Fractional Diffusion and Wave Equations

Yong Zhou

Fractional Diffusion and Wave Equations

Well-posedness and Inverse Problems

Springer

Yong Zhou
Macao Centre for Mathematical Sciences
Macau University of Science and
Technology
Macao, China

ISBN 978-3-031-74030-5 ISBN 978-3-031-74031-2 (eBook)
https://doi.org/10.1007/978-3-031-74031-2

Mathematics Subject Classification: 35A01, 35A24, 35Q93, 35L05, 35R11, 35R30

This Springer imprint is published by the registered company Springer Nature Switzerland AG
The registered company address is: Gewerbestrasse 11, 6330 Cham, Switzerland

Preface

The fractional calculus started more than three centuries ago. In the last years, the fractional calculus is playing a very important role in various scientific fields. In fact, it has been recognized as one of the best tools to describe long-memory processes. Fractional-order models are interesting not only for engineers and physicists but also for mathematicians. Among such models those described by partial differential equations (PDEs) containing fractional derivatives are of utmost importance. Their evolution was more complex than for the classical integer-order counterpart. Nonetheless, classical PDEs' methods are hardly applicable directly to fractional PDEs. Therefore, new theories and methods are required, with concepts and algorithms specifically developed for fractional PDEs. In the recent years, the theory of fractional PDEs has been highly developed and constitutes an important branch of differential equations.

This monograph gives an introduction to the theory for time-fractional diffusion and wave equations. Many of the basic results recently developed about this theory are presented, including the well-posedness, regularity, approximate controllability of initial value problems and the existence, regularity of terminal value problems. Some examples of applications relating to these equations are also discussed in detail. The materials in this monograph are based on the research work carried out by myself and some collaborators during the past five years. It is useful for researchers, graduates, or PhD students dealing with differential equations, applied analysis, and related areas of research.

I would like to thank Professors R.P. Agarwal, B. Ahmad, D. O'Regan, M. Fečkan, M. Kirane, V. Kiryakova, S.K. Ntouyas, G.M. N'Guerekata, J.J. Trujillo, and N.H. Tuan for their support. I also wish to express my appreciation to doctoral students J.W. He, L. Peng, J.N. Wang, X.X. Xi, and Y.L. Zhou for their help. Finally, I thank the editorial assistance of Springer, especially D. Chernyk.

I acknowledge with gratitude the support of National Natural Science Foundation of China (12071396) and the Macau Science and Technology Development Fund (0092/2022/A).

Macau, China Yong Zhou

Introduction

Fractional calculus has been attracting the attention of mathematicians and engineers from long time ago. The concept of fractional (or, more precisely, noninteger) differentiation appeared for the first time in a famous correspondence between L'Hospital and Leibniz, in 1695. Many mathematicians have further developed this area and we can mention the studies of Euler, Laplace, Abel, Liouville, and Riemann. However, the fractional calculus remained for centuries a purely theoretical topic, with little if any connections to practical problems of physics and engineering. In the past 30 years, the fractional calculus has been recognized as an effective modeling methodology for researchers. Fractional differential equations are generalizations of classical differential equations to an arbitrary (noninteger) order. Based on the wide applications in engineering and sciences such as physics, mechanics, electricity, chemistry, biology, economics, and many others, research on fractional differential equations is active and extensive around the world.

In the recent years, there has been a significant development in ordinary and partial differential equations involving fractional derivatives, see the monographs of Diethelm [52], Evangelista et al. [60], Hilfer [86], Jiao et al. [103], Kilbas et al. [115], Miller et al. [165], Podlubny [180], Povstenko [181], Umarov [212], Zhou [240, 241], and the references therein. A strong motivation for investigating this class of equations comes mainly from a compelling reason: the fractional order models of real systems are often more adequate than the classical integer order models, since the description of some systems is more accurate when the fractional derivative is used.

Diffusion equations with fractional time and space derivatives instead of the integer ones are widely used to describe anomalous diffusion processes where the mean squared displacement (MSD) scales as a power of time, $\langle x^2(t)\rangle \simeq t^\alpha$. Depending on the values of the anomalous diffusion exponent α one distinguishes the cases of subdiffusion for $0 < \alpha < 1$, normal Brownian diffusion for $\alpha = 1$, superdiffusion for $1 < \alpha < 2$, and ballistic motion for $\alpha = 2$. Well-known examples of anomalous transport include subdiffusion in artificially crowded systems and protein-crowded lipid bilayer membranes [99, 100, 203], subiffusive charge carrier motion in semiconductors [189], subdiffusive motion of submicron probes in living biological cells [72], superdiffusive tracer motion in chaotic laminar flows [197], diffusion in porous inhomogeneous media [239], and random search processes [217].

The time-fractional diffusion equation $\partial_t^\alpha u = \Delta u$, $\alpha \in (0, 1)$, can be used to model the anomalous diffusion exhibiting subdiffusive behavior, due to particle sticking and trapping phenomena (see [162, 163]). The fractional wave equation $\partial_t^\alpha u = \Delta u$, $\alpha \in (1, 2)$, governs the propagation of mechanical diffusive waves in viscoelastic media (see [149]). In the general case $\alpha \in (0, 2)$, we agree to refer to the equation $\partial_t^\alpha u = \Delta u$ as the fractional diffusion-wave equation.

This monograph introduces the theory for time-fractional diffusion equations and wave equations. Many of the basic results recently developed about this theory are presented, including the well-posedness (i.e., the existence, uniqueness, and continuous dependence), regularity, approximate controllability of initial value problems and the existence, regularity of terminal value problems (or called final value problem, initial inverse problem, backward problem). Some examples of applications relating to these equations are also discussed in detail. This fundamental theory should be the starting point for further research concerning the dynamics, numerical analysis, and applications of fractional partial differential equations.

This monograph is arranged and organized as follows.

In order to make the book self-contained, we devote the first chapter to a description of general information on fractional calculus, Mittag-Leffler functions, integral transforms, and semigroups.

The second chapter deals with initial value problems of time-fractional diffusion equations. In Sect. 2.1, we study a Cauchy problem for a space-time fractional diffusion equation with exponential nonlinearity. Based on the standard L^p-L^q estimates of strongly continuous semigroup generated by fractional Laplace operator, we investigate the existence of global solutions for initial data with small norm in the Orlicz space $\exp L^p(\mathbb{R}^d)$ and a time weighted $L^r(\mathbb{R}^d)$ space. In the framework of the Hölder's interpolation inequality, we also discuss the existence of local solutions without the Orlicz space. Section 2.2 is devoted to the study of a semilinear diffusion problem with distributed order fractional derivative on $\mathbb{R}^N$, which can be used to characterize the ultra-slow diffusion processes with time-dependent logarithmical-law attenuation. We use the resolvents approach to present the local well-posedness of mild solutions belonging to $L^r(\mathbb{R}^N)$ $(r > 2)$, in which the L^p-L^q estimates and continuity of the operator are first established. Then, under the assumption on the initial value belonging to $L^p(\mathbb{R}^N)$, the global well-posedness of mild solutions is derived. Moreover, a decay estimate in L^r-norm is included. Section 2.3 discusses an analysis of approximate controllability from the exterior of distributed order fractional diffusion problem with the fractional Laplace operator subject to the non-zero exterior condition. We first establish some well-posedness results, such as the existence, uniqueness, and regularity of the solutions allowing the weighted function μ that may be non-continuous. Especially, we show that the solutions can be represented by the series for the integral of a real-valued function. After giving the unique continuation property of the adjoint system, approximate controllability of the system is also included.

The third chapter deals with inverse problems of time-fractional diffusion equations of order $\alpha \in (0, 1)$. In Sect. 3.1, we study a backward problem for an inhomogeneous fractional diffusion equation in a bounded domain. By applying the properties of the Mittag-Leffler functions and the method of eigenvalue expansion, we establish some results about the existence, uniqueness, and regularity of the mild solutions as well as the classical solutions of the proposed problem in a weighted Hölder continuous function space. In Sect. 3.2, we consider a final value problem for a diffusion equation with time-space fractional differentiation on a bounded domain D of $\mathbb{R}^k$, $k \geq 1$, which includes the fractional power $\mathscr{L}^\beta$, $0 < \beta \leq 1$, of a symmetric uniformly elliptic operator $\mathscr{L}$ defined on $L^2(D)$. A representation of solutions is given by using the Laplace transform and the spectrum of $\mathscr{L}^\beta$. We present some existence and regularity results for our problem in both the linear and nonlinear case.

The fourth chapter is devoted to the study of the well-posedness and regularity for time-fractional wave equations of order $\alpha \in (1, 2)$. In Sect. 4.1, we study the well-posedness and regularity of mild solutions for a class of time-fractional damped wave equations. A concept of mild solutions is introduced to prove the existence for the linear problem, as well as the regularity of the solutions. We also establish a well-posedness result for nonlinear problem. As an application, we discuss a case of time-fractional telegraph equations. In Sect. 4.2, we study the semilinear time-fractional wave equation on a whole Euclidean space, also known as the super-diffusive equations. Based on the initial data taken in the fractional Sobolev spaces, and some known Sobolev embeddings, we prove the local/global well-posedness results of L^2-solutions for the linear and semilinear problems. In Sect. 4.3, we concern with an exponential nonlinearity for a fractional wave equation in the whole space, we establish the local existence of solutions in a dense subspace of the Orlicz classification. Moreover, we obtain the global existence of solutions for small initial data in lower dimension $1 \leq d \leq 3$. Our proofs are based on the analyticity of the Mittag-Leffler functions, the framework of prior estimates, and the type of exponential nonlinearity.

In fifth chapter, we discuss inverse problems of time-fractional wave equations of order $\alpha \in (1, 2)$. In Sect. 5.1, we concern with a backward problem (or called initial inverse problem) for a nonlinear time-fractional wave equation in a bounded domain. By applying the properties of the Mittag-Leffler functions and the method of eigenfunction expansion, we establish some results about the existence and uniqueness of mild solutions of the proposed problem based on the compact technique. Due to the ill-posedness of backward problem in the sense of Hadamard, a general filter regularization method is utilized to approximate the solution and further we prove the convergence rate for the regularized solutions. In Sect. 5.2, we consider the backward problem for an inhomogeneous time-fractional wave equation in a general bounded domain. We show that the backward problem is ill-posed, and we propose a regularizing scheme by using a fractional Landweber regularization method. We also present error estimates between the regularized solution and the exact solution under two parameter choice rules. In Sect. 5.3, we consider the terminal value problem of determining the initial value, in a general class of time-fractional wave equation with the Caputo derivative, from a given final value. We are concerned with the existence and regularity upon the terminal value data of the mild solutions. Under some assumptions of the nonlinear source function, we address and show the well-posedness for the terminal value problem. Some regularity results for the mild solutions and its derivatives of first and fractional orders are also derived. The effectiveness of our methods are shown by applying the results to two interesting models: time-fractional Ginzburg-Landau equation, and time-fractional Burgers equation, where time and spatial regularity estimates are obtained.

The materials in this monograph are based on the research work carried out by the author and some experts during the past five years. The Sect. 2.1 is taken from Peng, Zhou and He [179]. The material in Sect. 2.2 is taken from He, Zhou, Alsaedi, and Ahmad [85]. The results in Sect. 2.3 are taken from Peng and Zhou [178]. The material in Sect. 3.1 are adopted from Zhou, He, Ahmad and Tuan [245]. The contents in Sect. 3.2 are due to Tuan, Ngoc, Zhou and O'Regan [210]. The material in Sect. 4.1 is due to Zhou and He [242]. The results in Sect. 4.2 are taken from Zhou, He, Alsaedi and Ahmad [246]. The results in Sect. 4.3 are adopted from He and Zhou [83]. The contents of Sect. 5.1 are taken from He and Zhou [82]. The results in Sect. 5.2 are adopted from Huynh, Zhou, O'Regan and Tuan [91]. Section 5.3 is from Tuan, Caraballo, Ngoc and Zhou [211].

Keywords and Phrases Fractional calculus, fractional diffusion equations, fractional wave equations, damped wave equations, fractional Burgers equations, fractional Ginzburg-Landau equations, terminal value problem, final value problem, initial inverse problem, backward problem, mild solution, well-posedness, regularity, global existence, local existence, blow-up, uniqueness, compactness, continuation, controllability, C_0-Semigroup, analytic semigroup, integrated semigroup, Laplace transform, Fourier transform, Mittag-Leffler function, distributed order, error estimate, exponential growth, exponential nonlinearity.

Contents

Chapter 1
Preliminaries

1.1 Fractional Calculus

1.1.1 Definitions

A number of definitions for the fractional derivative have emerged over the years, and we refer the reader to Diethelm [52], Hilfer [86], Kilbas, Srivastava and Trujillo [115], Miller and Ross [165], and Podlubny [180]. In this book, we restrict our attention to the use of the Riemann–Liouville and Caputo fractional derivatives. In this section, we introduce some basic definitions and properties of the fractional integrals and fractional derivatives which are used further in this book. The materials in this section are taken from [115].

As usual $\mathbb{Z}$ denotes the set of integer numbers, $\mathbb{N}^+$ denotes the set of positive integer numbers, and $\mathbb{N}_0$ denotes the set of nonnegative integer numbers. $\mathbb{R}$ denotes the set of real numbers, $\mathbb{R}_+$ denotes the set of nonnegative reals and $\mathbb{R}^+$ the set of positive reals, and $\mathbb{R}^-$ denotes the set of nonpositive reals. Let $\mathbb{C}$ be the set of complex numbers.

Let $J = [a, b]$ $(-\infty < a < b < \infty)$ be a finite interval of $\mathbb{R}$. We assume that X is a Banach space with the norm $|\cdot|$. Denote $C(J, X)$ be the Banach space of all continuous functions from J into X with the norm $\|x\| = \sup_{t\in J} |x(t)|$, where $x \in C(J, X)$. $C^n(J, X)$ $(n \in \mathbb{N}_0)$ denotes the set of mappings having n times continuously differentiable on J, $AC(J, X)$ is the space of functions which are absolutely continuous on J, and $AC^n(J, X)$ $(n \in \mathbb{N}^+)$ is the space of functions f such that $f \in C^{n-1}(J, X)$ and $f^{(n-1)} \in AC(J, X)$. In particular, $AC^1(J, X) = AC(J, X)$.

Y. Zhou, *Fractional Diffusion and Wave Equations*,
https://doi.org/10.1007/978-3-031-74031-2_1

Let $1 \leq p \leq \infty$. $L^p(J, X)$ denotes the Banach space of all measurable functions $f : J \to X$. $L^p(J, X)$ is normed by

$$\|f\|_{L^pJ} = \begin{cases} \left(\int_J |f(t)|^p dt\right)^{\frac{1}{p}}, & 1 \leq p < \infty, \\ \inf\limits_{\mu(\bar{J})=0} \left\{ \sup\limits_{t \in J \setminus \bar{J}} |f(t)| \right\}, & p = \infty. \end{cases}$$

In particular, $L^1(J, X)$ is the Banach space of measurable functions $f : J \to X$ with the norm $\|f\|_{LJ} = \int_J |f(t)|dt$, and $L^\infty(J, X)$ is the Banach space of measurable functions $f : J \to X$ which are bounded, equipped with the norm $\|f\|_{L^\infty J} = \inf\{c > 0 | \, |f(t)| \leq c, \text{ a.e. } t \in J\}$.

The Gamma function $\Gamma(z)$ is defined by

$$\Gamma(z) = \int_0^\infty t^{z-1} e^{-t} dt \quad (Re(z) > 0),$$

where $t^{z-1} = e^{(z-1)\ln(t)}$. This integral is convergent for all complex $z \in \mathbb{C}$ $(Re(z) > 0)$.

For this function, the reduction formula

$$\Gamma(z+1) = z\Gamma(z) \quad (Re(z) > 0)$$

holds. In particular, if $z = n \in \mathbb{N}_0$, then

$$\Gamma(n+1) = n! \quad (n \in \mathbb{N}_0)$$

with (as usual) $0! = 1$.

Let us consider some of the starting points for a discussion of fractional calculus. One development begins with a generalization of repeated integration. Thus if f is locally integrable on (c, ∞), then the n-fold iterated integral is given by

$$\begin{aligned} {}_cD_t^{-n} f(t) &= \int_c^t ds_1 \int_c^{s_1} ds_2 \cdots \int_c^{s_{n-1}} f(s_n) ds_n \\ &= \frac{1}{(n-1)!} \int_c^t (t-s)^{n-1} f(s) ds \end{aligned}$$

for almost all t with $-\infty \leq c < t < \infty$ and $n \in \mathbb{N}^+$. Writing $(n-1)! = \Gamma(n)$, an immediate generalization is the integral of f of fractional order $\alpha > 0$,

$${}_cD_t^{-\alpha} f(t) = \frac{1}{\Gamma(\alpha)} \int_c^t (t-s)^{\alpha-1} f(s) ds \quad \text{(left hand)}$$

and similarly for $-\infty < t < d \leq \infty$

$$ {}_tD_d^{-\alpha} f(t) = \frac{1}{\Gamma(\alpha)} \int_t^d (s-t)^{\alpha-1} f(s) ds \quad \text{(right hand)} $$

both being defined for suitable f.

Definition 1.1 (Left and right Riemann–Liouville fractional integrals) Let $J = [a, b]$ $(-\infty < a < b < \infty)$ be a finite interval of $\mathbb{R}$. The left and right Riemann–Liouville fractional integrals ${}_aD_t^{-\alpha} f(t)$ and ${}_tD_b^{-\alpha} f(t)$ of order $\alpha \in \mathbb{R}^+$ are defined by

$$ {}_aD_t^{-\alpha} f(t) = \frac{1}{\Gamma(\alpha)} \int_a^t (t-s)^{\alpha-1} f(s) ds, \quad t > a, \ \alpha > 0 \tag{1.1} $$

and

$$ {}_tD_b^{-\alpha} f(t) = \frac{1}{\Gamma(\alpha)} \int_t^b (s-t)^{\alpha-1} f(s) ds, \quad t < b, \ \alpha > 0, \tag{1.2} $$

respectively, provided that the right-hand sides are pointwise defined on $[a, b]$. When $\alpha = n \in \mathbb{N}^+$, the definitions (1.1) and (1.2) coincide with the n-th integrals of the form

$$ {}_aD_t^{-n} f(t) = \frac{1}{(n-1)!} \int_a^t (t-s)^{n-1} f(s) ds $$

and

$$ {}_tD_b^{-n} f(t) = \frac{1}{(n-1)!} \int_t^b (s-t)^{n-1} f(s) ds. $$

Definition 1.2 (Left and right Riemann–Liouville fractional derivatives) The left and right Riemann–Liouville fractional derivatives ${}_aD_t^{\alpha} f(t)$ and ${}_tD_b^{\alpha} f(t)$ of order $\alpha \in \mathbb{R}_+$ are defined by

$$ \begin{aligned} {}_aD_t^{\alpha} f(t) &= \frac{d^n}{dt^n} {}_aD_t^{-(n-\alpha)} f(t) \\ &= \frac{1}{\Gamma(n-\alpha)} \frac{d^n}{dt^n} \left(\int_a^t (t-s)^{n-\alpha-1} f(s) ds \right), \quad t > a \end{aligned} $$

and

$$ \begin{aligned} {}_tD_b^{\alpha} f(t) &= (-1)^n \frac{d^n}{dt^n} {}_tD_b^{-(n-\alpha)} f(t) \\ &= \frac{1}{\Gamma(n-\alpha)} (-1)^n \frac{d^n}{dt^n} \left(\int_t^b (s-t)^{n-\alpha-1} f(s) ds \right), \quad t < b, \end{aligned} $$

respectively, where $n = [\alpha] + 1$, $[\alpha]$ means the integer part of α. In particular, when $\alpha = n \in \mathbb{N}_0$, then

$$_aD_t^0 f(t) = {}_tD_b^0 f(t) = f(t),$$

$$_aD_t^n f(t) = f^{(n)}(t) \quad \text{and} \quad {}_tD_b^n f(t) = (-1)^n f^{(n)}(t),$$

where $f^{(n)}(t)$ is the usual derivative of $f(t)$ of order n. If $0 < \alpha < 1$, then

$$_aD_t^\alpha f(t) = \frac{1}{\Gamma(1-\alpha)} \frac{d}{dt}\left(\int_a^t (t-s)^{-\alpha} f(s)ds \right), \quad t > a$$

and

$$_tD_b^\alpha f(t) = -\frac{1}{\Gamma(1-\alpha)} \frac{d}{dt}\left(\int_t^b (s-t)^{-\alpha} f(s)ds \right), \quad t < b.$$

Remark 1.1 If $f \in C([a, b]; \mathbb{R}^N)$, it is obvious that the Riemann–Liouville fractional integral of order $\alpha > 0$ exists on $[a, b]$. On the other hand, following Lemma 2.2 in [115], we know that the Riemann–Liouville fractional derivative of order $\alpha \in [n-1, n)$ exists almost everywhere on $[a, b]$ if $f \in AC^n([a, b]; \mathbb{R}^N)$.

The left and right Caputo fractional derivatives are defined via above the Riemann–Liouville fractional derivatives .

Definition 1.3 (Left and right Caputo fractional derivatives) The left and right Caputo fractional derivatives ${}_a^CD_t^\alpha f(t)$ and ${}_t^CD_b^\alpha f(t)$ of order $\alpha \in \mathbb{R}_+$ are defined by

$${}_a^CD_t^\alpha f(t) = {}_aD_t^\alpha\left(f(t) - \sum_{k=0}^{n-1} \frac{f^{(k)}(a)}{k!}(t-a)^k \right)$$

and

$${}_t^CD_b^\alpha f(t) = {}_tD_b^\alpha\left(f(t) - \sum_{k=0}^{n-1} \frac{f^{(k)}(b)}{k!}(b-t)^k \right),$$

respectively, where

$$n = [\alpha] + 1 \text{ for } \alpha \notin \mathbb{N}_0; \;\; n = \alpha \text{ for } \alpha \in \mathbb{N}_0. \tag{1.3}$$

In particular, when $0 < \alpha < 1$, then

$${}_a^CD_t^\alpha f(t) = {}_aD_t^\alpha (f(t) - f(a))$$

and

$$ {}_t^C D_b^\alpha f(t) = {}_t D_b^\alpha (f(t) - f(b)). $$

The Riemann–Liouville fractional derivative and the Caputo fractional derivative are connected with each other by the following relations.

Proposition 1.1

(i) *If $\alpha \notin \mathbb{N}_0$ and $f(t)$ is a function for which the Caputo fractional derivatives ${}_a^C D_t^\alpha f(t)$ and ${}_t^C D_b^\alpha f(t)$ of order $\alpha \in \mathbb{R}^+$ exist together with the Riemann–Liouville fractional derivatives ${}_a D_t^\alpha f(t)$ and ${}_t D_b^\alpha f(t)$, then*

$$ {}_a^C D_t^\alpha f(t) = {}_a D_t^\alpha f(t) - \sum_{k=0}^{n-1} \frac{f^{(k)}(a)}{\Gamma(k-\alpha+1)} (t-a)^{k-\alpha} $$

and

$$ {}_t^C D_b^\alpha f(t) = {}_t D_b^\alpha f(t) - \sum_{k=0}^{n-1} \frac{f^{(k)}(b)}{\Gamma(k-\alpha+1)} (b-t)^{k-\alpha}, $$

where $n = [\alpha] + 1$. In particular, when $0 < \alpha < 1$, we have

$$ {}_a^C D_t^\alpha f(t) = {}_a D_t^\alpha f(t) - \frac{f(a)}{\Gamma(1-\alpha)} (t-a)^{-\alpha} $$

and

$$ {}_t^C D_b^\alpha f(t) = {}_t D_b^\alpha f(t) - \frac{f(b)}{\Gamma(1-\alpha)} (b-t)^{-\alpha}. $$

(ii) *If $\alpha = n \in \mathbb{N}_0$ and the usual derivative $f^{(n)}(t)$ of order n exists, then ${}_a^C D_t^n f(t)$ and ${}_t^C D_b^n f(t)$ are represented by*

$$ {}_a^C D_t^n f(t) = f^{(n)}(t) \text{ and } {}_t^C D_b^n f(t) = (-1)^n f^{(n)}(t). \tag{1.4} $$

Proposition 1.2 *Let $\alpha \in \mathbb{R}_+$ and let n be given by (1.3). If $f \in AC^n([a,b]; \mathbb{R}^N)$, then the Caputo fractional derivatives ${}_a^C D_t^\alpha f(t)$ and ${}_t^C D_b^\alpha f(t)$ exist almost everywhere on $[a,b]$:*

(i) *If $\alpha \notin \mathbb{N}_0$, ${}_a^C D_t^\alpha f(t)$ and ${}_t^C D_b^\alpha f(t)$ are represented by*

$$ {}_a^C D_t^\alpha f(t) = \frac{1}{\Gamma(n-\alpha)} \left(\int_a^t (t-s)^{n-\alpha-1} f^{(n)}(s) ds \right) $$

and

$$ {}_t^C D_b^\alpha f(t) = \frac{(-1)^n}{\Gamma(n-\alpha)} \left(\int_t^b (s-t)^{n-\alpha-1} f^{(n)}(s) ds \right), $$

respectively, where $n = [\alpha] + 1$. *In particular, when* $0 < \alpha < 1$, $f \in AC([a,b];\mathbb{R}^N)$,

$$ {}_a^C D_t^\alpha f(t) = \frac{1}{\Gamma(1-\alpha)} \left(\int_a^t (t-s)^{-\alpha} f'(s) ds \right) \tag{1.5} $$

and

$$ {}_t^C D_b^\alpha f(t) = -\frac{1}{\Gamma(1-\alpha)} \left(\int_t^b (s-t)^{-\alpha} f'(s) ds \right). \tag{1.6} $$

(ii) *If* $\alpha = n \in \mathbb{N}_0$, *then* ${}_a^C D_t^\alpha f(t)$ *and* ${}_t^C D_b^\alpha f(t)$ *are represented by* (1.4). *In particular,*

$$ {}_a^C D_t^0 f(t) = {}_t^C D_b^0 f(t) = f(t). $$

Remark 1.2 If f is an abstract function with values in Banach space X, then integrals which appear in above definitions are taken in Bochner's sense.

The fractional integrals and derivatives, defined on a finite interval $[a,b]$ of $\mathbb{R}$, are naturally extended to whole axis $\mathbb{R}$.

Definition 1.4 (Left and right Liouville–Weyl fractional integrals on the real axis) The left and right Liouville–Weyl fractional integrals ${}_{-\infty}D_t^{-\alpha} f(t)$ and ${}_t D_{+\infty}^{-\alpha} f(t)$ of order $\alpha > 0$ on the whole axis $\mathbb{R}$ are defined by

$$ {}_{-\infty}D_t^{-\alpha} f(t) = \frac{1}{\Gamma(\alpha)} \int_{-\infty}^t (t-s)^{\alpha-1} f(s) ds \tag{1.7} $$

and

$$ {}_t D_{+\infty}^{-\alpha} f(t) = \frac{1}{\Gamma(\alpha)} \int_t^\infty (s-t)^{\alpha-1} f(s) ds, $$

respectively, where $t \in \mathbb{R}$ and $\alpha > 0$.

Definition 1.5 (Left and right Liouville–Weyl fractional derivatives on the real axis) The left and right Liouville–Weyl fractional derivatives ${}_{-\infty}D_t^{\alpha} f(t)$ and ${}_t D_{+\infty}^{\alpha} f(t)$ of order α on the whole axis $\mathbb{R}$ are defined by

$$
\begin{aligned}
{}_{-\infty}D_t^{\alpha} f(t) =& \frac{d^n}{dt^n}({}_{-\infty}D_t^{-(n-\alpha)} f(t)) \\
=& \frac{1}{\Gamma(n-\alpha)} \frac{d^n}{dt^n} \left(\int_{-\infty}^{t} (t-s)^{n-\alpha-1} f(s) ds \right)
\end{aligned}
$$

and

$$
\begin{aligned}
{}_tD_{+\infty}^{\alpha} f(t) =& (-1)^n \frac{d^n}{dt^n}({}_tD_{+\infty}^{-(n-\alpha)} f(t)) \\
=& \frac{1}{\Gamma(n-\alpha)} (-1)^n \frac{d^n}{dt^n} \left(\int_{t}^{\infty} (s-t)^{n-\alpha-1} f(s) ds \right),
\end{aligned}
$$

respectively, where $n = [\alpha] + 1$, $\alpha \geq 0$, and $t \in \mathbb{R}$.

In particular, when $\alpha = n \in \mathbb{N}_0$, then

$$
{}_{-\infty}D_t^{0} f(t) = {}_tD_{+\infty}^{0} f(t) = f(t),
$$

$$
{}_{-\infty}D_t^{n} f(t) = f^{(n)}(t) \quad \text{and} \quad {}_tD_{+\infty}^{n} f(t) = (-1)^n f^{(n)}(t),
$$

where $f^{(n)}(t)$ is the usual derivative of $f(t)$ of order n. If $0 < \alpha < 1$ and $t \in \mathbb{R}$, then

$$
\begin{aligned}
{}_{-\infty}D_t^{\alpha} f(t) =& \frac{1}{\Gamma(1-\alpha)} \frac{d}{dt} \left(\int_{-\infty}^{t} (t-s)^{-\alpha} f(s) ds \right) \\
=& \frac{\alpha}{\Gamma(1-\alpha)} \int_0^{\infty} \frac{f(t) - f(t-s)}{s^{\alpha+1}} ds
\end{aligned}
$$

and

$$
\begin{aligned}
{}_tD_{+\infty}^{\alpha} f(t) =& - \frac{1}{\Gamma(1-\alpha)} \frac{d}{dt} \left(\int_{t}^{\infty} (s-t)^{-\alpha} f(s) ds \right) \\
=& \frac{\alpha}{\Gamma(1-\alpha)} \int_0^{\infty} \frac{f(t) - f(t+s)}{s^{\alpha+1}} ds.
\end{aligned}
$$

Formulas (1.5) and (1.6) can be used for the definition of the Caputo fractional derivatives on the whole axis $\mathbb{R}$.

Definition 1.6 (Left and right Caputo fractional derivatives on the real axis) The left and right Caputo fractional derivatives ${}_{-\infty}^{\;\;C}D_t^{\alpha} f(t)$ and ${}_t^CD_{+\infty}^{\alpha} f(t)$ of order α (with $\alpha > 0$ and $\alpha \notin \mathbb{N}^+$) on the whole axis $\mathbb{R}$ are defined by

$$
{}_{-\infty}^{\;\;C}D_t^{\alpha} f(t) = \frac{1}{\Gamma(n-\alpha)} \left(\int_{-\infty}^{t} (t-s)^{n-\alpha-1} f^{(n)}(s) ds \right) \tag{1.8}
$$

and

$$ {}_{t}^{C}D_{+\infty}^{\alpha} f(t) = \frac{(-1)^n}{\Gamma(n-\alpha)} \left(\int_t^{\infty} (s-t)^{n-\alpha-1} f^{(n)}(s)ds \right), \tag{1.9} $$

respectively.

When $0 < \alpha < 1$, the relations (1.8) and (1.9) take the following forms:

$$ {}_{-\infty}^{\;\;C}D_t^{\alpha} f(t) = \frac{1}{\Gamma(1-\alpha)} \left(\int_{-\infty}^{t} (t-s)^{-\alpha} f'(s)ds \right) $$

and

$$ {}_{t}^{C}D_{+\infty}^{\alpha} f(t) = -\frac{1}{\Gamma(1-\alpha)} \left(\int_t^{\infty} (s-t)^{-\alpha} f'(s)ds \right). $$

1.1.2 Properties

We present here some properties of the fractional integral and fractional derivative operators that will be useful throughout this book.

Proposition 1.3 *If $\beta > 0$, then*

$$ {}_aD_t^{-\alpha}(t-a)^{\beta-1} = \frac{\Gamma(\beta)}{\Gamma(\beta+\alpha)}(t-a)^{\beta+\alpha-1} \quad (\alpha > 0), $$
$$ {}_aD_t^{\alpha}(t-a)^{\beta-1} = \frac{\Gamma(\beta)}{\Gamma(\beta-\alpha)}(t-a)^{\beta-\alpha-1} \quad (\alpha \geq 0) $$

and

$$ {}_tD_b^{-\alpha}(b-t)^{\beta-1} = \frac{\Gamma(\beta)}{\Gamma(\beta+\alpha)}(b-t)^{\beta+\alpha-1} \quad (\alpha > 0), $$
$$ {}_tD_b^{\alpha}(b-t)^{\beta-1} = \frac{\Gamma(\beta)}{\Gamma(\beta-\alpha)}(b-t)^{\beta-\alpha-1} \quad (\alpha \geq 0). $$

In particular, if $\beta = 1$ and $\alpha \geq 0$, then the Riemann–Liouville fractional derivatives of a constant are, in general, not equal to zero:

$$ {}_aD_t^{\alpha}1 = \frac{(t-a)^{-\alpha}}{\Gamma(1-\alpha)}, \quad {}_tD_b^{\alpha}1 = \frac{(b-t)^{-\alpha}}{\Gamma(1-\alpha)}. $$

On the other hand, for $j = 1, 2, \ldots, [\alpha]+1$,

$$ {}_aD_t^{\alpha}(t-a)^{\alpha-j} = 0, \quad {}_tD_b^{\alpha}(b-t)^{\alpha-j} = 0. $$

The semigroup properties of the fractional integral operators ${}_aD_t^{-\alpha}$ and ${}_tD_b^{-\alpha}$ are given by the following results.

Proposition 1.4 *If $\alpha > 0$ and $\beta > 0$, then the equations*

$$ {}_aD_t^{-\alpha}\left({}_aD_t^{-\beta} f(t)\right) = {}_aD_t^{-\alpha-\beta} f(t) \ \textit{and}\ {}_tD_b^{-\alpha}\left({}_tD_b^{-\beta} f(t)\right) = {}_tD_b^{-\alpha-\beta} f(t) \tag{1.10} $$

are satisfied at almost every point $t \in [a, b]$ for $f \in L^p(a, b; \mathbb{R}^N)$ $(1 \le p < \infty)$. If $\alpha + \beta > 1$, then the relations in (1.10) *hold at any point of* $[a, b]$.

Proposition 1.5

(i) *If $\alpha > 0$ and $f \in L^p(a, b; \mathbb{R}^N)$ $(1 \le p \le \infty)$, then the following equalities*

$$ {}_aD_t^{\alpha}\left({}_aD_t^{-\alpha} f(t)\right) = f(t) \ \textit{and}\ {}_tD_b^{\alpha}\left({}_tD_b^{-\alpha} f(t)\right) = f(t) $$

hold almost everywhere on $[a, b]$.

(ii) *If $\alpha > \beta > 0$, then, for $f \in L^p(a, b; \mathbb{R}^N)$ $(1 \le p \le \infty)$, the relations*

$$ {}_aD_t^{\beta}\left({}_aD_t^{-\alpha} f(t)\right) = {}_aD_t^{-\alpha+\beta} f(t) \ \textit{and}\ {}_tD_b^{\beta}\left({}_tD_b^{-\alpha} f(t)\right) = {}_tD_b^{-\alpha+\beta} f(t) $$

hold almost everywhere on $[a, b]$.

In particular, when $\beta = k \in \mathbb{N}^+$ and $\alpha > k$, then

$$ {}_aD_t^{k}\left({}_aD_t^{-\alpha} f(t)\right) = {}_aD_t^{-\alpha+k} f(t) \ \textit{and}\ {}_tD_b^{k}\left({}_tD_b^{-\alpha} f(t)\right) = (-1)^k {}_tD_b^{-\alpha+k} f(t). $$

To present the next property, we use the spaces of functions ${}_aD_t^{-\alpha}(L^p)$ and ${}_tD_b^{-\alpha}(L^p)$ defined for $\alpha > 0$ and $1 \le p \le \infty$ by

$$ {}_aD_t^{-\alpha}(L^p) = \{f : f = {}_aD_t^{-\alpha}\varphi,\ \varphi \in L^p(a, b; \mathbb{R}^N)\} $$

and

$$ {}_tD_b^{-\alpha}(L^p) = \{f : f = {}_tD_b^{-\alpha}\phi,\ \phi \in L^p(a, b; \mathbb{R}^N)\}, $$

respectively. The composition of the fractional integral operator ${}_aD_t^{-\alpha}$ with the fractional derivative operator ${}_aD_t^{\alpha}$ is given by the following results.

Proposition 1.6 *Let $\alpha > 0$, $n = [\alpha] + 1$ and let $f_{n-\alpha}(t) = {}_aD_t^{-(n-\alpha)} f(t)$ be the fractional integral* (1.1) *of order $n - \alpha$:*

(i) *If $1 \le p \le \infty$ and $f \in {}_aD_t^{-\alpha}(L^p)$, then*

$$ {}_aD_t^{-\alpha}\left({}_aD_t^{\alpha} f(t)\right) = f(t). $$

(ii) *If* $f \in L^1(a,b;\mathbb{R}^N)$ *and* $f_{n-\alpha} \in AC^n([a,b];\mathbb{R}^N)$, *then the equality*

$$ {}_aD_t^{-\alpha}\left({}_aD_t^{\alpha} f(t)\right) = f(t) - \sum_{j=1}^{n} \frac{f_{n-\alpha}^{(n-j)}(a)}{\Gamma(\alpha-j+1)}(t-a)^{\alpha-j} $$

holds almost everywhere on $[a,b]$.

Proposition 1.7 *Let* $\alpha > 0$ *and* $n = [\alpha]+1$. *Also let* $g_{n-\alpha}(t) = {}_tD_b^{-(n-\alpha)} g(t)$ *be the fractional integral* (1.2) *of order* $n-\alpha$:

(i) *If* $1 \le p \le \infty$ *and* $g \in {}_tD_b^{-\alpha}(L^p)$, *then*

$$ {}_tD_b^{-\alpha}\left({}_tD_b^{\alpha} g(t)\right) = g(t). $$

(ii) *If* $g \in L^1(a,b;\mathbb{R}^N)$ *and* $g_{n-\alpha} \in AC^n([a,b];\mathbb{R}^N)$, *then the equality*

$$ {}_tD_b^{-\alpha}\left({}_tD_b^{\alpha} g(t)\right) = g(t) - \sum_{j=1}^{n} \frac{(-1)^{n-j} g_{n-\alpha}^{(n-j)}(a)}{\Gamma(\alpha-j+1)}(b-t)^{\alpha-j} $$

holds almost everywhere on $[a,b]$.

In particular, if $0 < \alpha < 1$, *then*

$$ {}_tD_b^{-\alpha}\left({}_tD_b^{\alpha} g(t)\right) = g(t) - \frac{g_{1-\alpha}(a)}{\Gamma(\alpha)}(b-t)^{\alpha-1}, $$

where $g_{1-\alpha}(t) = {}_tD_b^{\alpha-1} g(t)$, *while for* $\alpha = n \in \mathbb{N}^+$, *the following equality holds:*

$$ {}_tD_b^{-n}\left({}_tD_b^{n} g(t)\right) = g(t) - \sum_{k=0}^{n-1} \frac{(-1)^k g^{(k)}(a)}{k!}(b-t)^k. $$

Proposition 1.8 *Let* $\alpha > 0$ *and let* $y \in L^\infty(a,b;\mathbb{R}^N)$ *or* $y \in C([a,b];\mathbb{R}^N)$. *Then*

$$ {}_a^CD_t^{\alpha}\left({}_aD_t^{-\alpha} y(t)\right) = y(t) \text{ and } {}_t^CD_b^{\alpha}\left({}_tD_b^{-\alpha} y(t)\right) = y(t). $$

Proposition 1.9 *Let* $\alpha > 0$ *and let* n *be given by* (1.3). *If* $y \in AC^n([a,b];\mathbb{R}^N)$ *or* $y \in C^n([a,b];\mathbb{R}^N)$, *then*

$$ {}_aD_t^{-\alpha}\left({}_a^CD_t^{\alpha} y(t)\right) = y(t) - \sum_{k=0}^{n-1} \frac{y^{(k)}(a)}{k!}(t-a)^k $$

and

$$ {}_tD_b^{-\alpha}\left({}_t^CD_b^\alpha y(t)\right) = y(t) - \sum_{k=0}^{n-1} \frac{(-1)^k y^{(k)}(b)}{k!}(b-t)^k. $$

In particular, if $0 < \alpha \le 1$ *and* $y \in AC([a,b];\mathbb{R}^N)$ *or* $y \in C([a,b];\mathbb{R}^N)$, *then*

$$ {}_aD_t^{-\alpha}\left({}_a^CD_t^\alpha y(t)\right) = y(t) - y(a) \text{ and } {}_tD_b^{-\alpha}\left({}_t^CD_b^\alpha y(t)\right) = y(t) - y(b). \tag{1.11} $$

1.2 Some Results from Analysis

1.2.1 Mittag–Leffler Function

Definition 1.7 ([165, 180]) The Mittag–Leffler function $E_{\alpha,\beta}$ is defined by

$$ E_{\alpha,\beta}(z) := \sum_{k=0}^{\infty} \frac{z^k}{\Gamma(\alpha k+\beta)} = \frac{1}{2\pi i}\int_{\Upsilon} \frac{\lambda^{\alpha-\beta}e^\lambda}{\lambda^\alpha - z} d\lambda, \quad \alpha > 0, \beta \in \mathbb{R}, \ z \in \mathbb{C}, $$

where Υ is a contour which starts and ends as $-\infty$ and encircles the disc $|\lambda| \le |z|^{1/\alpha}$ counter clockwise.

The function $E_{\alpha,\beta}(z)$ is an entire function, and so it is real analytic when restricted to the real line. Moreover, the approximation form of Mittag–Leffler function is given by

$$ E_{\alpha,\beta}(z) = -\sum_{k=1}^{N} \frac{1}{\Gamma(\beta-\alpha k)} \frac{1}{z^k} + O\left(\frac{1}{z^{N+1}}\right), $$

with $|z| \to \infty$, $\mu \le |\arg(z)| \le \pi$ for $\mu > 0$, and $N \in \mathbb{N}^+$. In particular,

$$ E_{\alpha,1}(z) = -\frac{1}{\Gamma(1-\alpha)}\frac{1}{z} + O\left(\frac{1}{z^2}\right), \tag{1.12} $$

with $|z| \to \infty$, $\mu \le |\arg(z)| \le \pi$ for $\mu > 0$.

For short, set

$$ E_\alpha(z) := E_{\alpha,1}(z), \quad e_\alpha(z) := E_{\alpha,\alpha}(z). $$

Then Mittag–Leffler functions have the following properties.

Proposition 1.10 ([165, 180]) *For $\alpha \in (0,1)$ and $t \in \mathbb{R}$:*

(i) $E_\alpha(t), e_\alpha(t) > 0$.
(ii) $(E_\alpha(t))' = \frac{1}{\alpha} e_\alpha(t)$.
(iii) $\lim_{t\to-\infty} E_\alpha(t) = \lim_{t\to-\infty} e_\alpha(t) = 0$.
(iv) ${}_0^C D_t^\alpha E_\alpha(\omega t^\alpha) = \omega E_\alpha(\omega t^\alpha)$, ${}_0D_t^{\alpha-1}(t^{\alpha-1} e_\alpha(\omega t^\alpha)) = E_\alpha(\omega t^\alpha)$, $\omega \in \mathbb{C}$.

Proposition 1.11 ([180, 207]) *Let $0 < \alpha_0 < \alpha_1 < 1$. Then there exist positive constants M_1, M_2 depending only on α_0, α_1 such that for all $\alpha \in [\alpha_0, \alpha_1]$,*

$$\frac{M_1}{1+x} \le E_\alpha(-x) \le \frac{M_2}{1+x}, \quad E_{\alpha,\beta}(-x) \le \frac{M_2}{1+x}, \quad \text{for all } x \ge 0,\ \beta \in \mathbb{R}.$$

Proposition 1.12 ([180]) *Let $0 < \alpha < 2$, and $\beta \in \mathbb{R}$ be arbitrary. Suppose that μ is such that $\pi\alpha/2 < \mu < \min\{\pi, \pi\alpha\}$. Then there exists a constant $M = M(\alpha, \beta, \mu) > 0$ such that*

$$|E_{\alpha,\beta}(z)| \le \frac{M}{1+|z|}, \quad \mu \le |\arg(z)| \le \pi.$$

Proposition 1.13 ([180]) *Let $0 < \alpha < 1$ and $\lambda > 0$. Then:*

(i) $\frac{d}{dt} E_\alpha(-\lambda t^\alpha) = -\lambda t^{\alpha-1} e_\alpha(-\lambda t^\alpha)$, *for $t > 0$.*
(ii) $\frac{d}{dt}\big(t^{\alpha-1} e_\alpha(-\lambda t^\alpha)\big) = t^{\alpha-2} E_{\alpha,\alpha-1}(-\lambda t^\alpha)$, *for $t > 0$.*
(iii) $\int_0^\infty e^{-st} E_\alpha(-\lambda t^\alpha) dt = \frac{s^{\alpha-1}}{s^\alpha + \lambda}$, *for $Re(s) > \lambda^{1/\alpha}$.*

Proposition 1.14 *Let $1 < \alpha < 2$ and $\lambda > 0$. Then:*

(i) $\frac{d}{dt} E_{\alpha,1}(-\lambda t^\alpha) = -\lambda t^{\alpha-1} E_{\alpha,\alpha}(-\lambda t^\alpha)$, *for $t > 0$.*
(ii) $\frac{d}{dt}(t^{\alpha-1} E_{\alpha,\alpha}(-\lambda t^\alpha)) = t^{\alpha-2} E_{\alpha,\alpha-1}(-\lambda t^\alpha)$, *for $t > 0$.*
(iii) ${}_0^C D_t^\alpha E_{\alpha,1}(-\lambda t^\alpha) = -\lambda E_{\alpha,1}(-\lambda t^\alpha)$, *for $t > 0$.*
(iv) ${}_0^C D_t^\alpha (t^{\alpha-1} E_{\alpha,\alpha}(-\lambda t^\alpha)) = -\lambda t^{\alpha-1} E_{\alpha,\alpha}(-\lambda t^\alpha)$, *for $t > 0$.*
(v) $\int_0^\infty e^{-st} E_{\alpha,1}(-\lambda t^\alpha) dt = \frac{s^{\alpha-1}}{s^\alpha + \lambda}$, *for $Re(s) > \lambda^{1/\alpha}$.*
(vi) $\lim_{t\to 0} E_{\alpha,1}(-\lambda t^\alpha) = 1$.
(vii) *The more accurate estimate $|E_{\alpha,\gamma}(-\lambda t^\alpha)| \le 1/\Gamma(\gamma)$ holds for $\gamma \in \{1, 2, \alpha\}$.*

It is well known that $E_{\alpha,1}(-t)$ is a positive and completely monotonic function for $\alpha \in (0,1)$, $t > 0$, that is, for all $t > 0$, $k \in \mathbb{N}_0$, we have

$$(-1)^k \Big(\frac{d}{dt}\Big)^k E_{\alpha,1}(-t) \ge 0.$$

Additionally, one can find that $\omega(t) := E_{\alpha,1}(\lambda t^\alpha)$ is a solution of equation ${}_0^C D_t^\alpha \omega(t) = \lambda\omega(t)$, $\lambda \in \mathbb{R}$, $\alpha \in (0, 2)$. We use the notation $a \lesssim b$ that stands for $a \le Cb$, with a positive constant C that does not depend on a, b. The following lemmas will be frequently used and can be found in [180].

Proposition 1.15 *For $\lambda > 0$, $\alpha > 0$, $\beta \in \mathbb{R}$, and any arbitrary positive number m, we have*

$$\left(\frac{d}{dt}\right)^m \left(t^{\beta-1} E_{\alpha,\beta}(-\lambda t^\alpha)\right) = t^{\beta-m-1} E_{\alpha,\beta-m}(-\lambda t^\alpha), \quad t > 0.$$

In particular,

$$\frac{d}{dt} E_{\alpha,1}(-\lambda t^\alpha) = -\lambda t^{\alpha-1} E_{\alpha,\alpha}(-\lambda t^\alpha), \quad t > 0.$$

Proposition 1.16 *If $0 < \alpha < 2$, $\beta \in \mathbb{R}$, $\pi\alpha/2 < \theta < \min\{\pi, \pi\alpha\}$, then*

$$\left|E_{\alpha,\beta}(z)\right| \lesssim \frac{1}{1+|z|}, \quad z \in \mathbb{C}, \quad \theta \le |\arg z| \le \pi.$$

Proposition 1.17 *If $0 < \alpha < 2$, $\beta \in \mathbb{R}$, θ is such that $\pi\alpha/2 < \theta < \min\{\pi, \pi\alpha\}$, then*

$$|E_{\alpha,\beta}(z)| \lesssim (1 + |z|)^{(1-\beta)/\alpha} \exp\left(Re(z^{1/\alpha})\right) + \frac{1}{1+|z|}, \quad z \in \mathbb{C}, \quad |\arg z| \le \theta.$$

By the fractional order term-by-term integration of the series, there is a more general relationship obtained as follows:

$$\frac{1}{\Gamma(\vartheta)} \int_0^t (t-s)^{\vartheta-1} s^{\beta-1} E_{\alpha,\beta}(\lambda s^\beta) ds = t^{\beta+\vartheta-1} E_{\alpha,\beta+\vartheta}(\lambda t^\beta), \ \vartheta > 0, \ \beta > 0, \ t > 0. \tag{1.13}$$

Proposition 1.18 ([10]) *Let $1 < \beta < 2$, $\beta' \in \mathbb{R}$, and $\lambda > 0$. Then the following estimates hold:*

(i) *Let $0 \le \alpha \le 1$, $0 < \beta' < \beta$. Then* $\left|\lambda^\alpha t^{\beta'} E_{\beta,\beta'}(-\lambda t^\beta)\right| \lesssim t^{\beta'-\beta\alpha}, \quad t > 0.$

(ii) *Let $0 \le \beta' \le 1$. Then* $\left|\lambda^{1-\beta'} t^{\beta-2} E_{\beta,\beta'}(-\lambda t^\beta)\right| \lesssim t^{\beta\beta'-2}, \quad t > 0.$

In what follows, let us state the definition and some properties of a function $\mathcal{M}_\alpha(\cdot)$ which is also called the Wright-type function. This function is a special case of the Wright function that plays an important role in different areas of fractional calculus, and it is introduced by Mainardi to characterize the solution of initial value problem for fractional diffusion-wave equations.

Definition 1.8 ([153]) The Wright-type function $\mathscr{M}_\alpha$ is defined by

$$\begin{aligned}\mathscr{M}_\alpha(z) &:= \sum_{n=0}^{\infty} \frac{(-z)^n}{n!\Gamma(-\alpha n + 1 - \alpha)}\\ &= \frac{1}{\pi}\sum_{n=1}^{\infty} \frac{(-z)^n}{(n-1)!}\Gamma(n\alpha)\sin(n\pi\alpha), \quad \text{for} \quad 0 < \alpha < 1, z \in \mathbb{C}.\end{aligned}$$

For $-1 < r < \infty$, $\lambda > 0$, the Wright-type function has the properties.

Proposition 1.19

(W1) $\mathscr{M}_\alpha(t) \geq 0$, $t > 0$.

(W2) $\int_0^\infty \frac{\alpha}{t^{\alpha+1}}\mathscr{M}_\alpha(\frac{1}{t^\alpha})e^{-\lambda t}dt = e^{-\lambda^\alpha}$.

(W3) $\int_0^\infty \mathscr{M}_\alpha(t)t^r dt = \frac{\Gamma(1+r)}{\Gamma(1+\alpha r)}$.

(W4) $\int_0^\infty \mathscr{M}_\alpha(t)e^{-zt}dt = E_\alpha(-z), \quad z \in \mathbb{C}$.

(W5) $\int_0^\infty \alpha t\mathscr{M}_\alpha(t)e^{-zt}dt = e_\alpha(-z), \quad z \in \mathbb{C}$.

1.2.2 Laplace and Fourier Transforms

In this subsection we present definitions and some properties of Laplace and Fourier transforms.

Definition 1.9 The Laplace transform of a function $f(t)$ of a real variable $t \in \mathbb{R}^+$ is defined by

$$(\mathscr{L}f)(s) = \mathscr{L}[f(t)](s) = \bar{f}(s) := \int_0^\infty e^{-st}f(t)dt \quad (s \in \mathbb{C}). \tag{1.14}$$

The inverse Laplace transform is given for $x \in \mathbb{R}^+$ by the formula

$$(\mathscr{L}^{-1}f)(x) = \mathscr{L}^{-1}[f(s)](x) := \frac{1}{2\pi i}\int_{\gamma-i\infty}^{\gamma+i\infty} e^{sx}f(s)ds \quad (\gamma = Re(s)). \tag{1.15}$$

Proposition 1.20 *Let* $f(t)$ *be defined on* $(0, \infty)$ *and* $0 < \alpha < 1$. *Then the Laplace transform of fractional integral and fractional differential operators satisfy:*

(i) $\overline{{}_0D_t^{-\alpha}f}(s) = s^{-\alpha}\bar{f}(s)$.

(ii) $\overline{{}_0D_t^{\alpha}f}(s) = s^{\alpha}\bar{f}(s) - ({}_0D_t^{\alpha-1}f)(0)$.

(iii) $\overline{{}_0^CD_t^{-\alpha}f}(s) = s^{\alpha}\bar{f}(s) - s^{\alpha-1}f(0)$.

Definition 1.10 The Fourier transform of a function $f(t)$ of a real variable $t \in \mathbb{R}$ is defined by

$$(\mathscr{F} f)(w) = \mathscr{F}[f(t)](w) = \hat{f}(w) := \int_{-\infty}^{\infty} e^{-it\cdot w} f(t)dt \quad (w \in \mathbb{R}). \tag{1.16}$$

The inverse Fourier transform is given by the formula

$$(\mathscr{F}^{-1} g)(w) = \mathscr{F}^{-1}[g(t)](w) = \frac{1}{2\pi}\hat{g}(-w) := \frac{1}{2\pi}\int_{-\infty}^{\infty} e^{it\cdot w} g(t)dt \quad (w \in \mathbb{R}). \tag{1.17}$$

The integrals in (1.16) and (1.17) converge absolutely for functions $f, g \in L^1(\mathbb{R})$ and in the norm of the space $L^2(\mathbb{R})$ for $f, g \in L^2(\mathbb{R})$.

Proposition 1.21 *Let $f(t)$ be defined on $(-\infty, \infty)$ and $0 < \alpha < 1$. Then the Fourier transform of Liouville–Weyl fractional integral and fractional differential operators satisfy:*

(i) $\widehat{{}_{-\infty}D_t^{-\alpha} f}(w) = (iw)^{-\alpha}\hat{f}(w)$.
(ii) $\widehat{{}_tD_{\infty}^{-\alpha} f}(w) = (-iw)^{-\alpha}\hat{f}(w)$.
(iii) $\widehat{{}_{-\infty}D_t^{\alpha} f}(w) = (iw)^{\alpha}\hat{f}(w)$.
(iv) $\widehat{{}_tD_{\infty}^{\alpha} f}(w) = (-iw)^{\alpha}\hat{f}(w)$.

1.3 Semigroups

1.3.1 C_0-Semigroup

Let us recall the definitions and properties of operator semigroups, for details see Pazy [175]. Let X be a Banach space and $\mathfrak{L}(X)$ be the Banach space of linear bounded operators with the norm $\|\cdot\|$.

Definition 1.11 A one parameter family $\{T(t)\}_{t\geq 0} \subset \mathfrak{L}(X)$ is a semigroup of bounded linear operators on X if:

(i) $T(t)T(s) = T(t+s)$, for $t, s \geq 0$.
(ii) $T(0) = I$; here, I denotes the identity operator in X.

Definition 1.12 A semigroup of bounded linear operators $\{T(t)\}_{t\geq 0}$ is uniformly continuous if

$$\lim_{t\to 0^+} \|T(t) - I\| = 0.$$

From the definition it is clear that if $\{T(t)\}_{t\geq 0}$ is a uniformly continuous semigroup of bounded linear operators, then

$$\lim_{s\to t} \|T(s) - T(t)\| = 0.$$

Definition 1.13 We say that the semigroup $\{T(t)\}_{t\geq 0}$ is strongly continuous (or a C_0-semigroup) if the mapping $t \to T(t)u$ is strongly continuous, for each $u \in X$, i.e.,

$$\lim_{t\to 0^+} T(t)u = u, \quad \forall\, u \in X.$$

Definition 1.14 Let $\{T(t)\}_{t\geq 0}$ be a C_0-semigroup defined on X. The linear operator A is the infinitesimal generator of $\{T(t)\}_{t\geq 0}$ defined by

$$Au = \lim_{t\to 0^+} \frac{T(t)u - u}{t}, \quad \text{for } u \in D(A),$$

where $D(A) = \Big\{u \in X : \lim_{t\to 0^+} \frac{T(t)u-u}{t} \text{ exists in } X\Big\}$.

Definition 1.15 The family $R(\lambda, A) = (\lambda I - A)^{-1}$, $\lambda \in \rho(A)$ of bounded linear operator is called of the resolvent of A, where $\rho(A)$ is the set of all complex numbers λ for which $\lambda I - A$ is invertible.

If there are $M \geq 0$ and $\nu \in \mathbb{R}$ such that $\|T(t)\| \leq Me^{\nu t}$, then

$$(\lambda I - A)^{-1}u = \int_0^\infty e^{-\lambda t}T(t)udt, \quad \text{Re}(\lambda) > \nu, \; u \in X. \tag{1.18}$$

A C_0-semigroup $\{T(t)\}_{t\geq 0}$ is called exponentially stable if there exist constants $M > 0$ and $\delta > 0$ such that

$$\|T(t)\| \leq Me^{-\delta t}, \qquad t \geq 0. \tag{1.19}$$

The growth bound ν_0 of $\{T(t)\}_{t\geq 0}$ is defined by

$$\nu_0 = \inf\{\delta \in \mathbb{R} : \text{ there exists } M_\delta > 0 \text{ such that } \|T(t)\| \leq M_\delta e^{\delta t}, \; \forall\, t \geq 0\}. \tag{1.20}$$

Furthermore, ν_0 can also be obtained by the following formula:

$$\nu_0 = \limsup_{t\to +\infty} \frac{\ln \|T(t)\|}{t}. \tag{1.21}$$

Definition 1.16 A C_0-semigroup $\{T(t)\}_{t\geq 0}$ is called uniformly bounded if there exists a constant $M > 0$ such that

$$\|T(t)\| \leq M, \quad t \geq 0. \tag{1.22}$$

Definition 1.17 A C_0-semigroup $\{T(t)\}_{t\geq 0}$ is called compact if $T(t)$ is compact for $t > 0$.

Proposition 1.22 *If $\{T(t)\}_{t>0}$ is compact, then $\{T(t)\}_{t>0}$ is equicontinuous for $t > 0$.*

Definition 1.18 A C_0-semigroup $\{T(t)\}_{t\geq 0}$ is called positive if $T(t)u \geq \theta$ for all $u \geq \theta$ and $t \geq 0$, where θ is the zero element.

1.3.2 Analytic Semigroup

Definition 1.19 Let $\Delta := \{z : \varphi_1 < \arg z < \varphi_2, \ \varphi_1 < 0 < \varphi_2\}$. The family $\{T(z)\}_{z\in\Delta} \subset \mathfrak{L}(X)$ is called an analytic semigroup in Δ if:

(i) $z \mapsto T(z)$ is analytic in Δ.
(ii) $T(0) = I$ and $\lim_{z\in\Delta, z\to 0} T(z)x = x$ for every $x \in X$.
(iii) $T(z_1 + z_2) = T(z_1)T(z_2)$ for $z_1, \ z_2 \in \Delta$.

A semigroup $T(t)$ is called analytic if it is analytic in some sector Δ containing the nonnegative real axis.

Theorem 1.1 *Let $\{T(t)\}_{t\geq 0}$ be a uniformly bounded C_0-semigroup. Let A be the infinitesimal generator of $\{T(t)\}_{t\geq 0}$ and assume $0 \in \rho(A)$. The following statements are equivalent:*

(i) *$T(t)$ can be extended to an analytic semigroup in a sector $\Sigma_\delta := \{z \in \mathbb{C} : |\arg z| < \delta\}$ and $\|T(z)\|$ is uniformly bounded in every closed subsector $\overline{\Sigma}_{\delta'}$, $\delta' < \delta$, of Σ_δ.*
(ii) *There exists a positive constant C such that for every $\sigma > 0$, $\tau \neq 0$,*

$$\|R(\sigma + i\tau, A)\| \leq \frac{C}{|\tau|}.$$

(iii) *There exist $0 < \delta < \frac{\pi}{2}$ and $M > 0$ such that*

$$\rho(A) \supset \Sigma := \{\lambda \in \mathbb{C} : |\arg \lambda| < \frac{\pi}{2} + \delta\} \cup \{0\}$$

and

$$\|R(\lambda, A)\| \leq \frac{M}{|\lambda|}, \quad \textit{for } \lambda \in \Sigma, \ \lambda \neq 0.$$

(iv) *$T(t)$ is differentiable for $t > 0$, and there is a constant C such that*

$$\|AT(t)\| \leq \frac{C}{t}, \quad \textit{for } t > 0.$$

1.3.3 Integrated Semigroup

Definition 1.20 Let X be a Banach space. An integrated semigroup is a family of bounded linear operators $\{S(t)\}_{t\geq 0}$ on X with the following properties:

(i) S(0)=0.
(ii) $t \mapsto S(t)$ is strongly continuous.
(iii) $S(s)S(t) = \int_0^s \big(S(t+r) - S(r)\big)dr$ for all $t, s \geq 0$.

Definition 1.21 An operator A is called a generator of an integrated semigroup if there exists $\omega \in \mathbb{R}$ such that $(\omega, \infty) \subset \rho(A)$ ($\rho(A)$ is the resolvent set of A) and there exists a strongly continuous exponentially bounded family $\{S(t)\}_{t\geq 0}$ of bounded operators such that $S(0) = 0$ and

$$R(\lambda, A) := (\lambda I - A)^{-1} = \lambda \int_0^\infty e^{-\lambda t} S(t)dt \text{ exists for all } \lambda \text{ with } \lambda > \omega.$$

Proposition 1.23 *Let A be the generator of an integrated semigroup $\{S(t)\}_{t\geq 0}$. Then for all $u \in X$ and $t \geq 0$,*

$$\int_0^t S(s)ds \in D(A) \ \textit{ and } S(t)u = A\int_0^t S(s)uds + tu.$$

Definition 1.22

(i) An integrated semigroup $\{S(t)\}_{t\geq 0}$ is called locally Lipschitz continuous if for all $\tau > 0$ there exists a positive constant L such that

$$\|S(t) - S(s)\| \leq L|t-s|, \quad t, s \in [0, \tau].$$

(ii) An integrated semigroup $\{S(t)\}_{t\geq 0}$ is called nondegenerate if $S(t)u = 0$ for all $t \geq 0$ implies that $u = 0$.

Definition 1.23 We say that the linear operator $A : D(A) \subset X \to X$ satisfies the Hille–Yosida condition if there exist two constants $\omega \in \mathbb{R}$ and $M > 0$ such that $(\omega, +\infty) \subset \rho(A)$ and

$$\|(\lambda I - A)^{-k}\| \leq \frac{M}{(\lambda - \omega)^k}, \quad \text{for all } \lambda > \omega, \ k \geq 1.$$

Theorem 1.2 *The following assertions are equivalent:*

(i) *A is the generator of a nondegenerate, locally Lipschitz continuous integrated semigroup.*
(ii) *A satisfies the Hille–Yosida condition.*

If A is the generator of an integrated semigroup $\{S(t)\}_{t\geq 0}$ which is locally Lipschitz, then $S(t)u$ is continuously differentiable if and only if $u \in \overline{D(A)}$ and $\{S'(t)\}_{t\geq 0}$ is a C_0-semigroup on $\overline{D(A)}$.

Chapter 2
Well-Posedness of Fractional Diffusion Equations

2.1 Diffusion Equation with Exponential Growth

2.1.1 Introduction

In this section, we consider a Cauchy problem for space-time fractional diffusion equation

$$\partial_t^\alpha u + (-\Delta)^\gamma u = f(u), \quad (t, x) \in (0, \infty) \times \mathbb{R}^d, \tag{2.1}$$

associated with an initial condition $u(0, x) = u_0(x)$, $x \in \mathbb{R}^d$, $d \geq 1$, where f is the exponential growth function, like asymptotic growth $f(u) \sim e^{4\pi|u|^2}$, and with a vanishing behavior at zero, and ∂_t^α stands for the Caputo fractional partial derivative of order $\alpha \in (0, 1)$ defined by

$$\partial_t^\alpha u(t, x) = \frac{1}{\Gamma(1-\alpha)} \int_0^t (t-s)^{-\alpha} \frac{\partial}{\partial s} u(s, x) ds, \quad (t, x) \in (0, \infty) \times \mathbb{R}^d.$$

In (2.1), $(-\Delta)^\gamma$ ($\gamma \in (0, 1)$) stands for the fractional Laplacian operator defined by

$$(-\Delta)^\gamma u(x) = C_{\gamma,d} \int_{\mathbb{R}^d} \frac{u(x) - u(y)}{|x-y|^{d+2\gamma}} dy,$$

with $C_{\gamma,d} = \gamma 2^{2\gamma} \Gamma((d + 2\gamma)/2)/(\pi^{d/2}\Gamma(1-\gamma))$.

Now we dwell on the literature dealing with the nonlinearity of exponential growth problems of diffusion and wave equations. The Cauchy problem for heat equation with exponential nonlinearity was studied by Ioku [94]. Inspired by this paper, Furioli et al. [65] discussed the asymptotic behavior and decay estimates for

Y. Zhou, *Fractional Diffusion and Wave Equations*,
https://doi.org/10.1007/978-3-031-74031-2_2

solutions of a nonlinear parabolic equation with exponential growth. Ioku et al. [95] obtained the existence and nonexistence results for a heat equation in the Orlicz space $\exp L(\mathbb{R}^2)$. Fino and Kirane [62] investigated the global solutions for heat equation with fractional Laplacian and exponential nonlinearity with small initial data and also the local solution in the Orlicz space. Regarding the wave equation, the global well-posed solutions with exponential growth type nonlinearity were studied in the critical Sobolev space in [170]. In order to overcome the issue of invalidity of the embedding $H^1(\mathbb{R}^2) \hookrightarrow L^\infty(\mathbb{R}^2)$, Ibrahim et al. [92] discussed the existence and asymptotic behavior of finite energy solutions for large time for the subcritical case of the wave operators via Trudinger-Moser-type inequality. On the supercritical regime of large energies for smooth and radially symmetric initial data, Struwe [199] established the global well-posedness of solutions for a nonlinear wave equation with nonlinearity $f(u) \sim ue^{u^2}$. Mahouachi and Saanouni [148] derived the well-posed and ill-posed results for a wave equation with exponential growth. Concerning the fractional derivatives, Bekkai et al. [17] discussed the local existence and blow-up of solutions for a space-time fractional diffusion equation with nonlocal nonlinearity of the form $f(u) \sim {}_0D_t^{1-\alpha}(e^u)$, where ${}_0D_t^{1-\alpha}$ represents the Riemann-Liouville fractional integral operator. Alsaedi et al. [9] proved the existence and uniqueness of the local mild solutions for a system of space-time fractional evolution equations with nonlocal nonlinearities of exponential growth. They also established a blow-up result by applying the Pokhozhaev capacity method and presented an estimate for the life span of blowing-up solutions under suitable conditions.

We note that there is no work concerning the existence of global solutions for space-time fractional diffusion equations with exponential growth data in $\mathbb{R}^d$. For this purpose, it is necessary to consider an appropriate space. Unlike the energy functional techniques used in previous work, this section will concern on the global solutions to the fractional Cauchy problem on the Orlicz spaces $\exp L^p(\mathbb{R}^d)$ via the subordinate principle and the semigroup theory. It is worthwhile to notice that $C_0^\infty(\mathbb{R}^d)$ is not dense in $\exp L^p(\mathbb{R}^d)$, but it works in $L^p(\mathbb{R}^d)$ for $1 \le p < \infty$. As it is difficult to consider a generic global space $C([0, \infty); \exp L^p(\mathbb{R}^d))$, we shall deal with this case in the sense of a weak topology. Furthermore, since $L^1(\mathbb{R}^d) \cap L^\infty(\mathbb{R}^d) \hookrightarrow \exp L^p(\mathbb{R}^d)$ and to solve equation (2.1) under the minimum required conditions, we will establish the local solutions in $L^1(\mathbb{R}^d) \cap L^\infty(\mathbb{R}^d)$.

Our main goal in this section is to investigate the existence of solutions for the problem (2.1). In the next subsection, we introduce the definition of the Orlicz spaces and the relevant solution operators of the problem (2.1), and then we study the space-time estimates in the frameworks of L^p-L^q and L^p-$\exp L^q$. In Sect. 2.1.3, we establish the existence of global solution for initial data with small norm in space $\exp L^p(\mathbb{R}^d)$ and the decay estimate of solutions. In Sect. 2.1.4, we prove the existence of local solutions in subspace $L^1(\mathbb{R}^d) \cap L^\infty(\mathbb{R}^d)$ of the Orlicz spaces.

2.1.2 Orlicz Spaces and Space-Time Estimates

Throughout this section, the notation $a \lesssim b$ stands for $a \leq Cb$, and $\sim$ stands for $a \leq Cb$ and $b \leq Ca$ for a positive generic constant C that does not depend on a, b. Symbols $\vee$ and $\wedge$ are expressed by $a \vee b = \max\{a, b\}$ and $a \wedge b = \min\{a, b\}$, respectively. It is well known that the fractional Laplace operator $(-\Delta)^\gamma$ can generate a strongly continuous semigroup $T_\gamma(t) = \exp(-t(-\Delta)^\gamma))$ on $L^p(\mathbb{R}^d)$ for $p \geq 1$, $d \geq 1$ with its Fourier transformation $(-\Delta)^\gamma u = \mathscr{F}^{-1}(|\xi|^{2\gamma}\mathscr{F}(u))$. Moreover, a space-time estimate of this semigroup is given by

$$\|T_\gamma(t)\phi\|_{L^p(\mathbb{R}^d)} \lesssim t^{-\frac{d}{2\gamma}\left(\frac{1}{q}-\frac{1}{p}\right)}\|\phi\|_{L^q(\mathbb{R}^d)},$$

for $t > 0$ and for all $\phi \in L^q(\mathbb{R}^d)$, $q \geq 1$, see [164].

For any $\alpha \in (0, 1)$, $\beta \in [0, \infty)$, $\gamma \in (0, 1)$, we introduce a subordinate operator $\mathscr{A}^\gamma_{\alpha,\beta}(\cdot)$ as

$$\mathscr{A}^\gamma_{\alpha,\beta}(t) = \int_0^\infty \alpha^\beta\theta^\beta \mathscr{M}_\alpha(\theta)T_\gamma(t^\alpha\theta)d\theta, \quad t \geq 0,$$

where $\mathscr{M}_\alpha(\cdot)$ is the Wright-type function defined in Definition 1.8.

Obviously, the operator $\mathscr{A}^\gamma_{\alpha,\beta}(\cdot)$ is well defined due to the estimate

$$\|\mathscr{A}^\gamma_{\alpha,\beta}(t)\phi\|_{L^p(\mathbb{R}^d)} \leq \int_0^\infty \alpha^\beta\theta^\beta \mathscr{M}_\alpha(\theta)\|T_\gamma(t^\alpha\theta)\|_{L^p(\mathbb{R}^d)}d\theta \lesssim \int_0^\infty \alpha^\beta\theta^\beta \mathscr{M}_\alpha(\theta)d\theta$$

and the properties of $\mathscr{M}_\alpha(\cdot)$ (see [240])

$$\mathscr{M}_\alpha(\cdot) \geq 0, \quad \int_0^\infty \theta^\delta \mathscr{M}_\alpha(\theta)d\theta = \frac{\Gamma(1+\delta)}{\Gamma(1+\alpha\delta)}, \quad \delta \in (-1, \infty).$$

Specially, if u is a solution of the problem (2.1), by using the strategy employed in [240], we have the following integral representation of the problem (2.1):

$$u(t) = \mathscr{A}^\gamma_{\alpha,0}(t)u_0 + \int_0^t (t-s)^{\alpha-1}\mathscr{A}^\gamma_{\alpha,1}(t-s)f(u(s))ds. \tag{2.2}$$

Recall that a space is Orlicz type if it can be expressed as

$$\exp L^p(\mathbb{R}^d) = \left\{u \in L^1_{loc}(\mathbb{R}^d) : \|u\|_{\exp L^p(\mathbb{R}^d)} < +\infty\right\},$$

endowed with the Luxemburg norm

$$\|u\|_{\exp L^p(\mathbb{R}^d)} := \inf\left\{\tau > 0 : \int_{\mathbb{R}^d} \left(\exp\left(|u(x)|^p/\tau^p\right) - 1\right) dx < 1 \text{for some } \tau > 0\right\}.$$

Clearly, an Orlicz space is a Banach space.

Lemma 2.1 ([94]) *For every* $1 \le p \le q < +\infty$, *the embedding* $\exp L^p(\mathbb{R}^d) \hookrightarrow L^q(\mathbb{R}^d)$ *holds, and moreover*

$$\|u\|_{L^q(\mathbb{R}^d)} \le (\Gamma(q/p+1))^{1/q}\, \|u\|_{\exp L^p(\mathbb{R}^d)}.$$

Lemma 2.2 ([62]) *For every* $1 \le q \le p < +\infty$, *the embedding* $L^q(\mathbb{R}^d) \cap L^\infty(\mathbb{R}^d) \hookrightarrow \exp L^p(\mathbb{R}^d)$ *holds, and moreover*

$$\|u\|_{\exp L^p(\mathbb{R}^d)} \le (\ln 2)^{-\frac{1}{p}}\left(\|u\|_{L^q(\mathbb{R}^d)} + \|u\|_{L^\infty(\mathbb{R}^d)}\right).$$

Lemma 2.3 *Let* $\alpha \in (0,1)$, $\beta \in [0,\infty)$, $\gamma \in (0,1)$, *and let* $\frac{d}{2\gamma}\left(\frac{1}{p} - \frac{1}{q}\right) < 1$ *for* $1 \le p \le q \le +\infty$, *for* $t > 0$, *then*

(i) $\|\mathscr{A}^\gamma_{\alpha,\beta}(t) f\|_{L^q(\mathbb{R}^d)} \lesssim t^{-\frac{\alpha d}{2\gamma}\left(\frac{1}{p}-\frac{1}{q}\right)} \|f\|_{L^p(\mathbb{R}^d)}, \quad for\ f \in L^p(\mathbb{R}^d)$,
(ii) $\|\mathscr{A}^\gamma_{\alpha,\beta}(t) f\|_{\exp L^p(\mathbb{R}^d)} \lesssim \|f\|_{\exp L^p(\mathbb{R}^d)}, \quad for\ f \in \exp L^p(\mathbb{R}^d)$,
(iii) $\|\mathscr{A}^\gamma_{\alpha,\beta}(t) f\|_{\exp L^q(\mathbb{R}^d)} \lesssim t^{-\frac{\alpha d}{2\gamma p}} (\ln(t^{-\frac{\alpha d}{2\gamma}} + 1))^{-\frac{1}{q}} \|f\|_{L^p(\mathbb{R}^d)}, \quad for\ f \in L^p(\mathbb{R}^d)$,
(iv) $\|\mathscr{A}^\gamma_{\alpha,\beta}(t) f\|_{\exp L^p(\mathbb{R}^d)} \lesssim t^{-\frac{\alpha d}{2\gamma q}} \|f\|_{L^q(\mathbb{R}^d)} + \|f\|_{L^p(\mathbb{R}^d)}, \quad for\ f \in L^p(\mathbb{R}^d) \cap L^q(\mathbb{R}^d)$.

Proof By the L^p-L^q estimates of semigroup $T_\gamma(t)$, for the operator $\mathscr{A}^\gamma_{\alpha,\beta}(\cdot)$, we obtain

$$\|\mathscr{A}^\gamma_{\alpha,\beta}(t) f\|_{L^q(\mathbb{R}^d)} \le \int_0^\infty \alpha^\beta \theta^\beta \mathscr{M}_\alpha(\theta) \|T_\gamma(t^\alpha\theta)\phi\|_{L^q(\mathbb{R}^d)} d\theta \lesssim t^{-\frac{\alpha d}{2\gamma r}} \|f\|_{L^p(\mathbb{R}^d)},$$

for $t > 0$, where $1/r := 1/p - 1/q$. Additionally, for any $\tau > 0$, $t > 0$, it follows by Taylor expansion combined with (i) that

$$\begin{aligned}
\int_{\mathbb{R}^d} \left(\exp\left(|\mathscr{A}^\gamma_{\alpha,\beta}(t) f|/\tau\right)^p - 1\right) dx &= \sum_{k=1}^\infty \frac{\|\mathscr{A}^\gamma_{\alpha,\beta}(t) f\|^{pk}_{L^{pk}(\mathbb{R}^d)}}{k!\tau^{pk}} \\
&\lesssim \sum_{k=1}^\infty \frac{\|f\|^{pk}_{L^{pk}(\mathbb{R}^d)}}{k!\tau^{pk}} = \int_{\mathbb{R}^d} \left(\exp\left(f/\tau\right)^p - 1\right) dx,
\end{aligned}$$

which implies that for $f \in \exp L^p(\mathbb{R}^d)$,

$$\|\mathscr{A}^{\gamma}_{\alpha,\beta}(t)f\|_{\exp L^p(\mathbb{R}^d)} \lesssim \|f\|_{\exp L^p(\mathbb{R}^d)}.$$

Next, by virtue of (i), for $t > 0$, we have

$$\begin{aligned}\int_{\mathbb{R}^d}\left(\exp\left(|\mathscr{A}^{\gamma}_{\alpha,\beta}(t)f|/\tau\right)^q - 1\right)dx &= \sum_{k=1}^{\infty}\frac{\|\mathscr{A}^{\gamma}_{\alpha,\beta}(t)f\|^{qk}_{L^{pk}(\mathbb{R}^d)}}{k!\tau^{qk}} \\ &\lesssim \sum_{k=1}^{\infty}\frac{t^{-\frac{\alpha d}{2\gamma}\left(\frac{1}{p}-\frac{1}{qk}\right)qk}\|f\|^{qk}_{L^p(\mathbb{R}^d)}}{k!\tau^{qk}},\end{aligned}$$

which implies that there exists a constant $C > 0$ such that

$$\int_{\mathbb{R}^d}\left(\exp\left(|\mathscr{A}^{\gamma}_{\alpha,\beta}(t)f|/\tau\right)^q - 1\right)dx \lesssim t^{\frac{\alpha d}{2\gamma}}\left(\exp\left(Ct^{-\frac{\alpha d}{2\gamma p}}\|f\|_{L^p(\mathbb{R}^d)}/\tau\right)^q - 1\right),$$

and then

$$\|\mathscr{A}^{\gamma}_{\alpha,\beta}(t)f\|_{\exp L^q(\mathbb{R}^d)} \lesssim t^{-\frac{\alpha d}{2\gamma p}}(\ln(t^{-\frac{\alpha d}{2\gamma}} + 1))^{-\frac{1}{q}}\|f\|_{L^p(\mathbb{R}^d)}.$$

The last inequality can easily be proved by using the standard L^p-L^q estimates of $\mathscr{A}^{\gamma}_{\alpha,\beta}(\cdot)$ and the embedding $L^p(\mathbb{R}^d)\cap L^\infty(\mathbb{R}^d) \hookrightarrow \exp L^p(\mathbb{R}^d)$ in Lemma 2.2. In consequence, we get

$$\begin{aligned}\|\mathscr{A}^{\gamma}_{\alpha,\beta}(t)f\|_{\exp L^p(\mathbb{R}^d)} &\lesssim \|\mathscr{A}^{\gamma}_{\alpha,\beta}(t)f\|_{L^p(\mathbb{R}^d)} + \|\mathscr{A}^{\gamma}_{\alpha,\beta}(t)f\|_{L^\infty(\mathbb{R}^d)} \\ &\lesssim t^{-\frac{\alpha d}{2\gamma q}}\|f\|_{L^q(\mathbb{R}^d)} + \|f\|_{L^p(\mathbb{R}^d)}.\end{aligned}$$

Hence, this completes the proof.

Observe that $C_0^\infty(\mathbb{R}^d)$ is not dense in $\exp L^p(\mathbb{R}^d)$ for $p \geq 1$; for more details, see [94, 95]. For the fractional version of $\mathscr{A}^{\gamma}_{\alpha,0}(\cdot)$, we have the following conclusion.

Remark 2.1 The operator $\mathscr{A}^{\gamma}_{\alpha,0}(t)$ is not strongly continuous at $t = 0$ in $\exp L^p(\mathbb{R}^d)$ for $p \geq 1$, that is, for any $\phi \in \exp L^p(\mathbb{R}^d)$, the following inequality holds:

$$\lim_{t\to 0}\|\mathscr{A}^{\gamma}_{\alpha,0}(t)\phi - \phi\|_{\exp L^p(\mathbb{R}^d)} \geq 1. \tag{2.3}$$

Proof In fact, for any $\lambda > 0$, let $\mu_v(\lambda) := |\{x \in \mathbb{R}^d : |v(x)| > \lambda\}|$ be a distribution function of v, let v^* be a nonincreasing rearrangement of v given by

$$v^*(r) := \inf\{\lambda > 0 : \mu_v(\lambda) \leq r\},$$

and the maximal function of v^* is denoted by v^{**} as follows:

$$v^{**}(r) = \frac{1}{r}\int_0^r v^*(s)ds.$$

We next use the rearrangement technique, that is,

$$\sup_{0<r<1} \frac{(\mathscr{A}^{\gamma}_{\alpha,0}\phi - \phi)^{**}(r)}{(\ln(e/r))^{1/p}} \lesssim \|\mathscr{A}^{\gamma}_{\alpha,0}(t)\phi - \phi\|_{\exp L^p(\mathbb{R}^d)},$$

for any $t \geq 0$, where

$$\|v\|_{\exp L^p(\mathbb{R}^d)} \lesssim \sup_{0<r<1} \frac{v^{**}(r)}{(\ln(e/r))^{1/p}} + \|v\|_{L^p(\mathbb{R}^d)} \lesssim \|v\|_{\exp L^p(\mathbb{R}^d)}.$$

The first inequality can be established as in [120, Theorem 3.4] and function $\chi(t) = (1+\ln t)/\ln(1+t)$ has a maximum value for $t \geq 1$, while the second inequality can be shown by the method employed in [94, Lemma 5.2].

Therefore, due to the triangle inequality for v^{**}, that is, $(f+g)^{**} \leq f^{**} + g^{**}$, we have

$$\frac{(\phi)^{**}(r) - (\mathscr{A}^{\gamma}_{\alpha,0}\phi)^{**}(r)}{(\ln(e/r))^{1/p}} \leq \frac{(\phi - \mathscr{A}^{\gamma}_{\alpha,0}(t)\phi)^{**}(r)}{(\ln(e/r))^{1/p}},$$

which implies from the nonnegative property of v^* ($v^*(r) \geq 0$ for any $r > 0$) that

$$\begin{aligned}\lim_{r\to 0} \frac{(\phi)^{**}(r) - (\mathscr{A}^{\gamma}_{\alpha,0}\phi)^{**}(r)}{(\ln(e/r))^{1/p}} &\lesssim \sup_{0<r<1} \frac{(\mathscr{A}^{\gamma}_{\alpha,0}(t)\phi - \phi)^{**}(r)}{(\ln(e/r))^{1/p}} \\ &\lesssim \|\mathscr{A}^{\gamma}_{\alpha,0}(t)\phi - \phi\|_{\exp L^p(\mathbb{R}^d)}.\end{aligned}$$

According to Lemmas 2.1 and 2.3, we get $\mathscr{A}^{\gamma}_{\alpha,0}(t)\phi \in L^\infty(\mathbb{R}^d)$ for all $t \geq \varepsilon$ with any $\varepsilon > 0$. This means that $(\mathscr{A}^{\gamma}_{\alpha,0}(t)\phi)^{**}(r) \in L^\infty(0,\infty)$ for all $t \geq \varepsilon$. Hence we have

$$\lim_{r\to 0} \frac{(\mathscr{A}^{\gamma}_{\alpha,0}(t)\phi)^{**}(r)}{(\ln(e/r))^{1/p}} = 0.$$

Let

$$\phi(x) = \begin{cases} \dfrac{p-1-p\ln(\omega_d|x|^d)}{p(1-\ln(\omega_d|x|^d))^{\frac{p-1}{p}}}, & 0<|x|<1, \\ 0, & \text{otherwise,} \end{cases}$$

where ω_d is the measure of the unit ball in $\mathbb{R}^d$. For $\psi(t) = e^{t^p} - 1$ with $p \geq 1$, it follows that

$$\int_{\mathbb{R}^d} \psi(|\phi(x)|)dx = \int_0^\infty \psi(|\phi^*(r)|)dr.$$

Furthermore, it is easy to check that $(\phi)^{**}(r) = (\ln(e/r))^{1/p}$ for $0 < r < \omega_d$. Consequently, we get

$$1 \lesssim \lim_{r \to 0} \frac{(\phi)^{**}(r)}{(\ln(e/r))^{1/p}} \lesssim \|\mathscr{A}^{\gamma}_{\alpha,0}(t)\phi - \phi\|_{\exp L^p(\mathbb{R}^d)},$$

which establishes (2.3). The proof is completed.

In view of Remark 2.1, we consider the Cauchy problem (2.1) in the following weak sense.

Definition 2.1 A function $u \in C((0,\infty); \exp L^p(\mathbb{R}^d))$ is a weak mild solution of (2.1) if it satisfies the integral equation (2.2) in $\exp L^p(\mathbb{R}^d)$ for almost all $t > 0$ and the initial data:

$$\lim_{t \to 0} u(t) \stackrel{*}{=} u_0, \text{ in } \exp L^p(\mathbb{R}^d).$$

Observe that $u(t) \to u_0$ as $t \to 0$ in weak* topology if and only if

$$\lim_{t \to 0} \int_{\mathbb{R}^d} (u(t,x)\phi(x) - u_0(x)\phi(x))d = 0,$$

for every $\phi \in L^1(\ln L)^{1/p}(\mathbb{R}^d)$, the pre-dual space of $\exp L^p(\mathbb{R}^d)$.

2.1.3 Global Existence

In this subsection, we show the global existence of solutions to the Cauchy problem (2.1). In order to achieve this aim, we assume that the exponential growth function f associated with $f(u) \sim |u|^\sigma$ near zero is given by

$$f(0) = 0, \quad |f(u) - f(v)| \lesssim |u - v|(|u|^{\sigma-1}e^{\lambda|u|^\varsigma} + |v|^{\sigma-1}e^{\lambda|v|^\varsigma}),$$

for some constants $\varsigma, \lambda > 0$ and $\sigma \geq 1$. A typical example satisfying previous nonlinearity is proposed $f(u) = u(e^{4\pi u^2} - 1)$ considering in [27].

Theorem 2.1 *Let $1 \leq d < 2\gamma p$, $p \geq 1$, $\gamma \in (0,1)$, and $\sigma \geq 1 + 2\gamma p/d$. If there exists a $\chi > 0$ such that, for all $u_0 \in \exp L^p(\mathbb{R}^d)$ with $\|u_0\|_{\exp L^p(\mathbb{R}^d)} \leq \chi$, then there exists a unique weak mild solution u of the Cauchy problem (2.1) satisfying*

$$\lim_{t\to 0}\|u(t)-\mathscr{A}^{\gamma}_{\alpha,0}(t)u_0\|_{\exp L^p(\mathbb{R}^d)}=0.$$

Moreover, for some $r>2\gamma p^2/d+p$, the decay estimate holds

$$\|u(t)\|_{L^r(\mathbb{R}^d)}\lesssim t^{-\frac{\alpha d}{2\gamma}\left(\frac{1}{p}-\frac{1}{r}\right)}\|u_0\|_{\exp L^p(\mathbb{R}^d)}.$$

Proof Let $\varrho=\frac{\alpha d}{2\gamma}(\frac{1}{p}-\frac{1}{r})$. For any $\varepsilon>0$, define a complete metric space X_ε by

$$\begin{aligned}X_\varepsilon:=\{&u\in C((0,\infty);\exp L^p(\mathbb{R}^d)):\\ &\sup_{t>0}t^{\varrho}\|u(t)\|_{L^r(\mathbb{R}^d)}+\|u\|_{L^\infty(0,\infty;\exp L^p(\mathbb{R}^d))}\le\varepsilon\},\end{aligned}$$

endowed with the distance $d(u,v)=\sup_{t>0}t^{\varrho}\|u(t)-v(t)\|_{L^r(\mathbb{R}^d)}$.

In the sequel, we set an operator Q by

$$Q(u)(t):=\mathscr{A}^{\gamma}_{\alpha,0}(t)\phi+\int_0^t(t-s)^{\alpha-1}\mathscr{A}^{\gamma}_{\alpha,1}(t-s)f(u(s))ds.$$

We solve the current problem with the exponential growth term by a contraction mapping argument by splitting the proof into four steps.

Step 1. Q is a contraction on X_ε. For any $u,v\in X_\varepsilon$, Lemma 2.3 (i) implies that

$$\|Q(u)(t)-Q(v)(t)\|_{L^r(\mathbb{R}^d)}\lesssim\int_0^t(t-s)^{\alpha-1-\frac{\alpha d}{2\gamma p}}\|f(u)-f(v)\|_{L^l(\mathbb{R}^d)}ds,$$

where $d(1/l-1/r)<2\gamma$. From the assumption of f, it follows by Taylor expansion that

$$\begin{aligned}&\int_0^t(t-s)^{\alpha-1-\frac{\alpha d}{2\gamma p}}\|f(u)-f(v)\|_{L^l(\mathbb{R}^d)}ds\\ \lesssim&\sum_{k=0}^{\infty}\frac{\lambda^k}{k!}\int_0^t(t-s)^{\alpha-1-\frac{\alpha d}{2\gamma p}}\||u-v|(|u|^{\sigma-1+\varsigma k}+|v|^{\sigma-1+\varsigma k})\|_{L^l(\mathbb{R}^d)}ds.\end{aligned}$$

By Hölder's inequality and Minkowski's inequality, for $1/l=1/r+1/p$, we have

$$\begin{aligned}&\||u-v|(|u|^{\sigma-1+\varsigma k}+|v|^{\sigma-1+\varsigma k})\|_{L^l(\mathbb{R}^d)}\\ \lesssim&\|u-v\|_{L^r(\mathbb{R}^d)}\|(|u|^{\sigma-1+\varsigma k}+|v|^{\sigma-1+\varsigma k})\|_{L^p(\mathbb{R}^d)}\\ \lesssim&\|u-v\|_{L^r(\mathbb{R}^d)}\left(\|u\|^{\sigma-1+\varsigma k}_{L^{p(\sigma-1+\varsigma k)}(\mathbb{R}^d)}+\|v\|^{\sigma-1+\varsigma k}_{L^{p(\sigma-1+\varsigma k)}(\mathbb{R}^d)}\right).\end{aligned}$$

For some $0 < c < \frac{(\sigma-1)(r-p)}{\gamma pr} \wedge \frac{2(\sigma-2)}{d(\sigma-1)}$, let

$$\zeta = \frac{(1-\vartheta)\gamma p^2 rc}{\vartheta(r-p-\gamma p^2 c)}, \qquad \vartheta = \frac{\gamma prc}{(\sigma-1+\varsigma k)(r-p)}, \qquad \frac{1}{\gamma p} = \frac{2}{d} - c.$$

Then $\zeta \geq p$, $0 < \vartheta \leq 1$ for each $k \in \mathbb{N}_0$ and $\alpha - \frac{\alpha d}{2\gamma p} - \varrho(\sigma - 1 + \varsigma k)\vartheta = 0$. Hence Hölder's interpolation inequality implies that

$$\|u\|_{L^{p(\sigma-1+\varsigma k)}(\mathbb{R}^d)} \leq \|u\|_{L^r(\mathbb{R}^d)}^{\vartheta} \|u\|_{L^\zeta(\mathbb{R}^d)}^{1-\vartheta},$$

where

$$\frac{1}{p(\sigma-1+\varsigma k)} = \frac{\vartheta}{r} + \frac{1-\vartheta}{\zeta}.$$

Additionally, for any $y \in \exp L^p(\mathbb{R}^d)$, Lemma 2.2 shows that

$$\|y\|_{L^\zeta(\mathbb{R}^d)}^{(\sigma-1+\varsigma k)(1-\vartheta)} \lesssim (\Gamma(\zeta/p+1))^{\frac{(\sigma-1+\varsigma k)(1-\vartheta)}{\zeta}} \|y\|_{\exp L^p(\mathbb{R}^d)}^{(\sigma-1+\varsigma k)(1-\vartheta)}.$$

By virtue of $\Gamma(x+1) \leq Cx^{x+1/2}$ for all $x \geq 1$ and for some constant $C > 0$, from Stirling's formula and the inequality $(\sigma - 1 + \varsigma k)(1 - \vartheta) \leq \zeta$, it follows that

$$\Gamma(\zeta/p+1)^{\frac{(\sigma-1+\varsigma k)(1-\vartheta)}{\zeta}} \leq C^{k-1}\Gamma(k+1).$$

Moreover, note the fact

$$\int_0^t (t-s)^{a-1}s^{b-1}ds = \frac{\Gamma(a)\Gamma(b)}{\Gamma(a+b)}t^{a+b-1}, \qquad \text{for } a, b > 0,\ t > 0,$$

combined with $d < 2p\gamma$, $0 < \alpha < 1$, and $\varrho + \varrho(\sigma - 1 + \varsigma k)\vartheta < 1$, and we obtain

$$\begin{aligned}
&\|Q(u)(t) - Q(v)(t)\|_{L^r(\mathbb{R}^d)} \\
&\lesssim \sum_{k=0}^{\infty} C^{k-1}\lambda^k \int_0^t (t-s)^{\alpha-1-\frac{\alpha d}{2\gamma p}} \|u-v\|_{L^r(\mathbb{R}^d)} \\
&\quad \times \left(\|u\|_{L^r(\mathbb{R}^d)}^{(\sigma-1+\varsigma k)\vartheta} \|u\|_{\exp L^p(\mathbb{R}^d)}^{(\sigma-1+\varsigma k)(1-\vartheta)} + \|v\|_{L^r(\mathbb{R}^d)}^{(\sigma-1+\varsigma k)\vartheta} \|v\|_{\exp L^p(\mathbb{R}^d)}^{(\sigma-1+\varsigma k)(1-\vartheta)}\right) ds \\
&\lesssim \sum_{k=0}^{\infty} C^{k-1}\lambda^k \varepsilon^{\sigma-1+\varsigma k} \int_0^t (t-s)^{\alpha-1-\frac{\alpha d}{2\gamma p}} s^{-\varrho-\varrho(\sigma-1+\varsigma k)\vartheta} ds\, d(u,v) \\
&\lesssim \sum_{k=0}^{\infty} C^{k-1}\lambda^k \varepsilon^{\sigma-1+\varsigma k} t^{-\varrho} d(u,v),
\end{aligned}$$

which means that there exists a constant $C_1 > 0$ for any $u \in X_\varepsilon$ such that

$$d(Q(u), Q(v)) \leq C_1 \sum_{k=0}^{\infty} C^{k-1} \lambda^k \varepsilon^{\sigma-1+\varsigma k} d(u, v).$$

For some small enough $\varepsilon > 0$ satisfying

$$C_1 \sum_{k=0}^{\infty} C^{k-1} \lambda^k \varepsilon^{\sigma-1+\varsigma k} \leq 1/4,$$

we deduce that Q is a contraction on X_ε.

Step 2. Q maps X_ε into itself. The continuity of Q follows from that of the operator $\mathscr{A}^{\gamma}_{\alpha,\beta}(t)$ and the strongly continuous behavior of semigroup $T_\gamma(t)$ for all $t \geq 0$. Since $g(x) = \ln(x+1) - x/2 = 0$ has two zeros, it follows by letting $a \in (2,4)$ with $a/2 = \ln(a+1)$ that $(\ln((t-s)^{-\frac{\alpha d}{2\gamma}} + 1))^{-\frac{1}{p}} \leq 2^{\frac{1}{p}} (t-s)^{\frac{\alpha d}{2\gamma p}}$, for $0 \leq s \leq t - a^{-\frac{2\gamma}{\alpha d}}$, and $(\ln((t-s)^{-\frac{\alpha d}{2\gamma}} + 1))^{-\frac{1}{p}} \leq 1$, for $t - a^{-\frac{2\gamma}{\alpha d}} \leq s \leq t$. Therefore, by Lemma 2.3 (iii), for some $1 \leq d < 2p\gamma$, we have

$$\begin{aligned}
&\left\| \int_0^t (t-s)^{\alpha-1} \mathscr{A}^{\gamma}_{\alpha,1}(t-s) f(u) ds \right\|_{\exp L^p(\mathbb{R}^d)} \\
&\lesssim \int_0^t (t-s)^{\alpha-1-\frac{\alpha d}{2p\gamma}} (\ln(t-s)^{-\frac{\alpha d}{2\gamma}} + 1)^{-\frac{1}{p}} \| f(u) \|_{L^p(\mathbb{R}^d)} ds \\
&\lesssim \int_0^{t-a^{-\frac{2\gamma}{\alpha d}}} (t-s)^{\alpha-1} \| f(u) \|_{L^p(\mathbb{R}^d)} ds + \int_{t-a^{-\frac{2\gamma}{\alpha d}}}^t (t-s)^{\alpha-1-\frac{\alpha d}{2p\gamma}} \| f(u) \|_{L^p(\mathbb{R}^d)} ds \\
&\lesssim \int_0^t (t-s)^{\alpha-1} \| f(u(s)) \|_{L^p(\mathbb{R}^d)} ds + \sup_{s>0} \| f(u(s)) \|_{L^p(\mathbb{R}^d)} =: I + II.
\end{aligned}$$

By Taylor expansion, we have

$$|f(u)| \leq \sum_{k=0}^{\infty} \frac{\lambda^k}{k!} |u|^{\varsigma k + \sigma}.$$

For $r > \frac{2\gamma p^2}{d} + p$ and $\sigma \geq 1 + 2\gamma p/d$, let

$$\theta = \frac{2\gamma p r}{d(r-p)(\varsigma k + \sigma)}, \quad \varpi = \frac{2\gamma p^2 r (1-\theta)}{\theta(d(r-p) - 2\gamma p^2)}.$$

Clearly, for each $k \in \mathbb{N}_0$, we have $\theta \in (0, 1]$ and $\alpha = (\varsigma k + \sigma)\theta \varrho$. Next, by virtue of Hölder's interpolation inequality, we get

$$\| u \|_{L^{(\varsigma k + \sigma)p}(\mathbb{R}^d)} \lesssim \| u \|^{\theta}_{L^r(\mathbb{R}^d)} \| u \|^{1-\theta}_{L^{\varpi}(\mathbb{R}^d)},$$

where

$$\frac{1}{(\varsigma k+\sigma)p}=\frac{\theta}{r}+\frac{1-\theta}{\varpi}.$$

Similar to the proof to Step 1, it follows by Lemma 2.1 that

$$\begin{aligned}&\int_0^t (t-s)^{\alpha-1}\|u(s)\|^{\varsigma k+\sigma}_{L^{(\varsigma k+\sigma)p}(\mathbb{R}^d)}ds\\ \lesssim&\Gamma(\varpi/p+1)^{\frac{(\varsigma k+\sigma)(1-\theta)}{\varpi}}\int_0^t (t-s)^{\alpha-1}\|u(s)\|^{(\varsigma k+\sigma)\theta}_{L^r(\mathbb{R}^d)}\|u(s)\|^{(\varsigma k+\sigma)(1-\theta)}_{\exp L^p(\mathbb{R}^d)}ds\\ \lesssim&\Gamma(k+1)\varepsilon^{\varsigma k+\sigma},\end{aligned}$$

for $u\in X_\varepsilon$ and $\Gamma(\varpi/p+1)\le C^k\Gamma(k+1)$ since $(\varsigma k+\sigma)(1-\theta)\le\varpi$. Therefore, by $\||u|^{\varsigma k+\sigma}\|_{L^p(\mathbb{R}^d)}=\|u\|^{\varsigma k+\sigma}_{L^{(\varsigma k+\sigma)p}(\mathbb{R}^d)}$, we obtain

$$I\le\sum_{k=0}^{\infty}\frac{\lambda^k}{k!}\int_0^t (t-s)^{\alpha-1}\||u(s)|^{\varsigma k+\sigma}\|_{L^p(\mathbb{R}^d)}ds\le\sum_{k=0}^{\infty}\frac{\lambda^k}{k!}\lesssim\sum_{k=0}^{\infty}\lambda^k\varepsilon^{\varsigma k+\sigma}. \tag{2.4}$$

Let us prove the second term. In fact, by the assumption of f, Hölder's inequality with $1/p=1/(a_1p)+1/(a_2p)$ for some constants $a_1\ge 1\vee 1/\varsigma$, $a_2\ge 1$, we have

$$\|f(u)\|_{L^p(\mathbb{R}^d)}\lesssim\|e^{\lambda|u|^\varsigma}-1\|_{L^{a_1p}(\mathbb{R}^d)}\|u\|^\sigma_{L^{\sigma a_2 p}(\mathbb{R}^d)}+\|u\|^\sigma_{L^{\sigma p}(\mathbb{R}^d)}. \tag{2.5}$$

For $a_1p\ge 1$, $\Gamma(x+1)\le Cx^{x+1/2}$, we obtain by the Stirling formula that

$$\begin{aligned}\|e^{\lambda|u|^\varsigma}-1\|_{L^{a_1p}(\mathbb{R}^d)}&\le\sum_{k=1}^{\infty}\frac{\lambda^k}{k!}\|u\|^{\varsigma k}_{L^{\varsigma k a_1 p}(\mathbb{R}^d)}\le\sum_{k=1}^{\infty}\frac{\lambda^k}{k!}\Gamma(\varsigma k a_1+1)^{\frac{1}{a_1p}}\|u\|^{\varsigma k}_{\exp L^p(\mathbb{R}^d)}\\ &\lesssim\sum_{k=0}^{\infty}\lambda^k\varepsilon^{\varsigma k}.\end{aligned}$$

In addition, we have

$$\sup_{t>0}\|f(u)\|_{L^p(\mathbb{R}^d)}\lesssim\sum_{k=0}^{\infty}\lambda^k\varepsilon^{\varsigma k}\|u\|^\sigma_{\exp L^p(\mathbb{R}^d)}+\|u\|^\sigma_{\exp L^p(\mathbb{R}^d)}\lesssim\sum_{k=0}^{\infty}\lambda^k\varepsilon^{\varsigma k+\sigma}+\varepsilon^\sigma. \tag{2.6}$$

By Lemma 2.3 (ii), we find that

$$\|\mathscr{A}^\gamma_{\alpha,0}(t)u_0\|_{\exp L^p(\mathbb{R}^d)}\lesssim\|u_0\|_{\exp L^p(\mathbb{R}^d)}. \tag{2.7}$$

Then it follows from (2.4), (2.6), and (2.7) that

$$\|Q(u)(t)\|_{\exp L^p(\mathbb{R}^d)} \lesssim \sum_{k=0}^{\infty} \lambda^k \varepsilon^{\varsigma k+\sigma} + \varepsilon^{\sigma} + \|u_0\|_{\exp L^p(\mathbb{R}^d)}. \tag{2.8}$$

Furthermore, by using Lemma 2.3 (i) and Lemma 2.1 and letting $f(v) = 0$ for $v = 0$ as in Step 1, we get

$$\begin{aligned}\|Q(u)(t)\|_{L^r(\mathbb{R}^d)} \leq & \|\mathscr{A}_{\alpha,0}^{\gamma}(t)u_0\|_{L^r(\mathbb{R}^d)} + \left\| \int_0^t (t-s)^{\alpha-1} \mathscr{A}_{\alpha,1}^{\gamma}(t-s) f(u) ds \right\|_{L^r(\mathbb{R}^d)} \\ \lesssim & t^{-\varrho} \|u_0\|_{\exp L^p(\mathbb{R}^d)} + t^{-\varrho} \sum_{k=0}^{\infty} \lambda^k \varepsilon^{\varsigma k+\sigma}.\end{aligned} \tag{2.9}$$

Consequently, by virtue of (2.8) and (2.9), there exists a constant $C > 0$ such that

$$\|Q(u)\|_{X_\varepsilon} \leq C\left(\|u_0\|_{\exp L^p(\mathbb{R}^d)} + \sum_{k=0}^{\infty} \lambda^k \varepsilon^{\varsigma k+\sigma} + \varepsilon^{\sigma} \right).$$

If we take $\varepsilon = 4C\chi$ for χ to be small enough such that $C\varepsilon^{\sigma} < \varepsilon/4$ and $C\sum_{k=0}^{\infty} \lambda^k \varepsilon^{\varsigma k} < 1$, then Q is a contraction from X_ε into itself. Thus, by the contraction mapping principle, there exists a unique solution to the Cauchy problem (2.1).

Step 3. Next we prove the continuity of solution at zero. According to Lemma 2.3 (iv), we have

$$\begin{aligned}\|u(t) - \mathscr{A}_{\alpha,0}^{\gamma}(t)u_0\|_{\exp L^p(\mathbb{R}^d)} \lesssim & \int_0^t (t-s)^{\alpha-1-\frac{\alpha d}{2\gamma p}} \|f(u)\|_{L^p(\mathbb{R}^d)} ds \\ & + \int_0^t \|f(u)\|_{L^p(\mathbb{R}^d)} ds.\end{aligned}$$

For the estimate of $\|f(u)\|_{L^p(\mathbb{R}^d)}$ given by (2.5) and for any $u \in X_\varepsilon$ with small $\varepsilon > 0$, we have

$$\begin{aligned}& \|u(t) - \mathscr{A}_{\alpha,0}^{\gamma}(t)u_0\|_{\exp L^p(\mathbb{R}^d)} \\ \lesssim & \int_0^t (t-s)^{\alpha-1-\frac{\alpha d}{2\gamma p}} \|u(s)\|_{\exp L^p(\mathbb{R}^d)} ds + \int_0^t \|u(s)\|_{\exp L^p(\mathbb{R}^d)} ds \\ \lesssim & t^{\alpha\left(1-\frac{d}{2\gamma p}\right)} \|u\|_{L^\infty(0,\infty;\ \exp L^p(\mathbb{R}^d))} + t\|u\|_{L^\infty(0,\infty;\ \exp L^p(\mathbb{R}^d))} \\ \to & 0, \quad as\ t \to 0.\end{aligned}$$

Step 4. Finally we check the weak* convergence at $t = 0$. Let $X = L^1(\ln L)^{1/p}(\mathbb{R}^d)$ be the pre-dual space of $\exp L^p(\mathbb{R}^d)$ (i.e., $X^* = \exp L^p(\mathbb{R}^d)$). Since X is a Banach space and $C_0^\infty(\mathbb{R}^d)$ is dense in X, it follows by the properties of $\mathscr{M}_\alpha(\cdot)$ that

$$\begin{aligned}
\langle \mathscr{A}^\gamma_{\alpha,0}(t)u_0 - u_0, \phi\rangle_{X^*,X} &= \int_{\mathbb{R}^d} (\mathscr{A}^\gamma_{\alpha,0}(t)u_0(x) - u_0(x))\phi(x)dx \\
&= \int_{\mathbb{R}^d}\int_0^\infty \mathscr{M}_\alpha(\theta)\left(\exp((-\Delta)^\gamma t^\alpha\theta)u_0(x) - u_0(x)\right) \\
&\quad \phi(x)d\theta dx \\
&= \int_0^\infty \mathscr{M}_\alpha(\theta)\int_{\mathbb{R}^d}\left(\exp((-\Delta)^\gamma t^\alpha\theta)u_0(x) - u_0(x)\right) \\
&\quad \phi(x)dxd\theta \\
&= \int_0^\infty \mathscr{M}_\alpha(\theta)\int_{\mathbb{R}^d}\left(\exp((-\Delta)^\gamma t^\alpha\theta)\phi(x) - \phi(x)\right) \\
&\quad u_0(x)dxd\theta \\
&= \int_{\mathbb{R}^d}\left(\mathscr{A}^\gamma_{\alpha,0}(t)\phi(x) - \phi(x)\right)u_0(x)dx,
\end{aligned}$$

which, by Hölder's inequality for the Orlicz space, implies that

$$|\langle \mathscr{A}^\gamma_{\alpha,0}(t)u_0 - u_0, \phi\rangle_{X^*,X}| \lesssim \|u_0\|_{X^*}\|\mathscr{A}^\gamma_{\alpha,0}(t)\phi - \phi\|_X.$$

By virtue of the density of $C_0^\infty(\mathbb{R}^d)$ in X, we have $\|\mathscr{A}^\gamma_{\alpha,0}(t)\phi - \phi\|_X \to 0$ as $t \to 0$. Consequently, the conclusions are achieved, and the proof is completed.

Remark 2.2 Notice that the solution of the problem (2.1) is in $\exp L^2(\mathbb{R})$ for $\gamma \in (1/4, 1/2]$, while the global solution may not exist for $\gamma \in (0, 1/4]$. If $\gamma \to 1$ in (2.1), then one can establish the global existence result for $\gamma \in (0, 1)$ by the same method by replacing the operator $(-\Delta)^\gamma$ with the Laplace operator.

Remark 2.3 Consider the embedding $H^{s,q}(\mathbb{R}^d) \hookrightarrow \exp L^\Psi(\mathbb{R}^d)$ together with Trudinger's inequality (see, e.g., [65]), where $\exp L^\Psi(\mathbb{R}^d)$ is an Orlicz space defined by the convex function

$$\Psi(t) := \exp(t^{q/(q-1)}) - \sum_{j=0}^{k-1} t^{jq/(q-1)}/j!,$$

k is the smallest integer satisfying $k \geq q - 1$, and $H^{s,q}(\mathbb{R}^d)$ is the Sobolev spaces for any $s \in \mathbb{R}$ and $1 < q < \infty$ defined by

$$H^{s,q}(\mathbb{R}^d) = \{\psi \in \mathscr{S}'(\mathbb{R}^d) : (1 - \Delta)^{s/2}\psi \in L^q(\mathbb{R}^d)\}.$$

Then the solution of the Cauchy problem (2.1) can be considered for initial data $u_0 \in H^{s,q}(\mathbb{R}^d)$. In particular, the growth of the nonlinearity at infinity is of the form $f(u) \sim e^{u^{q/(q-1)}}$ for $u_0 \in H^{n/q,q}(\mathbb{R}^d)$.

2.1.4 Local Existence

In this subsection, we set $X = L^1(\mathbb{R}^d) \cap L^\infty(\mathbb{R}^d)$ for each $d \geq 1$ and obtain the local solutions to the Cauchy problem (2.1) for a small initial data $u_0 \in X$. We are concerned with the local existence and uniqueness of mild solution of the problem (2.1). First, we give the definition of a mild solution to (2.1). Since $C_0^\infty(\mathbb{R}^d)$ is dense in $L^1(\mathbb{R}^d)$, by Lemma 2.2, it is clear that $X \hookrightarrow \exp L_0^p(\mathbb{R}^d)$ for all $p \geq 1$, where $\exp L_0^p(\mathbb{R}^d)$ is the closure of $C_0^\infty(\mathbb{R}^d)$ in $\exp L^p(\mathbb{R}^d)$ with respect to the same norm, for example, see [95]. So, it is natural to consider the local solution in X without the Orlicz space.

Definition 2.2 Let $u_0 \in X$ and $T > 0$. We say that $u \in C([0,T]; X)$ is a mild solution to (2.1) if (2.2) holds.

Theorem 2.2 *Let $u_0 \in X$ and $\sigma > 3/2$. Then the problem (2.1) has a unique mild solution on $[0, T_*]$ for some $T_* > 0$.*

Proof For given $T > 0$ and $R > 0$, we define a ball in Banach space $C([0,T]; X)$ by

$$B_R = \{u \in C([0,T]; X) : \ \|u\|_* \leq R\},$$

where the norm $\|u\|_* = \|u\|_{L^\infty(0,T;X)}$. Considering the operator Q be defined in Theorem 2.1, we shall show the existence of local solution by the fixed point theorem. We first verify that $Q(B_R) \subset B_R$.

In fact, for $\sigma > 3/2$, let $\theta = 1/(\varsigma k + \sigma)$ for each $k \in \mathbb{N}_0$. Clearly, $\theta \in (0,1)$. The Hölder interpolation inequality implies that

$$\|u\|_{L^{\varsigma k+\sigma}(\mathbb{R}^d)} \lesssim \|u\|_{L^1(\mathbb{R}^d)}^{\theta} \|u\|_{L^\infty(\mathbb{R}^d)}^{1-\theta}.$$

For any $u \in B_R$, we obtain

$$\int_0^t (t-s)^{\alpha-1} \|u(s)\|_{L^{\varsigma k+\sigma}(\mathbb{R}^d)}^{\varsigma k+\sigma} ds \lesssim \int_0^t (t-s)^{\alpha-1} \|u(s)\|_{L^1(\mathbb{R}^d)}^{(\varsigma k+\sigma)\theta}$$
$$\|u(s)\|_{L^\infty(\mathbb{R}^d)}^{(\varsigma k+\sigma)(1-\theta)} ds \lesssim T^\alpha R^{\varsigma k+\sigma},$$

which yields by Lemma 2.3 (i) that

$$\begin{aligned}\|Q(u)(t)\|_{L^1(\mathbb{R}^d)} &\lesssim \|u_0\|_{L^1(\mathbb{R}^d)} + \int_0^t (t-s)^{\alpha-1}\|f(u)(s)\|_{L^1(\mathbb{R}^d)}ds \\ &\lesssim \|u_0\|_X + \sum_{k=0}^{\infty}\int_0^t (t-s)^{\alpha-1}\||u|^{\varsigma k+\sigma}\|_{L^1(\mathbb{R}^d)}ds \\ &\lesssim \|u_0\|_X + \sum_{k=0}^{\infty}\frac{\lambda^k}{k!}\int_0^t (t-s)^{\alpha-1}\|u(s)\|^{\varsigma k+\sigma}_{L^{\varsigma k+\sigma}(\mathbb{R}^d)}ds.\end{aligned}$$

Therefore, for any $u \in B_R$, we have

$$\|Q(u)(t)\|_{L^1(\mathbb{R}^d)} \lesssim \|u_0\|_X + \sum_{k=0}^{\infty}\frac{\lambda^k}{k!}T^\alpha R^{\varsigma k+\sigma} \lesssim T^\alpha R^\sigma e^{\lambda R^\varsigma}.$$

In addition, for any $p \in (d/(2\gamma) \vee 1, \infty)$, we have

$$\begin{aligned}\|Q(u)(t)\|_{L^\infty(\mathbb{R}^d)} &\lesssim \|u_0\|_{L^\infty(\mathbb{R}^d)} + \int_0^t (t-s)^{\alpha-1-\frac{\alpha d}{2\gamma p}}\|f(u)(s)\|_{L^p(\mathbb{R}^d)}ds \\ &\lesssim \|u_0\|_X + \sum_{k=0}^{\infty}\int_0^t (t-s)^{\alpha-1-\frac{\alpha d}{2\gamma p}}\||u|^{\varsigma k+\sigma}\|_{L^p(\mathbb{R}^d)}ds \\ &\lesssim \|u_0\|_X + \sum_{k=0}^{\infty}\frac{\lambda^k}{k!}\int_0^t (t-s)^{\alpha-1-\frac{\alpha d}{2\gamma p}}\|u(s)\|^{\varsigma k+\sigma}_{L^{(\varsigma k+\sigma)p}(\mathbb{R}^d)}ds.\end{aligned}$$

Moreover, letting $r \in (1, \sigma p)$, we have that $\vartheta = r/((\varsigma k+\sigma)p) \in (0, 1)$ for each $k \in \mathbb{N}_0$. Hence the Hölder interpolation inequality yields

$$\|u\|_{L^{(\varsigma k+\sigma)p}(\mathbb{R}^d)} \lesssim \|u\|^{\vartheta}_{L^r(\mathbb{R}^d)}\|u\|^{1-\vartheta}_{L^\infty(\mathbb{R}^d)} \lesssim \|u\|^{\varpi\vartheta}_{L^1(\mathbb{R}^d)}\|u\|^{1-\varpi\vartheta}_{L^\infty(\mathbb{R}^d)},$$

for $\varpi = 1/r \in (0, 1)$. This shows that

$$\begin{aligned}&\|Q(u)(t)\|_{L^\infty(\mathbb{R}^d)} \\ &\lesssim \|u_0\|_X + \sum_{k=0}^{\infty}\frac{\lambda^k}{k!}\int_0^t (t-s)^{\alpha-1-\frac{\alpha d}{2\gamma p}}\|u(s)\|_X^{(\varsigma k+\sigma)\varpi\vartheta}\|u(s)\|^{(\varsigma k+\sigma)(1-\varpi\vartheta)}_{L^\infty(\mathbb{R}^d)}ds \\ &\lesssim \|u_0\|_X + \sum_{k=0}^{\infty}\frac{\lambda^k}{k!}T^{\alpha-\frac{\alpha d}{2\gamma p}}R^{\varsigma k+\sigma} \\ &\lesssim \|u_0\|_X + T^{\alpha-\frac{\alpha d}{2\gamma p}}R^\sigma e^{\lambda R^\varsigma}.\end{aligned}$$

Therefore, there exists a constant $C > 0$ such that

$$\|Q(u)(t)\|_X \leq C\|u_0\|_X + CT^{\alpha-\frac{\alpha d}{2\gamma p}} R^\sigma e^{\lambda R^\varsigma} + CT^\alpha R^\sigma e^{\lambda R^\varsigma}.$$

Consequently, letting $R = 2C\|u_0\|_X$ and choosing T small enough, we get

$$C\Big(T^\alpha + T^{\alpha-\frac{\alpha d}{2\gamma p}}\Big) R^{\sigma-1} e^{\lambda R^\varsigma} \leq 1/2. \tag{2.10}$$

Thus we deduce that $\|Q(u)\|_X \leq R$, and hence $Q(u) \in B_R$ for any $u \in B_R$.

Next we verify that Q is a contraction map. Let $\mu = 1/(2(\varsigma k + \sigma - 1))$ for $\sigma > 3/2$, $k \in \mathbb{N}_0$, and observe that $\mu \in (0, 1)$. Then, by Hölder's interpolation inequality, we have

$$\begin{aligned}
\| |u|^{\sigma-1} e^{\lambda |u|^\varsigma} \|_{L^2(\mathbb{R}^d)} &= \sum_{k=0}^{\infty} \frac{\lambda^k}{k!} \|u\|^{\varsigma k+\sigma-1}_{L^{2(\varsigma k+\sigma-1)}(\mathbb{R}^d)} \\
&\lesssim \sum_{k=0}^{\infty} \frac{\lambda^k}{k!} \|u\|^{\mu(\varsigma k+\sigma-1)}_{L^1(\mathbb{R}^d)} \|u\|^{(1-\mu)(\varsigma k+\sigma-1)}_{L^\infty(\mathbb{R}^d)} \\
&\lesssim R^{\sigma-1} e^{\lambda R^\varsigma},
\end{aligned}$$

for any $u \in B_R$. Letting $u, v \in B_R$, it follows by Lemma 2.3 (i) and the Hölder inequality that

$$\begin{aligned}
\|f(u)(s) - f(v)(s)\|_{L^1(\mathbb{R}^d)} &\lesssim \| |u - v| (|u|^{\sigma-1} e^{\lambda |u|^\varsigma} + |v|^{\sigma-1} e^{\lambda |v|^\varsigma}) \|_{L^1(\mathbb{R}^d)} \\
&\lesssim \|u - v\|_{L^2(\mathbb{R}^d)} \| |u|^{\sigma-1} e^{\lambda |u|^\varsigma} + |v|^{\sigma-1} e^{\lambda |v|^\varsigma} \|_{L^2(\mathbb{R}^d)} \\
&\lesssim R^{\sigma-1} e^{\lambda R^\varsigma} \|u - v\|^{1/2}_{L^1(\mathbb{R}^d)} \|u - v\|^{1/2}_{L^\infty(\mathbb{R}^d)} \\
&\lesssim R^{\sigma-1} e^{\lambda R^\varsigma} \|u - v\|_X.
\end{aligned}$$

Consequently, we get

$$\begin{aligned}
\|Q(u)(t) - Q(v)(t)\|_{L^1(\mathbb{R}^d)} &\lesssim \int_0^t (t-s)^{\alpha-1} \|f(u)(s) - f(v)(s)\|_{L^1(\mathbb{R}^d)} ds \\
&\lesssim T^\alpha R^{\sigma-1} e^{\lambda R^\varsigma} \|u - v\|_X.
\end{aligned}$$

On the other hand, for any fixed $p \in (d/(2\gamma) \vee 1, \infty)$, we have by the earlier argument that

$$\begin{aligned}
&\|f(u)(s) - f(v)(s)\|_{L^p(\mathbb{R}^d)} \\
&\lesssim \| |u - v| (|u|^{\sigma-1} e^{\lambda |u|^\varsigma} + |v|^{\sigma-1} e^{\lambda |v|^\varsigma}) \|_{L^p(\mathbb{R}^d)}
\end{aligned}$$

$$
\begin{aligned}
&\lesssim \|u-v\|_{L^{2p}(\mathbb{R}^d)} \| |u|^{\sigma-1}e^{\lambda|u|^\varsigma} + |v|^{\sigma-1}e^{\lambda|v|^\varsigma} \|_{L^{2p}(\mathbb{R}^d)} \\
&\lesssim \sum_{k=0}^{\infty} \frac{\lambda^k}{k!} \|u-v\|_{L^1(\mathbb{R}^d)}^{1/(2p)} \|u-v\|_{L^\infty(\mathbb{R}^d)}^{1-1/(2p)} \left(\|u\|_{L^{2(\varsigma k+\sigma-1)p}(\mathbb{R}^d)}^{\varsigma k+\sigma-1} + \|v\|_{L^{2(\varsigma k+\sigma-1)p}(\mathbb{R}^d)}^{\varsigma k+\sigma-1} \right) \\
&\lesssim \sum_{k=0}^{\infty} \frac{\lambda^k}{k!} \|u-v\|_X R^{\varsigma k+\sigma-1} \\
&\lesssim R^{\sigma-1} e^{\lambda R^\varsigma} \|u-v\|_X,
\end{aligned}
$$

which implies that

$$
\begin{aligned}
\|Q(u)(t)-Q(v)(t)\|_{L^\infty(\mathbb{R}^d)} &\lesssim \int_0^t (t-s)^{\alpha-1-\frac{\alpha d}{2\gamma p}} \|f(u)(s)-f(v)(s)\|_{L^p(\mathbb{R}^d)} ds \\
&\lesssim T^{\alpha-\frac{\alpha d}{2\gamma p}} R^{\sigma-1} e^{\lambda R^\varsigma} \|u-v\|_X.
\end{aligned}
$$

Thus, there exists a $C > 0$ (may be the same C given in (2.10)) such that

$$
\|Q(u)(t)-Q(v)(t)\|_X \le C T^{\alpha-\frac{\alpha d}{2\gamma p}} R^{\sigma-1} e^{\lambda R^\varsigma} \|u-v\|_X.
$$

Let T be small enough such that (2.10) holds; then Q is a contraction on B_R. Since $(-\Delta)^\gamma$ generates a strongly continuous semigroup $T_\gamma(t)$ on $L^1(\mathbb{R}^d)$, it is easy to check the continuity of Q. Hence, according to the Banach fixed point theorem, the problem (2.1) admits a local mild solution $u \in B_R$. The proof is completed.

2.2 Distributed Order Diffusion Problems

2.2.1 Introduction

One of the most important among anomalous diffusion processes is ultraslow diffusion where the mean squared variance grows only logarithmically with time, and then the corresponding mathematical and physical models lead to fractional diffusion equations with distributed order fractional derivative. These models have been the focus of many studies for their significant applications such as polymer physics and kinetics of particles moving in the quenched random force fields, see [160].

In this section, we consider the following semilinear distributed order fractional diffusion problems:

$$
\begin{cases}
D_t^{[\mu]} u - \Delta u = f(u), & t > 0,\ x \in \mathbb{R}^N, \\
u(0,x) = \varphi(x), & x \in \mathbb{R}^N,
\end{cases}
\tag{2.11}
$$

where Δ is the Laplacain operator, and $D_t^{[\mu]}u$ denotes the distributed order fractional derivative of u in time t (with respect to α) defined by

$$D_t^{[\mu]}u(t,x) = \int_0^1 \mu(\alpha)\partial_t^\alpha u d\alpha,$$

where ∂_t^α, $0 < \alpha < 1$, is the Caputo derivative of order α in t, defined by

$$\partial_t^\alpha u(t,x) = \frac{1}{\Gamma(1-\alpha)}\int_0^t (t-s)^{-\alpha}\frac{\partial}{\partial s}u(s,x)ds, \quad t>0,\ x\in\mathbb{R}^N,$$

provided the right-hand side of the equality is pointwise defined, and μ is a weighted function satisfying

$$\mu \in L^1(0,1), \quad \mu \geq 0, \ \mu \not\equiv 0. \tag{2.12}$$

From [124], we can know that there exists nonnegative $g \in L^1_{loc}[0,\infty)$ such that the fractional integration operator $I^{[\mu]}$, defined by the formula $I^{[\mu]}u = g * u$, satisfies

$$(D_t^{[\mu]}I^{[\mu]}u)(t) = u(t), \quad \text{for } u \in L^\infty(0,T),$$
$$(I^{[\mu]}D_t^{[\mu]}u)(t) = u(t) - u(0), \quad \text{for } u \in AC[0,T].$$

Furthermore, g satisfies $g(t) \leq c\max\{t^{\gamma-1},\ t^{-\gamma}\}$ for some positive constants c and $\gamma \in (0,\frac{1}{2})$.

It is generally known that Eq. (2.11) would resolve itself into time fractional diffusion equations when $\mu(\alpha) = \delta(\alpha-\alpha_0)$, where $\delta(\cdot)$ is the Dirac delta function and $\alpha_0 \in (0,1)$. It has attracted a growing interest due to its widespread applications in anomalous diffusion processes, the authors in [116] considered the problem with the Caputo derivative on $\mathbb{R}^N$ in L^q-framework, and then the uniqueness, existence, and $L^q(L^p)$-estimates of solutions were obtained. In addition, we can also refer to [30, 128, 135, 230] for the well-posedness, [139] for the asymptotic behavior, [51, 140, 237, 245] for the blow-up, and the references therein.

Compared with the extensive studies on the well-posedness for time fractional diffusion equations, the basic theoretical works for the distributed order case are far from sufficient. In [121], the author constructed fundamental solutions to the problem with distributed order Caputo derivative and established their positivity and subordination property. Later, the Fourier variable separation method was typically used to derive the representation of solutions to the problem, and then the existence, uniqueness, and regularity of solutions under the different assumptions on the weight function μ were derived in [132, 134]. Recently, Kubica and Ryszewska [124] used the Galerkin approximation method to study the parabolic-type problem with distributed order Caputo derivative, and the well-posedness of the solutions was obtained. For other results for the problem, we refer to [142] for the maximum principle, [123, 129] for the asymptotic behavior, [201] for the inverse source problems and the references therein.

On the other hand, since $D_t^{[\mu]}u$ can be represented as $\frac{d}{dt}(k * [u - u_0])$, where $k(t) = \int_0^1 \frac{t^{-\alpha}\mu(\alpha)}{\Gamma(1-\alpha)}d\alpha$, we know from the result [124, Proposition 6] that (k, g) is of type $\mathscr{PC}$ (see [110, 214, 215]). Thus the problem (2.11) is also a typical case of nonlocal in time problems. A very important result concerning the decay of solutions among them was found in [110]. The authors considered the equation

$$\frac{d}{dt}(k * [u - u_0])u - \Delta u = 0, \tag{2.13}$$

with the initial condition $u(0) = u_0$, and the kernel k is a given function of type $\mathscr{PC}$. The main results, given Theorem 5.1 and Corollary 5.1, claim that the L^r estimates of solutions to (2.13) are obtained by using Fourier multiplier methods and an estimation of relaxation functions and the decay behavior of solutions occurs a critical dimension phenomenon. In this section, we explore this problem for the distributed order fractional derivative in the semilinear case. Thus we are able to apply the results from [110] when $f = 0$. However, unlike the methods mentioned above, we develop our results based on the resolvents operator approach introduced in [84, 105, 182], and we give more straightforward argument than the one presented in [110] which shows that there are few restrictions to the dimension for deriving the resolvent operators. For other results for nonlocal in time problems, we also refer to [111, 214, 215].

In the papers mentioned above, it was shown that the properties of solutions are influenced by the assumption on μ. Under the assumption that μ satisfies (2.12), the polynomial decay was obtained in [110]. Further analysis was made in [123] if we additionally assume that $\mu(\alpha)/\alpha$ is integrable on $[0, 1]$. It was proved that they obtained the well-posedness and decay of solutions when $f = 0$ appeared in (2.11). Therefore, a question naturally arises: what is a factor that affects the existence and the decay of solutions for a distributed order fractional diffusion problem with a nonlinear term. In this section, we will present it in detail when μ satisfies (2.12). The primary contribution of the presented work is that we establish the L^p-L^q estimates and the continuity of the solution operator for $t > 0$ using the Gagliardo-Nirenberg inequality, which cannot be derived directly from the corresponding properties of the heat operator. Moreover, we for the first time point out the continuity of the solutions of the linear part at $t = 0$ in the sense of the space $L^p(\mathbb{R}^N)$.

The remainder of the section is organized as follows. In the next subsection, we introduce some notations and analytic properties of the Laplacian operator that will be used throughout the section. In Sect. 2.2.3, we establish the boundedness and the continuity of the operators derived from the initial value term and the nonlinear term. Finally, the local well-posedness and global well-posedness of mild solutions of the problem (2.11) are obtained, and then the decay of the solutions is also established which is influenced by the property of the Laplacian operator in space $L^p(\mathbb{R}^N)$.

2.2.2 Preliminaries

Let $(X, |\cdot|)$ be a Banach space and $\mathscr{L}(X)$ stand for the space of all linear bounded operators from X to itself with the norm $\|\cdot\|_{\mathscr{L}(X)}$. We denote by $C_b(I, X)$ the space of bounded continuous operators from I to X, $I \subset \mathbb{R}_+$, equipped with the norm $\sup_{t\in I} |\cdot|$. If A is a closed linear operator, we will denote by $D(A^\gamma)$ for $\gamma > 0$ the fractional power spaces associated with the operator A.

For $\varpi \in (0, \frac{\pi}{2})$, we define Σ_ϖ as

$$\Sigma_\varpi = \{z \in \mathbb{C} : z \neq 0, |\arg z| < \varpi\}.$$

In order to prove some crucial uniform estimates for the operators generated by the initial value term and the nonlinear term, respectively, we present the following technical lemma.

Lemma 2.4 *[11, Lemma 4.1.1] Given $\varpi \in (0, \pi/2)$, let Γ be an arbitrary piecewise smooth simple curve in $\Sigma_{\varpi+\pi/2}$ running from $\infty e^{-i(\varpi+\pi/2)}$ to $\infty e^{i(\varpi+\pi/2)}$, and let X be a Banach space. Suppose that the map $g : \Sigma_{\varpi+\pi/2} \times X \times \mathbb{R}^+ \to X$ has the following properties:*

(i) *$g(\cdot, x, t) : \Sigma_{\varpi+\pi/2} \to X$ is holomorphic for $(x, t) \in X \times \mathbb{R}^+$.*
(ii) *$g(z, \cdot, \cdot) \in C(X \times \mathbb{R}^+, X)$ for $z \in \Sigma_{\varpi+\pi/2}$.*
(iii) *There are $\kappa \in \mathbb{R}$ and $M > 0$ such that*

$$|g(z, x, t)| \leq M|z|^{\kappa-1}e^{tRe(z)}, \quad (z, x, t) \in \Sigma_{\varpi+\pi/2} \times X \times \mathbb{R}^+.$$

Then

$$(x, t) \mapsto \int_\Gamma g(z, x, t)dz \in C(X \times \mathbb{R}^+, X)$$

and

$$\left|\int_\Gamma g(z, x, t)dz\right| \leq M|t|^{-\kappa}, \quad (x, t) \in X \times \mathbb{R}^+.$$

Let $1 < p < \infty$. We consider the operator A in $L^p(\mathbb{R}^N)$ defined by $A = -\Delta$ with the domain

$$D(A) = \{u \in L^p(\mathbb{R}^N) : \Delta u \in L^p(\mathbb{R}^N)\}.$$

It follows from [31, Theorem 2.3.2] that $-A$ generates a bounded analytic semigroup of the spectral angle less than or equal to $\pi/2$. That is, for $\theta \in (0, \pi/2)$,

$$\|(z + A)^{-1}\|_{\mathscr{L}(X)} \leq M/|z|, \quad \text{for } z \in \Sigma_{\pi-\theta}. \tag{2.14}$$

For $\delta > 0$ and $\theta \in (0, \pi/2)$, we introduce the contour $\Gamma_{\delta,\theta}$ defined by

$$\Gamma_{\delta,\theta} = \{re^{-i\theta} : r \geq \delta\} \cup \{\delta e^{i\psi} : |\psi| \leq \theta\} \cup \{re^{i\theta} : r \geq \delta\},$$

where the circular arc is oriented counterclockwise, and the two rays are oriented with an increasing imaginary part. In the sequel, we introduce two linear operators $S_0(t)$ and $S_1(t)$ as

$$S_0(t) = \frac{1}{2\pi i}\int_{\Gamma_{\delta,\pi-\theta}} e^{zt}\omega(z)H(z)dz, \quad S_1(t) = \frac{1}{2\pi i}\int_{\Gamma_{\delta,\pi-\theta}} e^{zt}H(z)dz, \tag{2.15}$$

where

$$H(z) = (z\omega(z) + A)^{-1}, \quad \omega(z) = \int_0^1 z^{\alpha-1}\mu(\alpha)d\alpha.$$

2.2.3 *Technical Tools*

In this section, we denote $X = L^p(\mathbb{R}^N)$. Now we give some properties of $\omega(z)$. From [105, Lemma 2.3], we know that

$$z\omega(z) \in \Sigma_{\pi-\theta}, \quad \text{for } z \in \Sigma_{\pi-\theta}.$$

Using the generalization of the mean value theorem of integrals, it is easy to show that

$$|\omega(z)| \leq \|\mu\|_{L^1(0,1)}|z|^{\nu-1}, \quad \text{for } z \in \mathbb{C} \setminus \mathbb{R}^-, \tag{2.16}$$

where $\nu \in (0, 1)$.

Lemma 2.5 *Assume that μ satisfies* (2.12). *Then there exists a positive constant c, depending only on μ and γ such that*

$$|z\omega(z)| \geq c|z|^\rho, \quad \rho \in (\gamma, 1-\gamma), \quad \text{for } z \in \Sigma_{\pi-\theta}.$$

Proof From the proof of Proposition 6 in [124], we know that

$$|z\omega(z)| \geq c_0 \int_\gamma^{1-\gamma} \mu(\alpha)|z|^\alpha d\alpha, \quad \text{for } z \in \Sigma_{\pi-\theta},$$

where $c_0 = \max\left\{\cos(\frac{(1-\gamma)\pi}{2}), \min\{\sin(\gamma\pi), \sin(\frac{\gamma\pi}{2}), \frac{\sqrt{2}}{2}\}\right\}$.

Using the generalization of the mean value theorem of integrals again, it follows that

$$\int_{\gamma}^{1-\gamma} \mu(\alpha)|z|^{\alpha} d\alpha = \|\mu\|_{L^1(\gamma,1-\gamma)}|z|^{\rho},$$

where $\rho \in (\gamma, 1-\gamma)$. Then the conclusion holds.

Then we consider the relation between ν and ρ.

Remark 2.4 From (2.16) and Lemma 2.5, we know that $c|z|^{\rho} \le |z\omega(z)| \le c_1|z|^{\nu}$ for $z \in \Sigma_{\pi-\theta}$, where $c_1 > c > 0$. This implies $\rho = \nu$. Indeed, if we assume $\rho > \nu$, then $c \le \||z|^{1-\rho}\omega(z)| \le c_1|z|^{\nu-\rho}$. For any $r > 0$, we set $\vartheta(r) = r^{\nu-\rho}$. Thus $\vartheta(r) \ge \frac{c}{c_1}$ for $r > 0$, it is a contradiction. Similarly, the assumption $\rho < \nu$ is also impossible. Therefore $\rho = \nu$.

Next we state the uniform estimates of the operators $S_0(t)$ and $S_1(t)$ on $L^p(\mathbb{R}^N)$.

Lemma 2.6 *Assume that* μ *satisfies* (2.12). *The operator* $S_0(t)$ *defined by* (2.15) *is well defined,* $S_0(\cdot)x \in C([0,\infty); X)$ *and* $S_0(t)x \to x$ *as* $t \to 0$ *for any* $x \in X$. *Furthermore, there exists a constant* $M > 0$ *such that*

$$\|S_0(t)\|_{\mathscr{L}(X)} \le M, \quad \text{for } t \ge 0.$$

In addition, $AS_0(\cdot)x \in C(\mathbb{R}^+, X)$ *for* $x \in X$, *and there exists* $M^* > 0$ *such that*

$$\|AS_0(t)\|_{\mathscr{L}(X)} \le M^* t^{-\rho}, \quad \text{for } t > 0. \tag{2.17}$$

Proof In view of (2.14), we obtain

$$\|(z\omega(z) + A)^{-1}\|_{\mathscr{L}(X)} \le M/|z\omega(z)|, \quad z \in \Sigma_{\pi-\theta}, \tag{2.18}$$

and we deduce from (2.15) that $\|e^{zt}\omega(z)H(z)\|_{\mathscr{L}(X)} \le M|z|^{-1}e^{tRe(z)}$ for any $z \in \Sigma_{\pi-\theta}$. Then by taking $\varpi = \frac{\pi}{2} - \theta$ in Lemma 2.4, we conclude that $\|S_0(t)\|_{\mathscr{L}(X)} \le M$ for $t > 0$ and $S_0(\cdot)x \in C(\mathbb{R}^+, X)$. Consequently, the well-definition follows. To prove the limit of $S_0(t)x$ at $t = 0$, one can find the same way as in [182, Corollary 2.2], and it follows that $S_0(t)x \to x$ as $t \to 0$ for any $x \in X$.

Let $t > 0$, and we choose $\delta = t^{-1}$. By using the identity

$$A\omega(z)H(z) = \omega(z)\big(I - z\omega(z)H(z)\big)$$

and the inequality $|\omega(z)| \le \|\mu\|_{L^1(0,1)}|z|^{\rho-1}$ for $\rho = \nu$, we can deduce from (2.18) that

$$\|A\omega(z)H(z)\|_{\mathscr{L}(X)} \le (M+1)\|\mu\|_{L^1(0,1)}|z|^{\rho-1},$$

for any $z \in \Sigma_{\pi-\theta}$. Thus, we can estimate $AS_0(t)$ as

$$\begin{aligned}\|AS_0(t)\|_{\mathscr{L}(X)} \leq &\int_{\Gamma_{t^{-1},\pi-\theta}} e^{Re(z)t}\|A\omega(z)H(z)\|_{\mathscr{L}(X)}|dz| \\ \leq &(M+1)\|\mu\|_{L^1(0,1)}\left(2\int_{t^{-1}}^{\infty}\tau^{\rho-1}e^{-\tau t\cos(\theta)}d\tau\right. \\ &+\left.\int_{-\pi+\theta}^{\pi-\theta}t^{-\rho}e^{\cos(\psi)}d\psi\right) \\ \leq &M^*t^{-\rho},\end{aligned}$$

where $M^* > 0$ is a constant depending on M, θ, and $\|\mu\|_{L^1(0,1)}$. Therefore (2.17) holds.

Using Lemma 2.4 again, we see that $AS_0(\cdot)x \in C(\mathbb{R}^+, X)$ for $x \in X$. The proof is completed.

Remark 2.5 In fact, if we consider the Volterra equation

$$u(t) = \varphi(t) - \int_0^t g(t-s)Au(s)ds, \tag{2.19}$$

using Theorem 2.1 in Prüss [182], we show that (2.19) admits an analytic resolvent $S_0(t)$. This also ensures the results of Lemma 2.6 excluding formula (2.17).

Lemma 2.7 *Assume that μ satisfies* (2.12). *Then the operator $S_1(t)$ defined by* (2.15) *is well defined, $S_1(\cdot)x \in C(\mathbb{R}^+, X)$. Moreover, there exists a constant $M_1 > 0$ such that*

$$\|S_1(t)\|_{\mathscr{L}(X)} \leq M_1 t^{\rho-1}, \quad \textit{for } t > 0.$$

In addition, $AS_1(\cdot)x \in C(\mathbb{R}^+, X)$ for $x \in X$, and there exists $M_1^* > 0$ such that

$$\|AS_1(t)\|_{\mathscr{L}(X)} \leq \frac{M_1^*}{t}, \quad \textit{for } t > 0. \tag{2.20}$$

Proof As a similar approach in Lemma 2.6, we conclude that

$$\|e^{zt}H(z)\|_{\mathscr{L}(X)} \leq M|z\omega(z)|^{-1}e^{tRe(z)} \leq Mc_0|z|^{-\rho}e^{tRe(z)},$$

for any $z \in \Sigma_{\pi-\theta}$. Using Lemma 2.4 again, we find that $\|S_1(t)\|_{\mathscr{L}(X)} \leq M_1t^{\rho-1}$ and $S_1(\cdot)x \in C(\mathbb{R}^+, X)$.

Let $t > 0$ and $\delta = \frac{1}{t} > 0$. In view of the identity $AH(z) = I - z\omega(z)H(z)$, it follows from (2.18) that $\|AH(z)\|_{\mathscr{L}(X)} \leq (M+1)$ for any $z \in \Sigma_{\pi-\theta}$. Thus, we can estimate the upper bound on $AS_1(t)$ for $t \in \mathbb{R}^+$

$$\begin{aligned}\|AS_1(t)\|_{\mathscr{L}(X)} \leq &\int_{\Gamma_{\frac{1}{t},\pi-\theta}} e^{Re(z)t}\|AH(z)\|_{\mathscr{L}(X)}|dz|\\ \leq &(M+1)\left(2\int_{\frac{1}{t}}^{\infty} e^{-rt\cos(\theta)}dr + \int_{-\pi+\theta}^{\pi-\theta} e^{\cos(\psi)}t^{-1}d\psi\right)\\ \leq &M_1^* t^{-1},\end{aligned}$$

where $M_1^* > 0$ is a constant depending on M and θ. Similarly, we have $AS_1(\cdot)x \in C(\mathbb{R}^+, X)$ for $x \in X$. The proof is completed.

The inequalities (2.17) and (2.20) enable us to obtain some estimates of $S_0(t)x$ and $S_1(t)x$ in fractional power spaces.

Lemma 2.8 *Let $\omega \in (0, 1)$. Assume that μ satisfies (2.12). Then we have $A^\omega S_i(\cdot)x \in C(\mathbb{R}^+, X)$ $(i = 0, 1)$ for $x \in X$. Moreover there exists a constant $C_1 > 0$ such that*

$$|A^\omega S_0(t)x| \leq C_1 t^{-\rho\omega}|x| \quad \text{and} \quad |A^\omega S_1(t)x| \leq C_1 t^{\rho(1-\omega)-1}|x|,$$

for $t > 0$ and $x \in X$.

Proof Using the well-known inequality

$$|A^\omega x| \leq C|x|^{1-\omega}|Ax|^\omega, \quad \text{for } x \in D(A) \text{ and } \omega \in (0, 1),$$

we can immediately deduce the results from Lemma 2.6 and Lemma 2.7.

For convenience, we set

$$\beta_{pq} = \frac{N}{2}\left(\frac{1}{p} - \frac{1}{q}\right), \quad N \geq 1,\ 1 \leq p \leq q \leq \infty.$$

Our next lemma describes the L^p-L^q estimates and the continuity about the operators $S_0(t)$ and $S_1(t)$.

Lemma 2.9 *The operators $S_0(t)$ and $S_1(t)$ have the following properties:*

(i) *Assume that $N \geq 1$ and $v \in L^p(\mathbb{R}^N)$ for $1 \leq p \leq q \leq \infty$. Then we have $S_i(\cdot)v \in C(\mathbb{R}^+, L^q(\mathbb{R}^N))$ $(i = 0, 1)$. Moreover there exists a constant $C_2 > 0$ such that*

$$\|S_i(t)v\|_{L^q(\mathbb{R}^N)} \leq C_2 t^{-\rho\beta_{pq}+\rho i-i}\|v\|_{L^p(\mathbb{R}^N)}, \quad i = 0, 1,$$

for $t > 0$.

(ii) *Assume that $N > 1$ and $v \in L^p(\mathbb{R}^N)$ for $1 \leq q < \infty$, $1 < p < N$. Then we have $A^{1/2}S_i(\cdot)v \in C(\mathbb{R}^+, L^q(\mathbb{R}^N))$ $(i = 0, 1)$. Moreover there exists a constant $C_2 > 0$ such that*

$$\|A^{1/2}S_i(t)v\|_{L^q(\mathbb{R}^N)} \le C_2 t^{-\rho(\frac{1}{2}+\beta_{pq})+\rho i-i}\|v\|_{L^p(\mathbb{R}^N)}, \quad i=0,1,$$

for $t > 0$.

Proof (i) Let $N \ge 1$ and $1 \le p \le q \le \infty$. It follows from the Gagliardo-Nirenberg inequality that

$$\|S_0(t)v\|_{L^q(\mathbb{R}^N)} \le C\|AS_0(t)v\|^{\theta}_{L^p(\mathbb{R}^N)}\|S_0(t)v\|^{1-\theta}_{L^p(\mathbb{R}^N)}, \quad \text{for } v \in L^p(\mathbb{R}^N),$$

where $1/q = \theta(1/p - 2/N) + (1-\theta)/p$ for $\theta \in [0,1]$ and $C > 0$ is a constant. This together with Lemma 2.6 shows the continuity of $S_0(\cdot)v$ in $L^q(\mathbb{R}^N)$ for $t > 0$. Moreover, the following estimate holds:

$$\|S_0(t)v\|_{L^q(\mathbb{R}^N)} \le Ct^{-\rho\theta}\|v\|_{L^p(\mathbb{R}^N)}.$$

Taking the exponent of p, q into the above inequality, we immediately obtain the L^p-L^q estimate of operator $S_0(t)$.

We similarly consider the operator $S_1(t)$. It is easy to show the continuity and L^p-L^q estimate on $S_1(t)$ by the same arguments as above.

(ii) Using the fractional Gagliardo-Nirenberg inequality [78, Corollary 2.3.], Lemmas 2.6 and 2.8 lead to the estimate

$$\|A^{1/2}S_0(t)v\|_{L^q(\mathbb{R}^N)} \le \bar{C}\|AS_0(t)v\|^{\theta}_{L^p(\mathbb{R}^N)}\|A^{1/2}S_0(t)v\|^{1-\theta}_{L^p(\mathbb{R}^N)}, \text{ for } v \in L^p(\mathbb{R}^N),$$

where $1/q = \theta(1/p - 1/N) + (1-\theta)/p$ for $\theta \in (0,1)$ and $\bar{C} > 0$ is a constant. It yields the desired results. In addition, the results on the operator $(-\Delta)^{1/2}S_1(t)$ can be similarly derived, so we omit the proof.

Remark 2.6 It is noticed that when μ satisfies the assumption (2.12), from [110, Theorem 5.1] it just infers that

$$\|S_0(t)v\|_{L^q(\mathbb{R}^N)} \le [(1 * g)(t)]^{-\frac{N}{2}\left(1-\frac{1}{r}\right)}\|v\|_{L^p(\mathbb{R}^N)}, \quad \text{for } t > 0 \text{ and } v \in L^p(\mathbb{R}^N),$$

where $1 < r < \frac{N}{N-2}$, $1 < p, q < \infty$, $1 + \frac{1}{q} = \frac{1}{r} + \frac{1}{p}$. If we proceed to calculate the term $[(1 * g)(t)]^{-\frac{N}{2}\left(1-\frac{1}{r}\right)}$, we need to estimate the lower bound of $g(t)$ for $\frac{N}{2}\left(1 - \frac{1}{r}\right) > 0$. However, it is difficult to estimate the lower bound of $g(t)$. Therefore, it may not be easy to obtain the polynomial decay as in Lemma 2.9(i).

Set

$$W_f(t) = \int_0^t S_1(t-s)f(s)ds;$$

we next give the continuity of the function W_f in different spaces.

Lemma 2.10 *If $f \in L^p(0,T;X)$ for $\rho p > 1$, then $W_f \in C([0,T];X)$ for $0 < T < \infty$. If $f \in L^p(0,T;X)$ with $\rho p(1-\omega) > 1$ for some $0 < \omega < 1$, then $A^\omega W_f \in C([0,T];X)$ for $0 < T < \infty$. Furthermore, let $1 \leq p < \infty$; if $q \in [p,\infty]$ satisfies $\beta_{pq} < 1$ and there is $\xi \in [0,1)$ such that $\sup_{t\in[0,T)} t^\xi |f(t)| < \infty$ for $0 < T \leq \infty$, then $W_f \in C((0,T];L^q(\mathbb{R}^N))$. If $\xi < \rho(1-\beta_{pq})$, then $W_f \in C([0,T];L^q(\mathbb{R}^N))$.*

Proof For any $t_1, t_2 \in (0,T]$ with $t_1 < t_2$, we break the integral into the following two parts:

$$W_f(t_2) - W_f(t_1) = \int_{t_1}^{t_2} S_1(t_2-s)f(s)ds + \int_0^{t_1} (S_1(t_2-s) - S_1(t_1-s))f(s)ds.$$

Employing the condition of f and Lemma 2.7, we can obtain

$$\begin{aligned}\left|\int_{t_1}^{t_2} S_1(t_2-s)f(s)ds\right| &\leq M_1 \int_{t_1}^{t_2} (t_2-s)^{\rho-1}|f(s)|ds \\ &\leq M_1 \|f\|_{L^p(0,T;X)} \left(\frac{p-1}{\rho p-1}\right)^{1-\frac{1}{p}} (t_2-t_1)^{\rho-\frac{1}{p}} \\ &\to 0, \quad \text{as } t_2 \to t_1\end{aligned}$$

and

$$|(S_1(t_2-s) - S_1(t_1-s))f(s)| \leq 2M_1(t_1-s)^{\rho-1}|f(s)|, \quad s \in (0,t_1),$$

which is integrable in $L^1(0,t_1;X)$. From $S_1(t)f(\cdot) \in C((0,T];X)$ that is valid due to Lemma 2.7, it easily follows from the Lebesgue dominated convergence theorem that $W_f \in C((0,T];X)$.

For $t_1 = 0$ and $t_2 \in (0,T]$, we get

$$\begin{aligned}|W_f(t_2) - W_f(t_1)| &= \left|\int_0^{t_2} S_1(t_2-s)f(s)ds\right| \\ &\leq M_1 \|f\|_{L^p(0,T;X)} \left(\frac{p-1}{\rho p-1}\right)^{1-\frac{1}{p}} t_2^{\rho-\frac{1}{p}} \\ &\to 0, \quad \text{as } t_2 \to 0,\end{aligned}$$

which implies the continuity of the operator $W_f(t)$ at $t=0$. So $W_f \in C([0,T];X)$.

Next, from Lemma 2.8, it follows that

$$\left|\int_{t_1}^{t_2} A^\gamma S_1(t_2-s)f(s)ds\right| \leq C_1 \int_{t_1}^{t_2} (t_2-s)^{\rho(1-\gamma)-1}|f(s)|ds$$

and

$$\left|A^{\gamma}(S_1(t_2-s)-S_1(t_1-s))f(s)\right| \le 2C_1(t_1-s)^{\rho(1-\gamma)-1}|f(s)|, \quad s\in[0,t_1).$$

Because of $A^{\gamma}S_1(t)f(\cdot) \in C((0,T];X)$, we can conclude that $A^{\gamma}W_f \in C([0,T];X)$ as the same way as we derived $W_f \in C([0,T];X)$.

Finally, for $t_1, t_2 \in (0,T]$ with $t_1 < t_2$, we have from Lemma 2.9 that

$$\begin{aligned}
&\left\|\int_{t_1}^{t_2} S_1(t_2-s)f(s)ds\right\|_{L^q(\mathbb{R}^N)}\\
\le& C_2\int_{t_1}^{t_2}(t_2-s)^{\rho(1-\beta_{pq})-1}\|f(s)\|_{L^p(\mathbb{R}^N)}ds\\
\le& C_2\sup_{s\in[0,T]} s^{\xi}\|f(s)\|_{L^p(\mathbb{R}^N)}t_2^{\rho(1-\beta_{pq})-\xi}\int_{t_1/t_2}^{1}(1-s)^{\rho(1-\beta_{pq})-1}s^{-\xi}ds,
\end{aligned}$$

which tends to zero as $t_2 \to t_1$ by the properties of the incomplete Beta function. Moreover, Lemma 2.9 also yields that for $s \in (0,t_1)$,

$$\begin{aligned}
&\|(S_1(t_2-s)-S_1(t_1-s))f(s)\|_{L^q(\mathbb{R}^N)}\\
\le& 2C_2(t_1-s)^{\rho(1-\beta_{pq})-1}\|f(s)\|_{L^p(\mathbb{R}^N)}\\
\le& 2C_2(t_1-s)^{\rho(1-\beta_{pq})-1}s^{-\xi}\sup_{s\in[0,T]} s^{\xi}\|f(s)\|_{L^p(\mathbb{R}^N)},
\end{aligned}$$

which is integrable in $L^1(0,t_1)$. Therefore, the similar arguments show that $W_f \in C((0,T];L^q(\mathbb{R}^N))$. Furthermore, if $\xi < 1-\beta_{pq}$, then it is obvious to obtain

$$\|W_f(t)\|_{L^q(\mathbb{R}^N)} \le Ct^{\rho(1-\beta_{pq})-\xi},$$

where C is a positive constant. It also implies the continuity of $W_f(t)$ at $t=0$ in $L^q(\mathbb{R}^N)$, and then $W_f \in C([0,T];L^q(\mathbb{R}^N))$.

2.2.4 Well-Posedness and Decay of Solutions

In the subsection, the well-posedness and decay of local solutions and global solutions will be considered. For this, we start with giving a concept of mild solutions of the problem (2.11).

Let u be a solution of the problem (2.11). Applying the Laplace transform to (2.11) leads to

$$\bar{u}(z) = \omega(z)H(z)\varphi + H(z)\bar{f}(u)(z).$$

From the inverse Laplace transform, we deduce the following integral representation:

$$u(t) = S_0(t)\varphi + \int_0^t S_1(t-s)f(u(s))ds, \tag{2.21}$$

where the operators $S_0(t)$ and $S_1(t)$ are defined as in (2.15).

Definition 2.3 Let $p \geq 1$.

(i) If there exists $0 < T < \infty$ such that a continuous function $u : [0, T] \to L^p(\mathbb{R}^N)$ satisfies (2.21) for $t \in [0, T]$, we say that u is a local mild solution to the problem (2.11) in $L^p(\mathbb{R}^N)$.
(ii) A continuous function $u : [0, \infty) \to L^p(\mathbb{R}^N)$ satisfying (2.21) for $t \in [0, \infty)$ is called a global mild solution to the problem (2.11) in $L^p(\mathbb{R}^N)$.

Next, the hypothesis of the semilinear term is introduced.

(Hf) We suppose that $f(0) = 0$ and $f : L^r(\mathbb{R}^N) \to L^{r'}(\mathbb{R}^N)$, for some $r' > r$,

$$r \in \left[2, \frac{2N}{N-2}\right), \quad \text{if } N \geq 2, \quad (r \in [2, \infty], \text{ if } N = 1).$$

Additionally, we suppose that there exist $\sigma \geq 0$ and $K > 0$ such that

$$\|f(u) - f(v)\|_{L^{r'}(\mathbb{R}^N)} \leq K\left(\|u\|^{\sigma}_{L^r(\mathbb{R}^N)} + \|v\|^{\sigma}_{L^r(\mathbb{R}^N)}\right)\|u - v\|_{L^r(\mathbb{R}^N)},$$

for all $u, v \in L^r(\mathbb{R}^N)$.

We first consider the case $T < \infty$. Let r be such that

$$\begin{cases} r \in \left[2, \dfrac{2N}{N-2}\right) & \text{if } N \geq 2, \\ r \in [2, \infty] & \text{if } N = 1. \end{cases}$$

$Y_r[T]$ denotes the Banach space consisting of continuous functions $v : (0, T] \to L^r(\mathbb{R}^N)$ satisfying

$$t^{1-\rho(1-\beta_{r'r})}v \in C_b([0, T]; L^r(\mathbb{R}^N)), \quad \lim_{t\to 0} t^{1-\rho(1-\beta_{r'r})}v(t) = 0,$$

equipped with the norm

$$\|v\|_{Y_r[T]} = \sup_{t\in[0,T]} t^{1-\rho(1-\beta_{r'r})}\|v(t)\|_{L^r(\mathbb{R}^N)}.$$

Notice that $\beta_{r'r} < 1$ provided $r \in \left[2, \frac{2N}{N-2}\right)$ for $N \geq 2$ or $r \in [2, \infty]$ for $N = 1$. If $r' \leq p \leq r$, then $\beta_{r'p} < \beta_{r'r} < 1$.

Theorem 2.3 *Let $N \geq 1$ and $(1-\rho(1-\beta_{r'r}))(\sigma+1) < 1$. If $\varphi \in L^{r'}(\mathbb{R}^N)$ and (Hf) holds, then there exists $T_* > 0$ such that the problem (2.11) has a unique local mild solution u in $Y_r[T_*]$. Moreover,*

$$t^{1-\rho(1-\beta_{r'p})}u \in C_b([0,T_*]; L^p(\mathbb{R}^N)), \quad \textit{for } r' \leq p \leq r, \tag{2.22}$$

which values vanish at $t=0$.

Proof Let $R > 0$ and set

$$B(R) = \{u \in Y_r[T] : \ \|u\|_{Y_r[T]} \leq R\}.$$

It is easy to see that $B(R)$ is a closed ball of $Y_r[T]$ with center 0 and radius $R > 0$. Define the operator Φ in $B(R)$ as

$$\Phi(u)(t) = S_0(t)\varphi + \int_0^t S_1(t-s)f(u(s))ds. \tag{2.23}$$

By taking $p = r'$ and $q = r$ on Lemma 2.9, we obtain from (Hf) and the inequality $(1-\rho(1-\beta_{r'r}))(\sigma+1) < 1$ that

$$\begin{aligned}\|W_{f(u)}(t)\|_{L^r(\mathbb{R}^N)} \leq& C_2\int_0^t (t-s)^{\rho(1-\beta_{r'r})-1}\|f(u(s))\|_{L^{r'}(\mathbb{R}^N)}ds\\ \leq& C_2K\int_0^t (t-s)^{\rho(1-\beta_{r'r})-1}s^{-(1-\rho(1-\beta_{r'r}))(\sigma+1)}\\ &\times\Big(\sup_{s\in[0,T]} s^{1-\rho(1-\beta_{r'r})}\|u(s)\|_{L^r(\mathbb{R}^N)}\Big)^{\sigma+1}ds\\ \leq& C_2KB_0t^{\rho(1-\beta_{r'r})-(1-\rho(1-\beta_{r'r}))(\sigma+1)}\|u\|^{\sigma+1}_{Y_r[T]},\end{aligned}$$

where $B_0 = B(\rho(1-\beta_{r'r}), 1-(1-\rho(1-\beta_{r'r}))(\sigma+1))$; here $B(\cdot,\cdot)$ is the Beta function. It implies that

$$t^{1-\rho(1-\beta_{r'r})}\|W_{f(u)}(t)\|_{L^r(\mathbb{R}^N)} \leq C_2KB_0t^{1-(1-\rho(1-\beta_{r'r}))(\sigma+1)}R^{\sigma+1}, \quad \text{for } t \in (0,T]. \tag{2.24}$$

Then we choose $T_1 > 0$ small enough so that

$$2C_2KB_0T_1^{1-(1-\rho(1-\beta_{r'r}))(\sigma+1)}R^{\sigma} < 1. \tag{2.25}$$

Now, observe that $\varphi \in L^{r'}(\mathbb{R}^N)$. From Lemma 2.9, we find that

$$t^{1-\rho(1-\beta_{r'r})}\|S_0(t)\varphi\|_{L^r(\mathbb{R}^N)} \leq C_2t^{1-\rho}\|\varphi\|_{L^{r'}(\mathbb{R}^N)}. \tag{2.26}$$

Hence, there exists $T_2 > 0$ small enough such that

$$\|S_0(t)\varphi\|_{Y_r[T_2]} < \frac{R}{2}, \quad \forall\, t \in [0, T_2]. \tag{2.27}$$

Choose $T_* = \min\{T_1, T_2\}$. From (2.25) and (2.27), we see that

$$\|\Phi(u)\|_{Y_r[T_*]} \leq \|S_0(t)\varphi\|_{Y_r[T_*]} + \|W_{f(u)}(t)\|_{Y_r[T_*]} < R, \quad \text{for } u \in B(R).$$

Next, we prove the continuity of the operator $\Phi(u)$. Employing Lemmas 2.6 and 2.10 yields $\Phi(u) \in C([0, T_*]; L^{r'}(\mathbb{R}^N)) \cap C((0, T_*]; L^r(\mathbb{R}^N))$. Thus, it follows from the above arguments that $t^{1-\rho(1-\beta_{r'r})}\Phi(u) \in C_b([0, T_*]; L^r(\mathbb{R}^N))$. In addition, noting the inequalities (2.24) and (2.26), we obtain that $\lim_{t\to 0} t^{1-\rho(1-\beta_{r'r})}$ $\Phi(u)(t) = 0$. Therefore, the operator Φ maps $B(R)$ into itself.

Employing the assumption of f and Lemma 2.9, for any $u, v \in B(R)$, we further obtain that

$$\begin{aligned}
&\|\Phi(u)(t) - \Phi(v)(t)\|_{L^r(\mathbb{R}^N)}\\
\leq &\int_0^t \|S_1(t-s)(f(u(s)) - f(v(s)))\|_{L^r(\mathbb{R}^N)} ds\\
\leq &C_2 \int_0^t (t-s)^{\rho(1-\beta_{r'r})-1} \|f(u(s)) - f(v(s))\|_{L^{r'}(\mathbb{R}^N)} ds\\
\leq &C_2 K \int_0^t (t-s)^{\rho(1-\beta_{r'r})-1} \left(\|u(s)\|^\sigma_{L^r(\mathbb{R}^N)} + \|v(s)\|^\sigma_{L^r(\mathbb{R}^N)}\right) \|u(s) - v(s)\|_{L^r(\mathbb{R}^N)} ds\\
\leq &2C_2 K B_0 R^\sigma \|u - v\|_{Y_r[T_*]} t^{\rho(1-\beta_{r'r})-(1-\rho(1-\beta_{r'r}))(\sigma+1)},
\end{aligned}$$

which yields the inequality

$$\begin{aligned}
&t^{1-\rho(1-\beta_{r'r})} \|\Phi(u)(t) - \Phi(v)(t)\|_{L^r(\mathbb{R}^N)}\\
\leq &2C_2 K B_0 R^\sigma T_*^{1-(1-\rho(1-\beta_{r'r}))(\sigma+1)} \|u - v\|_{Y_r[T_*]}.
\end{aligned}$$

Set

$$\zeta = 2C_2 K B_0 R^\sigma T_*^{1-(1-\rho(1-\beta_{r'r}))(\sigma+1)}.$$

Using the inequality (2.25), we can ensure $\zeta \in (0, 1)$ that is valid for the choice of T_* and arrive at the estimate

$$\|\Phi(u) - \Phi(v)\|_{Y_r[T_*]} \leq \zeta \|u - v\|_{Y_r[T_*]}, \tag{2.28}$$

which implies that Φ is a strict contraction on $B(R)$. Thus, we arrive at the existence and uniqueness of a fixed point $u_* \in Y_r[T_*]$, and therefore u_* is the local mild solution of the problem (2.11) in $L^r(\mathbb{R}^N)$.

Finally, it remains to verify $t^{1-\rho(1-\beta_{r'p})}u_* \in C_b([0, T_*]; L^p(\mathbb{R}^N))$ vanishing at $t = 0$. Using the same arguments, which were employed for the contraction of Φ, we obtain

$$\begin{aligned}\|u_*(t)\|_{L^p(\mathbb{R}^N)} \leq& \|S_0(t)\varphi\|_{L^p(\mathbb{R}^N)} + \int_0^t \|S_1(t-s)f(u_*(s))\|_{L^p(\mathbb{R}^N)}ds\\ \leq& C_2 t^{-\rho\beta_{r'p}}\|\varphi\|_{L^{r'}(\mathbb{R}^N)}\\ &+ C_2 K B_1 t^{\rho(1-\beta_{r'p})-(1-\rho(1-\beta_{r'r}))(\sigma+1)}\|u_*\|^{\sigma+1}_{Y_r[T_*]},\end{aligned}$$

where $B_1 = B(\rho(1-\beta_{r'p}), 1-(1-\rho(1-\beta_{r'r}))(\sigma+1))$. Thus, the above inequality leads to the estimate

$$\begin{aligned}\sup_{0<t\leq T_*} t^{1-\rho(1-\beta_{r'p})}\|u_*(t)\|_{L^p(\mathbb{R}^N)} \leq& C_2 T_*^{1-\rho}\|\varphi\|_{L^{r'}(\mathbb{R}^N)}\\ &+ C_2 K B_1 T_*^{1-(1-\rho(1-\beta_{r'r}))(\sigma+1)}\|u_*\|^{\sigma+1}_{Y_r[T_*]}.\end{aligned}$$

Similar computations to that already done in Lemma 2.10 show the continuity of the nonlinear part. By Lemma 2.6, we assert that $S_0(t)\varphi \in C((0, T]; L^p(\mathbb{R}^N))$ for $\varphi \in L^{r'}(\mathbb{R}^N)$ and thus find that $t^{1-\rho(1-\beta_{r'p})}S_0(t)\varphi \in C_b([0, T]; L^p(\mathbb{R}^N))$. Moreover, the above inequality leads to the relation

$$\sup_{0<t\leq T_*} t^{1-\rho(1-\beta_{r'p})}\|u_*(t)\|_{L^p(\mathbb{R}^N)} \to 0, \quad \text{as } T_* \to 0,$$

which implies that $\lim_{t\to 0} t^{1-\rho(1-\beta_{r'p})}u(t) = 0$ in $L^p(\mathbb{R}^N)$ for $r' \leq p \leq r$. The proof is completed.

Corollary 2.1 *Under the conditions of Theorem 2.3, we suppose that*

$$1 < N < \frac{r}{r-2}. \tag{2.29}$$

Then the solution u_* of the problem (2.11) satisfies

$$t^{1-\rho(\frac{1}{2}-\beta_{r'p})}(-\Delta)^{1/2}u_* \in C_b([0, T_*]; L^p(\mathbb{R}^N)), \quad \textit{for } r' \leq p \leq r,$$

which values vanish at $t = 0$.

Proof Observe that (2.29), and we have $\beta_{r'p} \leq \beta_{r'r} < \frac{1}{2}$ for $r' \leq p \leq r$. Using Lemma 2.9 and the condition of f, we arrive at the inequality

$$
\begin{aligned}
&\|(-\Delta)^{1/2}\int_0^t S_1(t-s)f(u_*(s))ds\|_{L^p(\mathbb{R}^N)}\\
\leq & C_2\int_0^t (t-s)^{\rho(\frac{1}{2}-\beta_{r'p})-1}\|f(u_*(s))\|_{L^{r'}(\mathbb{R}^N)}ds\\
\leq & C_2K\int_0^t (t-s)^{\rho(\frac{1}{2}-\beta_{r'p})-1}s^{-(1-\rho(1-\beta_{r'r}))(\sigma+1)}\\
&\times\Big(\sup_{s\in[0,T_*]} s^{1-\rho(1-\beta_{r'r})}\|u_*(s)\|_{L^r(\mathbb{R}^N)}\Big)^{\sigma+1}ds\\
\leq & C_2KB_2\|u_*\|^{\sigma+1}_{Y_r[T_*]}t^{\rho(\frac{1}{2}-\beta_{r'p})-(1-\rho(1-\beta_{r'r}))(\sigma+1)},
\end{aligned}
$$

where $B_2 = B\Big(\rho(\frac{1}{2}-\beta_{r'p}), 1-(1-\rho(1-\beta_{r'r}))(\sigma+1)\Big)$; it follows that

$$
\begin{aligned}
t^{1-\rho(\frac{1}{2}-\beta_{r'p})}\|(-\Delta)^{1/2}\Phi(u_*)(t)\|_{L^p(\mathbb{R}^N)} \leq & C_2T_*^{(1-\rho)(\frac{1}{2}-\beta_{r'p})}\|\varphi\|_{L^{r'}(\mathbb{R}^N)}\\
&+C_2KB_2T_*^{1-(1-\rho(1-\beta_{r'r}))(\sigma+1)}R^{\sigma+1}
\end{aligned}
$$

and thus ensures

$$
\sup_{0<t\leq T_*} t^{1-\rho(1-\beta_{r'p})}\|u_*(t)\|_{L^p(\mathbb{R}^N)}\to 0, \quad \text{as } T_*\to 0.
$$

The continuity of the nonlinear operator evaluated in $(-\Delta)^{1/2}u_*$ follows by the same arguments as above. Also, by Lemma 2.8, we obtain the continuity of the linear part. The proof is completed.

Next, we consider the case $T=\infty$. For $1\leq r'\leq p<r\leq\infty$, Z_{pr} denotes the Banach space consisting of continuous functions $v:[0,\infty)\to L^p(\mathbb{R}^N)$ satisfying

$$
t^{\rho\beta_{pr}}v\in C_b([0,\infty);L^r(\mathbb{R}^N)), \quad \lim_{t\to 0}t^{\rho\beta_{pr}}v(t)=0,
$$

equipped with the norm

$$
\|v\|_{Z_{pr}}=\sup_{t\geq 0}\|v(t)\|_{L^p(\mathbb{R}^N)}+\sup_{t\geq 0}t^{\rho\beta_{pr}}\|v(t)\|_{L^r(\mathbb{R}^N)}.
$$

Theorem 2.4 *Let $1\leq N<\frac{2r}{r-2}$ and $\beta_{r'p}+\beta_{pr}(\sigma+1)=1$. Assume $\varphi\in L^p(\mathbb{R}^N)$. If (Hf) holds and there exists $\lambda>0$ such that $\|\varphi\|_{L^p(\mathbb{R}^N)}\leq\lambda$, then the problem (2.11) admits a unique global mild solution $u\in C_b([0,\infty);L^p(\mathbb{R}^N))$. Moreover, $t^{\rho\beta_{pr}}u\in C_b([0,\infty);L^r(\mathbb{R}^N))$ vanishing at $t=0$.*

Proof Let $\varepsilon > 0$. We define the operator Φ as in (2.23) in Ω_ε a closed ball of Z_{pr} with center 0 and radius 2ε. In fact, we only need to employ a contraction mapping technique, and thus the desired result follows.

Employing Lemma 2.9, we find that the estimates

$$\|S_0(t)\varphi\|_{L^p(\mathbb{R}^N)} \le C_2\|\varphi\|_{L^p(\mathbb{R}^N)} \quad \text{and} \quad t^{\rho\beta_{pr}}\|S_0(t)\varphi\|_{L^r(\mathbb{R}^N)} \le C_2\|\varphi\|_{L^p(\mathbb{R}^N)}$$

hold true.

Now we verify that $t^{\rho\beta_{pr}}\|S_0(t)\varphi\|_{L^r(\mathbb{R}^N)} \to 0$ as $t \to 0$ for $\varphi \in L^p(\mathbb{R}^N)$. Indeed, since $C_0^\infty(\mathbb{R}^N)$ is dense in $L^p(\mathbb{R}^N)$, there is a sequence $\varphi_n \in C_0^\infty(\mathbb{R}^N) \subset L^r(\mathbb{R}^N)$ such that $\varphi_n \to \varphi$ in $L^p(\mathbb{R}^N)$. Applying Lemma 2.9 gives

$$\begin{aligned} t^{\rho\beta_{pr}}\|S_0(t)\varphi\|_{L^r(\mathbb{R}^N)} \le& C_2 t^{\rho\beta_{pr}}\|S_0(t)(\varphi-\varphi_n)\|_{L^r(\mathbb{R}^N)} + t^{\rho\beta_{pr}}\|S_0(t)\varphi_n\|_{L^r(\mathbb{R}^N)} \\ \le& C_2\|\varphi-\varphi_n\|_{L^p(\mathbb{R}^N)} + t^{\rho\beta_{pr}}\|\varphi_n\|_{L^r(\mathbb{R}^N)} \\ \to& 0, \quad \text{as } t \to 0. \end{aligned}$$

Let us define $\lambda := \varepsilon/(2C_2)$ and notice that $\varphi \in L^p(\mathbb{R}^N)$ with $\|\varphi\|_{L^p(\mathbb{R}^N)} \le \lambda$; thus we deduce $S_0(t)\varphi \in Z_{pr}$.

On the other hand, the assumption $\beta_{r'p} + \beta_{pr}(\sigma+1) = 1$ implies that $\beta_{r'p} < 1$ and $\beta_{pr}(\sigma+1) < 1$. For any $u \in \Omega_\varepsilon$ and $t \ge 0$, using Lemma 2.9, we arrive at the estimate

$$\begin{aligned} &\|W_{f(u)}(t)\|_{L^p(\mathbb{R}^N)} \\ \le& C_2K\int_0^t (t-s)^{\rho(1-\beta_{r'p})-1}s^{-\rho\beta_{pr}(\sigma+1)}\big(\sup_{s\ge 0} s^{\rho\beta_{pr}}\|u(s)\|_{L^r(\mathbb{R}^N)}\big)^{\sigma+1}ds \\ \le& C_2KB(\rho(1-\beta_{r'p}), 1-\rho\beta_{pr}(\sigma+1))\|u\|_{Z_{pr}}^{\sigma+1}. \end{aligned}$$

This leads to the inequality

$$\|W_{f(u)}(t)\|_{L^p(\mathbb{R}^N)} \le C_2KB(\rho(1-\beta_{r'p}), 1-\rho\beta_{pr}(\sigma+1))(2\varepsilon)^{\sigma+1}.$$

Again, the assumption $1 \le N < \frac{2r}{r-2}$ ensures $\beta_{r'r} < 1$. The arguments similar to ones used for the estimation of $\|W_{f(u)}(t)\|_{L^p(\mathbb{R}^N)}$ lead to the relation

$$\begin{aligned} &\|W_{f(u)}(t)\|_{L^r(\mathbb{R}^N)} \\ \le& C_2K\int_0^t (t-s)^{\rho(1-\beta_{r'r})-1}s^{-\rho\beta_{pr}(\sigma+1)}\big(\sup_{s\ge 0} s^{\rho\beta_{pr}}\|u(s)\|_{L^r(\mathbb{R}^N)}\big)^{\sigma+1}ds \\ \le& C_2KB(\rho(1-\beta_{r'r}), 1-\rho\beta_{pr}(\sigma+1))\big(\sup_{0<s\le t} s^{\rho\beta_{pr}}\|u(s)\|_{L^r(\mathbb{R}^N)}\big)^{\sigma+1}t^{-\beta_{pr}}, \end{aligned}$$

which shows that

$$\sup_{t\geq 0} t^{\rho\beta_{pr}}\|W_{f(u)}(t)\|_{L^r(\mathbb{R}^N)} \leq C_2 K B(\rho(1-\beta_{r'r}), 1-\rho\beta_{pr}(\sigma+1))(2\varepsilon)^{\sigma+1}$$

and $\lim_{t\to 0} t^{\rho\beta_{pr}}\|\Phi(u)(t)\|_{L^r(\mathbb{R}^N)} = 0$. Because of $B(\rho(1-\beta_{r'p}), 1-\beta_{pr}(\sigma+1)) \leq B(\rho(1-\beta_{r'r}), 1-\beta_{pr}(\sigma+1)) =: B_3$, and choosing $\varepsilon \leq \left(\frac{1}{2^{\sigma+3}CKB_3}\right)^{1/\sigma}$, we derive that $\|\Phi(u)\|_{Z_{pr}} \leq 2\varepsilon$ for $u \in \Omega_\varepsilon$.

For the continuity of $\Phi(u)$. Using Lemma 2.6 and Lemma 2.10 again, we have $\Phi(u) \in C([0,\infty); L^p(\mathbb{R}^N)) \cap C((0,\infty); L^r(\mathbb{R}^N))$. The similar arguments easily ensure that $t^{\rho\beta_{pr}}\Phi(u) \in C_b([0,\infty); L^r(\mathbb{R}^N))$. Thus we proceed as in the proof of Theorem 2.3 to derive that Φ is a strict contraction on Ω_ε. Therefore, the desired results follow.

If we remove the condition that the initial value is small, then the existence of local mild solutions is immediately obtained. So we omit the proof.

Corollary 2.2 *Let $1 \leq N < \frac{2r}{r-2}$ and $\beta_{r'p} + \beta_{pr}(\sigma+1) \leq 1$. Assume $\varphi \in L^p(\mathbb{R}^N)$. If (Hf) holds, then there exists $T^* > 0$ such that the problem (2.11) admits a unique local mild solution $u \in C_b([0,T^*]; L^p(\mathbb{R}^N))$. Moreover, $t^{\rho\beta_{pr}}u \in C_b([0,T^*]; L^r(\mathbb{R}^N))$, vanishing at $t = 0$.*

2.3 Space-Time Fractional Diffusion Equations

2.3.1 Introduction

Let $\Omega \subset \mathbb{R}^N$ be a bounded domain with a Lipschitz continuous boundary $\partial\Omega$. $D_t^{[\mu]}u$ denotes the distributed order fractional derivative of u in time t (with respect to α). In the current section, we consider the exterior initial value problem of distributed order fractional diffusion equations

$$\begin{cases} D_t^{[\mu]}u + (-\Delta)^s u = 0 & \text{in } (0,T)\times\Omega, \\ u = g\chi_{(0,T)\times\mathcal{O}} & \text{in } (0,T)\times(\mathbb{R}^N\setminus\Omega), \\ u(0,\cdot) = u_0 & \text{in } \Omega, \end{cases} \tag{2.30}$$

where $s \in (0,1)$ and $(-\Delta)^s$ is the fractional Laplacain operator. In (2.30), $u = u(t,x)$ is the state to be controlled and $g = g(t,x)$ is the control function which is localized in a non-empty open subset $\mathcal{O}$ of $\mathbb{R}^N\setminus\Omega$.

This section discusses an analysis of approximate controllability from the exterior of distributed order fractional diffusion problem with the fractional Laplace operator subject to the nonzero exterior condition. In Sect. 2.3.2, we introduce some concepts of fractional calculus and properties of the fractional Laplacian operator that will be used throughout the section. In Sect. 2.3.3, two special functions are introduced, and their properties are discussed when $\mu \in L^1(0, 1; \mathbb{R}_+)$; then the properties of the corresponding operators are presented in detail. In Sect. 2.3.4, the existence, uniqueness, and regularity of solutions of the problem (2.30) and its associated adjoint system are obtained. Finally, we derive the unique continuation property of the adjoint system and approximate controllability of the problem (2.30).

2.3.2 Preliminaries

Let X denote a complex Banach space. Let J be a subinterval of $\mathbb{R}$ and $L^1(J, X)$ denote the Banach space of integral functions $f : J \to X$ with the norm $\|f\|_{L^1(J,X)}$. Especially, $\|f\|_{L^1(J,\mathbb{R})} = \|f\|_{L^1(J)}$.

We state the following characterization of functions that are holomorphic in a sector. Here $\bar{b}$ denotes the Laplace transform of a function b.

Lemma 2.11 ([182]) *Let $a : (0, +\infty) \to X$ and $\theta_0 \in (0, \pi/2]$. Then the following are equivalent:*

(i) *There is a function $b(z)$ holomorphic for $|\arg(z)| < \theta_0$ and bounded on each sector $|\arg(z)| \le \theta < \theta_0$ such that $a(s) = \bar{b}(s)$ for each $s > 0$.*
(ii) *$a(s)$ admits a holomorphic extension to the sector $|\arg(s)| < \frac{\pi}{2} + \theta_0$, and $sa(s)$ is bounded on each sector $|\arg(s)| \le \frac{\pi}{2} + \theta,\ \theta < \theta_0$.*

The following result shows that the inverse Laplace transform can be easily computed by means of the integral of a real-valued function, which was obtained in [20, 124].

Lemma 2.12 *Let F be an analytic function in $\mathbb{C}\backslash(-\infty, 0]$, satisfying the following assumptions:*

(i) *The limit $F^{\pm}(t) := \lim_{\phi\to\pi^-} F(te^{\pm i\phi})$ exists for a.a. $t > 0$ and $F^+ = \overline{F^-}$.*
(ii) *$|F(z)| = o(1)$ for $|z| \to \infty$, $|F(z)| = o(\frac{1}{|z|})$ for $|z| \to 0$ uniformly on $|\arg(z)| < \pi - \eta$, $\pi > \eta > 0$.*
(iii) *There exist $\varepsilon_0 \in (0, \frac{\pi}{2})$ and a function $a = a(r)$ such that $\varphi \in (\pi - \varepsilon_0, \pi)$ the estimate $|F(re^{\pm i\varphi})| \le a(r)$ holds, where $\frac{a(r)}{1+r} \in L^1(\mathbb{R})$.*

Then for $z \in \mathbb{C}$ such that $Re(z) > 0$, we have

$$F(z) = \int_0^\infty e^{-xz} f(x)dx, \quad \textit{where} \quad f(x) = \frac{1}{\pi}\int_0^\infty e^{-rx} Im(F^-(r))dr.$$

Next, let $T \in (0, \infty)$, and we define the left and right Riemann-Liouville integrals of $v \in L^1(0, T; \mathbb{R}^N)$:

$${}_0D_t^{-\alpha}v(t) = \int_0^t k_\alpha(t-\tau)v(\tau)d\tau, \quad {}_tD_T^{-\alpha}v(t) = \int_t^T k_\alpha(\tau - t)v(\tau)d\tau, \quad t > 0,$$

where $k_\alpha(t) = \frac{t^{\alpha-1}}{\Gamma(\alpha)}$. Furthermore, ${}_0^C D_t^\alpha v$ and ${}_t^L D_T^\alpha v$ stand for the left Caputo and right Riemann-Liouville fractional derivative operators of order α for $v \in AC[0,T]$, respectively; they are defined by

$$ {}_0^C D_t^\alpha v(t) = \int_0^t k_{1-\alpha}(t-\tau) D_\tau v(\tau) d\tau, \quad t > 0, $$

$$ {}_t^L D_T^\alpha v(t) = -D_t \Big(\int_t^T k_{1-\alpha}(\tau - t) v(\tau) d\tau \Big), \quad t > 0. $$

The following integration by parts formula is taken from [4]. Let $0 < \alpha \le 1$. Then

$$ \int_0^T u(t) {}_0^C D_t^\alpha v(t) dt = \int_0^T v(t) {}_t^L D_T^\alpha u(t) dt + \Big[v(t) {}_t D_T^{\alpha-1} u(t) \Big] \Big|_{t=0}^{t=T}, $$

provided that the left- and right-hand side expressions make sense. Furthermore, we have

$$ \int_0^T u(t) D_t^{[\mu]} v(t) dt = \int_0^T v(t) {}_t^L D_T^{[\mu]} u(t) dt + \Big[v(t) {}_t I_T^{[\mu]} u(t) \Big] \Big|_{t=0}^{t=T}, \tag{2.31} $$

where ${}_t^L D_T^{[\mu]} u = -\frac{\partial}{\partial t} {}_t I_T^{[\mu]} u$ and the distributed order right integral operator ${}_t I_T^{[\mu]} u$ is defined by

$$ {}_t I_T^{[\mu]} u = \int_0^1 \mu(\alpha) {}_t D_T^{\alpha-1} u d\alpha; $$

here μ is a weighted function.

Finally, we introduce fractional order Sobolev spaces and fractional Laplace operators. Given $0 < s < 1$, we recall that

$$ W^{s,2}(\Omega) = \Big\{ u \in L^2(\Omega) : \int_\Omega \int_\Omega \frac{|u(x) - u(y)|^2}{|x-y|^{N+2s}} dx dy < \infty \Big\} $$

with the natural norm

$$ \|u\|_{W^{s,2}(\Omega)} = \Big(\int_\Omega |u(x)|^2 dx + \int_\Omega \int_\Omega \frac{|u(x)-u(y)|^2}{|x-y|^{N+2s}} dx dy \Big)^{\frac{1}{2}} $$

and $W_0^{s,2}(\bar{\Omega}) = \{u \in W^{s,2}(\Omega) : u = 0 \text{ in } \mathbb{R}^N \setminus \Omega\}$. We let

$$ W_{loc}^{s,2}(\Omega) = \{u \in L^2(\Omega) : u\varphi \in W^{s,2}(\Omega), \ \forall \varphi \in \mathscr{D}(\Omega)\}, $$

where $\mathscr{D}(\Omega)$ denotes the space of all continuously infinitely differentiable functions with compact support in Ω. Furthermore, we let $W^{-s,2}(\mathbb{R}^N)$ and $W_0^{-s,2}(\bar{\Omega})$ denote the dual of $W^{s,2}(\mathbb{R}^N)$ and $W_0^{s,2}(\bar{\Omega})$, respectively.

The fractional Laplace operator $(-\Delta)^s$ is defined for $u \in \mathscr{L}_s^1(\mathbb{R}^N)$ by the formula

$$(-\Delta)^s u(x) := C_{N,s} P.V. \int_{\mathbb{R}^N} \frac{u(x) - u(y)}{|x - y|^{N+2s}} dy, \quad x \in \mathbb{R}^N,$$

where $C_{N,s} = \frac{s 2^{2s} \Gamma(\frac{2s+N}{2})}{\pi^{\frac{N}{2}} \Gamma(1-s)}$ and

$$\mathscr{L}_s^1(\mathbb{R}^N) = \left\{ u : \mathbb{R}^N \to \mathbb{R} \text{ measurable}, \int_{\mathbb{R}^N} \frac{|u(x)|}{1 + |x|^{N+2s}} dx < \infty \right\}.$$

For $u \in W^{s,2}(\mathbb{R}^N)$, we introduce the nonlocal normal derivative $\mathscr{N}_s$ given by

$$\mathscr{N}_s u(x) := C_{N,s} \int_{\Omega} \frac{u(x) - u(y)}{|x - y|^{N+2s}} dy, \quad x \in \mathbb{R}^N \setminus \bar{\Omega}. \tag{2.32}$$

By [67, Lemma 3.2], for every $u \in W^{s,2}(\mathbb{R}^N)$, we have $\mathscr{N}_s u \in W_{loc}^{s,2}(\mathbb{R}^N \setminus \Omega)$. This together with the definition of $W_{loc}^{s,2}(\mathbb{R}^N \setminus \Omega)$ shows that $\mathscr{N}_s u \in L^2(\mathbb{R}^N \setminus \Omega)$.

Let $u \in W^{s,2}(\mathbb{R}^N)$ be such that $(-\Delta)^s u \in L^2(\Omega)$. Then for every $v \in W^{s,2}(\mathbb{R}^N)$, the following identity holds:

$$\begin{aligned} &\frac{C_{N,s}}{2} \int\int_{\mathbb{R}^{2N} \setminus (\mathbb{R}^N \setminus \Omega)^2} \frac{(u(x) - u(y))(v(x) - v(y))}{|x - y|^{N+2s}} dxdy \\ &= \int_{\Omega} v(-\Delta)^s u dx + \int_{\mathbb{R}^N \setminus \Omega} v \mathscr{N}_s u dx, \end{aligned} \tag{2.33}$$

where $\mathbb{R}^{2N} \setminus (\mathbb{R}^N \setminus \Omega)^2 = (\Omega \times \Omega) \cup (\Omega \times (\mathbb{R}^N \setminus \Omega)) \cup ((\mathbb{R}^N \setminus \Omega) \times \Omega)$. The above integration by parts formula is contained in [53, 227]. For more details on the fractional Laplace operator, we refer the reader to [67] and the references therein.

The following unique continuation property has been presented in [227].

Lemma 2.13 *Let $\lambda > 0$ be a real number and $\mathscr{O} \subset \mathbb{R}^N \setminus \Omega$ a non-empty open set. If $\varphi \in D((-\Delta)_D^s)$ satisfies*

$$(-\Delta)_D^s \varphi = \lambda \varphi \ \text{ in } \Omega \ \text{ and } \ \mathscr{N}_s \varphi = 0 \ \text{ in } \mathscr{O},$$

then $\varphi = 0$ in $\mathbb{R}^N$.

Furthermore, we consider the operator $(-\Delta)_D^s$ defined by $(-\Delta)_D^s = (-\Delta)^s$ with the domain

$$D((-\Delta)^s_D) = \{u \in W_0^{s,2}(\bar{\Omega}) : (-\Delta)^s u \in L^2(\Omega)\},$$

which is the realization of $(-\Delta)^s u$ in $L^2(\Omega)$ with the zero exterior condition. Clearly, it is a self-adjoint operator in $L^2(\Omega)$. Let $\{\lambda_n\}$ be the set of eigenvalues of the operator $(-\Delta)^s_D$, and correspondingly, let $\{\varphi_n\}_{n=1}^{\infty}$ denote the complete orthonormal system of eigenfunctions which forms an orthogonal basis of $L^2(\Omega)$ such that

$$(-\Delta)^s \varphi_n = \lambda_n \varphi_n \text{ in } \Omega, \quad \varphi_n = 0 \text{ in } \mathbb{R}^N \setminus \Omega,$$

where $0 < \lambda_1 \le \lambda_2 \le \ldots \le \lambda_n \le \ldots$ and $\lim_{n\to\infty} \lambda_n = +\infty$. For more details on the fractional Laplace operator, we refer to [39, 192] and their references.

For all $\beta \ge 0$, the fractional power operator $[(-\Delta)^s_D]^\beta$ possesses the following representation:

$$[(-\Delta)^s_D]^\beta v = \sum_{n=1}^{\infty} \lambda_n^\beta (v, \varphi_n)\varphi_n,$$

$$D([(-\Delta)^s_D]^\beta) = \left\{ v \in L^2(\Omega) : \sum_{n=1}^{\infty} |\lambda_n^\beta (v, \varphi_n)|^2 < \infty \right\}.$$

Set $\mathbb{V}_{s,\beta} := D([(-\Delta)^s_D]^\beta)$. Then the norm is $\|u\|^2_{\mathbb{V}_{s,\beta}} = \sum_{n=1}^{\infty} |\lambda_n^\beta (v, \varphi_n)|^2$ for $v \in \mathbb{V}_{s,\beta}$. Notice that if $\beta \in (0, 1)$, we have

$$\mathbb{V}_{s,\beta} = [D((-\Delta)^s_D), L^2(\Omega)]_{1-\beta}.$$

It is well known that if $\beta \in (0, \frac{1}{4})$, then $W_0^{s,2}(\bar{\Omega}) \hookrightarrow \mathbb{V}_{s,\beta}$ and $\mathbb{V}_{s,1-\beta} \hookrightarrow W^{s,2}(\mathbb{R}^N)$.

2.3.3 Technical Tools

In this subsection, let $\mu \in L^1(0, 1; \mathbb{R}_+)$ and $\mu \not\equiv 0$. For $\theta \in (\frac{\pi}{2}, \pi)$, we define Σ_θ as

$$\Sigma_\theta = \{z \in \mathbb{C} : \ z \ne 0, \ |\arg z| < \theta\}.$$

Denote $\omega(z) := \int_0^1 z^{\alpha-1}\mu(\alpha)d\alpha$. Now we give some properties of $\omega(z)$. Using the generalization of the mean value theorem of integrals, it is easy to show that

$$|w(z)| \le \|\mu\|_{L^1(0,1)} |z|^{\nu-1}, \quad \text{for } z \in \mathbb{C} \setminus \mathbb{R}^-, \tag{2.34}$$

where $\nu \in (0, 1)$.

Lemma 2.14 *Let $\lambda > 0$ and $\mu \in L^1(0,1;\mathbb{R}_+)$ with $\mu \not\equiv 0$. The following properties hold:*

(i) *There exists a positive constant M such that*

$$|z\omega(z)+\lambda|^{-1} \le M \min\left\{\lambda^{-1},\ |z\omega(z)|^{-1}\right\},\ \ for\ z\in\Sigma_\theta. \tag{2.35}$$

(ii) *There exist a positive constant M_1 and $\gamma \in (0, \frac{1}{2})$ such that*

$$|z\omega(z)+\lambda|^{-1} \le M_1|z|^{-\rho},\quad \rho\in(\gamma, 1-\gamma),\ \ for\ z\in\Sigma_\theta. \tag{2.36}$$

Proof (i) For $z \in \Sigma_\theta$, from the proof of Lemma 2.1 in [134], we know the inequality $|z\omega(z)+\lambda| > \frac{\sin\theta}{2}\lambda$ also holds when $\mu \in L^1(0,1;\mathbb{R}_+)$. Furthermore, we write $z = re^{\pm i\phi}$ for $\phi \in [0,\theta)$. We have to consider two cases. If $\phi \in [0, \frac{\pi}{2}]$, then $\cos(\alpha\phi) > 0$ for $\alpha \in (0,1)$ and $Re(z\omega(z)+\lambda) > Re(z\omega(z)) > 0$. Thus

$$|z\omega(z)+\lambda| \ge |z\omega(z)| > 0.$$

If $\phi \in (\frac{\pi}{2}, \theta)$, we know that the imaginary part of $z\omega(z)$ is nonzero, and it follows from the arguments of [134, Lemma 2.1] that

$$|z\omega(z)+\lambda| \ge \frac{\sin\phi}{2}|z\omega(z)| \ge \frac{\sin\theta}{2}|z\omega(z)| > 0.$$

Therefore, the inequality (2.35) follows for $M = \frac{2}{\sin\theta}$.

(ii) Similar to the proof of proposition 6 in [124], we know that

$$|z\omega(z)+\lambda| \ge c_0 \int_\gamma^{1-\gamma} \mu(\alpha)|z|^\alpha d\alpha,\quad \text{for } z\in\Sigma_\theta \text{ and } \gamma\in(0,\frac{1}{2}),$$

where $c_0 = \max\left\{\cos(\frac{(1-\gamma)\pi}{2}), \min\{\sin(\gamma\pi), \sin(\frac{\gamma\pi}{2}), \frac{\sqrt{2}}{2}\}\right\}$.

Using the generalization of the mean value theorem of integrals again, it follows that

$$\int_\gamma^{1-\gamma} \mu(\alpha)|z|^\alpha d\alpha = \|\mu\|_{L^1(\gamma,1-\gamma)}|z|^\rho,$$

where $\rho \in (\gamma, 1-\gamma)$. The proof is finished.

For $\delta > 0$ and $\theta \in (0, \pi/2)$, we introduce the contour $\Gamma_{\delta,\theta}$ defined by

$$\Gamma_{\delta,\theta} = \{re^{-i\theta} : r\ge\delta\} \cup \{\delta e^{i\psi} : |\psi|\le\theta\} \cup \{re^{i\theta} : r\ge\delta\},$$

where the circular arc is oriented counterclockwise, and the two rays are oriented with an increasing imaginary part. For convenience, we define two functions: for $\lambda > 0$ and a.e. $t > 0$,

$$P(t,\lambda)=\frac{1}{2\pi i}\int_{\Gamma_{\delta,\theta}}\frac{\omega(z)e^{zt}}{z\omega(z)+\lambda}dz,\quad S(t,\lambda)=\frac{1}{2\pi i}\int_{\Gamma_{\delta,\theta}}\frac{e^{zt}}{z\omega(z)+\lambda}dz.$$

Moreover, we give several properties of $P(t,\lambda)$ and $S(t,\lambda)$ as follows.

Lemma 2.15 *Let $\mu\in L^1(0,1;\mathbb{R}_+)$ and $\mu\not\equiv 0$. For each $\lambda>0$, the functions $P(\cdot,\lambda)\in C((0,\infty);\mathbb{R})$ and $S(\cdot,\lambda)\in C((0,\infty);\mathbb{R})$. Moreover,*

$$|P(t,\lambda)|\le M^* \text{ for } t\ge 0 \text{ and } |S(t,\lambda)|\le M^*t^{\rho-1} \text{ for } t>0,$$

where $M^*=\max\{M,\ M_1\}$.

Proof We write $z=re^{i\phi}$. It follows from choosing the main branch of logarithm that

$$z\omega(z)+\lambda=\int_0^1 r^\alpha\cos(\alpha\phi)\mu(\alpha)d\alpha+\lambda+i\int_0^1 r^\alpha\sin(\alpha\phi)\mu(\alpha)d\alpha,$$

$$\omega(z)=\int_0^1 r^{\alpha-1}\cos((1-\alpha)\phi)\mu(\alpha)d\alpha-i\int_0^1 r^{\alpha-1}\sin((1-\alpha)\phi)\mu(\alpha)d\alpha$$

are analytic for $z\in\mathbb{C}\setminus(-\infty,0]$. Note that the function $z\omega(z)+\lambda$ has no zero in the main sheet of the Riemann surface including its boundaries on the cut, since the imaginary part of $z\omega(z)+\lambda$ is nonzero for $|\phi|\in(0,\pi)$ and the real part is positive for $\phi=0$. It also shows the analyticity of $\omega(z)(z\omega(z)+\lambda)^{-1}$ and $(z\omega(z)+\lambda)^{-1}$ for $z\in\mathbb{C}\setminus(-\infty,0]$.

From (2.35) and (2.36), we arrive at the following inequalities:

$$\left|\frac{\omega(z)e^{zt}}{z\omega(z)+\lambda}\right|\le M|z|^{-1}e^{tRe(z)} \text{ and } \left|\frac{e^{zt}}{z\omega(z)+\lambda}\right|\le M_1|z|^{-\rho}e^{tRe(z)},$$

for any $z\in\Sigma_\theta$. Thus from [11, Lemma 4.1.1], we conclude that $P(\cdot,\lambda)x\in C((0,\infty);$
$\mathbb{R})$ and $S(\cdot,\lambda)x\in C((0,\infty);\mathbb{R})$. Moreover, the operators have the uniform bound, i.e.,

$$|P(t,\lambda)|\le M \text{ and } |S(t,\lambda)|\le M_1t^{\rho-1},$$

for $t>0$. This together with the property of the Laplace transform shows that

$$P(0,\lambda)=\lim_{z\to+\infty}|z\mathscr{L}[P(\cdot,\lambda)](z)|=\lim_{z\to+\infty}\frac{1}{1+\frac{\lambda}{|z\omega(z)|}}=1.$$

Thus $P(0,\lambda)=1$.

Lemma 2.16 *Assume that* $\mu \in L^1(0, 1; \mathbb{R}_+)$ *and* $\mu \not\equiv 0$.

(i) *Then there exists a constant* M_c^* *depending only on* θ *and* $\|\mu\|_{L^1(0,1)}$ *such that*

$$|\lambda P(t, \lambda)| \le M_c^* t^{-\rho}, \ \textit{for}\ t > 0\ \textit{and for any}\ \lambda > 0.$$

(ii) *Then there exists a constant* M_1^* *depending only on* θ *and* $\|\mu\|_{L^1(0,1)}$ *such that*

$$|\lambda S(t, \lambda)| \le M_1^* t^{-1}, \ \textit{for}\ t > 0\ \textit{and for any}\ \lambda > 0.$$

Proof (i) Assume that $\mu \in L^1(0, 1; \mathbb{R}_+)$ and $\mu \not\equiv 0$. From Remark 3.1 in paper [179], we know that $\rho = \nu$. This together with (2.34) shows that the inequality

$$|w(z)| \le \|\mu\|_{L^1(0,1)} |z|^{\rho-1}$$

holds. Let $t > 0$ and $\delta = \frac{1}{t} > 0$. By using the identity

$$\lambda\omega(z)(z\omega(z) + \lambda)^{-1} = \omega(z)\big(I - z\omega(z)(z\omega(z) + \lambda)^{-1}\big),$$

we can deduce from (2.35) that

$$|\lambda\omega(z)(z\omega(z) + \lambda)^{-1}| \le (M + 1)\|\mu\|_{L^1(0,1)} |z|^{\rho-1},$$

for any $z \in \Sigma_\theta$ and $\lambda > 0$. Thus, we can estimate $\lambda P(t, \lambda)$ as

$$\begin{aligned}
|\lambda P(t, \lambda)| \le & \int_{\Gamma_{\delta,\theta}} e^{Re(z)t} |\lambda\omega(z)(z\omega(z) + \lambda)^{-1}||dz| \\
\le & (M + 1)\|\mu\|_{L^1(0,1)} \left(2\int_{t^{-1}}^{\infty} \tau^{\rho-1} e^{-\tau t \cos(\theta)} d\tau + \int_{-\pi+\theta}^{\pi-\theta} t^{-\rho} e^{\cos(\psi)} d\psi \right) \\
\le & M_c^* t^{-\rho},
\end{aligned}$$

where $M_c^* > 0$ is a constant depending on M, θ and $\|\mu\|_{L^1(0,1)}$.

(ii) Let $t > 0$ and $\delta = \frac{1}{t} > 0$. In view of the identity $\lambda(z\omega(z) + \lambda)^{-1} = I - z\omega(z)(z\omega(z)+\lambda)^{-1}$, it follows from (2.35) that $|\lambda(z\omega(z)+\lambda)^{-1}| \le (M+1)$ for any $z \in \Sigma_\theta$. Thus, we can estimate the upper bound on $\lambda S(t, \lambda)$ for $t \in \mathbb{R}_+$,

$$\begin{aligned}
|\lambda S(t, \lambda)| \le & \int_{\Gamma_{\frac{1}{t},\theta}} e^{Re(z)t} |\lambda(z\omega(z) + \lambda)^{-1}||dz| \\
\le & (M + 1) \left(2\int_{\frac{1}{t}}^{\infty} e^{-rt|\cos(\theta)|} dr + \int_{-\theta}^{\theta} e^{\cos(\psi)} t^{-1} d\psi \right) \\
\le & M_1^* t^{-1},
\end{aligned}$$

where M_1^* is a positive constant and it may be dependent on M and θ. Therefore the conclusion follows.

We establish the relation between $P(t, \lambda)$ and $S(t, \lambda)$ in the following lemma.

Lemma 2.17 *Assume that* $\mu \in L^1(0, 1; \mathbb{R}_+)$ *and* $\mu \not\equiv 0$. *For a.e.* $t \in (0, \infty)$ *and for* $\lambda > 0$, *we have*

$$\int_0^1 \mu(\alpha)_0 D_t^{\alpha-1} S(t, \lambda) d\alpha = P(t, \lambda) \quad \textit{and} \quad P'(t, \lambda) = -\lambda S(t, \lambda).$$

Proof Notice that

$$\begin{aligned}\mathscr{L}\left[\int_0^1 \mu(\alpha)_0 D_t^{\alpha-1} S(t, \lambda) d\alpha\right](z) =& \omega(z)\mathscr{L}[S(\cdot, \lambda)](z) \\ =& \frac{\omega(z)}{z\omega(z) + \lambda} \\ =& \mathscr{L}[P(\cdot, \lambda)](z).\end{aligned}$$

By the uniqueness of the inverse Laplace transform, we know that

$$\int_0^1 \mu(\alpha)_0 D_t^{\alpha-1} S(t, \lambda) d\alpha = P(t, \lambda).$$

From the arguments in Lemma 2.15, we know that $\frac{\omega(z)}{z\omega(z)+\lambda}$ is holomorphic in $\mathbb{C} \setminus \mathbb{R}^-$. In view of

$$|z\mathscr{L}[P(\cdot, \lambda)](z)| = \left|\frac{z\omega(z)}{z\omega(z) + \lambda}\right| \leq M, \quad \text{for } z \in \Sigma_\theta \text{ and } \lambda > 0,$$

it follows from Lemma 2.11 that $P(t, \lambda)$ is holomorphic for $|\arg(t)| < \theta_0$ ($\theta_0 \in (0, \pi/2]$). This also ensures the time analyticity of $P(t, \lambda)$ for $\lambda > 0$.

Observe that

$$\begin{aligned}\mathscr{L}[P'(\cdot, \lambda)](z) =& z\mathscr{L}[P(\cdot, \lambda)](z) - P(0, \lambda) \\ =& \frac{z\omega(z)}{z\omega(z) + \lambda} - 1 \\ =& -\lambda \frac{1}{z\omega(z) + \lambda} \\ =& -\lambda\mathscr{L}[S(\cdot, \lambda)](z).\end{aligned}$$

Then we have $P'(t, \lambda) = -\lambda S(t, \lambda)$ for $t > 0$.

Remark 2.7 Under the conditions of Lemma 2.15, the continuity of $P(t,\lambda)$ at $t=0$ is evident for each $\lambda>0$.

Proof We claim that $P(t,\lambda)\to 1$ as $t\to 0$. Indeed, for $t>0$, we use Lemma 2.15 and Lemma 2.17 to obtain that

$$|P(t,\lambda)-1| = \left|-\int_0^t P'(\tau,\lambda)d\tau\right| \le \lambda\int_0^t |S(\tau,\lambda)|d\tau$$
$$\le \lambda M\int_0^t \tau^{\rho-1}d\tau \to 0, \quad \text{as } t\to 0.$$

Then the continuity of the function $P(t,\lambda)$ at $t=0$ follows.

Lemma 2.18 *Assume that $\mu\in L^1(0,1;\mathbb{R}_+)$ and $\mu\not\equiv 0$. Then $S(t,\lambda)$ has the following explicit formula:*

$$S(t,\lambda)=\int_0^\infty e^{-rt}\mathscr{K}(r,\lambda)dr, \quad \textit{for } t>0 \textit{ and } \lambda>0, \tag{2.37}$$

where

$$\mathscr{K}(r,\lambda)=\frac{1}{\pi}\frac{\int_0^1 r^\alpha\sin(\alpha\pi)\mu(\alpha)d\alpha}{\left(\int_0^1 r^\alpha\cos(\alpha\pi)\mu(\alpha)d\alpha+\lambda\right)^2+\left(\int_0^1 r^\alpha\sin(\alpha\pi)\mu(\alpha)d\alpha\right)^2}. \tag{2.38}$$

Proof We denote $F(z)=\frac{1}{z\omega(z)+\lambda}$, and we prove that $F(z)$ satisfies assumptions of Lemma 2.12. From the arguments in Lemma 2.15, we know that $F(z)$ is holomorphic in $\mathbb{C}\setminus\mathbb{R}^-$.

Observe that $\int_0^1 r^\alpha\sin(\alpha\pi)\mu(\alpha)d\alpha$ is nonzero for $r>0$. Then $\lim_{\phi\to\pi^-}F(re^{\pm i\phi})$ exists. In view of

$$\int_0^1 r^\alpha e^{-i\alpha\pi}\mu(\alpha)d\alpha+\lambda=\int_0^1 r^\alpha\overline{e^{i\alpha\pi}}\mu(\alpha)d\alpha+\lambda=\overline{\int_0^1 r^\alpha e^{i\alpha\pi}\mu(\alpha)d\alpha+\lambda}$$

and $\lim_{\phi\to\pi^-}e^{\pm i\alpha\phi}=e^{\pm i\alpha\pi}$, one can obtain from the Lebesgue dominated convergence theorem that $F^+=\overline{F^-}$. On the other hand, it follows from (2.36) that $\lim_{|z|\to\infty}|F(z)|=0$ and $\lim_{|z|\to 0}|zF(z)|=0$. Furthermore, we also observe that

$$a(r)=Mr^{\gamma-1} \text{ for } r\le 1 \text{ and } a(r)=Mr^{-\gamma} \text{ for } r>1.$$

It is clear that $\frac{a(r)}{1+r}\in L^1(\mathbb{R})$. Therefore, using Lemma 2.12, we have

$$F(z)=\int_0^\infty e^{-tz}f(t)dt,$$

where

$$f(t) = \frac{1}{\pi}\int_0^\infty e^{-rt} Im(F^-(r))dr = \frac{1}{\pi}\int_0^\infty e^{-rt}\mathscr{K}(r,\lambda)dr,$$

for $\mathscr{K}$ defined as in (2.38). From the uniqueness of the inverse Laplace transform, it shows that $S(t,\lambda)$ has the form as (2.37).

As an immediate consequence of Lemma 2.17 and Lemma 2.18, we can infer the following results.

Lemma 2.19 *Let $\mu \in L^1(0,1;\mathbb{R}_+)$ and $\mu \not\equiv 0$. For $\lambda > 0$, the functions $S(t,\lambda)$ and $P(t,\lambda)$ have the following properties:*

(i) *$S(t,\lambda)$ is completely monotone for $t > 0$.*
(ii) *$0 < S(t,\lambda) \le M^* t^{\rho-1}$ for $t > 0$ and $0 < P(t,\lambda) \le 1$ for $t \ge 0$.*
(iii) *For any $\lambda_0 \ge \lambda$ and $t > 0$, $S(t,\lambda_0) \le S(t,\lambda)$.*

Lemma 2.20 *Assume that $\mu \in L^1(0,1;\mathbb{R}_+)$ and $\mu \not\equiv 0$. For fixed $t \ge 0$ and $\lambda > 0$, the function $\lambda \mapsto P(t,\lambda)$ is nonincreasing on $(0,\infty)$.*

Proof Notice that the Laplace transform of $P(t,\lambda)$ and $S(t,\lambda)$ is given by

$$\mathscr{L}[P(t,\lambda)](z) = \frac{\omega(z)}{z\omega(z)+\lambda}, \qquad \mathscr{L}[S(t,\lambda)](z) = \frac{1}{z\omega(z)+\lambda}.$$

Then we have

$$\frac{\partial}{\partial\lambda}\mathscr{L}[P(t,\lambda)](z) = -\frac{\omega(z)}{(z\omega(z)+\lambda)^2} = -\mathscr{L}[P(t,\lambda)](z)\mathscr{L}[S(t,\lambda)](z).$$

In view of the fact that $\mathscr{L}[S(t,\lambda)](z) = -\frac{1}{\lambda}\mathscr{L}[P'(t,\lambda)](z)$, we can deduce that

$$\frac{\partial}{\partial\lambda}\mathscr{L}[P(t,\lambda)](z) = \frac{1}{\lambda}\mathscr{L}[P(t,\lambda)](z)\mathscr{L}[P'(t,\lambda)](z).$$

Taking the inverse Laplace transform of the above equation, it follows from Lemma 2.17 and Lemma 2.19 that

$$\frac{\partial}{\partial\lambda}P(t,\lambda) = \frac{1}{\lambda}P(\cdot,\lambda) * P'(\cdot,\lambda) \le 0.$$

The desired assertion follows.

Next, we concern with the relaxation problem

$$\begin{cases} D_t^{[\mu]}\varpi(t) + \lambda\varpi(t) = f(t) & \text{in } (0,T), \\ \varpi(0) = \varpi_0, \end{cases} \tag{2.39}$$

where the unknown ϖ is a scalar function, λ is a positive parameter and $f \in L^1(0, T)$.

Lemma 2.21 *Assume that* $\mu \in L^1(0, 1; \mathbb{R}_+)$ *and* $\mu \not\equiv 0$. *The solution of* (2.39) *is given by*

$$\varpi(t) = P(t, \lambda)\varpi_0 + \int_0^t S(t-\tau, \lambda) f(\tau) d\tau. \tag{2.40}$$

Proof Applying the Laplace transform to the problem (2.39), we obtain

$$z\omega(z)\overline{\varpi} - \omega(z)\varpi_0 + \lambda\overline{\varpi} = \bar{f}.$$

Then

$$\overline{\varpi}(z) = \frac{\omega(z)}{z\omega(z) + \lambda}\varpi_0 + \frac{\bar{f}}{z\omega(z) + \lambda}.$$

Therefore, (2.40) follows from the inverse Laplace transform.

On the other hand, we can prove that the function ϖ given by (2.40) is the solution to the problem (2.39). Indeed, let $\mathscr{Q}[\varpi] = D_t^{[\mu]}\varpi(t) + \lambda\varpi(t)$; we have

$$\mathscr{Q}[\varpi] = \mathscr{Q}[P(\cdot, \lambda)]\varpi_0 + \mathscr{Q}[S(\cdot, \lambda) * f] = \mathscr{Q}[S(\cdot, \lambda) * f].$$

It remains to show that $\mathscr{Q}[S(\cdot, \lambda) * f] = f$. From Lemma 2.17, we obtain that

$$\begin{aligned} D_t^{[\mu]}(S(\cdot, \lambda) * f) =& k(\cdot) * \big(S(\cdot, \lambda) * f\big)' \\ =& \Big[k(\cdot) * \big(S(\cdot, \lambda) * f\big)\Big]' - k(t)\big(S(\cdot, \lambda) * f\big)(0) \\ =& [k(\cdot) * S(\cdot, \lambda) * f]' \\ =& [P(\cdot, \lambda) * f]' \\ =& f - \lambda[S(\cdot, \lambda) * f], \end{aligned}$$

where

$$k(t) = \int_0^1 \mu(\alpha) k_{1-\alpha}(t) d\alpha. \tag{2.41}$$

This implies

$$\mathscr{Q}[S(\cdot, \lambda) * f] = D_t^{[\mu]}(S(\cdot, \lambda) * f) + \lambda(S(\cdot, \lambda) * f) = f.$$

The proof is completed.

Now we give a representation of solutions to the following problem:

$$\begin{cases} D_t^{[\mu]}u + (-\Delta)^s u = f & \text{in } (0,T)\times\Omega, \\ u = 0 & \text{in } (0,T)\times(\mathbb{R}^N\setminus\Omega), \\ u(0,\cdot) = u_0 & \text{in } \Omega, \end{cases} \tag{2.42}$$

where $f \in L^1_{loc}(\mathbb{R}_+, L^2(\Omega))$ and $u_0 \in L^2(\Omega)$. Indeed, by applying eigenfunction expansion, we have

$$D_t^{[\mu]}u_n(t) + \lambda_n u_n(t) = f_n(t), \quad u_n(0) = u_{0n},$$

where $u_n(t) = (u(t,\cdot),\varphi_n)$, $f_n(t) = (f(t,\cdot),\varphi_n)$, and $u_{0n} = (u_0,\varphi_n)$. Then Lemma 2.21 shows

$$u_n(t) = P(t,\lambda_n)u_{0n} + \int_0^t S(t-\tau,\lambda_n)f_n(\tau)d\tau.$$

Thus, the solutions of the problem (2.42) can be written in the form

$$u(t) = P(t)u_0 + \int_0^t S(t-\tau)f(\tau)d\tau, \tag{2.43}$$

where

$$P(t)v = \sum_{n=1}^{\infty} P(t,\lambda_n)v_n\varphi_n \text{ and } S(t)v = \sum_{n=1}^{\infty} S(t,\lambda_n)v_n\varphi_n;$$

here, $v_n = (v(\cdot),\varphi_n)$ for $v \in L^2(\Omega)$.

In the sequel, the notation $\|\cdot\|$ is denoted as the norm in $L^2(\Omega)$. We collect some properties of the operators $P(t)$ and $S(t)$ in the following lemma.

Lemma 2.22 *Assume that* $\mu \in L^1(0,1;\mathbb{R}_+)$ *and* $\mu \not\equiv 0$. *For any* $v \in L^2(\Omega)$, *we have:*

(i) $\|P(t)v\| \le P(t,\lambda_1)\|v\|$ *for* $t \ge 0$. *In particular,* $\|P(t)\| \le 1$ *for* $t \ge 0$ *and* $P(0) = I$.
(ii) $\|S(t)v\| \le S(t,\lambda_1)\|v\| \le M^* t^{\rho-1}\|v\|$ *for* $t > 0$.
(iii) $P(\cdot)v \in C([0,T];L^2(\Omega))$ *and* $S(\cdot)v \in C((0,T];L^2(\Omega))$ *for* $T > 0$.

Proof Statements (i) and (ii) are immediately deduced by Lemma 2.19 and Lemma 2.20. We prove (iii) as follows. For $t \in (0,T]$ and $v \in L^2(\Omega)$, the continuity of $P(t)v$ and $S(t)v$ with respect to t follows by (i), (ii), and Lemma 2.15. We only need to show the continuity of $P(t)v$ in $L^2(\Omega)$ at $t = 0$. Indeed, let $v \in D((-\Delta)^s_D)$. Then

$$
\begin{aligned}
\|P(t)v - v\|^2 &= \sum_{n=1}^{\infty} |P(t, \lambda_n)v_n - v_n|^2 \\
&= \sum_{n=1}^{\infty} \left| \int_0^t S(\tau, \lambda) d\tau \right|^2 |\lambda_n v_n|^2 \\
&\leq \frac{M^{*2}}{\rho^2} t^{2\rho} \sum_{n=1}^{\infty} |\lambda_n v_n|^2 \\
&= \frac{M^{*2}}{\rho^2} t^{2\rho} \|v\|^2_{D((-\Delta)^s_D)} \to 0, \quad t \to 0.
\end{aligned}
$$

Since $D((-\Delta)^s_D)$ is dense in $L^2(\Omega)$, it also shows the continuity of $P(t)v$ at $t = 0$ for $v \in L^2(\Omega)$.

Lemma 2.23 *Let $\beta \in (0, 1)$. Assume that $\mu \in L^1(0, 1; \mathbb{R}_+)$ and $\mu \not\equiv 0$. There exists a constant $C > 0$ such that $P(t)v \in \mathbb{V}_{s,\beta}$ and $S(t)v \in \mathbb{V}_{s,\beta}$ for all $v \in L^2(\Omega)$ and for every $t > 0$. Moreover,*

$$
\|P(t)v\|_{\mathbb{V}_{s,\beta}} \leq Ct^{-\beta\rho}\|v\| \quad \textit{and} \quad \|S(t)v\|_{\mathbb{V}_{s,\beta}} \leq Ct^{\rho(1-\beta)-1}\|v\|, \quad \textit{for } t > 0.
$$

Proof Let $A = (-\Delta)^s_D$. By Lemma 2.16, it follows that $P(t)v$, $S(t)v \in D(A)$ for $v \in L^2(\Omega)$ and $t > 0$. Moreover, we can estimate $\|AP(t)v\|^2 = \sum_{n=1}^{\infty} |\lambda_n P(t, \lambda_n)v_n|^2 \leq M_c^* t^{-\rho}\|v\|$ and $\|AS(t)v\| \leq M_1^* t^{-1}\|v\|$. It together with Lemma 2.22 and the interpolation inequality shows that

$$
\|A^\beta P(t)v\| \leq Ct^{-\beta\rho}\|v\| \quad \text{and} \quad \|A^\beta S(t)v\| \leq Ct^{\rho(1-\beta)-1}\|v\|, \quad \text{for } t > 0.
$$

The proof is completed.

2.3.4 Well-Posedness Results

In this subsection, we investigate the well-posedness and regularity of solutions to the problem (2.30) and its associated adjoint system. To do this, let $u_0 \in L^2(\Omega)$, and let us split the system (2.30) into the following two systems:

$$
\begin{cases}
D_t^{[\mu]} v + (-\Delta)^s v = 0 & \text{in } (0, T) \times \Omega, \\
v = g & \text{in } (0, T) \times (\mathbb{R}^N \setminus \Omega), \\
v(0, \cdot) = 0 & \text{in } \Omega
\end{cases}
\tag{2.44}
$$

and

$$\begin{cases} D_t^{[\mu]}w + (-\Delta)^s w = 0 & \text{in } (0,T)\times\Omega, \\ w = 0 & \text{in } (0,T)\times(\mathbb{R}^N\setminus\Omega), \\ w(0,\cdot) = u_0 & \text{in } \Omega. \end{cases} \tag{2.45}$$

Then it is clear that $u = v + w$ solves the system (2.30). It suffices to consider the well-posedness and regularity of solutions to the system (2.44) and the system (2.45), respectively.

In the subsequent studies, we assume that $\mu \in L^1(0,1;\mathbb{R}_+)$ and $\mu \not\equiv 0$. First, we introduce the rigorous definition of strong solutions to the system (2.44).

Definition 2.4 Let g be a given function. We say that the function v is a strong solution of (2.44) if v satisfies the following properties:

(i) Regularity: $v \in C([0,T];L^2(\Omega))$, $D_t^{[\mu]}v \in C((0,T];W^{-s,2}(\mathbb{R}^N))$. Moreover, the variational identity

$$\langle D_t^{[\mu]}v, w\rangle + \langle(-\Delta)^s v, w\rangle = 0, \quad \text{for every } w \in W^{s,2}(\mathbb{R}^N) \text{ and a.e. } t \in (0,T)$$

holds.

(ii) Initial and exterior conditions: $v(0,\cdot) = 0$ in Ω and $v = g$ in $(0,T)\times(\mathbb{R}^N\setminus\Omega)$.

Remark 2.8 Since (2.45) can be rewritten as the problem

$$D_t^{[\mu]}w + (-\Delta)_D^s w = 0 \text{ in } (0,T)\times\Omega, \quad w(0,\cdot) = u_0 \text{ in } \Omega.$$

From [183] we can see that for every $u_0 \in D((-\Delta)_D^s)$, the system (2.45) has a unique strong solution $w \in C([0,T];D((-\Delta)_D^s))$ with $D_t^{[\mu]}w \in C([0,T];L^2(\Omega))$. Similarly, we can arrive at the conclusion that for every $u_0 \in L^2(\Omega)$, there exists a unique strong solution $w \in C([0,T];L^2(\Omega))$ of the system (2.45). Moreover, $D_t^{[\mu]}w \in C((0,T];W_0^{-s,2}(\Omega))$.

Now we are ready to state the following result.

Theorem 2.5 *For every $g \in \mathcal{D}((0,T)\times(\mathbb{R}^N\setminus\Omega))$, the system* (2.44) *has a unique strong solution $v \in C^\infty([0,T];W^{s,2}(\mathbb{R}^N))$ which can be represented in the form*

$$v(t,x) = -\sum_{n=1}^{\infty}\left(\int_0^t S(t-\tau,\lambda_n)(g(\tau,\cdot),\mathcal{N}_s\varphi_n)_{\mathbb{R}^N\setminus\Omega}d\tau\right)\varphi_n(x). \tag{2.46}$$

Moreover, we have for all $t \in [0,T]$ and $m \in \mathbb{N}^+$

$$\begin{aligned}&\|\partial_t^m v(t,\cdot)\|_{W^{s,2}(\mathbb{R}^N)}\\&\le C^*\Big(t^{1-\rho(1-\beta)}\|\partial_t^{m+1}g\|_{L^\infty((0,T);W^{s,2}(\mathbb{R}^N\setminus\Omega))}+\|\partial_t^m g(t,\cdot)\|_{W^{s,2}(\mathbb{R}^N\setminus\Omega)}\Big),\end{aligned} \tag{2.47}$$

where $C^*>0$ *is a constant and* $\beta\in(0,\frac{1}{4})$.

Proof First, we consider the Dirichlet exterior problem for fractional Laplacian:

$$\begin{cases}(-\Delta)^s\phi=0 & \text{in } \Omega,\\ \phi=g & \text{in } \mathbb{R}^N\setminus\Omega.\end{cases} \tag{2.48}$$

Using the theory of elliptic equations, we have that for every $g\in W^{s,2}(\mathbb{R}^N\setminus\Omega)$, the problem (2.48) has a unique solution $\phi\in W^{s,2}(\mathbb{R}^N)$. Moreover, the following estimate holds:

$$\|\phi\|_{W^{s,2}(\mathbb{R}^N)}\le C_1\|g\|_{W^{s,2}(\mathbb{R}^N\setminus\Omega)}, \tag{2.49}$$

where C_1 is a positive constant. For more details on this topic, we refer to [75] and the references therein. Clearly, (2.49) ensures $\|\partial_t^m\phi\|_{W^{s,2}(\mathbb{R}^N)}\le C_1\|\partial_t^m g\|_{W^{s,2}(\mathbb{R}^N\setminus\Omega)}$ for every $m\in\mathbb{N}^+$. This also implies that $\phi\in C^\infty([0,T];W^{s,2}(\mathbb{R}^N))$.

Next, we assume that v is a solution of (2.44). Let $w:=v-\phi$. Then we calculate

$$\begin{aligned}D_t^{[\mu]}w+(-\Delta)^s w&=D_t^{[\mu]}v-D_t^{[\mu]}\phi+(-\Delta)^s v-(-\Delta)^s\phi\\&=D_t^{[\mu]}v+(-\Delta)^s v-D_t^{[\mu]}\phi\\&=-D_t^{[\mu]}\phi\end{aligned}$$

in $(0,T)\times(\mathbb{R}^N)$. Moreover, $w=v-\phi=g-g=0$ in $(0,T)\times(\mathbb{R}^N\setminus\Omega)$ and $w(0,\cdot)=v(0,\cdot)-\phi(0,\cdot)=-g(0,\cdot)=0$ in Ω. Hence, a solution v of (2.44) will resolve into $v=\phi+w$, where w is a solution of the system

$$\begin{cases}D_t^{[\mu]}w+(-\Delta)^s w=-D_t^{[\mu]}\phi & \text{in } (0,T)\times\Omega,\\ w=0 & \text{in } (0,T)\times(\mathbb{R}^N\setminus\Omega),\\ w(0,\cdot)=0 & \text{in } \Omega.\end{cases} \tag{2.50}$$

We notice that $\phi\in\mathscr{D}((0,T);W^{s,2}(\mathbb{R}^N))$. Thus we can prove that $D_t^{[\mu]}\phi\in C^\infty([0,T];W^{s,2}(\mathbb{R}^N))$. Indeed, a simple calculation gives

$$\partial_t D_t^{[\mu]}\phi = \int_0^t k(t-\tau)\partial_{\tau\tau}\phi(\tau, x)d\tau,$$

where $k(\cdot)$ is defined as in (2.41). A fundamental argument shows $D_t^{[\mu]}\phi \in C^1([0,T]; W^{s,2}(\mathbb{R}^N))$. By induction, we know $D_t^{[\mu]}\phi \in C^\infty([0,T]; W^{s,2}(\mathbb{R}^N))$. Proceeding as the proof of [183, Theorem 3.2], we conclude that (2.50) has a unique solution w. From (2.43), one can infer that the solution w is given by

$$w(t,x) = -\sum_{n=1}^{\infty}\left(\int_0^t S(t-\tau,\lambda_n)(D_\tau^{[\mu]}\phi(\tau,\cdot),\varphi_n)d\tau\right)\varphi_n(x).$$

Using Lemma 2.17, we can rewrite $w(t,x)$ as the following form:

$$\begin{aligned} w(t,x) &= -\sum_{n=1}^{\infty}\int_0^t\left(\int_0^1 \mu(\alpha)d\alpha\right)S(t-\tau,\lambda_n)(I_\tau^{1-\alpha}\partial_\tau\phi(\tau,\cdot),\varphi_n)d\tau \\ &= -\sum_{n=1}^{\infty}\left(\int_0^1 \mu(\alpha)d\alpha\right)\int_0^t I_\tau^{1-\alpha}S(t-\tau,\lambda_n)(\partial_\tau\phi(\tau,\cdot),\varphi_n)d\tau \\ &= -\sum_{n=1}^{\infty}\int_0^t P(t-\tau,\lambda_n)(\partial_\tau\phi(\tau,\cdot),\varphi_n)d\tau. \end{aligned} \tag{2.51}$$

Notice that $P(0,\lambda_n) = 1$ and $\phi(0,\cdot) = 0$. Integrating (2.51) by parts leads to the representation

$$w(t,x) = -\sum_{n=1}^{\infty}(\phi(t,\cdot),\varphi_n)\varphi_n(x) - \sum_{n=1}^{\infty}\left(\int_0^t P'(t-\tau,\lambda_n)(\phi(\tau,\cdot),\varphi_n)d\tau\right)\varphi_n(x).$$

Using $\lambda_n(\phi,\varphi_n) = -(g,\mathcal{N}_s\varphi_n)_{\mathbb{R}^N\setminus\Omega}$ in Remark 15 of [227] and the fact $P'(t,\lambda_n) = -\lambda_n S(t,\lambda_n)$ for $t \in (0,T]$ by Lemma 2.17, the preceding identity yields that

$$w(t,x) = -\phi(t,x) - \sum_{n=1}^{\infty}\left(\int_0^t S(t-\tau,\lambda_n)(g(\tau,\cdot),\mathcal{N}_s\varphi_n)_{\mathbb{R}^N\setminus\Omega}d\tau\right)\varphi_n(x).$$

We thus get (2.46).

Employing (2.51), we obtain that for every $t \in [0,T]$,

$$
\begin{aligned}
\|(-\Delta)^s w(t,\cdot)\| =& \left\| \sum_{n=1}^{\infty} \lambda_n \int_0^t P(t-\tau,\lambda_n)(\partial_\tau \phi(\tau,\cdot),\varphi_n)\varphi_n(x)d\tau \right\| \\
\leq& \int_0^t \left\| \sum_{n=1}^{\infty} \lambda_n P(t-\tau,\lambda_n)(\partial_\tau \phi(\tau,\cdot),\varphi_n)\varphi_n(x) \right\| d\tau \\
\leq& \int_0^t \left(\sum_{n=1}^{\infty} |\lambda_n^{1-\beta} P(t-\tau,\lambda_n)|^2 |(\lambda_n^{\beta} \partial_\tau \phi(\tau,\cdot),\varphi_n)|^2 \right)^{\frac{1}{2}} d\tau.
\end{aligned}
\tag{2.52}
$$

From Lemma 2.23, we have $|\lambda_n^{1-\beta} P(t,\lambda_n)| \leq \|A^{1-\beta}P(t)\| \leq Ct^{-\rho(1-\beta)}$. Notice that $\|\partial_t\phi(t,\cdot)\|_{\mathbb{V}_{s,\beta}} \leq \|\partial_t\phi(t,\cdot)\|_{W_0^{s,2}(\overline{\Omega})} \leq \|\partial_t\phi(t,\cdot)\|_{W^{s,2}(\mathbb{R}^N)}$ for $\beta \in (0,\frac{1}{4})$. It follows from (2.49) that

$$
\begin{aligned}
\|(-\Delta)^s w(t,\cdot)\| \leq& C\int_0^t (t-\tau)^{-\rho(1-\beta)} \|\partial_\tau \phi(\tau,\cdot)\|_{W^{s,2}(\mathbb{R}^N)} d\tau \\
\leq& CC_1 \int_0^t (t-\tau)^{-\rho(1-\beta)} \|\partial_\tau g(\tau,\cdot)\|_{W^{s,2}(\mathbb{R}^N\setminus\Omega)} d\tau \\
\leq& C^* t^{1-\rho(1-\beta)} \|\partial_\tau g\|_{L^\infty(0,T;W^{s,2}(\mathbb{R}^N\setminus\Omega))}.
\end{aligned}
$$

This also shows that

$$
\begin{aligned}
\|v(t,\cdot)\|_{W^{s,2}(\mathbb{R}^N)} \leq& C^*\big(\|(-\Delta)^s w(t,\cdot)\| + \|\phi(t,\cdot)\|_{W^{s,2}(\mathbb{R}^N)}\big) \\
\leq& C^*\big(t^{1-\rho(1-\beta)}\|\partial_\tau g\|_{L^\infty(0,T;W^{s,2}(\mathbb{R}^N\setminus\Omega))} + \|g(t,\cdot)\|_{W^{s,2}(\mathbb{R}^N\setminus\Omega)}\big).
\end{aligned}
$$

Thus we get the inequality (2.47) for $m=0$. Proceeding by induction, we arrive at the desired estimate (2.47) for $m \in \mathbb{N}^+$.

Finally, it remains to verify the convergence of the series (2.46) in $C^\infty([0,T]; W^{s,2}(\mathbb{R}^N))$. Because of $\phi \in C^\infty([0,T];W^{s,2}(\mathbb{R}^N))$, we only need to show that w given by (2.51) converges in $C^\infty([0,T];W^{s,2}(\mathbb{R}^N))$. For this reason, we perform like in (2.52) to obtain that for $m_0, n_0 \in \mathbb{N}^+$ with $m_0 > n_0$,

$$
\begin{aligned}
&\left\| -\sum_{n=n_0}^{m_0} \left(\int_0^t P(t-\tau,\lambda_n)(\partial_\tau\phi(\tau,\cdot),\varphi_n)d\tau \right)\varphi_n(x) \right\|_{W^{s,2}(\mathbb{R}^N)} \\
\leq& \int_0^t \left\| \sum_{n=n_0}^{m_0} \lambda_n P(t-\tau,\lambda_n)(\partial_\tau\phi(\tau,\cdot),\varphi_n)\varphi_n(x) \right\| d\tau
\end{aligned}
$$

$$\leq C\int_0^t (t-\tau)^{-\rho(1-\beta)}\Big(\sum_{n=n_0}^{m_0}\lambda_n^{2\beta}|(\partial_\tau\phi(\tau,\cdot),\varphi_n)|^2\Big)^{\frac{1}{2}}d\tau$$

$$\leq Ct^{1-\rho(1-\beta)}\sup_{t\in[0,T]}\Big(\sum_{n=n_0}^{m_0}\lambda_n^{2\beta}|(\partial_\tau\phi(\tau,\cdot),\varphi_n)|^2\Big)^{\frac{1}{2}}.$$

In view of $\partial_t\phi \in C([0,T]; D((-\Delta)_D^s)) \subset C([0,T]; \mathbb{V}_{s,\beta})$, the above estimate leads to the relation

$$\sup_{t\in[0,T]}\Big\| -\sum_{n=n_0}^{m_0}\Big(\int_0^t P(t-\tau,\lambda_n)(\partial_\tau\phi(\tau,\cdot),\varphi_n)d\tau\Big)\varphi_n(x)\Big\|_{W^{s,2}(\mathbb{R}^N)}$$

$$\leq CT^{1-\rho(1-\beta)}\sup_{t\in[0,T]}\Big(\sum_{n=n_0}^{m_0}\lambda_n^{2\beta}|(\partial_\tau\phi(\tau,\cdot),\varphi_n)|^2\Big)^{\frac{1}{2}} \to 0,\quad \text{as } n_0, m_0\to\infty.$$

Consequently, it shows the uniform convergence of the series (2.51) for $t \in [0,T]$ in $W^{s,2}(\mathbb{R}^N)$. For the function $\partial_t^m w$, the arguments similar to ones used for estimation of w lead to the uniform convergence of the series

$$\partial_t^m w(t,x) = -\sum_{n=1}^{\infty}\Big(\int_0^t P(t-\tau,\lambda_n)(\partial_\tau^{m+1}\phi(\tau,\cdot),\varphi_n)d\tau\Big)\varphi_n(x),$$

for any $m \in \mathbb{N}^+$. This finishes the proof.

Next, we concern with the adjoint system associated with (2.44) derived from the integration by parts formula (see (2.31))

$$\begin{cases} {}_t^LD_T^{[\mu]}\psi + (-\Delta)^s\psi = 0 & \text{in } (0,T)\times\Omega,\\ \psi = 0 & \text{in } (0,T)\times(\mathbb{R}^N\setminus\Omega),\\ {}_tI_T^{[\mu]}\psi(T,\cdot) = \psi_0 & \text{in } \Omega, \end{cases} \tag{2.53}$$

where ${}_tI_T^{[\mu]}\psi(T,\cdot) = \lim_{t\to T^-} {}_tI_T^{[\mu]}\psi(t,\cdot)$.

We also introduce the following definition of strong solutions to the system (2.53).

Definition 2.5 Let $\psi_0 \in L^2(\Omega)$. We say that the function ψ is a strong solution of (2.53) if ψ satisfies the following properties.

(i) Regularity: ${}_tI_T^{[\mu]}\psi \in C([0,T]; L^2(\Omega))$, ${}_t^LD_T^{[\mu]}\psi \in C((0,T); W^{-s,2}(\bar{\Omega}))$. Moreover, the variational identity

$$\langle {}_t^L D_T^{[\mu]} \psi, w\rangle + \langle (-\Delta)^s \psi, w\rangle = 0, \quad \text{for every } w \in W_0^{s,2}(\bar{\Omega}) \text{ and a.e. } t \in (0, T)$$

holds.

(ii) Final conditions: ${}_t I_T^{[\mu]} \psi(T, \cdot) = \psi_0$ in Ω.

Furthermore, we show the existence and representation of solutions to the adjoint system.

Theorem 2.6 *Assume $\psi_0 \in L^2(\Omega)$. Then the adjoint system* (2.53) *has a unique strong solution ψ which can be represented in the form*

$$\psi(t, x) = \sum_{n=1}^{\infty} S(T - t, \lambda_n)\big(\psi_0, \varphi_n\big)\varphi_n(x). \tag{2.54}$$

In addition, the following assertions hold:

(i) ${}_t I_T^{[\mu]} \psi \in C([0, T]; L^2(\Omega))$, *and we have for all $t \in [0, T)$*

$$\|{}_t I_T^{[\mu]} \psi(t, \cdot)\| \leq \|\psi_0\|.$$

(ii) $\psi \in C([0, T); D((-\Delta)_D^s)) \cap L^1(0, T; L^2(\Omega))$ *and* ${}_t^L D_T^{[\mu]} \psi \in C((0, T); L^2(\Omega))$. *Moreover, there is a constant $\bar{M} > 0$ such that for all $t \in [0, T)$,*

$$\begin{aligned} \|\psi(t, \cdot)\| &\leq \bar{M}(T - t)^{\rho - 1} \|\psi_0\|, \\ \|{}_t^L D_T^{[\mu]} \psi(t, \cdot)\| &\leq \bar{M}(T - t)^{-1} \|\psi_0\|. \end{aligned}$$

(iii) *The mapping*

$$[0, T) \ni t \mapsto \mathcal{N}_s \psi(t, \cdot) \in L^2(\mathbb{R}^N \setminus \Omega)$$

can be analytically extended to the half-plane $\Sigma_T := \{\zeta \in \mathbb{C} : \mathrm{Re}(\zeta) < T\}$.

Proof First, we will prove the existence of solutions. For $m \geq 1$, we let

$$\psi_m(t, x) = \sum_{n=1}^{m} S(T - t, \lambda_n)\big(\psi_0, \varphi_n\big)\varphi_n(x).$$

For ψ given by (2.54), we claim that ${}_t I_T^{[\mu]} \psi \in C([0, T]; L^2(\Omega))$. For this, we integrate ψ_m with respect to t and use the formula $\int_0^1 \mu(\alpha)_0 D_t^{\alpha - 1} S(t, \lambda_n) d\alpha = P(t, \lambda_n)$ from Lemma 2.17 to obtain

$${}_t I_T^{[\mu]} \psi_m(t, x) = \sum_{n=1}^{m} P(T - t, \lambda_n)\big(\psi_0, \varphi_n\big)\varphi_n(x). \tag{2.55}$$

Using Lemma 2.19, we calculate that for every $t \in [0, T]$ and $m, \tilde{m} \in \mathbb{N}^+$ with $m > \tilde{m}$,

$$\|{}_tI_T^{[\mu]}\psi_{\tilde{m}}(t,\cdot) - {}_tI_T^{[\mu]}\psi_m(t,\cdot)\| = \Big(\sum_{n=\tilde{m}+1}^{m} |P(T-t,\lambda_n)\big(\psi_0,\varphi_n\big)|^2\Big)^{\frac{1}{2}}$$
$$\leq \Big(\sum_{n=\tilde{m}+1}^{m} |\big(\psi_0,\varphi_n\big)|^2\Big)^{\frac{1}{2}}.$$

The fact that the right-hand side of the above inequality tends to zero as $\tilde{m}, m \to \infty$ shows the series

$$\sum_{n=1}^{m} P(T-t,\lambda_n)\big(\psi_0,\varphi_n\big)\varphi_n(x) \to {}_tI_T^{[\mu]}\psi(t,x) \text{ uniformly for } t \in [0,T], \text{ as } m \to \infty$$

in $L^2(\Omega)$. Employing Lemma 2.22, we see that ${}_tI_T^{[\mu]}\psi \in C([0,T]; L^2(\Omega))$ and $\|{}_tI_T^{[\mu]}\psi(t,\cdot)\| \leq \|\psi_0\|$. Furthermore, ${}_tI_T^{[\mu]}\psi(T,x) = \sum_{n=1}^{\infty} P(0,\lambda_n)\big(\psi_0,\varphi_n\big)\varphi_n(x) = \psi_0$.

Next, we claim that ${}_t^LD_T^{[\mu]}\psi \in C([0,T); L^2(\Omega))$. Taking the derivative of (2.55) with respect to t, we infer from Lemma 2.17 that

$${}_t^LD_T^{[\mu]}\psi_m(t,x) = -\sum_{n=1}^{m} \lambda_n S(T-t,\lambda_n)\big(\psi_0,\varphi_n\big)\varphi_n(x).$$

For the function ${}_t^LD_T^{[\mu]}\psi_m$, the arguments similar to ones used for estimation of ${}_tI_T^{[\mu]}\psi_m$ lead to the relation

$$-\sum_{n=1}^{m} \lambda_n S(T-t,\lambda_n)\big(\psi_0,\varphi_n\big)\varphi_n(x) \to {}_t^LD_T^{[\mu]}\psi(t,x) \text{ in } L^2(\Omega), \text{ as } m \to \infty,$$

and the convergence is uniform in $t \in [0,T)$. Then for every $t \in [0,T)$, the estimate

$$\|{}_t^LD_T^{[\mu]}\psi(t,\cdot)\| \leq M_1^*(T-t)^{-1}\|\psi_0\|, \tag{2.56}$$

which can be easily derived from Lemma 2.16. Thereby, ${}_t^LD_T^{[\mu]}\psi \in C([0,T); L^2(\Omega))$. Moreover, in view of ${}_t^LD_T^{[\mu]}\psi = -(-\Delta)_D^s\psi$, we also have $\psi \in C([0,T); D((-\Delta)_D^s))$.

Using Lemma 2.19, a direct calculation yields

$$\begin{aligned}\|\psi(t,\cdot)\|^2 =& \sum_{n=1}^{\infty} |S(T-t,\lambda_n)|^2 |(\psi_0,\varphi_n)|^2 \\ \leq& M^*(T-t)^{2(\rho-1)} \sum_{n=1}^{\infty} (\psi_0,\varphi_n)|^2 \\ =& M^*(T-t)^{2(\rho-1)}\|\psi_0\|^2;\end{aligned}$$

we thus obtain the estimate $\|\psi(t,\cdot)\| \leq M^*(T-t)^{\rho-1}\|\psi_0\|$ and $\psi \in L^1(0,T;L^2(\Omega))$.

Second, we will prove the uniqueness. For this, assume that ψ is a solution of (2.53) with $\psi_0 = 0$. We take the inner product of (2.53) with φ_n and set $\psi_n(t) = (\psi(t,\cdot),\varphi_n)$ to find that

$$ {}_{t}^{L}D_{T}^{[\mu]}\psi_n(t) = -\lambda_n\psi_n(t), \quad \text{for a.e. } t \in (0,T). \tag{2.57}$$

Proceeding in the same way as the proof of existence, (2.54) yields $\psi = 0$ in $(0,T)\times\Omega$. The uniqueness of the solution follows.

Finally, we claim that the mapping $[0,T) \ni t \mapsto \mathcal{N}_s\psi(t,\cdot) \in L^2(\mathbb{R}^N\setminus\Omega)$ can be analytically extended to Σ_T. Notice the relation $\psi(t,\cdot) \in D((-\Delta)_D^s) \subset W^{s,2}(\mathbb{R}^N)$ for each $t \in [0,T)$; it yields that $\mathcal{N}_s\psi(t,\cdot)$ exists and $\mathcal{N}_s\psi(t,\cdot) \in L^2(\mathbb{R}^N \setminus \Omega)$.

We will prove that $\mathcal{N}_s\psi$ admits the representation

$$\mathcal{N}_s\psi(t,x) = \sum_{n=1}^{\infty} S(T-t,\lambda_n)(\psi_0,\varphi_n)\mathcal{N}_s\varphi_n(x), \tag{2.58}$$

and the series is convergent in $L^2(\mathbb{R}^N \setminus \Omega)$ for any $t \in [0,T)$. We first note that $\mathcal{N}_s : W^{s,2}(\mathbb{R}^N) \to L^2(\mathbb{R}^N \setminus \Omega)$ is bounded. For any $\delta > 0$ and $t \in [0,T-\delta]$, using Lemma 2.16 we find that

$$\begin{aligned}&\left\|\sum_{n=m+1}^{\infty} S(T-t,\lambda_n)(\psi_0,\varphi_n)\mathcal{N}_s\varphi_n(x)\right\|^2_{L^2(\mathbb{R}^N\setminus\Omega)} \\ \leq& C_2\left\|\sum_{n=m+1}^{\infty} S(T-t,\lambda_n)(\psi_0,\varphi_n)\varphi_n(x)\right\|^2_{D((-\Delta)_D^s))} \\ \leq& C_2\sum_{n=m+1}^{\infty} |\lambda_n S(T-t,\lambda_n)|^2 |(\psi_0,\varphi_n)|^2\end{aligned}$$

$$\leq C_2 M_1^*(T-t)^{-2} \sum_{n=m+1}^{\infty} |(\psi_0, \varphi_n)|^2$$

$$= \tilde{C}\delta^{-2} \sum_{n=m+1}^{\infty} |(\psi_0, \varphi_n)|^2 \to 0, \quad \text{as } m \to \infty,$$

where C_2 is a positive constant. Hence, $\mathcal{N}_s\psi$ is determined by (2.58), and we get the uniform convergence of the series in $L^2(\mathbb{R}^N \setminus \Omega)$ in any compact subset of $[0, T)$.

For $\zeta \in \mathbb{C}$, we consider $S(\zeta, \lambda_n)$ as defined by (2.37). Since $\zeta \mapsto e^{\zeta r}$ is analytic and the integral $\int_0^\infty e^{-r\zeta} \mathcal{K}(r, \lambda) dr$ is absolutely convergent for $|\arg(\zeta)| < \frac{\pi}{2}$, we can easily know the analyticity of $S(\zeta, \lambda_n)$ on the sector $|\arg(\zeta)| < \frac{\pi}{2}$. It follows that the function

$$\sum_{n=1}^{m} S(T-\zeta, \lambda_n)(\psi_0, \varphi_n)\mathcal{N}_s\varphi_n(x)$$

is analytic in Σ_T.

For arbitrary $\delta > 0$ and choosing $\zeta \in \mathbb{C}$ with $Re(\zeta) \leq T - \delta$, we apply a method similar to one we used for the above estimate to get

$$\left\| \sum_{n=m+1}^{\infty} S(T-\zeta, \lambda_n)(\psi_0, \varphi_n)\mathcal{N}_s\varphi_n(x) \right\|^2_{L^2(\mathbb{R}^N \setminus \Omega)}$$

$$\leq \tilde{C}\delta^{-2} \sum_{n=m+1}^{\infty} |(\psi_0, \varphi_n)|^2 \to 0,$$

which is valid as $m \to \infty$. This also indicates that

$$\mathcal{N}_s\psi(\zeta, x) = \sum_{n=1}^{\infty} S(T-\zeta, \lambda_n)(\psi_0, \varphi_n)\mathcal{N}_s\varphi_n(x), \tag{2.59}$$

and then we get the uniform convergence of the series in any compact subset of Σ_T. Consequently, $\mathcal{N}_s\psi$ determined by (2.59) is analytic in Σ_T as well. This finishes the proof.

2.3.5 Approximate Controllability Analysis

In this subsection, we provide the rigorous analysis of approximate controllability for the system (2.30). The definition of approximate controllability is introduced.

Definition 2.6 The system (2.30) is said to be approximately controllable at time $T > 0$ if for any $\tilde{u}_0 \in L^2(\Omega)$ and $\varepsilon > 0$, there is a control function g such that

$$\|u(T, \cdot) - \tilde{u}_0\|_{L^2(\Omega)} \leq \varepsilon,$$

where u is the unique strong solution of (2.30) corresponding to the control g.

The following unique continuation property of the system (2.53) is given. The following $\mathcal{N}_s$ denotes nonlocal normal derivative given by (2.32).

Theorem 2.7 *For $\psi_0 \in L^2(\Omega)$ and $\mathcal{O} \subset \mathbb{R}^N \setminus \Omega$ being an arbitrary non-empty open set, assume that ψ is the unique strong solution of* (2.53) *satisfying $\mathcal{N}_s\psi = 0$ in $(0, T) \times \mathcal{O}$. Then we have $\psi = 0$ in $(0, T) \times \Omega$.*

Proof If $\mathcal{N}_s\psi = 0$ in $(0, T) \times \mathcal{O}$ for any $\mathcal{O} \subset \mathbb{R}^N \setminus \Omega$ being non-empty open set, then we have

$$\mathcal{N}_s\psi(t, x) = \sum_{n=1}^{\infty} S(T - t, \lambda_n)\big(\psi_0, \varphi_n\big)\mathcal{N}_s\varphi_n(x) = 0, \quad \forall\, (t, x) \in (0, T) \times \mathcal{O}.$$

We first notice that $\mathcal{N}_s\psi$ can be analytically extended to the half-plane Σ_T, which can be derived from Theorem 2.6; it yields that

$$\mathcal{N}_s\psi(t, x) = \sum_{n=1}^{\infty} S(T - t, \lambda_n)\big(\psi_0, \varphi_n\big)\mathcal{N}_s\varphi_n(x) = 0, \quad \forall\, (t, x) \in (-\infty, T) \times \mathcal{O}. \tag{2.60}$$

Since $\{\lambda_n\}_{n\in\mathbb{N}^+}$ is the set of all eigenvalues of the operator $(-\Delta)^s_D$, we denote by $\{\varphi_{n_k}\}_{1\leq k\leq m_n}$ an orthonormal basis for $\ker(\lambda_n - (-\Delta)^s_D)$, and we deduce that $\mathcal{N}_s\psi$ given by (2.60) is also rewritten in the form

$$\mathcal{N}_s\psi(t, x) = \sum_{n=1}^{\infty}\sum_{k=1}^{m_n} \big(\psi_0, \varphi_{n_k}\big)\mathcal{N}_s\varphi_{n_k}(x)S(T - t, \lambda_n) = 0,$$
$$\forall\, (t, x) \in (-\infty, T) \times \mathcal{O}. \tag{2.61}$$

Choose $\zeta \in \mathbb{C}$ with $Re(\zeta) := \eta > 0$ and $m \in \mathbb{N}^+$. We set

$$\psi_m(t, x) = \sum_{n=1}^{m}\sum_{k=1}^{m_n} \big(\psi_0, \varphi_{n_k}\big)\mathcal{N}_s\varphi_{n_k}(x)e^{\zeta(t-T)}S(T - t, \lambda_n).$$

Recalling that φ_{n_k} are orthonormal for $1 \leq k \leq m_n$, $1 \leq n \leq m$, and $\mathcal{N}_s : \mathbb{V}_{s,1-\beta} \subset W^{s,2}(\mathbb{R}^N) \to L^2(\mathbb{R}^N \setminus \Omega)$ is bounded for $\beta \in (0, \frac{1}{4})$, we can conclude from Lemma 2.23 that for every $t \in (-\infty, T)$,

$$
\begin{aligned}
\|\psi_m(t,\cdot)\|^2_{L^2(\mathbb{R}^N\setminus\Omega)} \leq & C_3\left\|\sum_{n=1}^{\infty}\sum_{k=1}^{m_n}(\psi_0,\varphi_{n_k})\varphi_{n_k}(x)e^{\zeta(t-T)}S(T-t,\lambda_n)\right\|^2_{\mathbb{V}_{s,1-\beta}} \\
\leq & C_3\sum_{n=1}^{\infty}\sum_{k=1}^{m_n}\left|(\psi_0,\varphi_{n_k})\right|^2 e^{2\eta(t-T)}\left|\lambda_n^{1-\beta}S(T-t,\lambda_n)\right|^2 \\
\leq & C_3(T-t)^{2(\beta\rho-1)}e^{2\eta(t-T)}\|\psi_0\|^2.
\end{aligned}
$$

The last estimate leads to the inequality

$$
\|\psi_m(t,\cdot)\|_{L^2(\mathbb{R}^N\setminus\Omega)} \leq \bar{C}(T-t)^{\beta\rho-1}e^{\eta(t-T)}\|\psi_0\|,
$$

where $\bar{C} > 0$ is a positive constant. Moreover, the integrability of the right-hand side of the above inequality for $t \in (-\infty, T)$ can be easily derived, and

$$
\int_{-\infty}^{T}(T-t)^{\beta\rho-1}e^{\eta(t-T)}dt\|\psi_0\| = \frac{\Gamma(\beta\rho)}{\eta^{\beta\rho}}\|\psi_0\|.
$$

Then a direct calculation yields

$$
\begin{aligned}
&\int_{-\infty}^{T} e^{\zeta(t-T)}\sum_{n=1}^{\infty}\sum_{k=1}^{m_n}(\psi_0,\varphi_{n_k})\mathcal{N}_s\varphi_{n_k}(x)S(T-t,\lambda_n)dt \\
=&\sum_{n=1}^{\infty}\sum_{k=1}^{m_n}\frac{(\psi_0,\varphi_{n_k})}{\zeta\omega(\zeta)+\lambda_n}\mathcal{N}_s\varphi_{n_k}, \quad x\in\mathcal{O}, \quad Re(\zeta)>0,
\end{aligned}
$$

which together with (2.61) leads to the identity

$$
\sum_{n=1}^{\infty}\sum_{k=1}^{m_n}\frac{(\psi_0,\varphi_{n_k})}{\zeta\omega(\zeta)+\lambda_n}\mathcal{N}_s\varphi_{n_k} = 0, \quad x\in\mathcal{O}, \quad Re(\zeta)>0.
$$

Set $\xi := \zeta\omega(\zeta)$, and the above identity turns into the following form:

$$
\sum_{n=1}^{\infty}\sum_{k=1}^{m_n}\frac{(\psi_0,\varphi_{n_k})}{\xi+\lambda_n}\mathcal{N}_s\varphi_{n_k} = 0, \quad x\in\mathcal{O}, \quad Re(\xi)>0. \tag{2.62}
$$

Employing the analytic continuation in ξ, we see that the equality (2.62) is valid for every $\xi \in \mathbb{C}\setminus\{-\lambda_n\}$. Thus we choose a small circle including $-\lambda_l$, not including $\{-\lambda_n\}_{n\neq l}$, and integrate (2.62) over the circle to find that

$$
\sum_{k=1}^{m_l}(\psi_0,\varphi_{l_k})\mathcal{N}_s\varphi_{l_k} = 0, \quad x\in\mathcal{O}.
$$

Letting $\tilde{\psi}_l := \sum_{k=1}^{m_l} (\psi_0, \varphi_{l_k})\varphi_{l_k}$, we have $\mathcal{N}_s\tilde{\psi}_l = 0$ in $\mathcal{O}$. Therefore, we obtain that for every l, $\tilde{\psi}_l$ satisfies the following problem:

$$(-\Delta)^s_D\tilde{\psi}_l = \lambda_l\tilde{\psi}_l \text{ in } \Omega \text{ and } \mathcal{N}_s\tilde{\psi}_l = 0 \text{ in } \mathcal{O}.$$

Hence Lemma 2.13 implies that $\tilde{\psi}_l = 0$ in Ω for every l. Taking account of the linear independence among $\psi_{l_k}(1 \le k \le m_n)$ in $L^2(\Omega)$, we conclude that $(\psi_0, \varphi_{l_k}) = 0$ for $1 \le k \le m_n, n \in \mathbb{N}^+$. This also ensures $\psi_0 = 0$. We use the uniqueness of the solution to the adjoint system (2.53) and thus obtain that $\psi = 0$ in $(0, T) \times \Omega$. This finishes the proof.

Using the elementary arguments, we notice that the study of the approximate controllability of the system (2.30) can be reduced to the case $u_0 = 0$. Therefore, we only need to analyze the approximate controllability of the system (2.44).

Theorem 2.8 *Let $\mathcal{O} \subset \mathbb{R}^N \setminus \Omega$ be an arbitrary non-empty open set. Then for any $T > 0$ and every $g \in \mathcal{D}((0, T) \times \mathcal{O})$, the system* (2.44) *is approximately controllable, i.e.,*

$$\overline{\{v(T, \cdot) : g \in \mathcal{D}((0, T) \times \mathcal{O})\}}^{L^2(\Omega)} = L^2(\Omega),$$

where v denotes the unique strong solution to the system (2.44).

Proof For $g \in \mathcal{D}((0, T) \times \mathcal{O})$ and $\psi_0 \in L^2(\Omega)$, we assume that v and ψ are the unique strong solution to the system (2.44) and (2.53), respectively. As to v, we use Theorem 2.5 and thus obtain the relation $\partial_t v, (-\Delta)^s v \in L^\infty(0, T; L^2(\Omega))$. Moreover, because of the inequality $\int_0^1 \mu(\alpha)\frac{t^{1-\alpha}}{\Gamma(2-\alpha)}d\alpha \le 2\max\{1, t\}\|\mu\|_{L^1(0,1)}$ that is valid, we obtain that $D_t^{[\mu]}v \in L^\infty(0, T; L^2(\Omega))$. For ψ, using Theorem 2.6 we deduce that $\psi \in L^1(0, T; L^2(\Omega))$. In addition, we get $v(T, \cdot) \in L^2(\Omega)$ and ${}_tI_T^\mu\psi(T, \cdot) \in L^2(\Omega)$.

Let $\delta > 0$ be small; we use the formula (2.31) on $(0, T - \delta)$ and take the limit as $\delta \to 0^+$ to find that

$$\begin{aligned}
0 &= \int_0^T \int_\Omega \left(D_t^{[\mu]}v + (-\Delta)^s v\right)\psi dxdt \\
&= \int_0^T \int_\Omega v(t){}_t^LD_T^{[\mu]}\psi(t)dxdt + \int_\Omega v(T, \cdot){}_tI_T^{[\mu]}\psi(T, \cdot)dx \\
&\quad + \int_0^T \int_\Omega v(-\Delta)^s\psi dx + \int_0^T \int_{\mathbb{R}^N\setminus\Omega} g\mathcal{N}_s\psi dxdt \\
&= \int_0^T \int_\Omega v\left({}_t^LD_T^{[\mu]}\psi + (-\Delta)^s\psi\right)dxdt + \int_\Omega v(T, \cdot)\psi_0 dx \\
&\quad + \int_0^T \int_{\mathbb{R}^N\setminus\Omega} g\mathcal{N}_s\psi dxdt,
\end{aligned}$$

where we have used the formula (2.33) and $v(0,\cdot) = 0$. Thus the system (2.53) leads to the identity

$$\int_{\Omega} v(T,\cdot)\psi_0 dx + \int_0^T \int_{\mathbb{R}^N \setminus \Omega} g\mathcal{N}_s\psi dxdt = 0. \tag{2.63}$$

To show that $\overline{\{v(T,\cdot) : g \in \mathcal{D}((0,T)\times\mathcal{O})\}}^{L^2(\Omega)} = L^2(\Omega)$, we only need to verify the property that if $\psi_0 \in L^2(\Omega)$ admits $\int_{\Omega} v(T,\cdot)\psi_0 dx = 0$ for any $g \in \mathcal{D}((0,T)\times\mathcal{O})$, then $\psi_0 = 0$. In fact, if ψ_0 admits $\int_{\Omega} v(T,\cdot)\psi_0 dx = 0$, then the equality (2.63) yields that $\int_0^T \int_{\mathbb{R}^N\setminus\Omega} g\mathcal{N}_s\psi dxdt = 0$. Thus we can assert that $\mathcal{N}_s\psi = 0$ in $(0,T)\times\mathcal{O}$, which can be derived from the arbitrariness of g. Employing Theorem 2.7, we see that $\psi = 0$ in $(0,T)\times\mathcal{O}$. Finally, the uniqueness of the solution to the system (2.53) shows that $\psi_0 = 0$ in Ω. The proof is completed.

Chapter 3
Inverse Problems of Fractional Diffusion Equations

3.1 Backward Problem

3.1.1 Introduction

Let Ω be a bounded domain in $\mathbb{R}^N$ with the sufficiently smooth boundary $\partial\Omega$. We shall consider the following backward problem for fractional diffusion equations:

$$\begin{cases} \partial_t u(t,x) = \partial_t^{1-\alpha} Au(t,x) + f(t,x), & x \in \Omega,\ 0 < t < T, \\ u(t,x) = 0, & x \in \partial\Omega,\ 0 < t < T, \\ u(T,x) = g(x), & x \in \Omega, \end{cases} \tag{3.1}$$

where g is the terminal value status, f is a function representing kinetics, T is a backward finite time which ensures that some situation of phenomena can be measured at that point, $\partial_t = \partial/\partial t$, and $\partial_t^{1-\alpha}$ is the Riemann-Liouville fractional partial derivative of order $1-\alpha \in (0,1)$ defined by

$$\partial_t^{1-\alpha} u(t,x) = \frac{1}{\Gamma(\alpha)} \frac{\partial}{\partial t} \int_0^t (t-s)^{\alpha-1} u(s,x)ds, \quad t > 0,$$

where $\Gamma(\cdot)$ is the Gamma function. The operator A stands for the unbounded uniformly elliptic operator with the domain $D(A) := H_0^1(\Omega) \cap H^2(\Omega)$ defined by

$$Au(x) = -\sum_{i=1}^{N} \frac{\partial}{\partial x_i} \left(\sum_{j=1}^{N} A_{ij}(x) \frac{\partial}{\partial x_j} u(x) \right) + b(x)u(x),$$

Y. Zhou, *Fractional Diffusion and Wave Equations*,
https://doi.org/10.1007/978-3-031-74031-2_3

where $A_{ij} = A_{ji}$, $i, j = 1, 2, \ldots, N$, and there exists a constant $\mu > 0$ such that

$$\sum_{i,j=1}^{N} A_{ij}(x)\xi_i\xi_j \geq \mu \sum_{i=1}^{N} \xi_i^2, \quad x \in \overline{\Omega}, \ \xi_i \in \mathbb{R}^N,$$

$A_{ij} \in C^1(\overline{\Omega})$, $b \in C(\overline{\Omega})$, $b(x) \geq 0$ for all $x \in \overline{\Omega}$.

Diffusion models have been found a numerous of applications in chemical physics, biological cell dynamics and physiology, etc. Especially, fractional diffusion model is derived from a continuous time random walk model when the transport is dispersive and it can describe more precise for some models of anomalous diffusion in heterogeneous media. We refer to the books [180, 219] and the papers [5, 117, 155, 244].

The direct problem of fractional diffusion equations (3.1) is the typical heat equation for $\alpha = 1$ equipped with A being the Laplacian operator. Such equations were introduced by Schneider and Wyss [190], they pointed out that such equations can be described of diffusion in special types of porous media, and it can represent the "gray" noise in the fractional diffusion for $0 < \alpha < 1$, while the white noise for $\alpha = 1$. In addition, Seki et al. [191] derived such type fractional reaction-diffusion equations from a continuous time random walk model. More recently, Andrade and Viana [12] have developed the global well-posedness and spatiotemporal asymptotic behavior in the case of semilinear situation. Viana [216] studied the local well-posedness of such equations with perturbations behavior.

In many practical applications, we can just measure density of the diffusing substance at positive moment without knowing the initial density, and so, it reflects the advantages of the backward problem to deal with this kind of difficult situation. For more details of the fractional backward problem, one can see the papers [7, 206, 207, 226] and the references cited therein.

As far as we know, Liu and Yamamoto [138] studied the existence, regularity, and convergence analysis of a backward problem for time fractional diffusion equation in weighted continuous function space. Sakamoto and Yamamoto [186] investigated the existence, uniqueness, and regularity of solutions of a direct problem for fractional diffusion equation in space $C([0, T]; L^2(\Omega))$ with respect to the assumptions $a \in H_0^1(\Omega) \cap H^2(\Omega)$ and $f \in C^\theta([0, T]; L^2(\Omega))$. Mu et al. [168] established the existence and regularity of classical solutions on a weighted Hölder continuous function space, with respect to the assumption $f \in \mathscr{F}^{\beta,\theta}((0, T]; L^2(\Omega))$, which is weaker than Hölder continuous. It is natural to consider the existence, uniqueness, and regularity for a usual backward problem on continuous function space from the abovementioned points of view. Assuming that f is weighted Hölder continuous, the aim of this section is to provide some existence, uniqueness, and regularity results of mild/classical solutions for the proposed backward problem.

This section is organized as follows. In Sect. 3.1.2, we give some basic definitions and preliminaries results. In Sect. 3.1.3, the existence and uniqueness results for the fractional diffusion problem (3.1) are investigated. We also obtain some new estimations for regularity. In Sect. 3.1.4, two examples are given to illustrate the main results.

3.1.2 Preliminaries

It is well known that the operator A is introduced as above that can generate the following spectral problem:

$$Ae_n(x) = \lambda_n e_n(x), \quad x \in \Omega; \quad e_n(x) = 0, \quad x \in \partial\Omega, \ n \in \mathbb{N}^+, \tag{3.2}$$

where $\{\lambda_n\}_{n=1}^{\infty}$ denotes the set of eigenvalues satisfying

$$0 < \lambda_1 \le \lambda_2 \le \ldots \le \lambda_n \le \ldots,$$

and $\lim_{n\to\infty} \lambda_n = \infty$. The corresponding eigenfunctions are defined by $e_n \in H_0^1(\Omega) \cap H^2(\Omega)$ for every $n \in \mathbb{N}^+$. The eigenfunctions e_n are normalized so that $\{e_n\}_{n=1}^{\infty}$ is an orthonormal basis of $L^2(\Omega)$.

For all $s \ge 0$, the fractional power operator A^s possesses the following representation:

$$A^s u = \sum_{k=1}^{\infty} \lambda_k^s (u, e_k) e_k, \quad u \in D(A^s) = \left\{ u \in L^2(\Omega) : \sum_{k=1}^{\infty} \lambda_k^{2s} |(u, e_k)|^2 < \infty \right\}.$$

Let us define the norm on $D(A^s)$ by

$$\|u\|_{D(A^s)} = \left(\sum_{k=1}^{\infty} \lambda_k^{2s} |(u, e_k)|^2 \right)^{\frac{1}{2}}, \quad u \in D(A^s).$$

By duality, we can also set $D(A^{-s}) = (D(A^s))^*$ and use the so-called Gelfand triple. Then $D(A^{-s})$ is a Hilbert space endowed with the norm

$$\|u\|_{D(A^{-s})} = \left(\sum_{k=1}^{\infty} \lambda_k^{-2s} |\langle u, e_k \rangle|^2 \right)^{\frac{1}{2}}, \quad u \in D(A^{-s}),$$

where $\langle \cdot, \cdot \rangle$ denotes the duality bracket between $D(A^{-s})$ and $D(A^s)$. In particular, one has $D(A^s) \subset H^{2s}(\Omega)$ for $s > 0$, $D(A^0) = L^2(\Omega)$ and $D(A^{\frac{1}{2}}) = H_0^1(\Omega)$, see [186] and the references therein.

The space of Hölder continuous functions on $[0, T]$ with exponent $\theta \in (0, 1)$ is denoted by $C^\theta([0, T]; L^2(\Omega))$ which has the representation form

$$C^\theta([0, T]; L^2(\Omega)) = \Big\{ u \in C([0, T]; L^2(\Omega)) : \sup_{0 \le s < t \le T} \frac{\|u(t) - u(s)\|_{L^2(\Omega)}}{(t - s)^\theta} < \infty \Big\},$$

equipped with the norm

$$\|u\|_{C^\theta([0,T];L^2(\Omega))} = \sup_{0\le t\le T} \|u(t)\|_{L^2(\Omega)} + \sup_{0\le s<t\le T} \frac{\|u(t)-u(s)\|_{L^2(\Omega)}}{(t-s)^\theta}.$$

Let us introduce the space $\mathscr{F}^{\beta,\theta}((0,T];L^2(\Omega))$ by the weighted Hölder continuous function space of all continuous functions defined on $(0,T]$ for $0<\theta<\beta<1$ with satisfying properties:

(i) If $0<\beta<1$, $\lim_{t\to 0} t^{1-\beta}u(t)$ exists.
(ii) v is Hölder continuous with the exponent θ and with the weighted $s^{1-\beta+\theta}$, i.e.,

$$\sup_{0\le s<t\le T} \frac{s^{1-\beta+\theta}\|u(\cdot,t)-v(\cdot,s)\|_{L^2(\Omega)}}{(t-s)^\theta} < \infty. \tag{3.3}$$

(iii) As $t\to 0$, it yields

$$\omega_u(t) = \sup_{0\le s<t} \frac{s^{1-\beta+\theta}\|u(\cdot,t)-u(\cdot,s)\|_{L^2(\Omega)}}{(t-s)^\theta} \to 0. \tag{3.4}$$

Then the space $\mathscr{F}^{\beta,\theta}((0,T];L^2(\Omega))$ is a Banach space with the norm

$$\|u\|_{\mathscr{F}^{\beta,\theta}} = \sup_{0\le t\le T} t^{1-\beta}\|u(t)\|_{L^2(\Omega)} + \sup_{0\le s<t\le T} \frac{s^{1-\beta+\theta}\|u(\cdot,t)-u(\cdot,s)\|_{L^2(\Omega)}}{(t-s)^\theta}.$$

For $0<\theta<\sigma<\beta\le 1$, we have $\mathscr{F}^{\beta,\sigma}((0,T];L^2(\Omega)) \subset \mathscr{F}^{\beta,\theta}((0,T];L^2(\Omega))$ with continuous embedding; in addition, any function $v\in C^\theta([0,T];L^2(\Omega))$ belongs to $\mathscr{F}^{1,\theta}((0,T];L^2(\Omega))$ when $0<\theta<\beta=1$ (see, e.g., [231]).

Let $E_{\alpha,\beta}(z)$ be the Mittag-Leffler function as in Definition 1.7 and $\mathscr{M}_\varsigma(z)$ be the Wright-type function as in Definition 1.8, $\varsigma\in(0,1)$, $z\in\mathbb{C}$. The function $E_{\alpha,\beta}(z)$ is an entire function, and so it is real analytic when restricted to the real line. For convenience, we set $E_\alpha(z) := E_{\alpha,1}(z)$ for $z\in\mathbb{C}$.

3.1.3 Existence and Regularity

By using the separation of variables method, the formal solution to (3.1) can be expressed as

$$u(t,x) = \sum_{n=1}^{\infty} (u(t,x),e_n)e_n(x). \tag{3.5}$$

For convenience, we set $u_n(t) = (u(t,x), e_n)$, $g_n = (g(x), e_n)$, and $f_n(t) = (f(t,x), e_n)$. By applying $Ae_n = \lambda_n e_n$ for all $n \in \mathbb{N}^+$, let us take the Laplace transform into (3.1), and it can be rewritten as follows:

$$s\overline{u_n}(s) - u_n(0) = -\lambda_n s^{1-\alpha}\overline{u_n}(s) + \overline{f_n}(s),$$

where $\bar{v}$ denotes the Laplace transform of function v, and so

$$\overline{u_n}(s) = \frac{s^{\alpha-1}}{s^\alpha + \lambda_n} u_n(0) + \frac{s^{\alpha-1}}{s^\alpha + \lambda_n}\overline{f_n}(s).$$

It follows from Proposition 1.13 and the uniqueness of Laplace theorem that

$$u_n(t) = E_\alpha(-\lambda_n t^\alpha) u_n(0) + \int_0^t E_\alpha(-\lambda_n (t-s)^\alpha) f_n(s) ds. \tag{3.6}$$

By substituting $t = T$ into the above equation, it yields that

$$\begin{aligned} u_n(t) =& \frac{E_\alpha(-\lambda_n t^\alpha)}{E_\alpha(-\lambda_n T^\alpha)}\left(g_n - \int_0^T E_\alpha(-\lambda_n (T-s)^\alpha) f_n(s) ds\right) \\ &+ \int_0^t E_\alpha(-\lambda_n (t-s)^\alpha) f_n(s) ds. \end{aligned}$$

Let us consider the operators

$$\begin{aligned} (\mathscr{P}_1 v)(t,x) =& \sum_{n=1}^\infty \frac{E_\alpha(-\lambda_n t^\alpha)}{E_\alpha(-\lambda_n T^\alpha)} v_n e_n(x), \\ (\mathscr{P}_2 v)(t,x) =& \sum_{n=1}^\infty \left(-\int_0^T \frac{E_\alpha(-\lambda_n t^\alpha)}{E_\alpha(-\lambda_n T^\alpha)} E_\alpha(-\lambda_n (T-s)^\alpha) v_n(s) ds\right) e_n(x), \\ (\mathscr{P}_3 v)(t,x) =& \sum_{n=1}^\infty \left(\int_0^t E_\alpha(-\lambda_n (t-s)^\alpha) v_n(s) ds\right) e_n(x), \end{aligned}$$

for any $v \in L^2(\Omega)$. Then, it remains to find a solution of the problem (3.1) which can be transformed into the form

$$u(t,x) = (\mathscr{P}_1 g)(t,x) + (\mathscr{P}_2 f)(t,x) + (\mathscr{P}_3 f)(t,x). \tag{3.7}$$

From this point of view, we introduce the following definition of mild solution of the problem (3.1).

Definition 3.1 A function u is called a mild solution of the problem (3.1) if $u \in C((0,T]; L^2(\Omega))$, and it satisfies the equation

$$u(t,x) = \sum_{n=1}^{\infty} \frac{E_\alpha(-\lambda_n t^\alpha)}{E_\alpha(-\lambda_n T^\alpha)}$$

$$\left((g(x), e_n) - \int_0^T E_\alpha(-\lambda_n (T-s)^\alpha)(f(s,x), e_n) ds \right) e_n(x) \tag{3.8}$$

$$+ \sum_{n=1}^{\infty} \left(\int_0^t E_\alpha(-\lambda_n (t-s)^\alpha)(f(s,x), e_n) ds \right) e_n(x).$$

Remark 3.1 Note that the mild solution u is not continuous on $L^2(\Omega)$ if $t = 0$ and the given assumption of g belongs to $L^2(\Omega)$ or $H_0^1(\Omega)$. In fact, if u is a mild solution of the problem (3.1) and $\lim_{t \to 0} E_\alpha(-\lambda_n t^\alpha) = 1$, we just take into account the trivial case of $f \equiv 0$. In view of Proposition 1.11, one can see that

$$\begin{aligned} \|u(0,\cdot)\|_{L^2(\Omega)} &= \left(\sum_{n=1}^{\infty} \left| \frac{1}{E_\alpha(-\lambda_n T^\alpha)} \right|^2 |g_n|^2 \right)^{1/2} \\ &\geq M_2^{-1} \left(\sum_{n=1}^{\infty} (1 + \lambda_n T^\alpha)^2 |g_n|^2 \right)^{1/2} \\ &\geq M_2^{-1} \|g\|_{L^2(\Omega)}. \end{aligned}$$

This means that we cannot establish the continuous criterion of solution at $t = 0$ in the space $C([0,T]; L^2(\Omega))$ except for the higher assumptions of spatial regularity of given terminal value data g.

Remark 3.2 Let $v \in L^2(\Omega)$. The series $\sum_{n=1}^{\infty} \frac{E_\alpha(-\lambda_n t^\alpha)}{E_\alpha(-\lambda_n T^\alpha)} (v, e_n) e_n(x)$ is uniformly convergent for any $\delta > 0$ and $\delta \leq t \leq T$. Moreover, this series is also continuous on $(0, T]$. Indeed, from Proposition 1.11, we have

$$\left| \frac{E_{\alpha,\beta}(-\lambda_n t^\alpha)}{E_\alpha(-\lambda_n T^\alpha)} \right| \leq \frac{M_2}{M_1} \frac{1 + \lambda_n T^\alpha}{1 + \lambda_n t^\alpha} \leq \frac{M_2 T^\alpha}{M_1 t^\alpha}, \quad \text{for all } \beta \in \mathbb{R},\ t \in (0, T]. \tag{3.9}$$

On the other hand, for any $\varepsilon > 0$, there exists $N_1 > 0$ such that for all positive integers p, whereas $k > N_1$

$$\sum_{n=k+1}^{k+p} |(v, e_n)|^2 < \left(\frac{M_1 t^\alpha \sqrt{\varepsilon}}{M_2 T^\alpha} \right)^2, \quad \text{for all } t \in [\delta, T].$$

Let

$$S_k(t)v = \sum_{n=1}^{k} \frac{E_\alpha(-\lambda_n t^\alpha)}{E_\alpha(-\lambda_n T^\alpha)}(v, e_n)e_n(x).$$

Therefore, for all $t \in [\delta, T]$, we have

$$\begin{aligned}\|S_{k+p}(t)v - S_k(t)v\|^2_{L^2(\Omega)} &= \sum_{n=k+1}^{k+p} \left| \frac{E_\alpha(-\lambda_n t^\alpha)}{E_\alpha(-\lambda_n T^\alpha)}(v, e_n) \right|^2 \\ &\le \left(\frac{M_2 T^\alpha}{M_1 t^\alpha}\right)^2 \sum_{n=k+1}^{k+p} |(v, e_n)|^2 < \varepsilon.\end{aligned}$$

Hence, from the Cauchy convergence criterion, we obtain that the desired series is uniformly convergent on $[\delta, T]$, and thus it is continuous on $(0, T]$.

In this sequel, the existence, uniqueness, and regularity results of mild solutions for the proposed problem are investigated.

Theorem 3.1 *Assume that* $f \in \mathscr{F}^{\beta,\theta}((0, T]; L^2(\Omega))$ *with* $0 < \theta < \beta < 1$ *and* $g \in L^2(\Omega)$. *Then there is a unique mild solution* u *to the problem (3.1) which belongs to* $C((0, T]; L^2(\Omega))$. *Furthermore, there exists a positive constant* C *such that*

$$t^\alpha \|u(t, \cdot)\|_{L^2(\Omega)} \le C \left(\|g\|_{L^2(\Omega)} + \|f\|_{\mathscr{F}^{\beta,\theta}}\right). \tag{3.10}$$

Proof Firstly, in view of (3.9), for all $t \in (0, T]$, it is easy to check that

$$\|(\mathscr{P}_1 g)(t, \cdot)\|_{L^2(\Omega)} = \left(\sum_{n=1}^{\infty} \left| \frac{E_\alpha(-\lambda_n t^\alpha)}{E_\alpha(-\lambda_n T^\alpha)} g_n \right|^2\right)^{1/2} \le \frac{M_2 T^\alpha}{M_1 t^\alpha} \|g\|_{L^2(\Omega)}. \tag{3.11}$$

Furthermore, by Remark 3.2, we know that $\mathscr{P}_1 g$ is continuous on $(0, T]$. Therefore, we can deduce that $\mathscr{P}_1 g \in C((0, T]; L^2(\Omega))$.

Next, we show that $\mathscr{P}_2 f$ is continuous on $(0, T]$. With the aid of Proposition 1.11, for all $\kappa \in \mathbb{R}$ and $0 \le \tau < T$, we also have

$$\left| \frac{E_{\alpha,\kappa}(-\lambda_n (T-\tau)^\alpha)}{E_\alpha(-\lambda_n T^\alpha)} \right| \le \frac{M_2}{M_1} \frac{1 + \lambda_n T^\alpha}{1 + \lambda_n (T-\tau)^\alpha} \le \frac{M_2}{M_1} T^\alpha (T-\tau)^{-\alpha}. \tag{3.12}$$

Therefore, by the assumption of f and Proposition 1.11 again, one can see that

$$\begin{aligned}
&\|\mathscr{P}_2 f(t,\cdot)\|_{L^2(\Omega)} \\
\le &\int_0^T \left\| \sum_{n=1}^{\infty} \left(\frac{E_\alpha(-\lambda_n t^\alpha)}{E_\alpha(-\lambda_n T^\alpha)} E_\alpha(-\lambda_n (T-s)^\alpha) f_n(s) \right) e_n(x) \right\|_{L^2(\Omega)} ds \\
\le &\frac{M_2^2 T^\alpha}{M_1} \int_0^T (T-s)^{-\alpha} \|f(s,\cdot)\|_{L^2(\Omega)} ds \\
\le &\frac{M_2^2 T^\beta}{M_1} \frac{\Gamma(\beta)\Gamma(1-\alpha)}{\Gamma(1+\beta-\alpha)} \|f\|_{\mathscr{F}^{\beta,\theta}},
\end{aligned} \tag{3.13}$$

where we use the identity

$$\int_0^T (T-s)^{a-1} s^{b-1} ds = \frac{\Gamma(a)\Gamma(b)}{\Gamma(a+b)} T^{a+b-1}, \quad \text{for any } a, b > 0. \tag{3.14}$$

Furthermore, by applying (W3) and (W4) of Proposition 1.19, it follows from $1 - e^{-z} \le z$ and $ze^{-z} \le 1$ for $z \ge 0$ that

$$\begin{aligned}
|E_\alpha(-\lambda_n t^\alpha) - E_\alpha(-\lambda_n s^\alpha)| \le &\int_0^\infty \mathscr{M}_\alpha(\theta) \left| e^{-\lambda_n t^\alpha \theta} - e^{-\lambda_n s^\alpha \theta} \right| d\theta \\
= &\int_0^\infty \mathscr{M}_\alpha(\theta) e^{-\lambda_n s^\alpha \theta} \left| e^{-\lambda_n (t^\alpha - s^\alpha)\theta} - 1 \right| d\theta \\
\le &\lambda_n (t^\alpha - s^\alpha) \int_0^\infty \theta \mathscr{M}_\alpha(\theta) (\lambda_n s^\alpha \theta)^{-1} d\theta \\
\le &s^{-\alpha} (t-s)^\alpha,
\end{aligned} \tag{3.15}$$

where $0 < s < t \le T$, and we use the inequality

$$\xi_2^\rho - \xi_1^\rho \le (\xi_2 - \xi_1)^\rho, \quad \text{for any } 0 \le \xi_1 \le \xi_2,\ 0 \le \rho \le 1. \tag{3.16}$$

Therefore, by applying (3.15), for any $\varepsilon > 0$ and $t, s \in [\varepsilon, T]$ with $s < t$, it yields that

$$\begin{aligned}
&\|\mathscr{P}_2 f(t,\cdot) - \mathscr{P}_2 f(s,\cdot)\|_{L^2(\Omega)} \\
\le &\frac{M_2 T^\alpha}{M_1} \int_0^T (T-\tau)^{-\alpha} \left(\sum_{n=1}^{\infty} \left| E_\alpha(-\lambda_n t^\alpha) - E_\alpha(-\lambda_n s^\alpha) \right|^2 |f_n(\tau)|^2 \right)^{1/2} d\tau \\
\le &\frac{M_2 T^\beta}{M_1} \frac{\Gamma(\beta)\Gamma(1-\alpha)}{\Gamma(1+\beta-\alpha)} \varepsilon^{-\alpha} (t-s)^\alpha \|f\|_{\mathscr{F}^{\beta,\theta}},
\end{aligned}$$

which implies that $(\mathscr{P}_2 f)(t,\cdot) - (\mathscr{P}_2 f)(s,\cdot)$ tends to 0 when $t - s$ approaches 0. Hence, the desired results are obtained.

In the sequel, we show that $\mathscr{P}_3 f$ is continuous on $(0, T]$. By the assumption of f and Proposition 1.11, one has

$$\|(\mathscr{P}_3 f)(t,\cdot)\|_{L^2(\Omega)} \le M_2 \int_0^t \|f(s,\cdot)\|_{L^2(\Omega)} ds \le M_2 \beta^{-1} t^{\beta} \|f\|_{\mathscr{F}^{\beta,\theta}}. \tag{3.17}$$

On the other hand, it follows that

$$\begin{aligned}(\mathscr{P}_3 f)(t,\cdot) =& \sum_{n=1}^{\infty}\left(\int_0^t E_\alpha(-\lambda_n(t-s)^\alpha)(f_n(s)-f_n(t))ds\right)e_n(x)\\ &+\sum_{n=1}^{\infty}\left(\int_0^t E_\alpha(-\lambda_n(t-s)^\alpha)f_n(t)ds\right)e_n(x)\\ =:& l_1(t)+l_2(t).\end{aligned}$$

Therefore, for all $t, s \in (0, T]$ with $s < t$, it follows that

$$\begin{aligned} l_1(t)-l_1(s) =& \sum_{n=1}^{\infty}\left(\int_0^s \left(E_\alpha(-\lambda_n(t-\tau)^\alpha)\right.\right.\\ &\left.\left. - E_\alpha(-\lambda_n(s-\tau)^\alpha)\right)(f_n(\tau)-f_n(s))d\tau\right)e_n(x)\\ &+\sum_{n=1}^{\infty}\left(\int_0^s E_\alpha(-\lambda_n(t-\tau)^\alpha)(f_n(s)-f_n(t))d\tau\right)e_n(x)\\ &+\sum_{n=1}^{\infty}\left(\int_s^t E_\alpha(-\lambda_n(t-\tau)^\alpha)(f_n(\tau)-f_n(t))d\tau\right)e_n(x)\\ =:& L_1(t,s)+L_2(t,s)+L_3(t,s).\end{aligned} \tag{3.18}$$

By applying Proposition 1.13 and the assumption of f, we have

$$\begin{aligned} &\|L_1(t,s)\|_{L^2(\Omega)}\\ \le& \int_0^s \left\| \sum_{n=1}^{\infty} \left(E_\alpha(-\lambda_n(t-\tau)^\alpha)-E_\alpha(-\lambda_n(s-\tau)^\alpha)\right)(f_n(\tau)-f_n(s))e_n(x)\right\|_{L^2(\Omega)} d\tau\\ \le& \int_0^s (t-s)^\alpha (s-\tau)^{\theta-\alpha}\tau^{\beta-\theta-1} d\tau \sup_{0<\tau<s} \frac{\tau^{1-\beta+\theta}\|f(s,\cdot)-f(\tau,\cdot)\|_{L^2(\Omega)}}{(s-\tau)^\theta}\\ =& \frac{\Gamma(\beta-\theta)\Gamma(1-\alpha+\theta)}{\Gamma(1+\beta-\alpha)} s^{\beta-\alpha}(t-s)^\alpha \|f\|_{\mathscr{F}^{\beta,\theta}}.\end{aligned} \tag{3.19}$$

This means that

$$\|L_2(t,s)\|_{L^2(\Omega)} \le M_2 \int_0^s \|f(t) - f(s)\|_{L^2(\Omega)} d\tau \le M_2 s^{\beta-\theta}(t-s)^\theta \|f\|_{\mathscr{F}^{\beta,\theta}}. \tag{3.20}$$

Furthermore, in view of (3.16), we get the following result:

$$\begin{aligned} \|L_3(t,s)\|_{L^2(\Omega)} \le & M_2 \int_s^t (t-\tau)^\theta \tau^{\beta-\theta-1} d\tau \omega_f(t) \\ \le & \frac{M_2}{\theta+1}(t-s)^{\theta+1} s^{\beta-\theta-1} \|f\|_{\mathscr{F}^{\beta,\theta}}. \end{aligned} \tag{3.21}$$

Hence, we can deduce that $l_1(t) - l_1(s)$ tends to 0 when $t - s$ approaches 0. Furthermore, we know

$$\begin{aligned} l_2(t) - l_2(s) = & \sum_{n=1}^{\infty} \left(\int_0^t E_\alpha(-\lambda_n(t-\tau)^\alpha)(f_n(t) - f_n(s)) d\tau \right) e_n(x) \\ & + \sum_{n=1}^{\infty} \left(\int_0^s \left(E_\alpha(-\lambda_n(t-\tau)^\alpha) - E_\alpha(-\lambda_n(s-\tau)^\alpha) \right) f_n(s) d\tau \right) e_n(x) \\ & + \sum_{n=1}^{\infty} \left(\int_s^t E_\alpha(-\lambda_n(t-\tau)^\alpha) f_n(s) d\tau \right) e_n(x). \end{aligned}$$

By the similar way as in (3.19), (3.20), and (3.21), we can deduce that $l_2(t) - l_2(s)$ tends to 0 as $t - s \to 0$. Hence, it follows that $\mathscr{P}_3 f \in C((0,T]; L^2(\Omega))$. Consequently, we show that there exists a mild solution $u \in C((0,T]; L^2(\Omega))$ which can be expressed as in (3.8). Furthermore, together with (3.11), (3.13) and (3.17), the estimate (3.10) can be obtained.

Finally, we show the uniqueness. Let u and v be any two mild solutions of the problem (3.1) with the same terminal value data g and source function f, and set $w = u - v$. We have to prove that w is a trivial mild solution.

Since $e_n(x)$ is the eigenfunction of the eigenvalue problem (3.2), $w_n(t) = (w(t,\cdot), e_n)$ since $w \in C((0,T]; L^2(\Omega))$.

On the other hand, as the same procedure of solution v in (3.6), we obtain $w_n(T) = E_\alpha(-\lambda_n T^\alpha) w_n(0)$. Hence, it follows from $w_n(T) = 0$ and $E_\alpha(-\lambda_n T^\alpha) > 0$ that $w_n(0) = 0$, and the uniqueness result for the ordinary fractional differential equation implies that $w_n(t) = 0$ for all $n = 1, 2, 3, \ldots$. Since $\{e_n\}_{n=1}^{\infty}$ is also an orthonormal basis in $L^2(\Omega)$, we obtain that $w(t,\cdot) = 0$ in $(0,T) \times \Omega$. Thus, the desired conclusion is obtained, and the proof is completed.

Theorem 3.2 *Let* $0 < m < 1$, $0 < \theta < \beta \le 2\alpha m + \theta - 1$, *and* $\alpha m + \theta \le 1$. *Assume that* $f \in \mathscr{F}^{\beta,\theta}((0,T]; D(A^m))$ *and* $g \in D(A^{1+m})$. *Then the problem (3.1) possesses a unique mild solution u in the function space*

$$\partial_t u \in \mathscr{F}^{\beta,\theta}((0,T];L^2(\Omega)).$$

Furthermore, there exists a positive constant C such that

$$\|\partial_t u\|_{\mathscr{F}^{\beta,\theta}} \le C\left(\|g\|_{D(A^{1+m})} + \|f\|_{\mathscr{F}^{\beta,\theta}}\right). \tag{3.22}$$

Proof By applying the assumptions of functions g and f, it can be checked from Theorem 3.1 that there exists a unique mild solution u. Moreover, in view of Proposition 1.13, we obtain the following expression:

$$\begin{aligned}\partial_t u_n(t) =& \frac{-\lambda_n t^{\alpha-1}E_{\alpha,\alpha}(-\lambda_n t^\alpha)}{E_\alpha(-\lambda_n T^\alpha)}\left(g_n - \int_0^T E_\alpha(-\lambda_n(T-s)^\alpha)f_n(s)ds\right)\\ &+\int_0^t -\lambda_n(t-s)^{\alpha-1}E_{\alpha,\alpha}(-\lambda_n(t-s)^\alpha)f_n(s)ds + f_n(t).\end{aligned} \tag{3.23}$$

Let us consider the operators

$$\begin{aligned}(\mathscr{T}_1 g)(t,x) =& \sum_{n=1}^\infty \frac{-\lambda_n t^{\alpha-1}E_{\alpha,\alpha}(-\lambda_n t^\alpha)}{E_\alpha(-\lambda_n T^\alpha)} g_n e_n(x),\\ (\mathscr{T}_2 f)(t,x) =& \sum_{n=1}^\infty \left(\int_0^T \frac{\lambda_n t^{\alpha-1}E_{\alpha,\alpha}(-\lambda_n t^\alpha)}{E_\alpha(-\lambda_n T^\alpha)} E_\alpha(-\lambda_n(T-s)^\alpha)f_n(s)ds\right)e_n(x),\\ (\mathscr{T}_3 f)(t,x) =& \sum_{n=1}^\infty \left(-\int_0^t \lambda_n(t-s)^{\alpha-1}E_{\alpha,\alpha}(-\lambda_n(t-s)^\alpha)f_n(s)ds\right)e_n(x),\\ (\mathscr{T}_4 f)(t,x) =& \sum_{n=1}^\infty f_n(t)e_n(x).\end{aligned}$$

In order to check that $\partial_t u$ belongs to the space $\mathscr{F}^{\beta,\theta}((0,T];L^2(\Omega))$ and satisfies the estimate (3.22), for the definition of operators $\mathscr{T}_i$, $i=2,3,4$, we need to show that $\mathscr{T}_1 g$, $\mathscr{T}_i f \in \mathscr{F}^{\beta,\theta}((0,T];L^2(\Omega))$. Initially, by the assumption of f, it is easy to check that $\mathscr{T}_4 f \in \mathscr{F}^{\beta,\theta}((0,T];L^2(\Omega))$. Therefore, it remains to subdivide this proof into three steps.

Step 1. $\mathscr{T}_1 g \in \mathscr{F}^{\beta,\theta}((0,T];L^2(\Omega))$ for $g \in D(A^{1+m})$.

In view of Proposition 1.11, for all $\kappa \in \mathbb{R}$ and $t \in (0,T]$, we have

$$\begin{aligned}\left|\frac{E_{\alpha,\kappa}(-\lambda_n t^\alpha)}{E_\alpha(-\lambda_n T^\alpha)}\right| \le& \frac{M_2}{M_1}\left(\frac{1+\lambda_n T^\alpha}{1+\lambda_n t^\alpha}\right)^{1-m}\left(\frac{1+\lambda_n T^\alpha}{1+\lambda_n t^\alpha}\right)^m\\ \le& M_1^{-1}M_2 T^{\alpha(1-m)}t^{-\alpha(1-m)}(1+\lambda_n T^\alpha)^m\\ \le& M_1^{-1}M_2 T^{\alpha(1-m)}t^{-\alpha(1-m)}(1+\lambda_n^m T^{\alpha m}),\end{aligned} \tag{3.24}$$

where we use the inequality $(1+\xi)^\rho \le 1+\xi^\rho$ for any $\xi \ge 0$ and $0 \le \rho \le 1$. Therefore, for all $t \in (0, T]$, it yields that

$$
\begin{aligned}
\|(\mathscr{T}_1 g)(t,\cdot)\|^2_{L^2(\Omega)} &= \sum_{n=1}^{\infty} \left| \frac{-\lambda_n t^{\alpha-1} E_{\alpha,\alpha}(-\lambda_n t^\alpha)}{E_\alpha(-\lambda_n T^\alpha)} g_n \right|^2 \\
&\le \frac{2M_2^2}{M_1^2} T^{2\alpha(1-m)} t^{-2(1-\alpha m)} \sum_{n=1}^{\infty} \lambda_n^2 (1+\lambda_n^{2m} T^{2\alpha m}) \, |g_n|^2 ,
\end{aligned}
$$

where we use the inequality $(1+\xi)^2 \le 2(1+\xi^2)$ for all $\xi \in \mathbb{R}$. According to the definition of the fractional power space $D(A^s)$ for $s \ge 0$, we can find a positive constant C_1 which depends on λ_1 such that $\|g\|_{D(A)} \le C_1 \|g\|_{D(A^{1+m})}$ and then

$$
\|(\mathscr{T}_1 g)(t,\cdot)\|_{L^2(\Omega)} \le \frac{\mathscr{C}_1' M_2}{M_1} T^{\alpha(1-m)} t^{-(1-\alpha m)} \|g\|_{D(A^{1+m})}, \quad t \in (0, T],
$$

where $\mathscr{C}_1' = \left(2(C_1 + T^{2\alpha m})\right)^{1/2}$. Then, we conclude from the assumptions of $\beta \le 2\alpha m + \theta - 1$ and $\alpha m + \theta \le 1$ that $\lim_{t\to 0} t^{1-\beta} \|\mathscr{T}_1 g(t,\cdot)\|_{L^2(\Omega)}$ exists.

Furthermore, by applying (ii) in Proposition 1.13 and for $0 < s < t \le T$, we have

$$
\begin{aligned}
&\|(\mathscr{T}_1 g)(t,\cdot) - (\mathscr{T}_1 g)(s,\cdot)\|^2_{L^2(\Omega)} \\
&= \sum_{n=1}^{\infty} \lambda_n^2 \left| \frac{t^{\alpha-1} E_{\alpha,\alpha}(-\lambda_n t^\alpha) - s^{\alpha-1} E_{\alpha,\alpha}(-\lambda_n s^\alpha)}{E_\alpha(-\lambda_n T^\alpha)} \right|^2 |g_n|^2 \\
&= \sum_{n=1}^{\infty} \lambda_n^2 \left| \int_s^t \tau^{\alpha-2} \frac{E_{\alpha,\alpha-1}(-\lambda_n \tau^\alpha)}{E_\alpha(-\lambda_n T^\alpha)} d\tau \right|^2 |g_n|^2 \\
&\le \frac{2M_2^2}{M_1^2 (1-\alpha m)^2} \frac{T^{2\alpha(1-m)}}{(st)^{2(1-\alpha m)}} (t-s)^{2(1-\alpha m)} \sum_{n=1}^{\infty} \lambda_n^2 (1+\lambda_n^{2m} T^{2\alpha m}) |g_n|^2 \\
&\le \frac{\mathscr{C}_1'^2 M_2^2}{M_1^2 (1-\alpha m)^2} \frac{T^{2\alpha(1-m)}}{(st)^{2(1-\alpha m)}} (t-s)^{2(1-\alpha m)} \|g\|^2_{D(A^{1+m})},
\end{aligned}
$$

which implies that

$$
\begin{aligned}
&\frac{s^{1-\beta+\theta} \|(\mathscr{T}_1 g)(t,\cdot) - (\mathscr{T}_1 g)(s,\cdot)\|_{L^2(\Omega)}}{(t-s)^\theta} \\
&\le \frac{\mathscr{C}_1' M_2}{M_1 (1-\alpha m)} T^{\alpha(1-m)} s^{\alpha m - \beta} \|g\|_{D(A^{1+m})}
\end{aligned}
$$

$$\leq \frac{\mathscr{C}_1' M_2}{M_1(1-\alpha m)} T^{\alpha-\beta} \|g\|_{D(A^{1+m})}.$$

Therefore, $\mathscr{T}_1 g$ fulfills (3.3), and we have

$$\lim_{t\to 0} \sup_{0<s<t} \frac{s^{1-\beta+\theta} \|(\mathscr{T}_1 g)(t,\cdot) - (\mathscr{T}_1 g)(s,\cdot)\|_{L^2(\Omega)}}{(t-s)^\theta} = 0.$$

Hence, $\mathscr{T}_1 g$ also fulfills (3.4). This shows that $\mathscr{T}_1 g \in \mathscr{F}^{\beta,\theta}((0,T]; L^2(\Omega))$.

Step 2. $\mathscr{T}_2 f \in \mathscr{F}^{\beta,\theta}((0,T]; L^2(\Omega))$for $f \in \mathscr{F}^{\beta,\theta}((0,T]; D(A^m))$.

With the aid of (3.24) and the assumption of f, for all $0 < t \leq T$, one can see that

$$\begin{aligned}
&\|\mathscr{T}_2 f(t,\cdot)\|_{L^2(\Omega)} \\
&\leq \int_0^T \left\| \sum_{n=1}^\infty \lambda_n \left(\frac{t^{\alpha-1} E_{\alpha,\alpha}(-\lambda_n t^\alpha)}{E_\alpha(-\lambda_n T^\alpha)} E_\alpha(-\lambda_n (T-s)^\alpha) f_n(s) \right) e_n(x) \right\|_{L^2(\Omega)} ds \\
&\leq \frac{\sqrt{2} M_2^2}{M_1} T^{\alpha(1-m)} t^{\alpha m-1} \\
&\quad \times \int_0^T \left(\sum_{n=1}^\infty (1+\lambda_n^{2m} T^{2\alpha m}) \left(\frac{\lambda_n}{1+\lambda_n (T-s)^\alpha} \right)^2 |f_n(s)|^2 \right)^{1/2} ds.
\end{aligned} \tag{3.25}$$

It follows from the definition of space $D(A^m)$ that we can find a positive constant C_2 which may depend on λ_1 such that $\|f\|_{L^2(\Omega)} \leq C_2 \|f\|_{D(A^m)}$, and by (3.14) then

$$\begin{aligned}
\|\mathscr{T}_2 f(t,\cdot)\|_{L^2(\Omega)} &\leq \frac{\mathscr{C}_2' M_2^2}{M_1} T^{\alpha(1-m)} t^{\alpha m-1} \int_0^T (T-s)^{-\alpha} \|f(s,\cdot)\|_{D(A^m)} ds \\
&\leq \frac{\mathscr{C}_2' M_2^2}{M_1} \frac{\Gamma(\beta)\Gamma(1-\alpha)}{\Gamma(1+\beta-\alpha)} T^{\beta-\alpha m} t^{\alpha m-1} \|f\|_{\mathscr{F}^{\beta,\theta}},
\end{aligned} \tag{3.26}$$

in which $\mathscr{C}_2' = \left(2(C_2 + T^{2\alpha m})\right)^{1/2}$, and this implies that $\lim_{t\to 0} t^{1-\beta} \|\mathscr{T}_2 f(t,\cdot)\|_{L^2(\Omega)}$ exists. In addition, similarly to (3.25) and (3.26), for $0 < s < t \leq T$, we have

$$
\begin{aligned}
&\|(\mathscr{T}_2 f)(t,\cdot) - (\mathscr{T}_2 f)(s,\cdot)\|_{L^2(\Omega)} \\
&= \int_0^T \left(\sum_{n=1}^{\infty} \lambda_n^2 \left| \frac{t^{\alpha-1}E_{\alpha,\alpha}(-\lambda_n t^\alpha) - s^{\alpha-1}E_{\alpha,\alpha}(-\lambda_n s^\alpha)}{E_\alpha(-\lambda_n T^\alpha)} \right. \right. \\
&\quad \times E_\alpha(-\lambda_n (T-s)^\alpha) \bigg|^2 |f_n(s)|^2 \bigg)^{1/2} ds \\
&\le M_2 \int_0^T \left(\sum_{n=1}^{\infty} \left| \int_s^t \tau^{\alpha-2} \frac{E_{\alpha,\alpha-1}(-\lambda_n \tau^\alpha)}{E_\alpha(-\lambda_n T^\alpha)} d\tau \right|^2 \right. \\
&\quad \times \left(\frac{\lambda_n}{1+\lambda_n (T-s)^\alpha} \right)^2 |f_n(s)|^2 \bigg)^{1/2} ds \\
&\le \frac{\mathscr{C}_2' M_2^2}{M_1(1-\alpha m)} \frac{T^{\alpha(1-m)}}{(st)^{1-\alpha m}} (t-s)^{1-\alpha m} \int_0^T (T-s)^{-\alpha} \|f(s,\cdot)\|_{D(A^m)} ds \\
&\le \frac{\mathscr{C}_2' M_2^2}{M_1(1-\alpha m)} \frac{\Gamma(\beta)\Gamma(1-\alpha)}{\Gamma(1+\beta-\alpha)} \frac{T^{\beta-\alpha m}}{(st)^{1-\alpha m}} (t-s)^{1-\alpha m} \|f\|_{\mathscr{F}^{\beta,\theta}},
\end{aligned}
\tag{3.27}
$$

which implies that

$$
\begin{aligned}
&\frac{s^{1-\beta+\theta} \|(\mathscr{T}_2 f)(t,\cdot) - (\mathscr{T}_2 f)(s,\cdot)\|_{L^2(\Omega)}}{(t-s)^\theta} \\
&\le \frac{\mathscr{C}_2' M_2^2}{M_1(1-\alpha m)} \frac{\Gamma(\beta)\Gamma(1-\alpha)}{\Gamma(1+\beta-\alpha)} T^{\beta-\alpha m} s^{\alpha m-\beta} \|f\|_{\mathscr{F}^{\beta,\theta}} \\
&\le \frac{\mathscr{C}_2' M_2^2}{M_1(1-\alpha m)} \frac{\Gamma(\beta)\Gamma(1-\alpha)}{\Gamma(1+\beta-\alpha)} \|f\|_{\mathscr{F}^{\beta,\theta}}.
\end{aligned}
$$

Therefore, $\mathscr{T}_2 f$ fulfills (3.3). In a similar way, we have

$$
\limsup_{t\to 0} \sup_{0<s<t} \frac{s^{1-\beta+\theta} \|(\mathscr{T}_2 f)(t,\cdot) - (\mathscr{T}_2 f)(s,\cdot)\|_{L^2(\Omega)}}{(t-s)^\theta} = 0.
$$

Hence, $\mathscr{T}_2 f$ also fulfills (3.4). This shows that $\mathscr{T}_2 f \in \mathscr{F}^{\beta,\theta}((0,T];L^2(\Omega))$.

Step 3. $\mathscr{T}_3 f \in \mathscr{F}^{\beta,\theta}((0,T];L^2(\Omega))$for $f \in \mathscr{F}^{\beta,\theta}((0,T];D(A^m))$.

By Proposition 1.11 and (3.24), similarly to (3.25), for $0 < t \le T$, one has

$$
\begin{aligned}
&\|(\mathscr{T}_3 f)(t,\cdot)\|_{L^2(\Omega)} \\
&\le \int_0^t (t-s)^{\alpha-1}\left(\sum_{n=1}^{\infty}\left|\lambda_n \frac{E_{\alpha,\alpha}(-\lambda_n(t-s)^\alpha)}{E_\alpha(-\lambda_n T^\alpha)} E_\alpha(-\lambda_n T^\alpha) f_n(s)\right|^2\right)^{1/2} ds \\
&\le \frac{\mathscr{C}_2' M_2^2}{M_1} T^{-\alpha m}\int_0^t (t-s)^{\alpha m-1}\|f(s,\cdot)\|_{D(A^m)} ds \\
&\le \frac{\mathscr{C}_2' M_2^2}{M_1}\frac{\Gamma(\beta)\Gamma(\alpha m)}{\Gamma(\beta+\alpha m)} T^{-\alpha m} t^{\alpha m+\beta-1}\|f\|_{\mathscr{F}^{\beta,\theta}},
\end{aligned}
\tag{3.28}
$$

which implies that $\lim_{t\to 0} t^{1-\beta}\|\mathscr{T}_3 f(t,\cdot)\|_{L^2(\Omega)} = 0$. Furthermore, we have

$$
\begin{aligned}
&(\mathscr{T}_3 f)(t,\cdot) \\
&= \sum_{n=1}^{\infty}\left(\int_0^t -\lambda_n (t-s)^{\alpha-1} E_{\alpha,\alpha}(-\lambda_n(t-s)^\alpha)(f_n(s)-f_n(t))ds\right) e_n(x) \\
&+ \sum_{n=1}^{\infty}\left(\int_0^t -\lambda_n (t-s)^{\alpha-1} E_{\alpha,\alpha}(-\lambda_n(t-s)^\alpha) f_n(t)ds\right) e_n(x) \\
&=: J_1(t) + J_2(t).
\end{aligned}
$$

Therefore, for $0 < s < t \le T$, we obtain

$$
\begin{aligned}
&J_1(t) - J_1(s) \\
&= \sum_{n=1}^{\infty}\Bigg(\int_0^s -\lambda_n\big((t-\tau)^{\alpha-1} E_{\alpha,\alpha}(-\lambda_n(t-\tau)^\alpha) \\
&\quad - (s-\tau)^{\alpha-1} E_{\alpha,\alpha}(-\lambda_n(s-\tau)^\alpha)\big)(f_n(\tau)-f_n(t))d\tau\Bigg) e_n(x) \\
&+ \sum_{n=1}^{\infty} -\lambda_n \left(\int_0^s (t-\tau)^{\alpha-1} E_{\alpha,\alpha}(-\lambda_n(t-\tau)^\alpha)(f_n(s)-f_n(t))d\tau\right) e_n(x) \\
&+ \sum_{n=1}^{\infty} -\lambda_n \left(\int_s^t (t-\tau)^{\alpha-1} E_{\alpha,\alpha}(-\lambda_n(t-\tau)^\alpha)(f_n(\tau)-f_n(t))d\tau\right) e_n(x) \\
&=: \mathscr{J}_1(t,s) + \mathscr{J}_2(t,s) + \mathscr{J}_3(t,s).
\end{aligned}
$$

By applying Proposition 1.13, as in the same way of (3.27), one has

$$\begin{aligned}
&\|\mathscr{J}_1(t,s)\|_{L^2(\Omega)} \\
\le&\frac{M_2^2}{M_1}T^{-\alpha m}\int_0^s\left|\int_{s-\tau}^{t-\tau}\zeta^{\alpha m-2}d\zeta\right|\left(\sum_{n=1}^{\infty}(1+\lambda_n T^{\alpha})^{2m}|f_n(\tau)-f_n(t)|^2\right)^{1/2}d\tau \\
\le&\frac{\mathscr{C}_2' M_2^2 T^{-\alpha m}}{M_1(1-\alpha m)}(t-s)^{1-\alpha m}\int_0^s(t-\tau)^{\alpha m-1}(s-\tau)^{\alpha m-1}\|f(\tau,\cdot)-f(t,\cdot)\|_{D(A^m)}d\tau \\
\le&\frac{\mathscr{C}_2' M_2^2 T^{-\alpha m}}{M_1(1-\alpha m)}(t-s)^{1-\alpha m}\int_0^s(s-\tau)^{2(\alpha m-1)+\theta}\tau^{\beta-\theta-1}d\tau\omega_f(t) \\
\le&\frac{\mathscr{C}_2' M_2^2 T^{-\alpha m}}{M_1(1-\alpha m)}\frac{\Gamma(2\alpha m+\theta-1)\Gamma(\beta-\theta)}{\Gamma(2\alpha m+\beta-1)}(t-s)^{1-\alpha m}s^{2\alpha m+\beta-2}\|f\|_{\mathscr{F}^{\beta,\theta}},
\end{aligned} \tag{3.29}$$

where we use (3.16). By using the similar way as in (3.28), for $0<s<t\le T$, we obtain

$$\begin{aligned}
\|\mathscr{J}_2(t,s)\|_{L^2(\Omega)}\le&\frac{\mathscr{C}_2' M_2^2}{M_1}T^{-\alpha m}\int_0^s(t-\tau)^{\alpha m-1}\|f(s,\cdot)-f(t,\cdot)\|_{D(A^m)}d\tau \\
\le&\frac{\mathscr{C}_2' M_2^2}{M_1\alpha m}T^{-\alpha m}s^{\alpha m+\beta-\theta-1}(t-s)^{\theta}\|f\|_{\mathscr{F}^{\beta,\theta}}
\end{aligned} \tag{3.30}$$

and

$$\begin{aligned}
\|\mathscr{J}_3(t,s)\|_{L^2(\Omega)}\le&\frac{\mathscr{C}_2' M_2^2}{M_1}T^{-\alpha m}\int_s^t(t-\tau)^{\alpha m-1}\|f(\tau,\cdot)-f(t,\cdot)\|_{D(A^m)}d\tau \\
\le&\frac{\mathscr{C}_2' M_2^2}{M_1(\alpha m+\theta)}T^{-\alpha m}s^{\beta-\theta-1}(t-s)^{\alpha m+\theta}\|f\|_{\mathscr{F}^{\beta,\theta}}.
\end{aligned} \tag{3.31}$$

Therefore, together with (3.29), (3.30), and (3.31), there exists a positive constant C such that

$$\frac{s^{1-\beta+\theta}\|J_1(t)-J_1(s)\|_{L^2(\Omega)}}{(t-s)^{\theta}}\le C\|f\|_{\mathscr{F}^{\beta,\theta}}$$

and

$$\lim_{t\to 0}\sup_{0<s<t}\frac{s^{1-\beta+\theta}\|J_2(t)-J_2(s)\|_{L^2(\Omega)}}{(t-s)^{\theta}}=0.$$

On the other hand, we know

$$
\begin{aligned}
&J_2(t) - J_2(s)\\
&= \sum_{n=1}^{\infty} \left(\int_0^t -\lambda_n (t-\tau)^{\alpha-1} E_{\alpha,\alpha}(-\lambda_n (t-\tau)^{\alpha})(f_n(t) - f_n(s)) d\tau \right) e_n(x)\\
&+ \sum_{n=1}^{\infty} \left(\int_0^s -\lambda_n \big((t-\tau)^{\alpha-1} E_{\alpha,\alpha}(-\lambda_n (t-\tau)^{\alpha})\right.\\
&\left. - (s-\tau)^{\alpha-1} E_{\alpha,\alpha}(-\lambda_n (s-\tau)^{\alpha})\big) f_n(s) d\tau \right) e_n(x)\\
&+ \sum_{n=1}^{\infty} \left(\int_s^t -\lambda_n (t-\tau)^{\alpha-1} E_{\alpha,\alpha}(-\lambda_n (t-\tau)^{\alpha}) f_n(s) d\tau \right) e_n(x).
\end{aligned}
$$

Therefore, by using the similar way as in (3.25)–(3.28), there exists a constant C such that

$$
\frac{s^{1-\beta+\theta} \|J_2(t) - J_2(s)\|_{L^2(\Omega)}}{(t-s)^{\theta}} \leq C \|f\|_{\mathscr{F}^{\beta,\theta}}
$$

and

$$
\lim_{t \to 0} \sup_{0<s<t} \frac{s^{1-\beta+\theta} \|J_2(t) - J_2(s)\|_{L^2(\Omega)}}{(t-s)^{\theta}} = 0.
$$

The above arguments imply that $\mathscr{T}_3 f$ fulfills (3.3) and (3.4). Hence, we obtain that $\mathscr{T}_3 f \in \mathscr{F}^{\beta,\theta}((0,T]; L^2(\Omega))$. Consequently, combined with the above arguments, the estimate (3.22) holds, that is, the desired result is obtained.

Next, we introduce the following definition of classical solution of the problem (3.1).

Definition 3.2 A function u is called a classical solution of the problem (3.1) if $u \in L^2(\Omega)$ is continuous on [0,T], $\partial_t u$ exists and is continuous on $(0, T]$, $u(t) \in D(A)$ for $t \in (0, T]$, and (3.8) is satisfied.

Remark 3.3 It is obvious that a classical solution is a mild solution under the appropriate given data.

Theorem 3.3 *Let* $0 < m < 1$, $0 < \theta < \beta \leq 2\alpha m + \theta - 1$ *and* $\alpha m + \theta \leq 1$. *Assume that* $f \in \mathscr{F}^{\beta,\theta}((0,T]; D(A))$, *and* $g \in D(A^{1+m})$. *Then the problem (3.1) possesses a unique classical solution u in the function space*

$$\begin{cases} u \in C([0,T];L^2(\Omega)) \cap C((0,T];D(A)), \\ \partial_t u \in \mathcal{F}^{\beta,\theta}((0,T];L^2(\Omega)). \end{cases}$$

Furthermore, there exists a positive constant C such that

$$\|u(t,\cdot)\|_{L^2(\Omega)} + t^\alpha \|u(t,\cdot)\|_{D(A)} + \|\partial_t u\|_{\mathcal{F}^{\beta,\theta}} \le C\left(\|g\|_{D(A^{1+m})} + \|f\|_{\mathcal{F}^{\beta,\theta}}\right).$$

Proof With the aid of Proposition 1.11, we have

$$\left|\frac{E_{\alpha,\kappa}(-\lambda_n t^\alpha)}{E_\alpha(-\lambda_n T^\alpha)}\right| \le \frac{M_2}{M_1}\left(1+\lambda_n T^\alpha\right), \quad \text{for all } \kappa \in \mathbb{R},\ t \in [0,T].$$

Noting the definition of spaces $D(A^{1+m})$ and $D(A^m)$, it allows us to find a positive constant C_3 which may depend on λ_1 such that $\|g\|_{D(A^m)} \le C_3\|g\|_{D(A^{1+m})}$ and hence

$$\|(\mathcal{P}_1 g)(t,\cdot)\|_{L^2(\Omega)} \le \frac{M_2\mathcal{C}_3'}{M_1}\|g\|_{D(A^{1+m})},$$

where $\mathcal{C}_3' = \left(2(C_3 + T^{2\alpha})\right)^{1/2}$. For any $t, s \in [0,T]$ with $s < t$, by (i) in Proposition 1.13, (3.16), and (3.24), it follows that

$$\begin{aligned}
&\|(\mathcal{P}_1 g)(t,\cdot) - (\mathcal{P}_1 g)(s,\cdot)\|^2_{L^2(\Omega)} \\
=&\sum_{n=1}^{\infty}\left|\frac{E_\alpha(-\lambda_n t^\alpha) - E_\alpha(-\lambda_n s^\alpha)}{E_\alpha(-\lambda_n T^\alpha)}\right|^2 |g_n|^2 \\
=&\sum_{n=1}^{\infty}\left|\int_s^t \lambda_n \tau^{\alpha-1}\frac{E_\alpha(-\lambda_n \tau^\alpha)}{E_\alpha(-\lambda_n T^\alpha)}d\tau\right|^2 |g_n|^2 \\
\le&\frac{M_2^2}{M_1^2}T^{2\alpha(1-m)}\left|\int_s^t \tau^{\alpha m-1}d\tau\right|^2 \sum_{n=1}^{\infty}\lambda_n^2(1+\lambda_n^m T^{\alpha m})^2|g_n|^2 \\
\le&\left(\frac{\mathcal{C}_1' M_2}{M_1\alpha m}T^{\alpha(1-m)}(t-s)^{\alpha m}\right)^2 \|g\|^2_{D(A^{1+m})}.
\end{aligned}$$

It means that $(\mathcal{P}_1 g)(t,\cdot) - (\mathcal{P}_1 g)(s,\cdot)$ tends to 0 when $t-s$ approaches 0.

According to the definition of space $D(A)$, it follows that there exists a positive constant C_4 which may depend on λ_1 such that $\|f\|_{L^2(\Omega)} \le C_4\|f\|_{D(A)}$, and similarly to (3.13), we have

$$\|\mathcal{P}_2 f(t,\cdot)\|_{L^2(\Omega)} \le \frac{C_4 M_2^2 T^\beta}{M_1}\frac{\Gamma(\beta)\Gamma(1-\alpha)}{\Gamma(1+\beta-\alpha)}\|f\|_{\mathcal{F}^{\beta,\theta}}.$$

Moreover, for any $t, s \in [0, T]$ with $s < t$, it yields that

$$\begin{aligned}
&\|\mathscr{P}_2 f(t,\cdot) - \mathscr{P}_2 f(s,\cdot)\|_{L^2(\Omega)} \\
\le & \frac{M_2 T^\alpha}{M_1} \int_0^T (T-\tau)^{-\alpha} \left(\sum_{n=1}^\infty \left| E_\alpha(-\lambda_n t^\alpha) - E_\alpha(-\lambda_n s^\alpha) \right|^2 |f_n(\tau)|^2 \right)^{1/2} d\tau \\
\le & \frac{M_2 T^\alpha}{M_1} \int_0^T (T-\tau)^{-\alpha} \left(\sum_{n=1}^\infty \left| \int_s^t \lambda_n \tau^{\alpha-1} E_\alpha(-\lambda_n \tau^\alpha) d\tau \right|^2 |f_n(\tau)|^2 \right)^{1/2} d\tau \\
\le & \frac{M_2^2 T^\alpha}{M_1 \alpha} (t-s)^\alpha \int_0^T (T-\tau)^{-\alpha} \|f(\tau,\cdot)\|_{D(A)} d\tau \\
\le & \frac{M_2^2 T^\beta}{M_1 \alpha} \frac{\Gamma(\beta)\Gamma(1-\alpha)}{\Gamma(1+\beta-\alpha)} (t-s)^\alpha \|f\|_{\mathscr{F}^{\beta,\theta}},
\end{aligned}$$

which implies that $(\mathscr{P}_2 f)(t,\cdot) - (\mathscr{P}_2 f)(s,\cdot)$ tends to 0 when $t - s$ approaches 0.

On the other hand, similarly to (3.17), we obtain

$$\|(\mathscr{P}_3 f)(t,\cdot)\|_{L^2(\Omega)} \le C_4 M_2 \beta^{-1} t^\beta \|f\|_{\mathscr{F}^{\beta,\theta}}$$

and

$$\begin{aligned}
&\|(\mathscr{P}_3 f)(t,\cdot) - (\mathscr{P}_3 f)(s,\cdot)\|_{L^2(\Omega)} \\
\le & \int_s^t \left(\sum_{n=1}^\infty \left| E_\alpha(-\lambda_n (t-\tau)^\alpha) f_n(\tau) \right|^2 \right)^{1/2} d\tau \\
& + \int_0^s \left(\sum_{n=1}^\infty \left| \left(E_\alpha(-\lambda_n (t-\tau)^\alpha) - E_\alpha(-\lambda_n (s-\tau)^\alpha) \right) f_n(\tau) \right|^2 \right)^{1/2} d\tau \\
\le & C_4 M_2 \int_s^t \|f(\tau,\cdot)\|_{D(A)} d\tau \\
& + \int_0^s \left(\sum_{n=1}^\infty \left| \int_{s-\tau}^{t-\tau} \lambda_n \xi^{\alpha-1} E_{\alpha,\alpha}(-\lambda_n \xi^\alpha) d\xi \right|^2 |f_n(\tau)|^2 \right)^{1/2} d\tau \\
\le & C_4 M_2 \beta^{-1} (t^\beta - s^\beta) \|f\|_{\mathscr{F}^{\beta,\theta}} + M_2 \int_0^s \left| \int_{s-\tau}^{t-\tau} \xi^{\alpha-1} d\xi \right| \|f(\tau,\cdot)\|_{D(A)} d\tau \\
\le & C_4 M_2 \beta^{-1} (t-s)^\beta \|f\|_{\mathscr{F}^{\beta,\theta}} + \frac{M_2}{\beta\alpha} (t-s)^\alpha s^\beta \|f\|_{\mathscr{F}^{\beta,\theta}},
\end{aligned}$$

where we use (3.16). Hence we obtain that $(\mathscr{P}_3 f)(t,\cdot) - (\mathscr{P}_3 f)(s,\cdot)$ tends to 0 when $t - s$ approaches 0. Hence, this shows that $\mathscr{P}_3 f \in C([0,T]; L^2(\Omega))$. By applying the triangle inequality for Eq. (3.7), we have

$$\begin{aligned}&\|u(t,\cdot)-u(s,\cdot)\|_{L^2(\Omega)}\\ \leq&\|\mathscr{P}_1 g(t,\cdot)-\mathscr{P}_1 g(s,\cdot)\|_{L^2(\Omega)}+\|\mathscr{P}_2 f(t,\cdot)-\mathscr{P}_2 f(s,\cdot)\|_{L^2(\Omega)}\\ &+\|\mathscr{P}_3 f(t,\cdot)-\mathscr{P}_3 f(s,\cdot)\|_{L^2(\Omega)}\to 0,\quad \text{as } t-s\to 0.\end{aligned}$$

It means that $u \in C([0,T];L^2(\Omega))$.

We now consider the following operators:

$$\mathscr{Q}_1 g(t,\cdot) := \sum_{n=1}^{\infty}\frac{-\lambda_n E_\alpha(-\lambda_n t^\alpha)}{E_\alpha(-\lambda_n T^\alpha)} g_n e_n(x),$$

$$\mathscr{Q}_2 f(t,\cdot) := \sum_{n=1}^{\infty}\frac{\lambda_n E_\alpha(-\lambda_n t^\alpha)}{E_\alpha(\lambda_n T^\alpha)}\left(\int_0^T E_\alpha(-\lambda_n (T-s)^\alpha) f_n(s)ds\right) e_n(x),$$

$$\mathscr{Q}_3 f(t,\cdot) := \sum_{n=1}^{\infty}\left(\int_0^t -\lambda_n E_\alpha(-\lambda_n (t-s)^\alpha) f_n(s)ds\right) e_n(x).$$

In view of (3.9), as the same way we obtained in Remark 3.2, it is easy to check that $\mathscr{Q}_1 g$ is finite and continuous on $(0,T]$. Moreover, by using (3.24), (3.15), and (i) in Proposition 1.13, similarly to (3.25) and (3.27), we also obtain that $\mathscr{Q}_2 f$ is finite and continuous on $(0,T]$. The result $\mathscr{Q}_3 f \in C((0,T];L^2(\Omega))$ holds as well. Indeed, by (3.28), (3.24), and Proposition 1.11, it is easy to see that $\mathscr{Q}_3 f$ is finite. As the similar proof process in Theorem 3.1, by the embedded relationship between $D(A)$ and $L^2(\Omega)$, we can deduce that $(\mathscr{Q}_3 f)(t,\cdot)-(\mathscr{Q}_3 f)(s,\cdot)$ tend to 0 when $t-s$ approaches 0 for all $t,s\in(0,T]$. With the same argument as for Theorem 3.1 again, the uniqueness of the classical solution follows.

Since $e_n(x)$ is the eigenfunctions of the eigenvalue problem (3.2), it means that there exists a unique classical solution u such that $Au = Q_1 g + Q_2 f + Q_3 f$. Consequently, the result $\partial_t u \in \mathscr{F}^{\beta,\theta}((0,T];L^2(\Omega))$ can be verified as the same methods in Theorem 3.2 immediately. This shows that all desired results hold.

As an application of Theorems 3.1 and 3.2, we obtain the following conclusion.

Theorem 3.4 *Let* $0<m<1$, $0<\theta<\beta\leq 2\alpha m+\theta-1$ *and* $\alpha m+\theta\leq 1$. *Assume that* $f\in\mathscr{F}^{\beta,\theta}((0,T];L^2(\Omega))$, *and* $g\in D(A^m)$. *Then the problem (3.1) possesses a unique mild solution* u *in the function space*

$$\begin{cases} u\in C((0,T];L^2(\Omega))\cap C([0,T];D(A^{-1})),\\ \partial_t u\in\mathscr{F}^{\beta,\theta}((0,T];D(A^{-1})).\end{cases}$$

Furthermore, there exists a positive constant C *such that*

$$t^\alpha\|u(t,\cdot)\|_{L^2(\Omega)}+\|u(t,\cdot)\|_{D(A^{-1})}+\|\partial_t u\|_{\mathscr{F}^{\beta,\theta}}\leq C\left(\|g\|_{D(A^m)}+\|f\|_{\mathscr{F}^{\beta,\theta}}\right).$$

3.1.4 Example

We get two examples as follows to illustrate our main results.

Example 3.1 Let $\alpha \in (0, 1), \Omega = (0, 1), T = 1$, and $f(t, x) = t^{-\alpha}x(1-x)/\Gamma(1-\alpha)$ for $x \in \Omega$. Obviously, function f is not Hölder continuous on $[0, 1] \times \Omega$, but it is weighted Hölder continuous. Then, we shall consider the following backward problem for fractional diffusion equation:

$$\partial_t u(t, x) = \partial_t^{1-\alpha} u_{xx}(t, x) + f(t, x), \qquad 0 < t < 1,$$

associated with the boundary value conditions

$$u(t, 0) = u(t, 1) = 0, \qquad 0 < t < 1$$

and the finial value condition

$$u(1, x) = g(x), \qquad x \in \Omega.$$

For this case, we know that $\lambda_n = n^2\pi^2$ and $e_n(x) = \sin(n\pi x)$, $n \in \mathbb{N}^+$. Let $g(x) = \cos(x\pi/2)$. It is easy to check that

$$g_n = \frac{8\sqrt{\lambda_n}}{4\lambda_n - \pi^2} \text{ and } f_n(t) = \frac{4(1-(-1)^n)}{\lambda_n^{3/2}\Gamma(1-\alpha)} t^{-\alpha}.$$

Hence, $g \in D(A^m)$ for $0 < m < 1$ and $f(t, \cdot) \in D(A)$; then there exists a unique mild solution $u(t, x) = \sum_{n=1}^{\infty} u_n(t) \sin(n\pi x)$ that belongs to $C((0, T]; L^2(\Omega))$ by Theorem 3.1 in which

$$\begin{aligned} u_n(t) =& \frac{E_\alpha(-\lambda_n t^\alpha)}{E_\alpha(-\lambda_n)} \left(g_n - \frac{4}{\lambda_n^{3/2}}(1-(-1)^n)E_{\alpha,2-\alpha}(-\lambda_n) \right) \\ &+ \frac{4}{\lambda_n^{3/2}}(1-(-1)^n)t^{1-\alpha}E_{\alpha,2-\alpha}(-\lambda_n t^\alpha), \end{aligned}$$

and it further possesses the conclusions of Theorem 3.4 provided with $0 < \theta < \beta \le \min\{1-\alpha, 2\alpha m + \theta - 1\}$ and $\alpha m + \theta \le 1$.

Example 3.2 If we set $g(x) = x(1 - 2x^2 + x^3)$ in Example 3.1, for a simple calculation, we can deduce that $g_n = 24(1-(-1)^n)/\lambda_n^{5/2}$, $n \in \mathbb{N}^+$. Hence, $g \in D(A^{1+m})$ for all $0 < m < 1$. Then there exists a unique classical solution u defined as in Example 3.1 which satisfies all conclusions of Theorem 3.3, and there exists a mild solution which satisfies the conclusion of Theorem 3.2 both cases provided with $0 < \theta < \beta \le \min\{1-\alpha, 2m + \theta - 1\}$ and $\alpha m + \theta \le 1$.

3.2 Terminal Value Problem

3.2.1 Introduction

Nonlinear diffusion equations, an important class of parabolic equations, come from many diffuse phenomena that appear widely in nature. They are proposed as mathematical models of physical problems in many areas, such as filtering, phase transition, biochemistry, and dynamics of biological groups. Many new ideas and methods have been developed to consider some various kinds of nonlinear diffusion equations. We list some selected works in recent time, for example, Caffarelli et al. [49], Duzaar et al. [24, 25, 57], Vazquez et al. [22, 23, 89, 198], and the references therein.

We present existence and regularity estimates for the solutions to a final boundary value problem for a space-time fractional diffusion equation. Let D be an open and bounded domain in $\mathbb{R}^k$, $(k \geq 1)$ with boundary ∂D. Given $0 < \alpha < 1$ and $0 < \beta \leq 1$, a forcing (or source) function F, we consider the final value problem for the time fractional diffusion equation

$$ {}_0^C D_t^\alpha u(t,x) = -\mathscr{L}^\beta u(t,x) + F(t,x,u(t,x)), \quad (t,x) \in J \times D, \tag{3.32}$$

with the boundary condition

$$\mathscr{H} u(t,x) = 0, \quad (t,x) \in J \times \partial D \tag{3.33}$$

and the final condition

$$u(T,x) = \varphi(x), \quad x \in D, \tag{3.34}$$

where φ is a given function. Here J is the interval $(0,T)$, and the notation ${}_0^C D_t^\alpha$ for $0 < \alpha < 1$ represents the Caputo fractional derivative of order α which is defined as follows:

$$ {}_0^C D_t^\alpha u(t,x) := g_{1-\alpha}(t) * \frac{\partial}{\partial t} u(t,x) = \frac{1}{\Gamma(1-\alpha)} \int_0^t (t-s)^{-\alpha} \frac{\partial}{\partial s} u(s,x) ds, \quad t > 0;$$

here $g_\alpha(t) = \frac{1}{\Gamma(\alpha)} t^{\alpha-1}$, $t > 0$, and $*$ denotes the convolution. For $\alpha = 1$, we consider the usual time derivative $\frac{\partial u}{\partial t}$. The fractional power $\mathscr{L}^\beta (0 < \beta \leq 1)$ of the Laplacian operator $\mathscr{L}$ on D is defined by its spectrum. The symmetric uniformly elliptic operator is defined on the space $L^2(D)$ by

$$\mathscr{L} u(x) = -\sum_{i=1}^k \frac{\partial}{\partial x_i} \left(\sum_{j=1}^k \mathscr{L}_{ij}(x) \frac{\partial}{\partial x_j} u(x) \right) + b(x) u(x),$$

provided that $\mathscr{L}_{ij} \in C^1\left(\overline{\Omega}\right)$, $b \in C\left(\overline{\Omega}\right)$, $b(x) \geq 0$ for all $x \in \overline{\Omega}$, $\mathscr{L}_{ij} = \mathscr{L}_{ji}$, $1 \leq i, j \leq k$, and $\xi^T\left[\mathscr{L}_{ij}(x)\right]\xi \geq L_0|\xi|^2$ for some $L_0 > 0$, $x \in \overline{\Omega}$, $\xi = (\xi_1, \xi_2, \ldots, \xi_k) \in \mathbb{R}^k$. Equation (3.32) is equipped with $\mathscr{H}v = v$ or $\mathscr{H}v = (\overrightarrow{L}\nabla v) \cdot \overrightarrow{n}$, where $\overrightarrow{L} = [\mathscr{L}_{ij}(x)]_{i,j=1}^k$ is a $k \times k$ matrix and n is the outer normal vector of ∂D. Then the operator $\mathscr{L}$ is self-adjoint under this impedance boundary condition.

The time-fractional reaction diffusion equation arises in describing "memory" occurring in physics such as plasma turbulence [50], and it was introduced by Nigmatullin [171] to describe diffusion in media with fractal geometry, which is a special type of porous media and is applied in the flow in highly heterogeneous aquifer [19] and single-molecular protein dynamics [122]. In a physical model presented in [235], the fractional diffusion corresponds to a diverging jump length variance in the random walk, and a fractional time derivative arises when the characteristic waiting time diverges.

If the final condition (3.34) is replaced by the initial condition

$$u(0, x) = u_0(x), \quad x \in D, \tag{3.35}$$

then the problem (3.32)–(3.35) is called a forward problem (or an initial value problem) for time-space fractional diffusion equations; for applications of this type of equation see [66], and for the abstract form of (3.32) and (3.35) see [41]. Carvaho et al. [46] established a local theory of mild solutions for the problem (3.32) and (3.35) where $\mathscr{L}^\beta$ is a sectorial (nonpositive) operator. Guswanto [77] studied the existence and uniqueness of local mild solutions for a class of initial value problems for nonlinear fractional evolution equations, and the study of existence of initial value problems was considered by Warma et al. [66]. A significant number of papers were devoted to extend properties holding in the standard setting to the fractional one (see, e.g., [55, 69, 116, 134, 205]).

Numerical approximation for solutions for the problem (3.32)–(3.35) was studied by Jin et al. [104, 106], and for other works on fractional diffusion see [131, 172, 174, 243]. However, the literature on regularity of the initial value problem for fractional diffusion-wave equations is scarce; for the linear case see [43, 158, 172, 188], and for the nonlinear case see [10, 66, 112, 168]. Although there are many works on the direct problem, the results on inverse problem for fractional diffusion are scarce. We can list some papers of Yamamoto and his group [102, 127, 130, 145, 166, 221], of Kaltenbacher et al. [107, 108], of Rundell et al. [183, 184], of Janno [97, 98], etc.

In practice, the initial data of some problems may not be known since many phenomena cannot be measublack at the initial time. Phenomena can be observed at a final time $t = T$, such as, in the image processing area. A picture is not processed at the capturing time $t = 0$. Instead, one wishes to recover the original information of the picture from its blurry form. Hence, inverse problems or terminal value problems or final value problems (IPs/FVPs), i.e., the fractional differential equations (FDEs) equipped with final value data, have been considered.

IPs/FVPs are important in engineering in detecting the previous status of physical fields from its present information. If $b = 0$ and $F(u(x,t)) = F(x,t)$, Tuan et al. [206] showed that problem (3.32)–(3.34) has a unique weak solution when $\varphi \in H^2(\Omega)$ and $F \in L^\infty(0,T; H^2(\Omega))$, and other works on the homogeneous case for the problem (3.32)–(3.34) can be found in [101, 206]. Wei and her group [224–226, 228] studied some regularization methods for homogeneous backward problem, and Yamamoto et al. [138] considered a backward problem in time for a time-fractional partial differential equation in the one-dimensional case. When $\alpha = 1$, systems (3.32) and (3.34) are reduced to the backward problem for classical reaction diffusion equations and were studied in [15, 80, 209].

To the best of the authors' knowledge, [210] was the first paper that analyzed the problem (3.32)–(3.34). The authors presented existence and uniqueness results and derived regularity estimates in both time and space. In what follows, the authors analyzed the difficulties of this problem. By letting $u(t,\cdot) = \mathscr{O}(t)(f,\varphi)$, the solution operator $\mathscr{O}(0)$ is really not bounded in $L^2(D)$. Hence continuity of mild solutions does not hold at the initial time $t = 0$. In addition, since the fractional derivative is nonlocally defined, if we put $v(t) = u(T-t)$, then ${}_0^C D_t^\alpha u(s)|_{s=T-t}$ does not equal ${}_0^C D_t^\alpha v(t)$, so the problem cannot be changed to an initial value problem. As a result we need new techniques to deal with the FVP (3.32)–(3.34). To the best of our knowledge, the work on the final value problem is still limited.

The main results in this section can be split into two parts, linear and nonlinear source functions. Linear models are sometimes good approximations of the real problems under consideration and provide mathematical tools needed to study nonlinear phenomena, especially for semi-linear and quasi-linear equations. In part 1, we consider the regularity property of the solutions in the linear case F. We seek to address the following question: If the data is regular, how regular is the solution? Our task in this part is to find a suitable Banach space for the given data (φ, F) in order to obtain regularity results for the corresponding solution. In part 2, we discuss existence, uniqueness, and regularity for the solutions to (3.32)–(3.34) for the nonlinear problem. Our main motivation for deriving regularity results is that one needs it for a rigorous study of a numerical scheme to approximate the solution. To the best of our knowledge, regularity results on inverse initial value problems (final value problems) for fractional diffusion are still unavailable in the literature. For initial value problems, McLean et al. [158], Jin et al. [106], and Mu et al. [168] considered existence and regularity results of the solutions in $C([0,T]; L^2(D))$. However it seems that the techniques in [106, 168] cannot be applied for our problems (it is impossible to apply some well-known fixed point theorems with some spaces in [106] for establishing unique solution). To overcome this we need data φ in a suitable space, and we will use a Picard iteration argument and then develop some new techniques to obtain existence and regularity of the solutions.

The rest of this section is organized as follows. In Sect. 3.2.2, we give basic notations and preliminaries, and we propose a mild solution of our problem. In Sect. 3.2.3, we give some regularity results of the linear inhomogeneous problem. Section 3.2.4 is devoted to existence and regularity for nonlinear problems, and Sect. 3.2.5 considers global existence.

3.2.2 Notations and Preliminaries

3.2.2.1 Functional Space

In this subsection, we introduce some functional spaces for solutions of FVP (3.32)–(3.34). By $\{m_j\}_{j\geq 1}$ and $\{e_j(x)\}_{j\geq 1}$, we denote the spectrum and sequence of eigenfunctions of $\mathscr{L}$ which satisfy $e_j \in \{v \in H^2(D) : \mathscr{H}v = 0\}$, $\mathscr{L}e_j(x) = m_j e_j(x)$, $0 < m_1 \leq m_2 \leq \ldots \leq m_j \leq \ldots$, and $\lim_{j\to\infty} m_j = \infty$. The sequence $\{e_j(x)\}_{j\geq 1}$ forms an orthonormal basis of the space $L^2(D)$. For a given real number $p \geq 0$, the Hilbert scale space $H^{2p}(D)$ is defined by

$$\left\{ v \in L^2(D) : \|v\|^2_{H^{2p}(D)} := \sum_{j=1}^{\infty} (v, e_j)^2 m_j^{2p} < \infty \right\},$$

where $(\cdot,\cdot)$ is the usual inner product of $L^2(D)$. The fractional power $\mathscr{L}^\beta$, $\beta \geq 0$, of the Laplacian operator $\mathscr{L}$ on D is defined by

$$\mathscr{L}^\beta v(x) := \sum_{j=1}^{\infty} (v, e_j) m_j^\beta e_j(x). \tag{3.36}$$

Then, $\{m_j^\beta\}_{j\geq 1}$ is the spectrum of the operator $\mathscr{L}^\beta$. We denote by $\mathbf{V}_\beta$ the domain of $\mathscr{L}^\beta$, and then

$$\mathbf{V}_\beta = \{v \in L^2(D) : \|\mathscr{L}^\beta v\| < \infty\},$$

where $\|\cdot\|$ is the usual norm of $L^2(D)$, and $\mathbf{V}_\beta$ is a Banach space with respect to the norm $\|v\|_{\mathbf{V}_\beta} = \|\mathscr{L}^\beta v\|$. Moreover, the inclusion $\mathbf{V}_\beta \subset H^{2\beta}(D)$ holds for $\beta > 0$. We identify the dual space $\left(L^2(D)\right)' = L^2(D)$ and define the domain $\mathbf{V}_{-\beta} := D(\mathscr{L}^{-\beta})$ by the dual space of $\mathbf{V}_\beta$, i.e., $\mathbf{V}_{-\beta} = \left(\mathbf{V}_\beta\right)'$. Then, $\mathbf{V}_{-\beta}$ is a Hilbert space endowed with the norm

$$\|v\|_{\mathbf{V}_{-\beta}} := \left\{ \sum_{j=1}^{\infty} (v, e_j)^2_{-\beta,\beta} m_j^{-2\beta} \right\}^{1/2},$$

where $(\cdot,\cdot)_{-\beta,\beta}$ denotes the dual inner product between $\mathbf{V}_{-\beta}$ and $\mathbf{V}_\beta$. We note that the Sobolev embedding $\mathbf{V}_\beta \hookrightarrow L^2(D) \hookrightarrow \mathbf{V}_{-\beta}$ holds for $0 < \beta < 1$, and $(\tilde{v}, v)_{-\beta,\beta} = (\tilde{v}, v)$, for $\tilde{v} \in L^2(D)$, $v \in \mathbf{V}_\beta$. Hence, we have

$$(e_i, e_j)_{-\beta,\beta} = (e_i, e_j) = \delta_{ij}, \tag{3.37}$$

where δ_{ij} is the Kronecker delta for $i, j \in \mathbb{N}^+$. Moreover, for given $p_1 \geq 1$ and $0 < \eta < 1$, we denote by $\mathscr{X}_{p_1,\eta}(J \times D)$ the set of all functions f from J to $L^{p_1}(D)$ such that

$$|||f|||_{\mathscr{X}_{p_1,\eta}} := \operatorname{esssup}_{0\leq t\leq T} \int_0^t \|f(\tau,\cdot)\|_{p_1} (t-\tau)^{\eta-1} d\tau < \infty, \tag{3.38}$$

where $\|.\|_{p_1}$ is the norm of $L^{p_1}(D)$. Note that, for fixed $t > 0$, Hölder's inequality shows that

$$\int_0^t \|f(\tau,\cdot)\|_{p_1} (t-\tau)^{\eta-1} d\tau \leq \left[\int_0^t \|f(\tau,\cdot)\|_{p_1}^{p_2} d\tau\right]^{\frac{1}{p_2}} \left[\int_0^t (t-\tau)^{\frac{p_2(\eta-1)}{p_2-1}} d\tau\right]^{\frac{p_2-1}{p_2}}.$$

In the above inequality, we note that the function $\tau \to (t-\tau)^{\frac{p_2(\eta-1)}{p_2-1}}$ is integrable for $p_2 > \frac{1}{\eta}$. Therefore, if we let $L^{p_2}(0, T; L^{p_1}(D))$, $p_1, p_2 \geq 1$, be the space of all Bochner's measurable functions f from J to $L^{p_1}(D)$ such that

$$\|f\|_{L^{p_2}(0,T;L^{p_1}(D))} := \left[\int_0^t \|f(\tau,\cdot)\|_{p_1}^{p_2} d\tau\right]^{\frac{1}{p_2}} < \infty,$$

then the following inclusion holds

$$L^{p_2}(0, T; L^{p_1}(D)) \subset \mathscr{X}_{p_1,\eta}(J \times D), \quad \text{for } p_2 > \frac{1}{\eta}, \tag{3.39}$$

and there exists a positive constant $C > 0$ such that

$$|||f|||_{\mathscr{X}_{p_1,\eta}} \leq C \|f\|_{L^{p_2}(0,T;L^{p_1}(D))}; \tag{3.40}$$

here, C depends only on p_2, η, and T. Moreover, for a given number s such that $0 < s < \eta$, we have $\mathscr{X}_{p_1,\eta-s}(J \times D) \subset \mathscr{X}_{p_1,\eta}(J \times D)$ since $|||f|||_{\mathscr{X}_{p_1,\eta}} \leq T^s |||f|||_{\mathscr{X}_{p_1,\eta-s}}$. Let B be a Banach space, and we denote by $C([0, T]; B)$ the space of all continuous functions from $[0, T]$ to B endowed with the norm $\|v\|_{C([0,T];B)} := \sup_{0\leq t\leq T} \|v(t)\|_B$ and by $C^\theta([0, T]; B)$, $0 < \theta \leq 1$, the subspace of $C([0, T]; B)$ which includes all Hölder continuous functions and is equipped with the norm

$$|||v|||_{C^\theta([0,T];B)} := \sup_{0\leq t_1<t_2\leq T} \frac{\|v(t_2) - v(t_1)\|_B}{|t_2 - t_1|^\theta}.$$

In some cases, a given function might not be continuous at $t = 0$. Hence, it is useful to consider the set $C((0, T]; B)$ which consists of all continuous functions from $(0, T]$ to B. We define by $C_w^\rho((0, T]; B)$ the weighted Banach space of all functions v in $C((0, T]; B)$ such that

$$\|v\|_{C_w^\rho((0,T];B)} := \sup_{0<t\leq T} t^\rho \|v(t)\|_B < \infty.$$

Now, we discuss solutions of the FVP for the fractional ordinary equation

$$ {}_0^C D_t^\alpha v(t) = g(t, v(t)) - mv(t), \ t \in J \quad \text{and} \quad v(T) = v_T, \tag{3.41}$$

where m and v_T are given real numbers. Here, we wish to find a representation formula for v in terms of the given function g and the final value data v_T. By applying the fractional integral ${}_0D_t^{-\alpha}$ on both sides of the equation (3.41), we obtain

$$v(t) = v(0) + {}_0D_t^{-\alpha} \left[g(t, v(t)) - mv(t)\right].$$

The Laplace transform yields that

$$\overline{v} = \frac{\lambda^{\alpha-1}}{\lambda^\alpha + m} v(0) + \frac{1}{\lambda^\alpha + m} \overline{g(v)},$$

where $\overline{v}$ is the Laplace transform of v. Hence, the inverse Laplace transform implies

$$v(t) = v(0) E_{\alpha,1}(-mt^\alpha) + g(t, v(t)) * \left[t^{\alpha-1} E_{\alpha,\alpha}(-mt^\alpha)\right]. \tag{3.42}$$

Here, $E_{\alpha,1}$ and $E_{\alpha,\alpha}$ are the Mittag-Leffler functions which are defined in Definition 1.7. Now, a representation of the solution of FVP (3.41) can be obtained by substituting $t = T$ into (3.42) and using the final value data $v(T) = v_T$, i.e.,

$$\begin{aligned} v(t) =& g(t, v(t)) * \widetilde{E}_{\alpha,\alpha}(-mt^\alpha) \\ &+ \left[v_T - \left(g(r, v(r)) * \widetilde{E}_{\alpha,\alpha}(-mr^\alpha)\right)\Big|_{r=T}\right] \frac{E_{\alpha,1}(-mt^\alpha)}{E_{\alpha,1}(-mT^\alpha)}, \end{aligned}$$

where

$$\widetilde{E}_{\alpha,\alpha}(-mt^\alpha) := t^{\alpha-1} E_{\alpha,\alpha}(-mt^\alpha). \tag{3.43}$$

3.2.2.2 Mild Solutions of FVP and Unboundedness of Solution Operators

A representation of solutions and the definition of mild solutions are given in this subsection, and then we analyze the unboundedness of solution operators. By the definition (3.36) of $\mathscr{L}^\beta$, the identity $\mathscr{L}^\beta e_j(x) = m_j^\beta e_j(x)$ holds. Hence, in view of the Fourier expansion $u(t, x) = \sum_{j=1}^\infty u_j(t) e_j(x)$, where $u_j(t) = (u(t, \cdot), e_j)$, Eq. (3.32) can be rewritten as

$$\left({}_0^C D_t^\alpha \sum_{j=1}^\infty u_j(t)e_j, e_j\right) = -\left(\mathscr{L}^\beta \sum_{j=1}^\infty u_j(t)e_j, e_j\right) + \left(F(t,x,u(t,x)), e_j\right), \quad t \in J.$$

This is equivalent to the equation

$${}_0^C D_t^\alpha u_j(t) = F_j(t, u(t)) - m_j^\beta u_j(t), \qquad F_j(t, u(t)) = (F(t,x,u(t,x)), e_j).$$

By applying the method of solutions of FVPs for fractional ordinary equations in Subsection 3.2.2.1 and using the final value data (3.34), we derive

$$\begin{aligned} u_j(t) =& F_j(t, u(t)) * \widetilde{E}_{\alpha,\alpha}(-m_j^\beta t^\alpha) \\ &+ \left[\varphi_j - \left(F_j(r, u(r)) * \widetilde{E}_{\alpha,\alpha}(-m_j^\beta r^\alpha)\right)\Big|_{r=T}\right] \frac{E_{\alpha,1}(-m_j^\beta t^\alpha)}{E_{\alpha,1}(-m_j^\beta T^\alpha)}, \end{aligned} \tag{3.44}$$

where $\varphi_j = (\varphi, e_j)$. Therefore, we obtain a spectral representation for u as follows:

$$\begin{aligned} u(t,x) =& \sum_{j=1}^\infty F_j(t, u(t)) * \widetilde{E}_{\alpha,\alpha}(-m_j^\beta t^\alpha) e_j(x) \\ &+ \sum_{j=1}^\infty \left[\varphi_j - \left(F_j(r, u(r)) * \widetilde{E}_{\alpha,\alpha}(-m_j^\beta r^\alpha)\right)\Big|_{r=T}\right] \frac{E_{\alpha,1}(-m_j^\beta t^\alpha)}{E_{\alpha,1}(-m_j^\beta T^\alpha)} e_j(x). \end{aligned}$$

For $g \in L^2(0,T; L^2(D))$ and $v \in L^2(D)$, let us denote by $\mathscr{O}_n$, $1 \le n \le 3$, the following operators

$$(\mathscr{O}_1 g)(t,x) := \sum_{j=1}^\infty g_j(t) * \widetilde{E}_{\alpha,\alpha}(-m_j^\beta t^\alpha) e_j(x),$$

$$(\mathscr{O}_2(t) v)(x) := \sum_{j=1}^\infty v_j \frac{E_{\alpha,1}(-m_j^\beta t^\alpha)}{E_{\alpha,1}(-m_j^\beta T^\alpha)} e_j(x),$$

and $(\mathscr{O}_3 g)(t) := -\mathscr{O}_2(t)(\mathscr{O}_1 g)(T)$ on $L^2(D)$, for $t \in J$. Then, the solution u can be represented as

$$u(t,x) = (\mathscr{O}_1 F)(t,x) + (\mathscr{O}_2(t)\varphi)(x) + (\mathscr{O}_3 F)(t,x), \tag{3.45}$$

for $(t,x) \in J \times D$.

One of the most important things, when we consider the well-posedness of a PDE, is the boundedness of solution operators. Corresponding to the initial value problem (3.32), (3.33), (3.35), the solution operators are usually bounded in $L^2(D)$, see, e.g., [112, 114, 158, 168, 174]. Unfortunately, some solution operators of FVP (3.32) and (3.34) are not bounded on $L^2(D)$ at $t = 0$. For this purpose, we recall that, for $0 < \alpha < 1$ and $z < 0$, there exist positive constants $c_\alpha, \widehat{c_\alpha}$ such that

$$\frac{c_\alpha}{1+|z|} \le |E_{\alpha,1}(z)| \le \frac{\widehat{c_\alpha}}{1+|z|}, \quad |E_{\alpha,\alpha}(z)| \le \min\left\{\frac{\widehat{c_\alpha}}{1+|z|}, \frac{\widehat{c_\alpha}}{1+|z|^2}\right\}, \tag{3.46}$$

see, for example, [52, 180, 187]. Now, let v_0 be defined by $v_{0,j} := (v_0, e_j) = j^{-1/2} m_j^{-\beta}$, $j \ge 1$. Then, it is easy to see that v_0 belongs to $\mathbf{V}_{\beta\gamma}$ for $0 \le \gamma < 1$ and does not for $\gamma \ge 1$. Using the inequalities (3.46), we have

$$\begin{aligned} \|\mathscr{O}_2(0)v_0\|^2 =& \sum_{j=1}^{\infty} \frac{v_{0,j}^2}{E_{\alpha,1}^2(-m_j^\beta T^\alpha)} \ge c_\alpha^{-2} \sum_{j=1}^{\infty} v_{0,j}^2 (1 + m_j^\beta T^\alpha)^2 \\ \ge& c_\alpha^{-2} T^{2\alpha} \sum_{j=1}^{\infty} v_{0,j}^2 m_j^{2\beta} = c_\alpha^{-2} T^{2\alpha} \sum_{j=1}^{\infty} j^{-1} = \infty, \end{aligned}$$

which shows the unboundedness of $\mathscr{O}_2(0)$ on $L^2(D)$. Similarly, the unboundedness of $\mathscr{O}_3(0)$ on $L^2(D)$ can be shown.

3.2.3 *Final Value Problem with a Linear Source*

In this subsection, we study the regularity of mild solutions of final value problem (FVP) (3.32)–(3.34) corresponding to the linear source function F, i.e., $F(t, x, u(t, x)) = F(t, x)$ which does not include u. We will investigate the regularity of the following FVP:

$$\begin{cases} {}_0^C D_t^\alpha u(t,x) = -\mathscr{L}^\beta u(t,x) + F(t,x), & (t,x) \in J \times D, \\ \mathscr{H} u(t,x) = 0, & (t,x) \in J \times \partial D, \\ u(T,x) = \varphi(x), & x \in D, \end{cases} \tag{3.47}$$

where φ, F will be specified later. In order to consider this problem, it is necessary to give a definition of mild solutions based on (3.45) as follows.

Definition 3.3 If a function u belongs to $L^p(0, T; L^q(D))$, for some $p, q \ge 1$, and satisfies the equation

$$u(t,x) = (\mathscr{O}_1 F)(t,x) + (\mathscr{O}_2(t)\varphi)(x) + (\mathscr{O}_3 F)(t,x), \tag{3.48}$$

then u is said to be a mild solution of FVP (3.47).

In what follows, we introduce some assumptions on the final value data φ and the linear source function F.

(R1) $0 < p, q < 1$ such that $p + q = 1$.
(R2) $0 < r \le \frac{1-\alpha q}{\alpha q}$.
(R3) $0 < s < \min\{\alpha q, 1 - \alpha q\}$.
(R4) $0 < p' \le p - \frac{s}{\alpha}, \quad q' = 1 - p', \quad 0 < r \le \frac{1-\alpha q'}{\alpha q'}$.
(R5) $0 \le \widehat{q} \le \min\{p, q, \frac{s}{\alpha}\}, \quad \widehat{p} = 1 - \widehat{q}, \quad 0 < \widehat{r} \le \frac{1-\alpha}{\alpha}$.

In the following lemma, we will show that solutions of FVP (3.47) must be bounded by a power function $t^{-\alpha q}$, for some appropriate number q, i.e.,

$$\|u(t,\cdot)\| \lesssim t^{-\alpha q}, \quad \text{for all} \quad 0 < t \le T.$$

Lemma 3.1 *Let p, q be defined by (R1), and u satisfies (3.48). If $\varphi \in \mathbf{V}_{\beta p}$ and $F \in \mathscr{X}_{2,\alpha q}(J \times D)$, then there exists a constant $C_0 > 0$ such that*

$$\|u(t,\cdot)\| \le C_0 \left(\|\varphi\|_{\mathbf{V}_{\beta p}} + |||F|||_{\mathscr{X}_{2,\alpha q}} \right) t^{-\alpha q}. \tag{3.49}$$

Proof The inequality (3.46) shows that

$$E_{\alpha,\alpha}(-m_j^\beta (t-\tau)^\alpha) \le \widehat{c}_\alpha \left[1 + m_j^\beta (t-\tau)^\alpha\right]^{-p} \le \widehat{c}_\alpha m_j^{-\beta p} (t-\tau)^{-\alpha p}. \tag{3.50}$$

Combined with the definition $(\mathscr{O}_1 F)(t,x) = \sum_{j=1}^\infty F_j(t) * \widetilde{E}_{\alpha,\alpha}(-m_j^\beta t^\alpha) e_j(x)$, we have that

$$\begin{aligned} \|(\mathscr{O}_1 F)(t,\cdot)\| &\le \int_0^t \left\| \sum_{j=1}^\infty F_j(\tau) \widetilde{E}_{\alpha,\alpha}(-m_j^\beta (t-\tau)^\alpha) e_j \right\| d\tau \\ &= \int_0^t \left\{ \sum_{j=1}^\infty F_j^2(\tau) E_{\alpha,\alpha}^2(-m_j^\beta (t-\tau)^\alpha)(t-\tau)^{2\alpha-2} \right\}^{1/2} d\tau \\ &\le \widehat{c}_\alpha \int_0^t \left\{ \sum_{j=1}^\infty F_j^2(\tau) m_j^{-2\beta p} (t-\tau)^{-2\alpha p} (t-\tau)^{2\alpha-2} \right\}^{1/2} d\tau. \end{aligned} \tag{3.51}$$

Hence, we obtain the following estimate:

$$\|(\mathscr{O}_1 F)(t,\cdot)\| \leq \widehat{c}_\alpha m_1^{-\beta p} \int_0^t \|F(\tau,\cdot)\| (t-\tau)^{\alpha q-1} d\tau \leq M_1 t^{-\alpha q} |||F|||_{\mathscr{X}_{2,\alpha q}}, \tag{3.52}$$

by noting (3.38) and letting $M_1 = \widehat{c}_\alpha m_1^{-\beta p} T^{\alpha q}$. In addition, the norm $\|\mathscr{O}_2(t)\varphi\|$ can be estimated as

$$\|\mathscr{O}_2(t)\varphi\| = \left\{ \sum_{j=1}^{\infty} \varphi_j^2 \frac{E_{\alpha,1}^2(-m_j^\beta t^\alpha)}{E_{\alpha,1}^2(-m_j^\beta T^\alpha)} \right\}^{1/2} \leq \widehat{c}_\alpha c_\alpha^{-1} \left\{ \sum_{j=1}^{\infty} \varphi_j^2 \left[\frac{1+m_j^\beta T^\alpha}{1+m_j^\beta t^\alpha} \right]^2 \right\}^{1/2}.$$

The ratio $\frac{1+m_j^\beta T^\alpha}{1+m_j^\beta t^\alpha}$ is clearly bounded by both $1+m_j^\beta T^\alpha$ and $\frac{T^\alpha}{t^\alpha}$. Moreover, the increasing property of the sequence $\{m_j\}_{j\geq 1}$ shows $1 \leq m_1^{-\beta} m_j^\beta$. Thus, we have $1+m_j^\beta T^\alpha \leq (m_1^{-\beta}+T^\alpha) m_j^\beta$. By noting $p+q=1$, one can deduce that the ratio is bounded by the product of $T^{\alpha q} t^{-\alpha q}$ and $(1+m_j^\beta T^\alpha)^p$. Bring the above arguments together, and this leads to

$$\|\mathscr{O}_2(t)\varphi\| \leq \widehat{c}_\alpha c_\alpha^{-1} \left\{ \sum_{j=1}^{\infty} \varphi_j^2 \frac{T^{2\alpha q}}{t^{2\alpha q}} (1+m_j^\beta T^\alpha)^{2p} \right\}^{1/2} \leq M_2 t^{-\alpha q} \|\varphi\|_{\mathbf{V}_{\beta p}}, \tag{3.53}$$

where

$$M_2 = \widehat{c}_\alpha c_\alpha^{-1} T^{\alpha q} (m_1^{-\beta} + T^\alpha)^p.$$

Now, we proceed to estimate $\|(\mathscr{O}_3 F)(t,\cdot)\|$ by using the same techniques as in (3.51) and (3.53). As a consequence of

$$\begin{aligned}(\mathscr{O}_3 F)(t,x) &= -\mathscr{O}_2(t)(\mathscr{O}_1 F)(T)(x)\\ &= -\sum_{j=1}^{\infty} \left(F_j(r) * \widetilde{E}_{\alpha,\alpha}(-m_j^\beta r^\alpha)\right)\Big|_{r=T} \frac{E_{\alpha,1}(-m_j^\beta t^\alpha)}{E_{\alpha,1}(-m_j^\beta T^\alpha)} e_j(x),\end{aligned}$$

we can obtain the following estimates:

$$\begin{aligned}&\|(\mathscr{O}_3 F)(t,\cdot)\|\\ &\leq \int_0^T \left\| \sum_{j=1}^{\infty} F_j(\tau) E_{\alpha,\alpha}(-m_j^\beta (T-\tau)^\alpha)(T-\tau)^{\alpha-1} \frac{E_{\alpha,1}(-m_j^\beta t^\alpha)}{E_{\alpha,1}(-m_j^\beta T^\alpha)} e_j \right\| d\tau\end{aligned}$$

$$=\int_0^T\left\{\sum_{j=1}^{\infty}F_j^2(\tau)E_{\alpha,\alpha}^2(-m_j^\beta(T-\tau)^\alpha)(T-\tau)^{2\alpha-2}\frac{E_{\alpha,1}^2(-m_j^\beta t^\alpha)}{E_{\alpha,1}^2(-m_j^\beta T^\alpha)}\right\}^{1/2}d\tau$$

$$\leq \widehat{c}_\alpha^2 c_\alpha^{-1}\int_0^T\left\{\sum_{j=1}^{\infty}F_j^2(\tau)m_j^{-2\beta p}(T-\tau)^{2\alpha q-2}\frac{T^{2\alpha q}}{t^{2\alpha q}}(1+m_j^\beta T^\alpha)^{2p}\right\}^{1/2}d\tau.$$

A simple computation shows that

$$\|(\mathscr{O}_3F)(t,\cdot)\| \leq M_3t^{-\alpha q}\int_0^T\|F(\tau,\cdot)\|\,(T-\tau)^{\alpha q-1}d\tau \leq M_3t^{-\alpha q}|||F|||_{\mathscr{X}_{2,\alpha q}}, \tag{3.54}$$

where we let $M_3=\widehat{c}_\alpha M_2$. Finally, it follows from (3.52)–(3.54) and the identity (3.48) that

$$\begin{aligned}\|u(t,\cdot)\| &\leq \|(\mathscr{O}_3F)(t,\cdot)\| + \|\mathscr{O}_2(t)\varphi\| + \|(\mathscr{O}_3F)(t,\cdot)\|\\ &\leq\Big(\sum_{1\leq n\leq 3}M_n\Big)\Big(\|\varphi\|_{\mathbf{V}_{\beta p}}+|||F|||_{\mathscr{X}_{2,\alpha q}}\Big)t^{-\alpha q}.\end{aligned}$$

The inequality (3.49) is proved by letting $C_0=\sum_{1\leq n\leq 3}M_n$.

Based on Lemma 3.1, we consider existence, uniqueness, and regularity of solutions in the following theorem which is divided into two parts. In the first part, we obtain the existence and uniqueness of a mild solution in the space $L^{\frac{1}{\alpha q}-r}(0,T;L^2(D))$ for some suitable numbers q,r and for the given assumptions on φ and F as in Lemma 3.1. In the second part, we improve the smoothness of the mild solution by considering the spatial-fractional derivative $\mathscr{L}^{\beta(p-p')}$. It is very important to investigate the continuity of the mild solution. We first show that the mild solution is continuous from $(0,T]$ to $L^2(D)$. Moreover, we establish the continuity on the closed interval $[0,T]$ which corresponds to lower spatial smoothness, $V_{-\beta q'}$, for a relevant number q'.

Theorem 3.5

(a) *Let p,q,r be defined by (R1), (R2). If $\varphi\in\mathbf{V}_{\beta p}$ and $F\in\mathscr{X}_{2,\alpha q}(J\times D)$, then FVP (3.47) has a unique solution u in $L^{\frac{1}{\alpha q}-r}(0,T;L^2(D))$. Moreover, there exists a positive constant C_1 such that*

$$\|u\|_{L^{\frac{1}{\alpha q}-r}(0,T;L^2(D))}\leq C_1\,\|\varphi\|_{\mathbf{V}_{\beta p}}+C_1|||F|||_{\mathscr{X}_{2,\alpha q}}. \tag{3.55}$$

(b) *Let p,q,s,r,p',q' be defined by (R1), (R3), (R4). If $\varphi\in\mathbf{V}_{\beta p}$, and $F\in\mathscr{X}_{2,\alpha q-s}(J\times D)$, then FVP (3.47) has a unique solution u such that*

$$u \in L^{\frac{1}{\alpha q'}-r}(0,T;\mathbf{V}_{\beta(p-p')}) \cap C_w^{\alpha q}((0,T];L^2(D)) \cap C^s([0,T];\mathbf{V}_{-\beta q'}).$$

Moreover, there exists a positive constant C_2 such that

$$\begin{aligned}&\|u\|_{L^{\frac{1}{\alpha q'}-r}(0,T;\mathbf{V}_{\beta(p-p')})} + \|u\|_{C_w^{\alpha q}((0,T];L^2(D))} + |||u|||_{C^s([0,T];\mathbf{V}_{-\beta q'})}\\ \leq & C_2\,\|\varphi\|_{\mathbf{V}_{\beta p}} + C_2|||F|||_{\mathscr{X}_{2,\alpha q-s}}.\end{aligned} \tag{3.56}$$

Proof The proof of part (a) can be easily obtained from Lemma 3.1. Indeed, the inequality (3.49) leads to

$$\|u\|_{L^{\frac{1}{\alpha q}-r}(0,T;L^2(D))} \leq C_0\left(\|\varphi\|_{\mathbf{V}_{\beta p}} + |||F|||_{\mathscr{X}_{2,\alpha q}}\right)\left\{\int_0^T t^{-\alpha q\left(\frac{1}{\alpha q}-r\right)}dt\right\}^{\frac{\alpha q}{1-\alpha qr}}.$$

Since $-\alpha q\left(\frac{1}{\alpha q}-r\right) > -1$, the integral in the above inequality exists, i.e., $t^{-\alpha q}$ belongs to $L^{\frac{1}{\alpha q}-r}(0,T;\mathbb{R})$. Hence, FVP (3.47) has a solution u in $L^{\frac{1}{\alpha q}-r}(0,T;L^2(D))$. The uniqueness of u depends on the uniqueness of the ODE (3.41). For the uniqueness of this ODE, see [115] (Theorem 3.25). Moreover, the inequality (3.55) is derived by letting $C_1 = C_0\left\|t^{-\alpha q}\right\|_{L^{\frac{1}{\alpha q}-r}(0,T;\mathbb{R})}$. Now, we proceed to prove part (b) which will be presented in the following steps.

Step 1. We prove $u \in L^{\frac{1}{\alpha q'}-r}(0,T;\mathbf{V}_{\beta(p-p')})$.

Firstly, by the same argument as in the proof of (3.51), we derive the following chain of inequalities:

$$\begin{aligned}&\|(\mathscr{O}_1F)(t,\cdot)\|_{\mathbf{V}_{\beta(p-p')}}\\ \leq&\int_0^t\left\|\mathscr{L}^{\beta(p-p')}\sum_{j=1}^{\infty}F_j(\tau)\widetilde{E}_{\alpha,\alpha}(-m_j^{\beta}(t-\tau)^{\alpha})e_j\right\|d\tau,\\ \leq&\widehat{c}_{\alpha}\int_0^t\left\{\sum_{j=1}^{\infty}F_j^2(\tau)m_j^{-2\beta p}(t-\tau)^{-2\alpha p}(t-\tau)^{2\alpha-2}m_j^{2\beta(p-p')}\right\}^{1/2}d\tau\\ \leq&\widehat{c}_{\alpha}m_1^{-\beta p'}\int_0^t\|F(\tau,\cdot)\|\,(t-\tau)^{\alpha q-1}d\tau \leq M_4t^{-\alpha q'}|||F|||_{\mathscr{X}_{2,\alpha q-s}},\end{aligned} \tag{3.57}$$

for $M_4 = \widehat{c}_{\alpha}m_1^{-\beta p'}T^{\alpha q'+s}$, where the inequality $|||F|||_{\mathscr{X}_{2,\alpha q}} \leq T^s|||F|||_{\mathscr{X}_{2,\alpha q-s}}$ holds. Similarly, from $\|\mathscr{O}_2(t)\varphi\|_{\mathbf{V}_{\beta(p-p')}} = \left\|\mathscr{L}^{\beta(p-p')}\mathscr{O}_2(t)\varphi\right\|$ and the same way as in the proof of (3.53), we have

$$\|\mathscr{O}_2(t)\varphi\|_{\mathbf{V}_{\beta(p-p')}} \le \widehat{c}_\alpha c_\alpha^{-1}\left\{\sum_{j=1}^{\infty}\varphi_j^2\frac{T^{2\alpha q'}}{t^{2\alpha q'}}(1+m_j^\beta T^\alpha)^{2p'}m_j^{2\beta(p-p')}\right\}^{1/2}$$
$$\le M_5 t^{-\alpha q'}\|\varphi\|_{\mathbf{V}_{\beta p}}, \tag{3.58}$$

where we let $M_5 = \widehat{c}_\alpha c_\alpha^{-1} T^{\alpha q'}(m_1^{-\beta}+T^\alpha)^{p'}$. Now, we will estimate the norm $\|(\mathscr{O}_3 F)(t,\cdot)\|_{\mathbf{V}_{\beta(p-p')}}$ which will use the same estimate for the fraction $\frac{E_{\alpha,1}(-m_j^\beta t^\alpha)}{E_{\alpha,1}(-m_j^\beta T^\alpha)}$ as in (3.58). Indeed, noting that $(1+m_j^\beta T^\alpha)^{p'} \le (m_1^{-\beta}+T^\alpha)^{p'}m_j^{\beta p'}$, by using (3.46), we see that

$$\|(\mathscr{O}_3 F)(t,\cdot)\|_{\mathbf{V}_{\beta(p-p')}}$$
$$\le \int_0^T \left\| \mathscr{L}^{\beta(p-p')}\sum_{j=1}^{\infty} F_j(\tau)E_{\alpha,\alpha}(-m_j^\beta(T-\tau)^\alpha)(T-\tau)^{\alpha-1}\frac{E_{\alpha,1}(-m_j^\beta t^\alpha)}{E_{\alpha,1}(-m_j^\beta T^\alpha)}e_j\right\| d\tau$$
$$= \int_0^T \left\{\sum_{j=1}^{\infty} F_j^2(\tau)E_{\alpha,\alpha}^2(-m_j^\beta(T-\tau)^\alpha)(T-\tau)^{2\alpha-2}\frac{E_{\alpha,1}^2(-m_j^\beta t^\alpha)}{E_{\alpha,1}^2(-m_j^\beta T^\alpha)}m_j^{2\beta(p-p')}\right\}^{1/2} d\tau$$
$$\le \widehat{c}_\alpha M_5 \int_0^T \left\{\sum_{j=1}^{\infty} F_j^2(\tau)m_j^{-2\beta p}(T-\tau)^{-2\alpha p}(T-\tau)^{2\alpha-2}t^{-2\alpha q'}m_j^{2\beta p'}m_j^{2\beta(p-p')}\right\}^{1/2} d\tau.$$

Thus, by some simple computations, one can get

$$\|(\mathscr{O}_3 F)(t,\cdot)\|_{\mathbf{V}_{\beta(p-p')}} \le \widehat{c}_\alpha M_5 t^{-\alpha q'}\int_0^T \|F(t,\cdot)\|\,(T-\tau)^{\alpha q-1}d\tau$$
$$\le M_6 t^{-\alpha q'}|||F|||_{\mathscr{X}_{2,\alpha q-s}}, \tag{3.59}$$

with $M_6 = \widehat{c}_\alpha M_5 T^s$, where the inequality $|||F|||_{\mathscr{X}_{2,\alpha q}} \le T^s|||F|||_{\mathscr{X}_{2,\alpha q-s}}$ has been used. Bring (3.57)–(3.59) and (3.48) together, and we have

$$\|u(t,\cdot)\|_{\mathbf{V}_{\beta(p-p')}} \le M_7\left(\|\varphi\|_{V_{\beta p}} + |||F|||_{\mathscr{X}_{2,\alpha q-s}}\right)t^{-\alpha q'},$$

where $M_7 = \sum_{4\le n\le 6} M_n$. Since the function $t \to t^{-\alpha q'}$ is clearly contained in the space $L^{\frac{1}{\alpha q'}-r}(0,T;\mathbb{R})$, we can take the $L^{\frac{1}{\alpha q'}-r}(0,T;\mathbb{R})$-norm on both sides of the above inequality, namely

$$\|u\|_{L^{\frac{1}{\alpha q'}-r}(0,T;\mathbf{V}_{\beta(p-p')})} \leq M_8 \|\varphi\|_{\mathbf{V}_{\beta p}} + M_8 |||F|||_{\mathscr{X}_{2,\alpha q-s}}, \tag{3.60}$$

where $M_8 = M_7 \left\| t^{-\alpha q'} \right\|_{L^{\frac{1}{\alpha q'}-r}(0,T;\mathbb{R})}$.

Step 2. We prove $u \in C_w^{\alpha q}((0,T]; L^2(D))$.

Let us consider $0 < t_1 < t_2 \leq T$. By (3.48), the difference $u(t_2,x) - u(t_1,x)$ can be calculated as

$$\begin{aligned}
&u(t_2,x) - u(t_1,x) \\
&= \sum_{j=1}^{\infty} F_j(t) * \widetilde{E}_{\alpha,\alpha}(-m_j^{\beta} t^{\alpha}) \Big|_{t=t_1}^{t=t_2} e_j(x) + \sum_{j=1}^{\infty} \varphi_j \left. \frac{E_{\alpha,1}(-m_j^{\beta} t^{\alpha})}{E_{\alpha,1}(-m_j^{\beta} T^{\alpha})} \right|_{t=t_1}^{t=t_2} e_j(x) \\
&\quad - \sum_{j=1}^{\infty} \left(F_j(r) * \widetilde{E}_{\alpha,\alpha}(-m_j^{\beta} r^{\alpha})\right)\Big|_{r=T} \left. \frac{E_{\alpha,1}(-m_j^{\beta} t^{\alpha})}{E_{\alpha,1}(-m_j^{\beta} T^{\alpha})} \right|_{t=t_1}^{t=t_2} e_j(x).
\end{aligned}$$

By using the differentiation identities

$$\frac{d}{d\omega} E_{\alpha,1}(-m_j^{\beta}\omega^{\alpha}) = -m_j^{\beta} \widetilde{E}_{\alpha,\alpha}(-m_j^{\beta}\omega^{\alpha})$$

and

$$\frac{d}{d\omega}\left(\widetilde{E}_{\alpha,\alpha}(-m_j^{\beta}\omega^{\alpha})\right) = \omega^{\alpha-2} E_{\alpha,\alpha-1}(-m_j^{\beta}\omega^{\alpha}),$$

see, for example, [52, 180, 187], we have

$$\begin{aligned}
&F_j(t) * \widetilde{E}_{\alpha,\alpha}(-m_j^{\beta} t^{\alpha}) \Big|_{t=t_1}^{t=t_2} \\
&= \int_0^{t_1} F_j(\tau)\, \widetilde{E}_{\alpha,\alpha}(-m_j^{\beta}\omega^{\alpha}) \Big|_{\omega=t_1-\tau}^{\omega=t_2-\tau} d\tau + \int_{t_1}^{t_2} F_j(\tau) \widetilde{E}_{\alpha,\alpha}(-m_j^{\beta}(t_2-\tau)^{\alpha}) d\tau \\
&= \int_0^{t_1} \int_{t_1-\tau}^{t_2-\tau} F_j(\tau)\omega^{\alpha-2} E_{\alpha,\alpha-1}(-m_j^{\beta}\omega^{\alpha}) d\omega d\tau \\
&\quad + \int_{t_1}^{t_2} F_j(\tau) \widetilde{E}_{\alpha,\alpha}(-m_j^{\beta}(t_2-\tau)^{\alpha}) d\tau
\end{aligned}$$

and

$$E_{\alpha,1}(-m_j^{\beta} t^{\alpha}) \Big|_{t=t_1}^{t=t_2} = -m_j^{\beta} \int_{t_1}^{t_2} \widetilde{E}_{\alpha,\alpha}(-m_j^{\beta}\omega^{\alpha}) d\omega.$$

Combining the above arguments gives

$$
\begin{aligned}
&u(t_2,x)-u(t_1,x)\\
=&\sum_{j=1}^{\infty}\int_0^{t_1}\int_{t_1-\tau}^{t_2-\tau}F_j(\tau)\omega^{\alpha-2}E_{\alpha,\alpha-1}(-m_j^\beta\omega^\alpha)d\omega d\tau e_j(x)\\
&+\sum_{j=1}^{\infty}\int_{t_1}^{t_2}F_j(\tau)\widetilde{E}_{\alpha,\alpha}(-m_j^\beta(t_2-\tau)^\alpha)d\tau e_j(x)\\
&-\mathscr{L}^\beta\sum_{j=1}^{\infty}\varphi_j\int_{t_1}^{t_2}\frac{\widetilde{E}_{\alpha,\alpha}(-m_j^\beta\omega^\alpha)}{E_{\alpha,1}(-m_j^\beta T^\alpha)}d\omega e_j(x)\\
&+\mathscr{L}^\beta\sum_{j=1}^{\infty}\int_0^T\int_{t_1}^{t_2}F_j(\tau)\widetilde{E}_{\alpha,\alpha}(-m_j^\beta(T-\tau)^\alpha)\frac{\widetilde{E}_{\alpha,\alpha}(-m_j^\beta\omega^\alpha)}{E_{\alpha,1}(-m_j^\beta T^\alpha)}d\omega d\tau e_j(x)\\
=:&\mathscr{I}_1+\mathscr{I}_2+\mathscr{I}_3+\mathscr{I}_4.
\end{aligned}
\tag{3.61}
$$

Now, we will establish estimates for $\mathscr{I}_j$, $j=1,2,3,4$, and show that $\mathscr{I}_j$ tends to 0 as $t_2-t_1\to 0$. Firstly, by the inequality (3.46), we see that the absolute value of $E_{\alpha,\alpha-1}(-m_j^\beta\omega^\alpha)$ is bounded by $\widehat{c}_\alpha m_j^{-\beta p}\omega^{-\alpha p}$. This implies

$$
\omega^{\alpha-2}\left|E_{\alpha,\alpha-1}(-m_j^\beta\omega^\alpha)\right|\le\widehat{c}_\alpha\omega^{\alpha q-2}m_j^{-\beta p}.
$$

Moreover, for $0<\tau<t_1$, we have

$$
\int_{t_1-\tau}^{t_2-\tau}\omega^{\alpha q-2}d\omega=\frac{1}{1-\alpha q}\frac{(t_2-\tau)^{1-\alpha q}-(t_1-\tau)^{1-\alpha q}}{(t_1-\tau)^{1-\alpha q}(t_2-\tau)^{1-\alpha q}},
$$

where we note that the estimates $(t_2-\tau)^{1-\alpha q}-(t_1-\tau)^{1-\alpha q}\le(t_2-t_1)^{1-\alpha q}$ and $(t_2-\tau)^{1-\alpha q}\ge(t_1-\tau)^s(t_2-t_1)^{1-\alpha q-s}$ can be showed easily from $0<\alpha q<1$ and $1-\alpha q-s>0$. Hence, we deduce

$$
\begin{aligned}
\|\mathscr{I}_1\|\le&\int_0^{t_1}\left\|\sum_{j=1}^{\infty}\int_{t_1-\tau}^{t_2-\tau}F_j(\tau)\omega^{\alpha-2}E_{\alpha,\alpha-1}(-m_j^\beta\omega^\alpha)d\omega e_j\right\|d\tau\\
\le&\widehat{c}_\alpha m_1^{-\beta p}\int_0^{t_1}\int_{t_1-\tau}^{t_2-\tau}\omega^{\alpha q-2}d\omega\,\|F(\tau,\cdot)\|\,d\tau\\
\le&\frac{\widehat{c}_\alpha m_1^{-\beta p}}{1-\alpha q}\int_0^{t_1}\|F(\tau,\cdot)\|\,(t_1-\tau)^{\alpha q-s-1}d\tau(t_2-t_1)^s.
\end{aligned}
$$

This leads to

$$\|\mathscr{I}_1\| \le M_9 |||F|||_{\mathscr{X}_{2,\alpha q-s}} (t_2 - t_1)^s, \tag{3.62}$$

where we let $M_9 = \frac{\widehat{c}_\alpha m_1^{-\beta p}}{1-\alpha q}$. Secondly, an estimate for the term $\mathscr{I}_2$ can be shown by using (3.46) as follows:

$$\begin{aligned}\|\mathscr{I}_2\| \le& \int_{t_1}^{t_2} \left\| \sum_{j=1}^{\infty} F_j(\tau) E_{\alpha,\alpha}(-m_j^\beta (t_2-\tau)^\alpha) e_j \right\| (t_2-\tau)^{\alpha-1} d\tau \\ \le& \widehat{c}_\alpha \int_{t_1}^{t_2} \|F(\tau,\cdot)\| (t_2-\tau)^{\alpha q-s-1} (t_2-\tau)^{\alpha p+s} d\tau \\ \le& M_{10} |||F|||_{\mathscr{X}_{2,\alpha q-s}} (t_2-t_1)^s,\end{aligned} \tag{3.63}$$

where we let $M_{10} = \widehat{c}_\alpha T^{\alpha p}$. Thirdly, we will estimate the term $\mathscr{I}_3$. We have

$$\|\mathscr{I}_3\| = \left\| \mathscr{L}^\beta \sum_{j=1}^{\infty} \varphi_j \int_{t_1}^{t_2} \frac{\widetilde{E}_{\alpha,\alpha}(-m_j^\beta \omega^\alpha)}{E_{\alpha,1}(-m_j^\beta T^\alpha)} d\omega e_j \right\|.$$

Here, the fraction can be estimated as follows:

$$\left| \frac{\widetilde{E}_{\alpha,\alpha}(-m_j^\beta \omega^\alpha)}{E_{\alpha,1}(-m_j^\beta T^\alpha)} \right| \le \widehat{c}_\alpha c_\alpha^{-1} \left[\frac{1+m_j^\beta T^\alpha}{1+m_j^\beta \omega^\alpha} \right]^p \left[\frac{1+m_j^\beta T^\alpha}{1+m_j^{2\beta} \omega^{2\alpha}} \right]^q \omega^{\alpha-1}, \tag{3.64}$$

by using (3.46) and $\widetilde{E}_{\alpha,\alpha}(-m_j^\beta \omega^\alpha) = E_{\alpha,\alpha}(-m_j^\beta \omega^\alpha)\omega^{\alpha-1}$. Moreover, we can see that

$$\frac{1+m_j^\beta T^\alpha}{1+m_j^{2\beta} \omega^{2\alpha}} \le (m_1^{-\beta} + T^\alpha) m_j^\beta m_j^{-2\beta} \omega^{-2\alpha} = (m_1^{-\beta} + T^\alpha) m_j^{-\beta} \omega^{-2\alpha}. \tag{3.65}$$

Taking these estimates together, we thus obtain the following chain of the inequalities:

$$\begin{aligned}\|\mathscr{I}_3\| \le& \left\{ \sum_{j=1}^{\infty} \varphi_j^2 m_j^{2\beta} \left[\int_{t_1}^{t_2} \left| \frac{\widetilde{E}_{\alpha,\alpha}(-m_j^\beta \omega^\alpha)}{E_{\alpha,1}(-m_j^\beta T^\alpha)} \right| d\omega \right]^2 \right\}^{1/2} \\ \le& M_{11} \left\{ \sum_{j=1}^{\infty} \varphi_j^2 m_j^{2\beta} \left[\int_{t_1}^{t_2} \omega^{-\alpha p} m_j^{-\beta q} \omega^{-2\alpha q} \omega^{\alpha-1} d\omega \right]^2 \right\}^{1/2}\end{aligned}$$

$$=M_{11}\left\{\sum_{j=1}^{\infty}\varphi_j^2 m_j^{2\beta p}\left[\int_{t_1}^{t_2}\omega^{-\alpha q-1}d\omega\right]^2\right\}^{1/2},$$

which implies that

$$\|\mathscr{I}_3\| \le M_{12}t_1^{-2\alpha q}(t_2-t_1)^{\alpha q}\|\varphi\|_{\mathbf{V}_{\beta p}}, \tag{3.66}$$

where $M_{11}=\widehat{c}_\alpha c_\alpha^{-1}T^{\alpha p}(m_1^{-\beta}+T^\alpha)^q$ and $M_{12}=M_{11}[\alpha q]^{-1}$. Fourthly, we proceed to estimate $\mathscr{I}_4$. According to (3.64), we have $\left|\frac{\widetilde{E}_{\alpha,\alpha}(-m_j^\beta\omega^\alpha)}{E_{\alpha,1}(-m_j^\beta T^\alpha)}\right| \le M_{11}m_j^{-\beta q}\omega^{-\alpha q-1}$. Moreover, $\widetilde{E}_{\alpha,\alpha}(-m_j^\beta(T-\tau)^\alpha)\le\widehat{c}_\alpha m_j^{-\beta p}(T-\tau)^{\alpha q-1}$ can be established by using the inequalities (3.46). Hence, we obtain

$$\begin{aligned}
&\|\mathscr{I}_4\|\\
\le&\int_0^T\left\|\mathscr{L}^\beta\sum_{j=1}^{\infty}\int_{t_1}^{t_2}F_j(\tau)\widetilde{E}_{\alpha,\alpha}(-m_j^\beta(T-\tau)^\alpha)\frac{\widetilde{E}_{\alpha,\alpha}(-m_j^\beta\omega^\alpha)}{E_{\alpha,1}(-m_j^\beta T^\alpha)}d\omega e_j\right\|d\tau\\
\le&\int_0^T\left\{\sum_{j=1}^{\infty}m_j^{2\beta}F_j^2(\tau)\left[\int_{t_1}^{t_2}\left|\widetilde{E}_{\alpha,\alpha}(-m_j^\beta(T-\tau)^\alpha)\frac{\widetilde{E}_{\alpha,\alpha}(-m_j^\beta\omega^\alpha)}{E_{\alpha,1}(-m_j^\beta T^\alpha)}\right|d\omega\right]^2\right\}^{1/2}d\tau\\
\le&\widehat{c}_\alpha M_{11}\int_0^T\left\{\sum_{j=1}^{\infty}m_j^{2\beta}F_j^2(\tau)\left[\int_{t_1}^{t_2}m_j^{-\beta p}(T-\tau)^{\alpha q-1}m_j^{-\beta q}\omega^{-\alpha q-1}d\omega\right]^2\right\}^{1/2}d\tau\\
=&\widehat{c}_\alpha M_{12}\frac{t_2^{\alpha q}-t_1^{\alpha q}}{t_1^{2\alpha q}}\int_0^T\|F(\tau,\cdot)\|(T-\tau)^{\alpha q-1}d\tau,
\end{aligned}$$

and we arrive at

$$\|\mathscr{I}_4\|\le\widehat{c}_\alpha M_{13}t_1^{-2\alpha q}(t_2-t_1)^{\alpha q}|||F|||_{\mathscr{X}_{2,\alpha q-s}}, \tag{3.67}$$

where $M_{13}=M_{12}T^s$.

We deduce from (3.62), (3.63), (3.66), and (3.67) that $\left\|\sum_{1\le j\le 4}\mathscr{I}_j\right\|$ tends to 0 as t_2-t_1 tends to 0 for $0<t_1<t_2\le T$. Thus, u belongs to the set $C((0,T];L^2(D))$. On the other hand, by assumption (R3), we have $0<\alpha q-s<\alpha q$ and $\mathscr{X}_{2,\alpha q-s}(J\times D)\subset\mathscr{X}_{2,\alpha q}(J\times D)$. Therefore, the assumptions on φ and F in this theorem also fulfill Lemma 3.1. Hence, the inequality (3.49) holds, i.e.,

$$t^{\alpha q}\|u(t,\cdot)\|\le C_0\left(\|\varphi\|_{\mathbf{V}_{\beta p}}+|||F|||_{\mathscr{X}_{2,\alpha q}}\right),\quad t>0. \tag{3.68}$$

This implies u belongs to $C_w^{\alpha q}((0,T];L^2(D))$. Moreover, by taking the supremum on both sides of (3.68) on $(0,T]$, we obtain

$$\|u\|_{C_w^{\alpha q}((0,T];L^2(D))} \le C_0 \|\varphi\|_{\mathbf{V}_{\beta p}} + T^s C_0 |||F|||_{\mathscr{X}_{2,\alpha q-s}}. \tag{3.69}$$

Step 3. We prove $u \in C^s([0,T];\mathbf{V}_{-\beta q'})$.

In this step, we establish the continuity of the solution on the closed interval $[0,T]$. Now, we consider $0 \le t_1 < t_2 \le T$. If $t_1 = 0$, then $\mathscr{I}_1 = 0$. If $t_1 > 0$, then combining (3.37) in the same way as in (3.62) gives

$$\begin{aligned}
&\|\mathscr{I}_1\|_{\mathbf{V}_{-\beta q'}} \\
&\le \int_0^{t_1} \left\| \sum_{j=1}^{\infty} \int_{t_1-\tau}^{t_2-\tau} F_j(\tau)\omega^{\alpha-2} E_{\alpha,\alpha-1}(-m_j^\beta \omega^\alpha) d\omega e_j \right\|_{\mathbf{V}_{-\beta q'}} d\tau \\
&\le \int_0^{t_1} \left\{ \sum_{j=1}^{\infty} m_j^{-2\beta q'} \left(\int_{t_1-\tau}^{t_2-\tau} F_j(\tau)\omega^{\alpha-2} E_{\alpha,\alpha-1}(-m_j^\beta \omega^\alpha) d\omega e_j, e_j \right)^2_{-\beta q',\beta q'} \right\}^{1/2} d\tau \\
&\le \int_0^{t_1} \left\{ \sum_{j=1}^{\infty} m_j^{-2\beta q'} F_j^2(\tau) \left| \int_{t_1-\tau}^{t_2-\tau} \omega^{\alpha-2} \left| E_{\alpha,\alpha-1}(-m_j^\beta \omega^\alpha) \right| d\omega \right|^2 \right\}^{1/2} d\tau
\end{aligned}$$

and so

$$\|\mathscr{I}_1\|_{\mathbf{V}_{-\beta q'}} \le \frac{\widehat{c}_\alpha m_1^{-\beta q'-\beta p}}{1-\alpha q} (t_2-t_1)^s |||F|||_{\mathscr{X}_{2,\alpha q-s}}. \tag{3.70}$$

On the other hand, the inequality (3.63) also holds for all $0 \le t_1 < t_2 \le T$. Hence, the same way as in the proof (3.63) shows that

$$\begin{aligned}
\|\mathscr{I}_2\|_{\mathbf{V}_{-\beta q'}} &\le \int_{t_1}^{t_2} \left\| \sum_{j=1}^{\infty} F_j(\tau) E_{\alpha,\alpha}(-m_j^\beta (t_2-\tau)^\alpha) e_j \right\|_{\mathbf{V}_{-\beta q'}} (t_2-\tau)^{\alpha-1} d\tau \\
&\le m_1^{-\beta q'} \widehat{c}_\alpha (t_2-t_1)^{\alpha p+s} |||F|||_{\mathscr{X}_{2,\alpha q-s}} \\
&\le m_1^{-\beta q'} \widehat{c}_\alpha T^{\alpha p} (t_2-t_1)^s |||F|||_{\mathscr{X}_{2,\alpha q-s}}.
\end{aligned}$$

Now, we will establish estimates for $\|\mathscr{I}_3\|_{\mathbf{V}_{-\beta q'}}$ and $\|\mathscr{I}_4\|_{\mathbf{V}_{-\beta q'}}$. Indeed, we have

$$\left| \frac{\widetilde{E}_{\alpha,\alpha}(-m_j^\beta \omega^\alpha)}{E_{\alpha,1}(-m_j^\beta T^\alpha)} \right| \le \widehat{c}_\alpha c_\alpha^{-1} \left[\frac{1+m_j^\beta T^\alpha}{1+m_j^\beta \omega^\alpha} \right]^{p-p'} \left[\frac{1+m_j^\beta T^\alpha}{1+m_j^\beta \omega^\alpha} \right]^{q+p'} \omega^{\alpha-1},$$

$$\leq \widehat{c}_\alpha c_\alpha^{-1}(m_1^{-\beta} + T^\alpha)^{p-p'} m_j^{\beta(p-p')} T^{\alpha(q+p')} \omega^{-\alpha(q+p')} \omega^{\alpha-1}.$$

Thus, we can derive

$$\left| \frac{\widetilde{E}_{\alpha,\alpha}(-m_j^\beta \omega^\alpha)}{E_{\alpha,1}(-m_j^\beta T^\alpha)} \right| \leq M_{14} m_j^{\beta(p-p')} \omega^{\alpha(p-p')-1}, \tag{3.71}$$

where we let

$$M_{14} = \widehat{c}_\alpha c_\alpha^{-1}(m_1^{-\beta} + T^\alpha)^{p-p'} T^{\alpha(q+p')}.$$

This leads to

$$\begin{aligned}
\|\mathscr{I}_3\|_{\mathbf{V}_{-\beta q'}} \leq & \left\{ \sum_{j=1}^\infty \varphi_j^2 m_j^{2\beta} m_j^{-2\beta q'} \left[\int_{t_1}^{t_2} \left| \frac{\widetilde{E}_{\alpha,\alpha}(-m_j^\beta \omega^\alpha)}{E_{\alpha,1}(-m_j^\beta T^\alpha)} \right| d\omega \right]^2 \right\}^{1/2} \\
\leq & M_{14} \left\{ \sum_{j=1}^\infty \varphi_j^2 m_j^{2\beta} m_j^{-2\beta q'} \left[\int_{t_1}^{t_2} m_j^{\beta(p-p')} \omega^{\alpha(p-p')-1} d\omega \right]^2 \right\}^{1/2} \\
\leq & M_{14} \left\{ \sum_{j=1}^\infty \varphi_j^2 m_j^{2\beta p} \left[\int_{t_1}^{t_2} \omega^{\alpha(p-p')-1} d\omega \right]^2 \right\}^{1/2}.
\end{aligned}$$

Thus, by letting $M_{15} = \frac{M_{14}}{\alpha(p-p')} T^{\alpha(p-p')-s}$, we obtain the estimate

$$\begin{aligned}
\|\mathscr{I}_3\|_{\mathbf{V}_{-\beta q'}} \leq & \frac{M_{14}}{\alpha(p-p')} \|\varphi\|_{\mathbf{V}_{\beta p}} \left(t_2^{\alpha(p-p')} - t_1^{\alpha(p-p')} \right) \\
\leq & M_{15} \|\varphi\|_{\mathbf{V}_{\beta p}} (t_2 - t_1)^s,
\end{aligned} \tag{3.72}$$

where we have used that

$$t_2^{\alpha(p-p')} - t_1^{\alpha(p-p')} \leq (t_2 - t_1)^{\alpha(p-p')} \leq T^{\alpha(p-p')-s} (t_2 - t_1)^s,$$

for $p' \leq p - \frac{s}{\alpha}$. By employing (3.71) and following the same way as in (3.72), we have

$$\begin{aligned}
& \|\mathscr{I}_4\|_{\mathbf{V}_{-\beta q'}} \\
\leq & \int_0^T \left\| \mathscr{L}^\beta \sum_{j=1}^\infty \int_{t_1}^{t_2} F_j(\tau) \widetilde{E}_{\alpha,\alpha}(-m_j^\beta (T-\tau)^\alpha) \frac{\widetilde{E}_{\alpha,\alpha}(-m_j^\beta \omega^\alpha)}{E_{\alpha,1}(-m_j^\beta T^\alpha)} d\omega e_j \right\|_{\mathbf{V}_{-\beta q'}} d\tau
\end{aligned}$$

$$\leq \int_0^T \left\{ \sum_{j=1}^{\infty} m_j^{2\beta p'} F_j^2(\tau) \left[\int_{t_1}^{t_2} \left| \widetilde{E}_{\alpha,\alpha}(-m_j^\beta (T-\tau)^\alpha) \frac{\widetilde{E}_{\alpha,\alpha}(-m_j^\beta \omega^\alpha)}{E_{\alpha,1}(-m_j^\beta T^\alpha)} \right| d\omega \right]^2 \right\}^{1/2} d\tau$$

$$\leq \widehat{c}_\alpha M_{14} \int_0^T \left\{ \sum_{j=1}^{\infty} m_j^{2\beta p'} F_j^2(\tau) \left[\int_{t_1}^{t_2} (T-\tau)^{\alpha q-1} m_j^{-\beta p'} \omega^{\alpha(p-p')-1} d\omega \right]^2 \right\}^{1/2} d\tau$$

$$\leq \widehat{c}_\alpha \frac{M_{14}}{\alpha(p-p')} \left(t_2^{\alpha(p-p')} - t_1^{\alpha(p-p')} \right) \int_0^T \|F(\tau,\cdot)\| (T-\tau)^{\alpha q-1} d\tau,$$

where

$$|\widetilde{E}_{\alpha,\alpha}(-m_j^\beta (T-\tau)^\alpha)| \leq m_j^{-\beta p} (T-\tau)^{\alpha q-1}.$$

This implies that

$$\|\mathscr{I}_4\|_{\mathbf{V}_{-\beta q'}} \leq M_{16}(t_2-t_1)^s |||F|||_{\mathscr{X}_{2,\alpha q-s}}, \tag{3.73}$$

where $M_{16} = \widehat{c}_\alpha T^s M_{15}$. Combining the above arguments guarantees that u belongs to $C^s([0,T];\mathbf{V}_{-\beta q'})$. Moreover, there also exists a positive constant M_{17} such that

$$|||u|||_{C^s([0,T];\mathbf{V}_{-\beta q'})} \leq M_{17} \|\varphi\|_{\mathbf{V}_{\beta p}} + M_{17}|||F|||_{\mathscr{X}_{2,\alpha q-s}}. \tag{3.74}$$

Finally, the inequality (3.56) is obtained by taking the inequalities (3.60), (3.69), and (3.74) together. The proof is completed.

In the next theorem, we will investigate the time-space fractional derivative of the mild solution u. More specifically, we investigate ${}_0^C D_t^\alpha \mathscr{L}^{-\beta(q-\widehat{q})} u$, for a suitably chosen number $\widehat{q} \leq q$. We also establish the continuity of ${}_0^C D_t^\alpha \mathscr{L}^{-\beta q} u$ on the interval $(0,T]$ without establishing it at $t=0$ since this requires a strong assumption of F, for example, F must be continuous on whole interval $[0,T]$.

Theorem 3.6 *Let $p, q, s, p', q', \widehat{p}, \widehat{q}, r, \widehat{r}$ be defined by (R1), (R3), (R4), (R5).*

(a) *If $\varphi \in \mathbf{V}_{\beta(p+\widehat{q})}$ and $F \in L^{\frac{1}{\alpha q-s}+\widehat{r}}(0,T;L^2(D))$, then FVP (3.47) has a unique solution u such that*

$$u \in L^{\frac{1}{\alpha q'}-r}(0,T;\mathbf{V}_{\beta(p-p')}) \cap C_w^{\alpha q}((0,T];L^2(D)) \cap C^s([0,T];\mathbf{V}_{-\beta q'}),$$

$${}_0^C D_t^\alpha u \in L^{\frac{1}{\alpha}-\widehat{r}}(0,T;\mathbf{V}_{-\beta(q-\widehat{q})}).$$

Moreover, there exists a positive constant C_3 such that

$$\|{}_0^C D_t^\alpha u\|_{L^{\frac{1}{\alpha}-\widehat{r}}(0,T;\mathbf{V}_{-\beta(q-\widehat{q})})} \leq C_3 \|F\|_{L^{\frac{1}{\alpha q-s}+\widehat{r}}(0,T;L^2(D))} + C_3 \|\varphi\|_{\mathbf{V}_{\beta(p+\widehat{q})}}. \tag{3.75}$$

(b) *If* $\varphi \in \mathbf{V}_{\beta(p+\widehat{q})}$ *and* $F \in L^{\frac{1}{\alpha q-s}+\widehat{r}}(0,T;L^2(D)) \cap C_w^\alpha((0,T];\mathbf{V}_{-\beta q})$*, then FVP (3.47) has a unique solution u such that*

$$u \in L^{\frac{1}{\alpha q'}-r}(0,T;\mathbf{V}_{\beta(p-p')}) \cap C_w^{\alpha q}((0,T];L^2(D)) \cap C^s([0,T];\mathbf{V}_{-\beta q'}),$$

$${}_0^C D_t^\alpha u \in L^{\frac{1}{\alpha}-\widehat{r}}(0,T;\mathbf{V}_{-\beta(q-\widehat{q})}) \cap C_w^\alpha((0,T];\mathbf{V}_{-\beta q}).$$

Moreover, there exists a positive constant C_4 *such that*

$$\|{}_0^C D_t^\alpha u\|_{L^{\frac{1}{\alpha}-\widehat{r}}(0,T;\mathbf{V}_{-\beta(q-\widehat{q})})} + \|{}_0^C D_t^\alpha u\|_{C_w^\alpha((0,T];\mathbf{V}_{-\beta q})}$$
$$\leq C_4 \|\varphi\|_{\mathbf{V}_{\beta(p+\widehat{q})}} + C_4 \|F\|_{L^{\frac{1}{\alpha q-s}+\widehat{r}}(0,T;L^2(D))} + C_4 \|F\|_{C_w^\alpha((0,T];\mathbf{V}_{-\beta q})}.$$

Proof (a) By (R3), we have $0 < \alpha q - s < 1$, and $\frac{1}{\alpha q-s} + \widehat{r} > \frac{1}{\alpha q-s} > 1$. Thus, one can deduce from (3.40) that

$$L^{\frac{1}{\alpha q-s}+\widehat{r}}(0,T;L^2(D)) \subset \mathscr{X}_{2,\alpha q-s}(J \times D), \tag{3.76}$$

where we have used the inclusion (3.39). Moreover, the Sobolev embedding $\mathbf{V}_{\beta(p+\widehat{q})} \hookrightarrow \mathbf{V}_{\beta p}$ holds. Therefore, the assumptions of this theorem also fulfill part (b) of Theorem 3.5. Hence, FVP (3.47) has a unique solution

$$u \in L^{\frac{1}{\alpha q'}-r}(0,T;\mathbf{V}_{\beta(p-p')}) \cap C_w^{\alpha q}((0,T];L^2(D)) \cap C^s([0,T];\mathbf{V}_{-\beta q'}).$$

Now, we prove that ${}_0^C D_t^\alpha u$ exists and belongs to $L^{\frac{1}{\alpha}-\widehat{r}}(0,T;\mathbf{V}_{-\beta(q-\widehat{q})})$. It follows from the identities

$${}_0^C D_t^\alpha E_{\alpha,1}(-m_j^\beta t^\alpha) = -m_j^\beta E_{\alpha,1}(-m_j^\beta t^\alpha),$$
$${}_0^C D_t^\alpha \widetilde{E}_{\alpha,\alpha}(-m_j^\beta t^\alpha) = -m_j^\beta \widetilde{E}_{\alpha,\alpha}(-m_j^\beta t^\alpha),$$

see, for example, [52, 180, 187], and Eq. (3.44) that

$${}_0^C D_t^\alpha u_j(t) = {}_0^C D_t^\alpha \left[F_j(t) * \widetilde{E}_{\alpha,\alpha}(-m_j^\beta t^\alpha)\right]$$
$$+ \left[\varphi_j - \left(F_j(r) * \widetilde{E}_{\alpha,\alpha}(-m_j^\beta r^\alpha)\right)\Big|_{r=T}\right] \frac{{}_0^C D_t^\alpha E_{\alpha,1}(-m_j^\beta t^\alpha)}{E_{\alpha,1}(-m_j^\beta T^\alpha)}$$

$$
\begin{aligned}
&= F_j(t) - m_j^\beta F_j(t) * \widetilde{E}_{\alpha,\alpha}(-m_j^\beta t^\alpha) - \varphi_j \frac{m_j^\beta E_{\alpha,1}(-m_j^\beta t^\alpha)}{E_{\alpha,1}(-m_j^\beta T^\alpha)} \\
&\quad + \left(F_j(r) * \widetilde{E}_{\alpha,\alpha}(-m_j^\beta r^\alpha)\right)\Big|_{r=T} \frac{m_j^\beta E_{\alpha,1}(-m_j^\beta t^\alpha)}{E_{\alpha,1}(-m_j^\beta T^\alpha)} \\
&=: F_j(t) + \psi_j^{(1)}(t) + \psi_j^{(2)}(t) + \psi_j^{(3)}(t),
\end{aligned}
$$

for all $j \in \mathbb{N}^+$. Firstly, let us consider the sum $\sum_{n_1 \le j \le n_2} \psi_j^{(1)}(t) e_j$, for $n_1, n_2 \in \mathbb{N}^+$, $1 \le n_1 < n_2$. By the definition of the dual space $\mathbf{V}_{-\beta(q-\widehat{q})}$ of $\mathbf{V}_{\beta(q-\widehat{q})}$ and the identity (3.37) of their dual inner product, we have

$$
\begin{aligned}
&\left\| \sum_{n_1 \le j \le n_2} \psi_j^{(1)}(t) e_j \right\|_{\mathbf{V}_{-\beta(q-\widehat{q})}} \\
&\le \int_0^t \left\| \sum_{n_1 \le j \le n_2} m_j^\beta F_j(\tau) \widetilde{E}_{\alpha,\alpha}(-m_j^\beta (t-\tau)^\alpha) e_j \right\|_{\mathbf{V}_{-\beta(q-\widehat{q})}} d\tau \\
&= \int_0^t \left\{ \sum_{i=1}^\infty m_i^{2\beta(p+\widehat{q})} \left(\sum_{n_1 \le j \le n_2} F_j(\tau) \widetilde{E}_{\alpha,\alpha}(-m_j^\beta (t-\tau)^\alpha) e_j, e_i \right)^2_{-\beta(q-\widehat{q}),\beta(q-\widehat{q})} \right\}^{1/2} d\tau \\
&= \int_0^t \left\{ \sum_{n_1 \le j \le n_2} m_j^{2\beta(p+\widehat{q})} F_j^2(\tau) \widetilde{E}_{\alpha,\alpha}^2(-m_j^\beta (t-\tau)^\alpha) \right\}^{1/2} d\tau.
\end{aligned}
$$

Assumption (R5) shows that $0 < p+\widehat{q} < 1$. Hence, by using the inequalities (3.46), we have $|\widetilde{E}_{\alpha,\alpha}(-m_j^\beta (t-\tau)^\alpha)| \le \widehat{c}_\alpha m_j^{-\beta(p+\widehat{q})} (t-\tau)^{-\alpha(p+\widehat{q})} (t-\tau)^{\alpha-1}$. This together with the above argument gives

$$
\left\| \sum_{n_1 \le j \le n_2} \psi_j^{(1)}(t) e_j \right\|_{\mathbf{V}_{-\beta(q-\widehat{q})}} \le M_{18} t^{-\alpha} \int_0^t (t-\tau)^{\alpha(q-\widehat{q})-1} \left\{ \sum_{n_1 \le j \le n_2} F_j^2(\tau) \right\}^{1/2} d\tau, \tag{3.77}
$$

where $M_{18} = \widehat{c}_\alpha T^\alpha$. Secondly, we proceed to establish an estimate for the sum $\sum_{n_1 \le j \le n_2} \psi_j^{(2)}(t) e_j$. Using the inequality (3.46), the absolute value of $\frac{E_{\alpha,1}(-m_j^\beta t^\alpha)}{E_{\alpha,1}(-m_j^\beta T^\alpha)}$ is bounded by $\widehat{c}_\alpha c_\alpha^{-1} T^\alpha t^{-\alpha}$. Therefore, we derive

$$
\begin{aligned}
\left\| \sum_{n_1 \le j \le n_2} \psi_j^{(2)}(t) e_j \right\|_{\mathbf{V}_{-\beta(q-\widehat{q})}} &= \left\{ \sum_{n_1 \le j \le n_2} m_j^{2\beta(p+\widehat{q})} \varphi_j^2 \frac{E_{\alpha,1}^2(-m_j^\beta t^\alpha)}{E_{\alpha,1}^2(-m_j^\beta T^\alpha)} \right\}^{1/2} \\
&\le \widehat{c}_\alpha c_\alpha^{-1} T^\alpha \left\{ \sum_{n_1 \le j \le n_2} m_j^{2\beta(p+\widehat{q})} \varphi_j^2 t^{-2\alpha} \right\}^{1/2},
\end{aligned}
$$

which shows that

$$\left\| \sum_{n_1 \le j \le n_2} \psi_j^{(2)}(t) e_j \right\|_{\mathbf{V}_{-\beta(q-\widehat{q})}} \le M_{19} t^{-\alpha} \left\{ \sum_{n_1 \le j \le n_2} m_j^{2\beta(p+\widehat{q})} \varphi_j^2 \right\}^{1/2}, \tag{3.78}$$

where $M_{19} = \widehat{c}_\alpha c_\alpha^{-1} T^\alpha$. Thirdly, we proceed to establish an estimate for the sum $\sum_{n_1 \le j \le n_2} \psi_j^{(3)}(t) e_j$. By a similar argument as in (3.78), we have

$$\begin{aligned}
&\left\| \sum_{n_1 \le j \le n_2} \psi_j^{(3)}(t) e_j \right\|_{\mathbf{V}_{-\beta(q-\widehat{q})}} \\
&\le \int_0^T \left\| \sum_{n_1 \le j \le n_2} F_j(\tau) \widetilde{E}_{\alpha,\alpha}(-m_j^\beta (T-\tau)^\alpha) \frac{m_j^\beta E_{\alpha,1}(-m_j^\beta t^\alpha)}{E_{\alpha,1}(-m_j^\beta T^\alpha)} e_j \right\|_{\mathbf{V}_{-\beta(q-\widehat{q})}} d\tau \\
&\le \int_0^T \left\{ \sum_{n_1 \le j \le n_2} m_j^{2\beta(p+\widehat{q})} F_j^2(\tau) \widetilde{E}_{\alpha,\alpha}^2(-m_j^\beta (T-\tau)^\alpha) \frac{E_{\alpha,1}^2(-m_j^\beta t^\alpha)}{E_{\alpha,1}^2(-m_j^\beta T^\alpha)} \right\}^{1/2} d\tau \\
&\le \widehat{c}_\alpha M_{19} t^{-\alpha} \int_0^T \left\{ \sum_{n_1 \le j \le n_2} m_j^{2\beta(p+\widehat{q})} F_j^2(\tau) m_j^{-2\beta(p+\widehat{q})} (T-\tau)^{2\alpha(q-\widehat{q})-2} \right\}^{1/2} d\tau.
\end{aligned}$$

Therefore, we obtain the following estimate:

$$\begin{aligned}
&\left\| \sum_{n_1 \le j \le n_2} \psi_j^{(3)}(t) e_j \right\|_{\mathbf{V}_{-\beta(q-\widehat{q})}} \\
&\le \widehat{c}_\alpha M_{19} t^{-\alpha} \int_0^T (T-\tau)^{\alpha(q-\widehat{q})-1} \left\{ \sum_{n_1 \le j \le n_2} F_j^2(\tau) \right\}^{1/2} d\tau.
\end{aligned} \tag{3.79}$$

For almost every τ in the interval $(0, T)$, by (3.76), we have that $F(\tau, \cdot)$ belongs to $L^2(D)$. This implies $\sum_{1 \le j \le n} F_j(\tau) e_j$ is a Cauchy sequence in $L^2(D)$. This together with the embedding

$$L^2(D) \hookrightarrow \mathbf{V}_{-\beta(q-\widehat{q})}$$

implies that $\sum_{1 \le j \le n} F_j(\tau) e_j$ is also a Cauchy sequence in $\mathbf{V}_{-\beta(q-\widehat{q})}$. On the other hand, it follows from $\varphi \in \mathbf{V}_{\beta(p+\widehat{q})}$ that

$$\lim_{n_1,n_2\to\infty}\sum_{n_1\le j\le n_2}\varphi_j^2 m_j^{2\beta(p+\widehat{q})}=0.$$

By (R5), $0\le\widehat{q}\le\frac{s}{\alpha}$, and we obtain the inclusion $\mathscr{X}_{2,\alpha q-s}(J\times D)\subset\mathscr{X}_{2,\alpha(q-\widehat{q})}(J\times D)$. We deduce that $F\in\mathscr{X}_{2,\alpha(q-\widehat{q})}(J\times D)$, and

$$\lim_{n_1,n_2\to\infty}\int_0^T (T-\tau)^{\alpha(q-\widehat{q})-1}\left\{\sum_{n_1\le j\le n_2}F_j^2(\tau)\right\}^{1/2}d\tau=0,$$

by the dominated convergence theorem. Combining these with the estimates (3.77), (3.78), and (3.79), we have that

$$\lim_{n_1,n_2\to\infty}\left\|\sum_{n_1\le j\le n_2}{}_0^C D_t^\alpha u_j(t)e_j\right\|_{\mathbf{V}_{-\beta(q-\widehat{q})}}=0.$$

Hence $\sum_{j=1}^n {}_0^C D_t^\alpha u_j(t)e_j$ is a Cauchy sequence and a convergent sequence in $\mathbf{V}_{-\beta(q-\widehat{q})}$. We then conclude that ${}_0^C D_t^\alpha u(t,\cdot)=\sum_{j=1}^\infty {}_0^C D_t^\alpha u_j(t)e_j$ finitely exists in the space $\mathbf{V}_{-\beta(q-\widehat{q})}$. Moreover, by taking the inequalities (3.77), (3.78), and (3.79), there exist constants $M_{20}>0$ and $M_{20}'>0$ such that

$$\begin{aligned}&\|{}_0^C D_t^\alpha u(t,\cdot)\|_{\mathbf{V}_{-\beta(q-\widehat{q})}}\\ \le& M_{20}\,\|F(t,\cdot)\|_{\mathbf{V}_{-\beta(q-\widehat{q})}}+M_{20}\left(\|\varphi\|_{\mathbf{V}_{\beta(p+\widehat{q})}}+|||F|||_{\mathscr{X}_{2,\alpha(q-\widehat{q})}}\right)t^{-\alpha}\\ \le& M_{20}'\,\|F(t,\cdot)\|+M_{20}\left(\|\varphi\|_{\mathbf{V}_{\beta(p+\widehat{q})}}+|||F|||_{\mathscr{X}_{2,\alpha(q-\widehat{q})}}\right)t^{-\alpha}.\end{aligned}\tag{3.80}$$

Now, it follows from $0<\widehat{r}\le\frac{1-\alpha}{\alpha}$ and $0<\alpha q-s<\alpha$ that $1\le\frac{1}{\alpha}-\widehat{r}<\frac{1}{\alpha q-s}+\widehat{r}$. This implies the following Sobolev embedding:

$$L^{\frac{1}{\alpha q-s}+\widehat{r}}(0,T;\mathbf{V}_{-\beta(q-\widehat{q})})\hookrightarrow L^{\frac{1}{\alpha}-\widehat{r}}(0,T;\mathbf{V}_{-\beta(q-\widehat{q})}).$$

Moreover, by the assumption (R5), $\widehat{q}<\frac{s}{\alpha}$, we have $\frac{1}{\alpha q-s}+\widehat{r}>\frac{1}{\alpha(q-\widehat{q})}$. This implies that there exists a constant $C_*>0$ such that

$$|||F|||_{\mathscr{X}_{2,\alpha(q-\widehat{q})}}\le C_*\,\|F\|_{L^{\frac{1}{\alpha q-s}+\widehat{r}}(0,T;L^2(D))}.\tag{3.81}$$

Hence, we deduce from (3.80) that there exists a constant $M_{21}>0$ satisfying

$$\|{}_0^C D_t^\alpha u\|_{L^{\frac{1}{\alpha}-\widehat{r}}(0,T;\mathbf{V}_{-\beta(q-\widehat{q})})}\le M_{21}\,\|F\|_{L^{\frac{1}{\alpha q-s}+\widehat{r}}(0,T;L^2(D))}+M_{21}\,\|\varphi\|_{\mathbf{V}_{\beta(p+\widehat{q})}},$$

where we note that $||t^{-\alpha}||_{L^{\frac{1}{\alpha}-\hat{r}}(0,T;\mathbb{R})} < \infty$. The inequality (3.75) is proved by letting $C_3 = M_{21}$.

(b) It is clear that the assumptions of this part also satisfy part (a). Therefore, by part (a), it is necessary to prove ${}_0^C D_t^\alpha u \in C_w^\alpha((0,T];\mathbf{V}_{-\beta q})$, i.e.,

$$\lim_{t_2-t_1\to 0} ||{}_0^C D_t^\alpha u(t_2,\cdot) - {}_0^C D_t^\alpha u(t_1,\cdot)||_{\mathbf{V}_{-\beta q}} = 0, \tag{3.82}$$

where we note $0 < t_1 < t_2 \le T$. After some simple computations, we find that

$${}_0^C D_t^\alpha u(t_2,x) - {}_0^C D_t^\alpha u(t_1,x) = F(t_2,x) - F(t_1,x) + \sum_{1\le n\le 4} \mathscr{J}_n, \tag{3.83}$$

where $\mathscr{J}_n = -\mathscr{L}^\beta \mathscr{I}_n$, and $\mathscr{I}_n$ is defined by (3.61). Since F is in $C_w^\alpha((0,T];\mathbf{V}_{-\beta q})$, we have just to prove $||\mathscr{J}_n||_{\mathbf{V}_{-\beta q}}$ approaches 0 as $t_2 - t_1$ approaches 0. Let us first consider $||\mathscr{J}_1||_{\mathbf{V}_{-\beta q}}$. The inequality (3.46) yields that

$$\begin{aligned}
&||\mathscr{J}_1||_{\mathbf{V}_{-\beta q}} \\
&\le \int_0^{t_1} \left\| \mathscr{L}^\beta \sum_{j=1}^\infty \int_{t_1-\tau}^{t_2-\tau} F_j(\tau)\omega^{\alpha-2} E_{\alpha,\alpha-1}(-m_j^\beta \omega^\alpha) d\omega e_j \right\|_{\mathbf{V}_{-\beta q}} d\tau \\
&\le \int_0^{t_1} \left\{ \sum_{j=1}^\infty m_j^{2\beta} m_j^{-2\beta q} F_j^2(\tau) \left| \int_{t_1-\tau}^{t_2-\tau} \omega^{\alpha-2} \left| E_{\alpha,\alpha-1}(-m_j^\beta \omega^\alpha)\right| d\omega \right|^2 \right\}^{1/2} d\tau \\
&\le \widehat{c_\alpha} \int_0^{t_1} \left\{ \sum_{j=1}^\infty m_j^{2\beta} m_j^{-2\beta q} F_j^2(\tau) \left| \int_{t_1-\tau}^{t_2-\tau} \omega^{\alpha-2} m_j^{-\beta p} \omega^{-\alpha p} d\omega \right|^2 \right\}^{1/2} d\tau.
\end{aligned}$$

We recall that, by (3.76), F belongs to $\mathscr{X}_{2,\alpha q-s}(J\times D)$. Thus, we can deduce from the above inequality that

$$\begin{aligned}
||\mathscr{J}_1||_{\mathbf{V}_{-\beta q}} &\le \widehat{c_\alpha} \int_0^{t_1} \|F(\tau,\cdot)\| \left| \int_{t_1-\tau}^{t_2-\tau} \omega^{\alpha q-2} d\omega \right| d\tau \\
&\le \frac{\widehat{c_\alpha}}{1-\alpha q} |||F|||_{\mathscr{X}_{2,\alpha q-s}} (t_2-t_1)^s,
\end{aligned} \tag{3.84}$$

where we have used the same argument as in the estimate (3.62). Let us secondly consider $||\mathscr{J}_2||_{\mathbf{V}_{-\beta q}}$. We have

$$
\begin{aligned}
||\mathscr{J}_2||_{\mathbf{V}_{-\beta q}} &\leq \int_{t_1}^{t_2} \left\| \mathscr{L}^{\beta} \sum_{j=1}^{\infty} F_j(\tau) E_{\alpha,\alpha}(-m_j^{\beta}(t_2-\tau)^{\alpha}) e_j \right\|_{\mathbf{V}_{-\beta q}} (t_2-\tau)^{\alpha-1} d\tau \\
&\leq \widehat{c}_{\alpha} \int_{t_1}^{t_2} \left\{ \sum_{j=1}^{\infty} m_j^{2\beta} m_j^{-2\beta q} F_j^2(\tau) m_j^{-2\beta p} (t_2-\tau)^{-2\alpha p} \right\}^{1/2} (t_2-\tau)^{\alpha-1} d\tau \\
&\leq \widehat{c}_{\alpha} \int_{t_1}^{t_2} \|F(\tau,\cdot)\| (t_2-\tau)^{\alpha q-s-1} (t_2-\tau)^{s} d\tau \leq \widehat{c}_{\alpha} |||F|||_{\mathscr{X}_{2,\alpha q-s}} (t_2-t_1)^{s},
\end{aligned}
\tag{3.85}
$$

where (3.46) has been used. Thirdly, we consider the norm $||\mathscr{J}_3||_{\mathbf{V}_{-\beta q}}$. It is clear that

$$
\mathscr{J}_3 = \mathscr{L}^{\beta} \mathscr{I}_3 = \mathscr{L}^{2\beta} \sum_{j=1}^{\infty} \varphi_j \int_{t_1}^{t_2} \frac{\widetilde{E}_{\alpha,\alpha}(-m_j^{\beta}\omega^{\alpha})}{E_{\alpha,1}(-m_j^{\beta}T^{\alpha})} d\omega e_j .
$$

Hence, we deduce

$$
||\mathscr{J}_3||_{\mathbf{V}_{-\beta q}} = \left\{ \sum_{j=1}^{\infty} m_j^{4\beta} m_j^{-2\beta q} \varphi_j^2 \left| \int_{t_1}^{t_2} \frac{\widetilde{E}_{\alpha,\alpha}(-m_j^{\beta}\omega^{\alpha})}{E_{\alpha,1}(-m_j^{\beta}T^{\alpha})} d\omega \right|^2 \right\}^{1/2} .
$$

By applying (3.64) and (3.65), we obtain

$$
\begin{aligned}
||\mathscr{J}_3||_{\mathbf{V}_{-\beta q}} &\leq \widehat{c}_{\alpha} c_{\alpha}^{-1} (m_1^{-\beta} + T^{\alpha}) \left\{ \sum_{j=1}^{\infty} m_j^{4\beta} m_j^{-2\beta q} \varphi_j^2 \left| \int_{t_1}^{t_2} m_j^{-\beta} \omega^{-2\alpha} \omega^{\alpha-1} d\omega \right|^2 \right\}^{1/2} \\
&\leq \frac{M_{23}}{\alpha} t_1^{-2\alpha} \|\varphi\|_{\mathbf{V}_{\beta p}} \left(t_2^{\alpha} - t_1^{\alpha} \right),
\end{aligned}
\tag{3.86}
$$

where $M_{23} = \widehat{c}_{\alpha} c_{\alpha}^{-1} (m_1^{-\beta} + T^{\alpha})$. Finally, we can look at $||\mathscr{J}_4||_{\mathbf{V}_{-\beta q}}$ as follows:

$$
\begin{aligned}
&||\mathscr{J}_4||_{\mathbf{V}_{-\beta q}} \\
&\leq \int_0^T \left\| \mathscr{L}^{2\beta} \sum_{j=1}^{\infty} \int_{t_1}^{t_2} F_j(\tau) \widetilde{E}_{\alpha,\alpha}(-m_j^{\beta}(T-\tau)^{\alpha}) \frac{\widetilde{E}_{\alpha,\alpha}(-m_j^{\beta}\omega^{\alpha})}{E_{\alpha,1}(-m_j^{\beta}T^{\alpha})} d\omega e_j \right\|_{\mathbf{V}_{-\beta q}} d\tau \\
&\leq \int_0^T \left\{ \sum_{j=1}^{\infty} m_j^{4\beta} m_j^{-2\beta q} F_j^2(\tau) \left| \int_{t_1}^{t_2} \widetilde{E}_{\alpha,\alpha}(-m_j^{\beta}(T-\tau)^{\alpha}) \frac{\widetilde{E}_{\alpha,\alpha}(-m_j^{\beta}\omega^{\alpha})}{E_{\alpha,1}(-m_j^{\beta}T^{\alpha})} d\omega \right|^2 \right\}^{1/2} d\tau
\end{aligned}
$$

$$\leq M_{24}\int_0^T \left\{ \sum_{j=1}^{\infty} m_j^{4\beta} m_j^{-2\beta q} F_j^2(\tau) \left| \int_{t_1}^{t_2} m_j^{-\beta p}(T-\tau)^{\alpha q-1} m_j^{-\beta}\omega^{-\alpha-1} d\omega \right|^2 \right\}^{1/2} d\tau,$$

where the fraction $\frac{\widetilde{E}_{\alpha,\alpha}(-m_j^{\beta}\omega^{\alpha})}{E_{\alpha,1}(-m_j^{\beta}T^{\alpha})}$ can be estimated in the same way as in the proof of (3.86) and $M_{24} = \widehat{c}_{\alpha} M_{23}$. This leads to

$$||\mathscr{J}_4||_{\mathbf{V}_{-\beta q}} \leq \frac{M_{24}}{\alpha} t_1^{-2\alpha}(t_2^{\alpha} - t_1^{\alpha}) \int_0^T \|F(\tau,\cdot)\| \,(T-\tau)^{\alpha q-1} d\tau,$$

which shows that

$$||\mathscr{J}_4||_{\mathbf{V}_{-\beta q}} \leq \frac{M_{24}}{\alpha} t_1^{-2\alpha} |||F|||_{\mathscr{X}_{2,\alpha q}}(t_2^{\alpha} - t_1^{\alpha}). \tag{3.87}$$

This implies (3.82) by taking (3.83)–(3.87) together. Thus, ${}_0^C D_t^{\alpha} u$ is contained in $C((0,T];\mathbf{V}_{-\beta q})$.

On the other hand, it is easy to see that the estimates (3.77)–(3.79) also hold for $\widehat{q} = 0$. Hence, we deduce from (3.80) and (3.81) that

$$t^{\alpha}||{}_0^C D_t^{\alpha} u(t,\cdot)||_{\mathbf{V}_{-\beta q}} \leq M_{20} t^{\alpha} \, \|F(t,\cdot)\|_{\mathbf{V}_{-\beta q}} + M_{20}\left(\|\varphi\|_{\mathbf{V}_{\beta p}} + C_* \, \|F\|_{L^{\frac{1}{\alpha q - s}+\widehat{r}}(0,T;L^2(D))} \right).$$

Now ${}_0^C D_t^{\alpha} u \in C_w^{\alpha}((0,T];\mathbf{V}_{-\beta q})$. In addition, there exists a positive constant $C' > 0$ such that

$$||{}_0^C D_t^{\alpha} u||_{C_w^{\alpha}((0,T];\mathbf{V}_{-\beta q})} \leq M_{20} \, \|F\|_{C_w^{\alpha}((0,T];\mathbf{V}_{-\beta q})} + C' M_{20} \, \|\varphi\|_{\mathbf{V}_{\beta(p+\widehat{q})}} + C_* M_{20} \, \|F\|_{L^{\frac{1}{\alpha q - s}+\widehat{r}}(0,T;L^2(D))}.$$

We can complete the proof by taking (3.75) and the above inequality together.

3.2.4 *Final Value Problem with a Nonlinear Source*

In this subsection, we study the existence, uniqueness, and regularity of mild solutions of FVP (3.32)–(3.34) corresponding to the nonlinear source function $F(t,x,u(t,x))$. It is suitable considering assumption that $u(t,\cdot)$ and $F(t,\cdot,u(t,\cdot))$ belong to the same spatial space H. In view of most considerations of PDEs, we let $H = L^2(D)$.

We introduce the following assumptions on the numbers $p, q, p', q', \widehat{p}, \widehat{q}, r, \widehat{r}$:

(R1b) $0 < q < p < 1$ such that $p + q = 1$.
(R4b) $0 < p' < p, \quad q' = 1 - p', \quad 0 < r \le \frac{1-\alpha q'}{\alpha q'}$.
(R4c) $0 < p' \le p - q, \quad q' = 1 - p', \quad 0 < r \le \frac{1-\alpha q'}{\alpha q'}$.
(R5b) $0 \le \widehat{q} < q, \quad \widehat{p} = 1 - \widehat{q}, \quad 0 < \widehat{r} \le \frac{1-\alpha}{\alpha}$.

In our work, we will assume on $F(t, \cdot, u(t, \cdot))$ the following assumptions:

(A1) $F(t, \cdot, \mathbf{0}) = \mathbf{0}$, and there exists a constant $K > 0$ such that, for all $v_1, v_2 \in L^2(D)$ and $t \in J$,

$$\|F(t, \cdot, v_1) - F(t, \cdot, v_2)\| \le K \|v_1 - v_2\| .$$

(A2) $F(t, \cdot, \mathbf{0}) = \mathbf{0}$, and there exists a constant $K_* > 0$ such that, for all $v_1, v_2 \in L^2(D)$ and $t_1, t_2 \in J$,

$$||F(t_1, \cdot, v_1) - F(t_2, \cdot, v_2)|| \le K_* (|t_1 - t_2| + ||v_1 - v_2||) .$$

Note that the assumption (A1) and (A2) imply that, for $v \in L^2(D)$,

$$||F(t, \cdot, v)|| \le K \|v\| . \tag{3.88}$$

We try to develop the ideas of the linear FVP (3.47) to deal with the nonlinear FVP (3.32) and (3.34). In Sect. 3.2.3, for the linear function $F(t, x)$, we assume that

$$F \in \mathscr{X}_{2,\alpha q}(J \times D), \text{ or } F \in \mathscr{X}_{2,\alpha q-s}(J \times D), \text{ or } F \in L^{\frac{1}{\alpha q - s}+\widehat{r}}(0, T; L^2(D)), \tag{3.89}$$

where $p, q, s, \widehat{r}$ are defined by (R1), (R3), (R5). However, we cannot suppose that the nonlinear source function $F(t, x, u(t, x))$ satisfies the same assumptions as in (3.89) and then find the solution u. A natural idea might be to combine the idea of Lemma 3.1 with the inequality (3.88), i.e., we predict the solution u may be contained in the set

$$\mathbf{W}^{\rho}_{\gamma,\eta}(J \times D) := \left\{ w \in \mathscr{X}_{2,\eta}(J \times D) : \|w(t, \cdot)\| \le \rho t^{-\gamma}, \text{ for } 0 < t \le T \right\},$$

for $\rho > 0, 0 < \gamma \le \eta < 1$.

The prediction will be proved in the next lemma. However, it is necessary to give some useful notes on $\mathbf{W}^{\rho}_{\gamma,\eta}(J \times D)$ as follows. For $w \in \mathbf{W}^{\rho}_{\gamma,\eta}(J \times D)$, we see

$$\text{esssup}_{0 \le t \le T} \int_0^t \|w(\tau, \cdot)\| (t - \tau)^{\eta-1} d\tau \le \rho \text{esssup}_{0 \le t \le T} \int_0^t \tau^{-\gamma} (t - \tau)^{\eta-1} d\tau.$$

The function $\tau \to \tau^{-\gamma}(t-\tau)^{\eta-1}$ is integrable on $(0, t)$ since both numbers $-\gamma$ and $\eta-1$ are greater than -1. In addition, we have $\int_0^t \tau^{-\gamma}(t-\tau)^{\eta-1}d\tau = t^{\eta-\gamma}B(\eta, 1-\gamma)$, where $B(\cdot,\cdot)$ is the Beta function, see, for example, [52, 180, 187]. Hence, we have

$$|||w|||_{\mathscr{X}_{2,\eta}} \le \rho T^{\eta-\gamma}B(\eta, 1-\gamma). \tag{3.90}$$

Moreover, if $\gamma < \eta$, then there always exists a real number p such that $1 < \frac{1}{\eta} < p < \frac{1}{\gamma}$. This implies that the function $t^{-\gamma}$ belongs to $L^p(0, T; L^2(D))$. Therefore, we can obtain the following inclusions:

$$\mathbf{W}^{\rho}_{\gamma,\eta}(J \times D) \subset L^p(0, T; L^2(D)) \subset \mathscr{X}_{2,\eta}(J \times D).$$

In the following lemma, we will consider the case $\gamma = \eta$, which we will denote by $\mathbf{W}^{\rho}_{\gamma}(J \times D) := \mathbf{W}^{\rho}_{\gamma,\eta}(J \times D)$.

Now, the Sobolev embedding $\mathbf{V}_{\beta p} \hookrightarrow L^2(D)$ shows that there exists a positive constant C_D depending on D, β, q such that $\|v\| \le C_D \|v\|_{\mathbf{V}_{\beta p}}$ for all $v \in \mathbf{V}_{\beta p}$. In this subsection, we let

$$k_0(T) = KB(\alpha q, 1-\alpha q)T^{\alpha q}\left[\widehat{c}_\alpha m_1^{-\beta p} + \widehat{c}_\alpha^2 c_\alpha^{-1}(m_1^{-\beta} + T^\alpha)^p\right]$$

and

$$M_0 = C_D T^{\alpha q} + \widehat{c}_\alpha c_\alpha^{-1} T^{\alpha q}(m_1^{-\beta} + T^\alpha)^p.$$

Lemma 3.2 *Let p, q be defined by (R1) and ϕ be a function on D. Let $\{w_{(n)}\}_{n\ge 0}$ be defined by $w_{(0)} = \phi$,*

$$\begin{aligned} w_{(n)}(t,x) &= (\mathscr{O}w_{(n-1)})(t,x) \\ &= (\mathscr{O}_1 F(w_{(n-1)}))(t,x) + \mathscr{O}_2(t)\phi(x) + (\mathscr{O}_3 F(w_{(n-1)}))(t,x), \quad n \in \mathbb{N}^+. \end{aligned} \tag{3.91}$$

If ϕ belongs to $\mathbf{V}_{\beta p}$, F satisfies (A1), and $k_0(T) < 1$, then

$$\{w_{(n)}\}_{n\ge 0} \subset \mathbf{W}^{\widehat{C}_0}_{\alpha q}(J \times D), \tag{3.92}$$

where $\widehat{C}_0 := \widetilde{C}_0 \|\phi\|_{\mathbf{V}_{\beta p}}$ and $\widetilde{C}_0 = \frac{M_0}{1-k_0(T)}$.

Proof First, we have

$$||w_{(0)}|| = ||\phi|| \le C_D||\phi||_{\mathbf{V}_{\beta p}} \le M_0 \|\phi\|_{\mathbf{V}_{\beta p}} t^{-\alpha q} \le \widehat{C}_0 t^{-\alpha q}.$$

Hence, inequality (3.90) and $\alpha q < 1$ imply $w_{(0)} \in \mathbf{W}_{\alpha q}^{\widehat{C}_0}(J \times D)$. Now, we assume that $w_{(n-1)}$ belongs to $\mathbf{W}_{\alpha q}^{\widehat{C}_0}(J \times D)$ for some $n \geq 1$. Then, by using (3.90), we have

$$|||w_{(n-1)}|||_{\mathscr{X}_{2,\alpha q}} \leq \widehat{C}_0 B(\alpha q, 1-\alpha q). \tag{3.93}$$

By induction, the inclusion (3.92) will be proved by showing that $w_{(n)}$ belongs to $\mathbf{W}_{\alpha q}^{\widehat{C}_0}(J \times D)$. Indeed, by using the same arguments as in the proof of (3.52), we have

$$\begin{aligned} \|(\mathscr{O}_1 F(w_{(n-1)}))(t,\cdot)\| &\leq \widehat{c}_\alpha m_1^{-\beta p} \int_0^t \|F(\tau,\cdot,w_{(n-1)}(\tau,\cdot))\|(t-\tau)^{\alpha q-1} d\tau, \\ &\leq K\widehat{c}_\alpha m_1^{-\beta p} |||w_{(n-1)}|||_{\mathscr{X}_{2,\alpha q}} \leq M_{26}\widehat{C}_0 t^{-\alpha q}, \end{aligned} \tag{3.94}$$

where we have used (3.88) and (3.93), and let $M_{26} = K\widehat{c}_\alpha m_1^{-\beta p} B(\alpha q, 1-\alpha q) T^{\alpha q}$. On the other hand, the norm $\|\mathscr{O}_2(t)\phi\|$ is estimated by (3.53), i.e.,

$$\|\mathscr{O}_2(t)\phi\| \leq M_0 \|\phi\|_{\mathbf{V}_{\beta p}} t^{-\alpha q},$$

where we note that

$$M_2 = \widehat{c}_\alpha c_\alpha^{-1} T^{\alpha q}(m_1^{-\beta} + T^\alpha)^p \leq M_0.$$

The norm $\|(\mathscr{O}_3 F(w_{(n-1)}))(t,\cdot)\|$ can be estimated in the same way as in the proof of (3.54); that is,

$$\begin{aligned} \|(\mathscr{O}_3 F(w_{(n-1)}))(t,\cdot)\| &\leq M_3 t^{-\alpha q} \int_0^T \|F(\tau,\cdot,w_{(n-1)}(\tau,\cdot))\|(T-\tau)^{\alpha q-1} d\tau \\ &\leq K M_3 t^{-\alpha q} |||w_{(n-1)}|||_{\mathscr{X}_{2,\alpha q}} \leq M_{27}\widehat{C}_0 t^{-\alpha q}, \end{aligned} \tag{3.95}$$

where (3.88) and (3.93) have been used. Here,

$$M_{27} = K M_3 B(\alpha q, 1-\alpha q), \quad M_3 = \widehat{c}_\alpha^2 c_\alpha^{-1} T^{\alpha q}(m_1^{-\beta} + T^\alpha)^p.$$

We deduce from the above arguments and $w_{(n)}(t,\cdot) = (\mathscr{O} w_{(n-1)})(t,\cdot)$ that

$$\begin{aligned} \|w_{(n)}(t,\cdot)\| &\leq \|(\mathscr{O}_1 F(w_{(n-1)}))(t,\cdot)\| + \|\mathscr{O}_2(t)\phi\| + \|(\mathscr{O}_3 F(w_{(n-1)}))(t,\cdot)\| \\ &\leq k_0(T)\widehat{C}_0 t^{-\alpha q} + M_0 \|\phi\|_{\mathbf{V}_{\beta p}} t^{-\alpha q}, \end{aligned} \tag{3.96}$$

by noting that $k_0(T) = M_{26} + M_{27}$. Since $\widehat{C}_0 = \frac{M_0}{1-k_0(T)}\|\phi\|_{\mathbf{V}_{\beta p}}$, the above inequality implies that $\|w_{(n)}(t,\cdot)\| \le \widehat{C}_0 t^{-\alpha q}$. Therefore, from $\alpha q < 1$, we obtain the inclusion (3.92).

Next, it is necessary to give a definition of mild solutions of FVP (3.32) and (3.34).

Definition 3.4 Let F be defined by (A1) or (A2). If a function u belongs to $L^{p_2}(0,T;L^{p_1}(D))$, for some $p_1, p_2 \ge 1$, and satisfies equation (3.45) where F stands for the nonlinear source function $F(t,x,u(t,x))$, then u is said to be a mild solution of FVP (3.32) and (3.34).

The following theorem presents existence, uniqueness, and regularity of a mild solution of FVP (3.32) and (3.34).

Theorem 3.7

(a) *Let p, q, r, p', q' be defined by (R1), (R4b). If φ belongs to $\mathbf{V}_{\beta p}$, F satisfies (A1), and $k_0(T) < 1$, then FVP* (3.32)–(3.34) *has a unique solution*

$$u \in L^{\frac{1}{\alpha q'}-r}(0,T;\mathbf{V}_{\beta(p-p')}) \cap C_w^{\alpha q}((0,T];L^2(D)),$$

and there exists a positive constant C_5 such that

$$\|u\|_{L^{\frac{1}{\alpha q'}-r}(0,T;\mathbf{V}_{\beta(p-p')})} + \|u\|_{C_w^{\alpha q}((0,T];L^2(D))} \le C_5\,\|\varphi\|_{\mathbf{V}_{\beta p}}.$$

(b) *Let p, q, r, p', q' be defined by (R1b), (R4c). If φ belongs to $\mathbf{V}_{\beta p}$, F satisfies (A1), and $k_0(T) < 1$, then FVP* (3.32) *and* (3.34) *has a unique solution*

$$u \in L^{\frac{1}{\alpha q'}-r}(0,T;\mathbf{V}_{\beta(p-p')}) \cap C_w^{\alpha q}((0,T];L^2(D)) \cap C^{\alpha q}([0,T];\mathbf{V}_{-\beta q'}),$$

and there exists a positive constant C_6 such that

$$\|u\|_{L^{\frac{1}{\alpha q'}-r}(0,T;\mathbf{V}_{\beta(p-p')})} + \|u\|_{C_w^{\alpha q}((0,T];L^2(D))} + |||u|||_{C_w^{\alpha q}([0,T];\mathbf{V}_{-\beta q'})}$$
$$\le C_6\,\|\varphi\|_{\mathbf{V}_{\beta p}}.$$

Proof

(a) We divide the proof of this part into the following steps.

Step 1. We prove the existence and uniqueness of a mild solution.
In order to prove the existence of a mild solution of FVP (3.32) and (3.34), we will construct a convergent sequence in $L^{\frac{1}{\alpha q}-r}(0,T;L^2(D))$ whose limit will be a mild solution of the problem. Here, r is defined by (R2). Let $\{w_{(n)}\}_{n\ge 0}$ be a

sequence defined by Lemma 3.2 with respect to $\phi = \varphi \in \mathbf{V}_{\beta p}$; then $\{w_{(n)}\}_{n\geq 0} \subset \mathbf{W}_{\alpha q}^{\widehat{C}_0}(J \times D)$, where $\widehat{C}_0 := \frac{M_0}{1-k_0(T)} \|\varphi\|_{\mathbf{V}_{\beta p}}$. Therefore,

$$\|w_{(n)}(t,\cdot)\| \leq \widehat{C}_0 t^{-\alpha q}, \quad 0 < t \leq T,$$

for all $n \geq 1$. This together with $t^{-\alpha q}$ belonging to $L^{\frac{1}{\alpha q}-r}(0, T; \mathbb{R})$ implies that $\{w_{(n)}\}_{n\geq 0}$ is a bounded sequence in $L^{\frac{1}{\alpha q}-r}(0, T; L^2(D))$. Now, we will show that $\{w_{(n)}\}_{n\geq 0}$ is convergent by proving that it is also a Cauchy sequence. For fixed $n \geq 1$ and $k \geq 1$, the definition (3.91) of $\{w_{(n)}\}_{n\geq 0}$ yields that

$$\begin{aligned} w_{(n+k)} - w_{(n)} =& \mathscr{O}_1 \left[F(w_{(n-1+k)}) - F(w_{(n-1)})\right] \\ &+ \mathscr{O}_3 \left[F(w_{(n-1+k)}) - F(w_{(n-1)})\right]. \end{aligned}$$

Since F satisfies Lemma 3.2, the latter equation shows that we can apply the same arguments as in Lemma 3.2 with $\phi = \mathbf{0}$. Hence, one can deduce

$$\begin{aligned} &\|w_{(1+k)}(t,\cdot) - w_{(1)}(t,\cdot)\| \\ &\leq \left\|\left(\mathscr{O}_1 \left[F(w_{(0+k)}) - F(w_{(0)})\right]\right)(t,\cdot)\right\| + \left\|\left(\mathscr{O}_3 \left[F(w_{(0+k)}) - F(w_{(0)})\right]\right)(t,\cdot)\right\| \\ &\leq \widehat{c}_\alpha m_1^{-\beta p} \int_0^t \left\|F(\tau,\cdot, w_{(0+k)}(\tau,\cdot)) - F(\tau,\cdot, w_{(0)}(\tau,\cdot))\right\| (t-\tau)^{\alpha q-1} d\tau \\ &+ M_3 t^{-\alpha q} \int_0^T \left\|F(\tau,\cdot, w_{(0+k)}(\tau,\cdot)) - F(\tau,\cdot, w_{(0)}(\tau,\cdot))\right\| (T-\tau)^{\alpha q-1} d\tau, \end{aligned}$$

where we combined the estimates (3.94) and (3.95). From $\{w_{(n)}\}_{n\geq 0} \subset \mathbf{W}_{\alpha q}^{\widehat{C}_0}(J \times D)$, we have

$$\left\|w_{(0+k)}(\tau,\cdot) - w_{(0)}(\tau,\cdot)\right\| \leq 2\widehat{C}_0 t^{-\alpha q}.$$

Thus, by noting the identity

$$\int_0^t (t-\tau)^{a-1}\tau^{b-1} d\tau = t^{a+b-1} B(a,b),$$

where $a, b > 0$ and $B(\cdot,\cdot)$ is the Beta function, we find that

$$\begin{aligned} &\left\|w_{(1+k)}(t,\cdot) - w_{(1)}(t,\cdot)\right\| \\ &\leq \widehat{c}_\alpha m_1^{-\beta p} K \int_0^t \left\|w_{(0+k)}(\tau,\cdot) - w_{(0)}(\tau,\cdot)\right\| (t-\tau)^{\alpha q-1} d\tau \\ &+ M_3 K t^{-\alpha q} \int_0^T \left\|w_{(0+k)}(\tau,\cdot) - w_{(0)}(\tau,\cdot)\right\| (T-\tau)^{\alpha q-1} d\tau \end{aligned}$$

$$\leq \widehat{c}_\alpha m_1^{-\beta p} K(2\widehat{C}_0) \int_0^t \tau^{-\alpha q}(t-\tau)^{\alpha q-1} d\tau + M_3 K(2\widehat{C}_0) t^{-\alpha q}$$

$$\int_0^T \tau^{-\alpha q}(T-\tau)^{\alpha q-1} d\tau$$

$$\leq \left(\widehat{c}_\alpha m_1^{-\beta p} K T^{\alpha q} B(\alpha q, 1-\alpha q) + M_3 K B(\alpha q, 1-\alpha q)\right)(2\widehat{C}_0) t^{-\alpha q}.$$

From the definition of $k_0(T)$, we conclude that

$$\left\| w_{(1+k)}(t,\cdot) - w_{(1)}(t,\cdot) \right\| \leq k_0(T)(2\widehat{C}_0) t^{-\alpha q}.$$

Iterating this method n times shows

$$\left\| w_{(n+k)}(t,\cdot) - w_{(n)}(t,\cdot) \right\| \leq k_0^n(T)(2\widehat{C}_0) t^{-\alpha q}.$$

Taking the $L^{\frac{1}{\alpha q}-r}(0,T;\mathbb{R})$-norm of both sides of the above inequality directly implies

$$\left\| w_{(n+k)} - w_{(n)} \right\|_{L^{\frac{1}{\alpha q}-r}(0,T;L^2(D))} \leq k_0^n(T)(2\widehat{C}_0) \left\| t^{-\alpha q} \right\|_{L^{\frac{1}{\alpha q}-r}(0,T;\mathbb{R})}. \tag{3.97}$$

Here, we emphasize that the constants in (3.97) also do not depend on (n,k). Therefore, by letting n go to infinity, we obtain

$$\lim_{n\to\infty} \left\| w_{(n+k)} - w_{(n)} \right\|_{L^{\frac{1}{\alpha q}-r}(0,T;L^2(D))} = 0,$$

i.e., $\left\{w_{(n)}\right\}_{n\geq 0}$ is a bounded Cauchy sequence in $L^{\frac{1}{\alpha q}-r}(0,T;L^2(D))$. Hence, there exists a function u in $L^{\frac{1}{\alpha q}-r}(0,T;L^2(D))$ such that

$$u = \lim_{n\to\infty} w_{(n)} \text{ in } L^{\frac{1}{\alpha q}-r}(0,T;L^2(D)),$$

and u satisfies equation (3.45), i.e., u is a mild solution of FVP (3.32) and (3.34). Moreover, the boundedness (3.96) of $\left\{w_{(n)}\right\}_{n\geq 0}$ gives

$$\|u(t,\cdot)\| \leq \widetilde{C}_0 \|\varphi\|_{\mathbf{V}_{\beta p}} t^{-\alpha q}, \tag{3.98}$$

and so that

$$\|u\|_{L^{\frac{1}{\alpha q}-r}(0,T;L^2(D))} \leq M_{28} \|\varphi\|_{\mathbf{V}_{\beta p}},$$

where $M_{28} = \widetilde{C}_0 \left\| t^{-\alpha q} \right\|_{L^{\frac{1}{\alpha q}-r}(0,T;\mathbb{R})}$.

Now, we show the uniqueness of the solution u. Assume that $\widetilde{u}$ is another solution of FVP (3.32) and (3.34). Then, by applying the same argument as in (3.97), we also have

$$\|u-\widetilde{u}\|_{L^{\frac{1}{\alpha q}-r}(0,T;L^2(D))} \leq k_0^n(T)(2\widehat{C}_0)\left\|t^{-\alpha q}\right\|_{L^{\frac{1}{\alpha q}-r}(0,T;\mathbb{R})},$$

for all $n \in \mathbb{N}^+$. Thus $\|u-\widetilde{u}\|_{L^{\frac{1}{\alpha q}-r}(0,T;L^2(D))} = 0$ by letting n go to infinity. Hence, $u = \widetilde{u}$ in $L^{\frac{1}{\alpha q}-r}(0, T; L^2(D))$.

Step 2. We prove that $u \in L^{\frac{1}{\alpha q'}-r}(0, T; \mathbf{V}_{\beta(p-p')})$.

This will be proved by using the inequality (3.98). We now apply the same arguments as in the proofs of (3.57) and (3.59) to estimate $\|u(t,\cdot)\|_{\mathbf{V}_{\beta(p-p')}}$ as follows. First, we have

$$\begin{aligned}
&\|(\mathscr{O}_1 F(u))(t,\cdot)\|_{\mathbf{V}_{\beta(p-p')}}\\
\leq&\widehat{c}_\alpha \int_0^t \left\{\sum_{j=1}^{\infty} F_j^2(\tau,u(\tau)) m_j^{2\beta(p-p')} m_j^{-2\beta p}(t-\tau)^{-2\alpha p}(t-\tau)^{2\alpha-2}\right\}^{1/2} d\tau\\
\leq&\widehat{c}_\alpha m_1^{-\beta p'} \int_0^t \|F(\tau,\cdot,u(\tau,\cdot))\|(t-\tau)^{\alpha q-1} d\tau\\
\leq&\widehat{c}_\alpha m_1^{-\beta p'} K \int_0^t \|u(\tau,\cdot)\|(t-\tau)^{\alpha q-1} d\tau\\
\leq&\widehat{c}_\alpha m_1^{-\beta p'} K\widehat{C}_0 \int_0^t \tau^{-\alpha q}(t-\tau)^{\alpha q-1} d\tau \leq M_{29}\|\varphi\|_{\mathbf{V}_{\beta p}} t^{-\alpha q'},
\end{aligned}$$

where we let

$$M_{29} = \widehat{c}_\alpha m_1^{-\beta p'} K\widetilde{C}_0 B(\alpha q, 1-\alpha q) T^{\alpha q'}.$$

Secondly,

$$\begin{aligned}
\|(\mathscr{O}_3 F(u))(t,\cdot)\|_{\mathbf{V}_{\beta(p-p')}} \leq& M_6 t^{-\alpha q'} \int_0^T \|F(\tau,\cdot,u(\tau,\cdot))\|(T-\tau)^{\alpha q-1} d\tau\\
\leq& M_6 t^{-\alpha q'} K \int_0^T \|u(\tau,\cdot)\|(T-\tau)^{\alpha q-1} d\tau\\
\leq& M_6 t^{-\alpha q'} K\widehat{C}_0 \int_0^t \tau^{-\alpha q}(t-\tau)^{\alpha q-1} d\tau\\
\leq& M_{30}\|\varphi\|_{\mathbf{V}_{\beta p}} t^{-\alpha q'},
\end{aligned}$$

where we let $M_{30} = M_6 K\widetilde{C}_0 B(\alpha q, 1-\alpha q)$. We recall that $\|\mathscr{O}_2(t)\varphi\|_{\mathbf{V}_{\beta(p-p')}}$ have been estimated by (3.58). According to the above arguments, we arrive at the estimate

$$\begin{aligned}\|u(t,\cdot)\|_{\mathbf{V}_{\beta(p-p')}} \leq& \|(\mathscr{O}_1 F(u))(t,\cdot)\|_{\mathbf{V}_{\beta(p-p')}} + \|\mathscr{O}_2(t)\varphi\|_{\mathbf{V}_{\beta(p-p')}} \\ &+ \|(\mathscr{O}_3 F(u))(t,\cdot)\|_{\mathbf{V}_{\beta(p-p')}} \\ \leq& M_{31}\,\|\varphi\|_{\mathbf{V}_{\beta p}}\, t^{-\alpha q'},\end{aligned}$$

for $M_{31} = M_{29} + M_5 + M_{30}$. By taking the $L^{\frac{1}{\alpha q'}-r}(0,T;\mathbb{R})$-norm, then the latter inequalities can be transformed into the following estimate:

$$\|u\|_{L^{\frac{1}{\alpha q'}-r}(0,T;\mathbf{V}_{\beta(p-p')})} \leq M_{32}\,\|\varphi\|_{\mathbf{V}_{\beta p}}, \tag{3.99}$$

where $M_{32} = M_{31}\left\|t^{-\alpha q'}\right\|_{L^{\frac{1}{\alpha q'}-r}(0,T;\mathbb{R})}$.

Step 3. We prove that $u \in C_w^{\alpha q}((0,T];L^2(D))$.

Let us consider $0 < t_1 < t_2 \leq T$. By the same arguments as in (3.61), we have

$$\begin{aligned}&u(t_2,x) - u(t_1,x)\\ =&\sum_{j=1}^{\infty}\int_0^{t_1}\int_{t_1-\tau}^{t_2-\tau} F_j(\tau,u(\tau))\omega^{\alpha-2}E_{\alpha,\alpha-1}(-m_j^{\beta}\omega^{\alpha})d\omega d\tau e_j(x)\\ &+\sum_{j=1}^{\infty}\int_{t_1}^{t_2} F_j(\tau,u(\tau))\widetilde{E}_{\alpha,\alpha}(-m_j^{\beta}(t_2-\tau)^{\alpha})d\tau e_j(x)\\ &-\mathscr{L}^{\beta}\sum_{j=1}^{\infty}\varphi_j\int_{t_1}^{t_2}\frac{\widetilde{E}_{\alpha,\alpha}(-m_j^{\beta}\omega^{\alpha})}{E_{\alpha,1}(-m_j^{\beta}T^{\alpha})}d\omega e_j(x)\\ &+\mathscr{L}^{\beta}\sum_{j=1}^{\infty}\int_0^T\int_{t_1}^{t_2} F_j(\tau,u(\tau))\widetilde{E}_{\alpha,\alpha}(-m_j^{\beta}(T-\tau)^{\alpha})\frac{\widetilde{E}_{\alpha,\alpha}(-m_j^{\beta}\omega^{\alpha})}{E_{\alpha,1}(-m_j^{\beta}T^{\alpha})}d\omega d\tau e_j(x)\\ =:&\mathscr{I}_1^N + \mathscr{I}_2^N + \mathscr{I}_3^N + \mathscr{I}_4^N.\end{aligned} \tag{3.100}$$

Here, by (3.66), $\left\|\mathscr{I}_3^N\right\|$ tends to 0 as $t_2 - t_1$ tends to 0. In what follows, we will establish the convergence for $\left\|\mathscr{I}_n^N\right\|$, $n = 1, 2, 4$, which can be treated similarly as in (3.62), (3.63), and (3.67) based on the Lipschitzian assumption (A1). We first see that

$$
\begin{aligned}
\left\|\mathscr{I}_1^N\right\| \leq & \widehat{c}_\alpha m_1^{-\beta p} \int_0^{t_1} \left\{ \sum_{j=1}^{\infty} F_j^2(\tau, u(\tau)) \left| \int_{t_1-\tau}^{t_2-\tau} \omega^{\alpha-2} \omega^{-\alpha p} d\omega \right|^2 \right\}^{1/2} d\tau \\
\leq & \frac{\widehat{c}_\alpha m_1^{-\beta p}}{1-\alpha q} \int_0^{t_1} \|F(\tau, \cdot, u(\tau, \cdot))\| \left[(t_1-\tau)^{\alpha q-1} - (t_2-\tau)^{\alpha q-1} \right] d\tau \\
\leq & \frac{\widehat{c}_\alpha m_1^{-\beta p}}{1-\alpha q} K \int_0^{t_1} \|u(\tau, \cdot)\| \left[(t_1-\tau)^{\alpha q-1} - (t_2-\tau)^{\alpha q-1} \right] d\tau \\
\leq & \frac{\widehat{c}_\alpha m_1^{-\beta p}}{1-\alpha q} K \widetilde{C}_0 \|\varphi\|_{\mathbf{V}_{\beta p}} \int_0^{t_1} \tau^{-\alpha q} \left[(t_1-\tau)^{\alpha q-1} - (t_2-\tau)^{\alpha q-1} \right] d\tau .
\end{aligned}
$$

Note that

$$
\int_0^{t_1} \tau^{-\alpha q} (t_1-\tau)^{\alpha q-1} d\tau = B(\alpha q, 1-\alpha q).
$$

Thus, due to the substitution $\tau = t_2 \mu$, we have

$$
\begin{aligned}
& \int_0^{t_1} \tau^{-\alpha q} \left[(t_1-\tau)^{\alpha q-1} - (t_2-\tau)^{\alpha q-1} \right] d\tau \\
= & B(\alpha q, 1-\alpha q) - \int_0^{t_1/t_2} \mu^{-\alpha q} (1-\mu)^{\alpha q-1} d\mu \\
= & B(\alpha q, 1-\alpha q) - \left[B(\alpha q, 1-\alpha q) - \int_{t_1/t_2}^{1} \mu^{-\alpha q} (1-\mu)^{\alpha q-1} d\mu \right] \qquad (3.101) \\
\leq & \left(\frac{t_2}{t_1} \right)^{\alpha q} \int_{t_1/t_2}^{1} (1-\mu)^{\alpha q-1} d\mu \\
= & \frac{1}{\alpha q} \left(\frac{t_2}{t_1} - 1 \right)^{\alpha q} .
\end{aligned}
$$

As a consequence, $\lim_{t_2-t_1 \to 0} \int_0^{t_1} \tau^{-\alpha q} \left[(t_1-\tau)^{\alpha q-1} - (t_2-\tau)^{\alpha q-1} \right] d\tau = 0$, and so $\lim_{t_2-t_1 \to 0} \|\mathscr{I}_1^N\| = 0$. Secondly we proceed to deal with $\mathscr{I}_2^N$. Now $0 < p_0 < p$, and by (3.46), we have

$$
|E_{\alpha,\alpha}(-m_j^\beta (t_2-\tau)^\alpha)| \leq \widehat{c}_\alpha m_1^{-\beta(p-p_0)} (t_2-\tau)^{-\alpha(p-p_0)}.
$$

We deduce the following chain of estimates:

$$
\begin{aligned}
\left\| \mathscr{I}_2^N \right\| \le & \int_{t_1}^{t_2} \left\| \sum_{j=1}^{\infty} F_j(\tau, u(\tau)) E_{\alpha,\alpha}(-m_j^\beta (t_2-\tau)^\alpha) e_j \right\| (t_2-\tau)^{\alpha-1} d\tau \\
\le & \widehat{c}_\alpha m_1^{-\beta(p-p_0)} \int_{t_1}^{t_2} \| F(\tau, \cdot, u(\tau, \cdot)) \| (t_2-\tau)^{\alpha q - 1 + \alpha p_0} d\tau \\
\le & \widehat{c}_\alpha m_1^{-\beta(p-p_0)} K \int_{t_1}^{t_2} \| u(\tau, \cdot) \| (t_2-\tau)^{\alpha q-1} d\tau (t_2-t_1)^{\alpha p_0} \\
\le & \widehat{c}_\alpha m_1^{-\beta(p-p_0)} K \widetilde{C}_0 \|\varphi\|_{\mathbf{V}_{\beta p}} \int_0^{t_2} \tau^{-\alpha q} (t_2-\tau)^{\alpha q-1} d\tau (t_2-t_1)^{\alpha p_0} \\
\le & \widehat{c}_\alpha m_1^{-\beta(p-p_0)} K \widetilde{C}_0 \|\varphi\|_{\mathbf{V}_{\beta p}} B(\alpha q, 1-\alpha q) (t_2-t_1)^{\alpha p_0}.
\end{aligned}
\tag{3.102}
$$

This implies $\lim_{t_2-t_1\to 0} \left\| \mathscr{I}_2^N \right\| = 0$. Next, we thirdly proceed to consider $\mathscr{I}_4^N$. The same argument as in (3.67) gives

$$
\begin{aligned}
\left\| \mathscr{I}_4^N \right\| \le & \widehat{c}_\alpha M_{11} \frac{t_2^{\alpha q} - t_1^{\alpha q}}{t_1^{2\alpha q}} \int_0^T \| F(\tau, \cdot, u(\tau, \cdot)) \| (T-\tau)^{\alpha q-1} d\tau \\
\le & \widehat{c}_\alpha M_{11} K \widetilde{C}_0 \|\varphi\|_{\mathbf{V}_{\beta p}} \frac{t_2^{\alpha q} - t_1^{\alpha q}}{t_1^{2\alpha q}} \int_0^T \tau^{-\alpha q} (T-\tau)^{\alpha q-1} d\tau \\
\le & \widehat{c}_\alpha M_{11} K \widetilde{C}_0 B(\alpha q, 1-\alpha q) \|\varphi\|_{\mathbf{V}_{\beta p}} \frac{t_2^{\alpha q} - t_1^{\alpha q}}{t_1^{2\alpha q}},
\end{aligned}
$$

and we arrive at $\lim_{t_2-t_1\to 0} \left\| \mathscr{I}_4^N \right\| = 0$. The above arguments prove $u \in C((0,T]; L^2(D))$. This combines with (3.98) so that $u \in C_w^{\alpha q}((0,T]; L^2(D))$ and

$$
\|u\|_{C_w^{\alpha q}((0,T]; L^2(D))} \le \widetilde{C}_0 \|\varphi\|_{\mathbf{V}_{\beta p}}. \tag{3.103}
$$

We complete step 1 by combining the inequalities (3.99) and (3.103).

(b) According to part (a), we have just to prove that $u \in C^{\alpha q}([0,T]; \mathbf{V}_{-\beta q'})$. In this part, we consider $0 \le t_1 < t_2 \le T$. Let us first show

$$
\left\| \mathscr{I}_1^N \right\|_{\mathbf{V}_{-\beta q'}} \le M_{33} \|\varphi\|_{\mathbf{V}_{\beta p}} (t_2-t_1)^{\alpha q}, \tag{3.104}
$$

for some positive constant M_{33}, where the case $t_1 = 0$ is trivial. It is necessary to prove (3.104) for $t_1 > 0$. From the proof of the estimate (3.70), we have

$$
\begin{aligned}
&\left\| \mathscr{I}_1^N \right\|_{\mathbf{V}_{-\beta q'}} \\
\leq & \int_0^{t_1} \left\{ \sum_{j=1}^{\infty} m_j^{-2\beta q'} F_j^2(\tau, u(\tau)) \left| \int_{t_1-\tau}^{t_2-\tau} \omega^{\alpha-2} \left| E_{\alpha,\alpha-1}(-m_j^{\beta} \omega^{\alpha}) \right| d\omega \right|^2 \right\}^{1/2} d\tau.
\end{aligned}
$$

In addition, the inequalities (3.46) yield that $|E_{\alpha,\alpha-1}(-m_j^{\beta}\omega^{\alpha})| \leq \widehat{c}_{\alpha} m_j^{-\beta p'} \omega^{-\alpha p'}$. This associates with $\alpha q' - 2 = \alpha q - 2 + \alpha(p - p')$ so that

$$
\begin{aligned}
&\left\| \mathscr{I}_1^N \right\|_{\mathbf{V}_{-\beta q'}} \\
\leq & \widehat{c}_{\alpha} m_1^{-\beta} \int_0^{t_1} \| F(\tau, \cdot, u(\tau, \cdot)) \| \left| \int_{t_1-\tau}^{t_2-\tau} \omega^{\alpha q - 2 + \alpha(p-p')} d\omega \right| d\tau \\
\leq & \frac{\widehat{c}_{\alpha} m_1^{-\beta}}{1 - \alpha q'} K \int_0^{t_1} \| u(\tau, \cdot) \| \left[(t_1 - \tau)^{\alpha q - 1 + \alpha(p-p')} - (t_2 - \tau)^{\alpha q - 1 + \alpha(p-p')} \right] d\tau \\
\leq & \frac{\widehat{c}_{\alpha} m_1^{-\beta}}{1 - \alpha q'} K \widehat{C}_0 \int_0^{t_1} \tau^{-\alpha q} (t_1 - \tau)^{\alpha(p-p')} \left[(t_1 - \tau)^{\alpha q - 1} - (t_2 - \tau)^{\alpha q - 1} \right] d\tau \\
\leq & \frac{\widehat{c}_{\alpha} m_1^{-\beta}}{1 - \alpha q'} K \widehat{C}_0 t_1^{\alpha(p-p')} \frac{1}{\alpha q} \left(\frac{t_2}{t_1} - 1 \right)^{\alpha q} \\
\leq & M_{33} \| \varphi \|_{\mathbf{V}_{\beta p}} (t_2 - t_1)^{\alpha q},
\end{aligned}
$$

where

$$
M_{33} = \frac{\widehat{c}_{\alpha} m_1^{-\beta}}{(1 - \alpha q')\alpha q} K \widetilde{C}_0 T^{\alpha(p-p') - \alpha q}.
$$

Here, we have used (3.61), (3.101), and $\alpha(p - p') \geq \alpha q$ by (R4b). Secondly, we are going to consider $\mathscr{I}_2^N$. The Sobolev embedding $L^2(D) \hookrightarrow \mathbf{V}_{-\beta q'}$ yields that there exists a positive constant M_{34} such that

$$
\begin{aligned}
\left\| \mathscr{I}_2^N \right\|_{\mathbf{V}_{-\beta q'}} \leq & M_{34} \left\| \mathscr{I}_2^N \right\| \\
\leq & M_{34} \widehat{c}_{\alpha} m_1^{-\beta p'} K \widetilde{C}_0 B(\alpha q, 1 - \alpha q) \| \varphi \|_{\mathbf{V}_{\beta p}} (t_2 - t_1)^{\alpha(p-p')} \\
\leq & M_{34} \widehat{c}_{\alpha} m_1^{-\beta p'} T^{\alpha(p-p') - \alpha q} K \widetilde{C}_0 B(\alpha q, 1 - \alpha q) \| \varphi \|_{\mathbf{V}_{\beta p}} (t_2 - t_1)^{\alpha q},
\end{aligned}
$$

where we applied (3.102) with respect to $0 < p_0 = p - p' < p$. Thirdly, we now consider $\mathscr{I}_3^N$. By applying the same arguments as in (3.72), one can get

$$\begin{aligned}\left\|\mathscr{I}_3^N\right\|_{\mathbf{V}_{-\beta q'}} &\leq \frac{M_{14}}{\alpha(p-p')}\|\varphi\|_{\mathbf{V}_{\beta p}}\left(t_2^{\alpha(p-p')}-t_1^{\alpha(p-p')}\right)\\ &\leq \frac{M_{14}}{\alpha(p-p')}T^{\alpha(p-p')-\alpha q}\|\varphi\|_{\mathbf{V}_{\beta p}}(t_2-t_1)^{\alpha q}.\end{aligned}$$

Finally, the arguments proving (3.73) give

$$\begin{aligned}\left\|\mathscr{I}_4^N\right\|_{\mathbf{V}_{-\beta q'}} \leq& \widehat{C}_\alpha\frac{M_{14}}{\alpha(p-p')}\left(t_2^{\alpha(p-p')}-t_1^{\alpha(p-p')}\right)\\ &\int_0^T\|F(\tau,\cdot,u(\tau,\cdot))\|(T-\tau)^{\alpha q-1}d\tau\\ \leq& \widehat{C}_\alpha\frac{M_{14}}{\alpha(p-p')}(t_2-t_1)^{\alpha(p-p')}K\int_0^T\|u(\tau,\cdot)\|(T-\tau)^{\alpha q-1}d\tau\\ \leq& \widehat{C}_\alpha\frac{M_{14}}{\alpha(p-p')}T^{\alpha(p-p')-\alpha q}K\widetilde{C}_0B(\alpha q,1-\alpha q)\|\varphi\|_{\mathbf{V}_{\beta p}}(t_2-t_1)^{\alpha q}.\end{aligned}$$

Taking the above estimates for $\mathscr{I}_n^N$, $1\leq n\leq 4$, together, we conclude that u belongs to $C^{\alpha q}([0,T];\mathbf{V}_{-\beta q'})$. Moreover, there exists a positive constant M_{35} such that

$$|||u|||_{C^{\alpha q}([0,T];\mathbf{V}_{-\beta q'})}\leq M_{35}\|\varphi\|_{\mathbf{V}_{\beta p}}.$$

By combining this inequality with (3.99) and (3.103), we complete the proof.

Theorem 3.8 *Let $p,q,p',q',\widehat{p},\widehat{q},r,\widehat{r}$ be defined by (R1), (R4b), (R5b). If φ belongs to $\mathbf{V}_{\beta(p+\widehat{q})}$, F satisfies the assumption (A2), and $k_0(T)<1$, then FVP* (3.32)–(3.34) *has a unique solution u satisfying that*

$$\begin{aligned}&u\in L^{\frac{1}{\alpha q'}-r}(0,T;\mathbf{V}_{\beta(p-p')})\cap C_w^{\alpha q}((0,T];L^2(D)),\\ &{}_0^CD_t^\alpha u\in L^{\frac{1}{\alpha q}-\widehat{r}}(0,T;\mathbf{V}_{-\beta(q+\widehat{p})})\cap C_w^{\alpha}((0,T];\mathbf{V}_{-\beta q}).\end{aligned}$$

Moreover, there exists a constant $C_7>0$ such that

$$\left\|{}_0^CD_t^\alpha u\right\|_{L^{\frac{1}{\alpha}-\widehat{r}}(0,T;\mathbf{V}_{-\beta(q-\widehat{q})})}+\left\|{}_0^CD_t^\alpha u(t,\cdot)\right\|_{C_w^{\alpha}((0,T];\mathbf{V}_{-\beta q})}\leq C_7\|\varphi\|_{\mathbf{V}_{\beta(p+\widehat{q})}}.\tag{3.105}$$

Proof Since F satisfies (A2), F also satisfies (A1) with respect to the Lipschitz constant K_*. In addition, the Sobolev embedding $\mathbf{V}_{\beta(p+\widehat{q})}\hookrightarrow\mathbf{V}_{\beta p}$ shows that φ belongs to $\mathbf{V}_{\beta p}$. Hence, by Theorem 3.7, FVP (3.32) and (3.34) has a unique solution

$$u \in L^{\frac{1}{\alpha q'}-r}(0, T; \mathbf{V}_{\beta(p-p')}) \cap C_w^{\alpha q}((0, T]; L^2(D)).$$

Moreover, the inequality (3.98) also holds. We deduce that, for $0 < t \leq T$,

$$\|F(t, \cdot, u(t, \cdot))\| \leq K_* \widetilde{C}_0 \|\varphi\|_{\mathbf{V}_{\beta p}} t^{-\alpha q} \leq M_{36} K_* \widetilde{C}_0 \|\varphi\|_{\mathbf{V}_{\beta(p+\widehat{q})}} t^{-\alpha q}. \tag{3.106}$$

The remainder of this proof falls naturally into two steps as follows.

Step 1. We prove that ${}_0^C D_t^\alpha u$ finitely exists and belongs to $L^{\frac{1}{\alpha}-\widehat{r}}(0, T; \mathbf{V}_{-\beta(q-\widehat{q})})$. By the same way as in part (a) of Theorem 3.6, we have

$$\begin{aligned}
{}_0^C D_t^\alpha u_j(t) =& F_j(t, u(t)) - m_j^\beta F_j(t, u(t)) * \widetilde{E}_{\alpha,\alpha}(-m_j^\beta t^\alpha) - \varphi_j \frac{m_j^\beta E_{\alpha,1}(-m_j^\beta t^\alpha)}{E_{\alpha,1}(-m_j^\beta T^\alpha)} \\
&+ \left(F_j(r, u(r)) * \widetilde{E}_{\alpha,\alpha}(-m_j^\beta r^\alpha)\right)\Big|_{r=T} \frac{m_j^\beta E_{\alpha,1}(-m_j^\beta t^\alpha)}{E_{\alpha,1}(-m_j^\beta T^\alpha)} \\
=:& F_j(t, u(t)) + \psi_j^{N,1}(t) + \psi_j^{N,2}(t) + \psi_j^{N,3}(t),
\end{aligned}$$

for all $j \in \mathbb{N}^+$. In view of (3.106), $F(t, \cdot, u(t, \cdot))$ is contained in $L^2(D)$ for $0 < t \leq T$. This associates with the Sobolev embedding $L^2(D) \hookrightarrow \mathbf{V}_{-\beta(q-\widehat{q})}$ that $F(t, \cdot, u(t, \cdot))$ is contained in $\mathbf{V}_{-\beta(q-\widehat{q})}$, namely $\sum_{j=1}^\infty F_j(t, u(t)) e_j$ is contained in $\mathbf{V}_{-\beta(q-\widehat{q})}$. On the other hand, $\psi_j^{N,2} = \psi_j^{(2)}$, and the norm $\left\|\sum_{j=1}^\infty \psi_j^{N,2}(t) e_j\right\|_{\mathbf{V}_{-\beta(q-\widehat{q})}}$ exists finitely by (3.78). Now, we consider $\left\|\sum_{j=1}^\infty \psi_j^{N,n}(t) e_j\right\|_{\mathbf{V}_{-\beta(q-\widehat{q})}}$, $n = 1, 3$. According to the estimates (3.77) and (3.79), the following ones hold:

$$\begin{aligned}
&\left\|\sum_{n_1 \leq j \leq n_2} \psi_j^{N,1}(t) e_j\right\|_{\mathbf{V}_{-\beta(q-\widehat{q})}} \\
\leq& \widehat{c}_\alpha T^\alpha t^{-\alpha} \int_0^t (t-\tau)^{\alpha(q-\widehat{q})-1} \left\{\sum_{n_1 \leq j \leq n_2} F_j^2(\tau, u(\tau))\right\}^{1/2} d\tau, \\
&\left\|\sum_{n_1 \leq j \leq n_2} \psi_j^{N,3}(t) e_j\right\|_{\mathbf{V}_{-\beta(q-\widehat{q})}} \\
\leq& \widehat{c}_\alpha M_{19} t^{-\alpha} \int_0^T (T-\tau)^{\alpha(q-\widehat{q})-1} \left\{\sum_{n_1 \leq j \leq n_2} F_j^2(\tau, u(\tau))\right\}^{1/2} d\tau.
\end{aligned}$$

For $0 < \tau < T$, we have $F(\tau, \cdot, u(\tau, \cdot))$ belonging to $L^2(D)$. This follows that the sequence $\{G_n(\tau)\}$, which is defined by $G_n(\tau) = \left\{\sum_{j\geq n} F_j^2(\tau, u(\tau))\right\}^{1/2}$, converges pointwise to 0 as n goes to infinity. Moreover, by (3.106), we have

$$\left|(t-\tau)^{\alpha(q-\widehat{q})-1} G_n(\tau)\right| \leq M_{36} K_* \widetilde{C}_0 \|\varphi\|_{\mathbf{V}_{\beta(p+\widehat{q})}} (t-\tau)^{\alpha(q-\widehat{q})-1} \tau^{-\alpha q}.$$

The function $\tau \to (t-\tau)^{\alpha(q-\widehat{q})-1}\tau^{-\alpha q}$ is integrable on the open interval $(0, t)$, $t > 0$, since

$$\int_0^t (t-\tau)^{\alpha(q-\widehat{q})-1}\tau^{-\alpha q} d\tau = t^{-\alpha\widehat{q}} B(\alpha(q-\widehat{q}), 1-\alpha q).$$

Therefore, the dominated convergence theorem yields that

$$\lim_{n\to\infty} \int_0^t (t-\tau)^{\alpha(q-\widehat{q})-1} G_n(\tau) d\tau = 0.$$

This together with $\left\{\sum_{n_1\leq j\leq n_2} F_j^2(\tau, u(\tau))\right\}^{1/2} \leq G_{n_1}(\tau)$ gives

$$\lim_{n_1, n_2\to\infty} \int_0^t (t-\tau)^{\alpha(q-\widehat{q})-1} \left\{ \sum_{n_1\leq j\leq n_2} F_j^2(\tau, u(\tau)) \right\}^{1/2} d\tau = 0.$$

Similarly, we also have

$$\lim_{n_1, n_2\to\infty} \int_0^T (T-\tau)^{\alpha(q-\widehat{q})-1} \left\{ \sum_{n_1\leq j\leq n_2} F_j^2(\tau, u(\tau)) \right\}^{1/2} d\tau = 0.$$

We deduce $\left\|\sum_{j=1}^{\infty} \psi_j^{N,n}(t) e_j\right\|_{\mathbf{V}_{-\beta(q-\widehat{q})}}$, $n = 1, 3$, exist finitely. Taking all the above arguments together, we conclude that $\left\|\sum_{j=1}^{\infty} {}_0^C D_t^{\alpha} u_j(t) e_j\right\|_{\mathbf{V}_{-\beta(q-\widehat{q})}}$ finitely exists. In addition, the Sobolev embedding $L^2(D) \hookrightarrow \mathbf{V}_{-\beta(q-\widehat{q})}$ yields that there exists a positive constant M_{37} such that

$$\|F(t, \cdot, u(t, \cdot))\|_{\mathbf{V}_{-\beta(q-\widehat{q})}} \leq M_{37} \|F(t, \cdot, u(t, \cdot))\|.$$

Hence,

$$
\begin{aligned}
&\left\| {}_0^C D_t^\alpha u(t,\cdot) \right\|_{\mathbf{V}_{-\beta(q-\widehat{q})}} \\
&\le \|F(t,\cdot,u(t,\cdot))\|_{\mathbf{V}_{-\beta(q-\widehat{q})}} + \sum_{1\le n\le 3} \left\| \sum_{j=1}^{\infty} \psi_j^{N,n}(t) e_j \right\|_{\mathbf{V}_{-\beta(q-\widehat{q})}} \\
&\le M_{37} \|F(t,\cdot,u(t,\cdot))\| + \widehat{c}_\alpha \int_0^t (t-\tau)^{\alpha(q-\widehat{q})-1} \|F(\tau,\cdot,u(\tau,\cdot))\| \, d\tau \\
&\quad + M_{19} t^{-\alpha} \|\varphi\|_{\mathbf{V}_{\beta(p+\widehat{q})}} + \widehat{c}_\alpha M_{19} t^{-\alpha} \int_0^T (T-\tau)^{\alpha(q-\widehat{q})-1} \|F(\tau,\cdot,u(\tau,\cdot))\| \, d\tau.
\end{aligned}
$$

We now note that

$$
\int_0^t (t-\tau)^{\alpha(q-\widehat{q})-1} \tau^{-\alpha q} d\tau \le T^{\alpha-\alpha\widehat{q}} B(\alpha(q-\widehat{q}), 1-\alpha q) t^{-\alpha}
$$

and

$$
\int_0^T (T-\tau)^{\alpha(q-\widehat{q})-1} \tau^{-\alpha q} d\tau = T^{-\alpha\widehat{q}} B(\alpha(q-\widehat{q}), 1-\alpha q).
$$

This combines with (3.106), and there exists a constant $M_{38} > 0$ such that

$$
\left\| {}_0^C D_t^\alpha u(t,\cdot) \right\|_{\mathbf{V}_{-\beta(q-\widehat{q})}} \le M_{38} \|\varphi\|_{\mathbf{V}_{\beta(p+\widehat{q})}} t^{-\alpha}, \tag{3.107}
$$

which leads to

$$
\left\| {}_0^C D_t^\alpha u \right\|_{L^{\frac{1}{\alpha}-\widehat{r}}(0,T;\mathbf{V}_{-\beta(q-\widehat{q})})} \le M_{38} \left\| t^{-\alpha} \right\|_{L^{\frac{1}{\alpha}-\widehat{r}}(0,T;\mathbb{R})} \|\varphi\|_{\mathbf{V}_{\beta(p+\widehat{q})}}. \tag{3.108}
$$

Step 2. We prove ${}_0^C D_t^\alpha u \in C_w^\alpha((0,T];\mathbf{V}_{-\beta q})$.
We consider $0 < t_1 < t_2 \le T$. A similar argument as in (3.83) yields

$$
{}_0^C D_t^\alpha u(t_2,x) - {}_0^C D_t^\alpha u(t_1,x) = F(t_2,x,u(t_2,x)) - F(t_1,x,u(t_1,x)) + \sum_{1\le n\le 4} \mathscr{J}_n^N,
$$

where $\mathscr{J}_n^N = -\mathscr{L}^\beta \mathscr{I}_n^N$ and $\mathscr{I}_n^N$ is defined by (3.100). By applying the Sobolev embedding $L^2(D) \hookrightarrow \mathbf{V}_{-\beta q}$, there exists a positive constant M_{39} such that

$$
\begin{aligned}
&\lim_{t_2-t_1\to 0} \|F(t_2,\cdot,u(t_2,\cdot)) - F(t_1,\cdot,u(t_1,\cdot))\|_{\mathbf{V}_{-\beta q}} \\
&\le \lim_{t_2-t_1\to 0} M_{39} \|F(t_2,\cdot,u(t_2,\cdot)) - F(t_1,\cdot,u(t_1,\cdot))\| \\
&\le \lim_{t_2-t_1\to 0} M_{39} K_* \big(|t_2-t_1| + \|u(t_2,\cdot) - u(t_1,\cdot)\|\big) = 0,
\end{aligned}
$$

where we note that $u \in C_w^{\alpha q}((0,T]; L^2(D))$. From (3.84) and (3.101), we have

$$\begin{aligned}\left\|\mathscr{I}_1^N\right\|_{\mathbf{V}_{-\beta q}} \leq & \widehat{c}_\alpha \int_0^{t_1} \|F(\tau,\cdot,u(\tau,\cdot))\| \left|\int_{t_1-\tau}^{t_2-\tau} \omega^{\alpha q-2} d\omega\right| d\tau \\ \leq & \frac{\widehat{c}_\alpha M_{36} K_* \widetilde{C}_0}{1-\alpha q} \|\varphi\|_{\mathbf{V}_{\beta(p+\widehat{q})}} \int_0^{t_1} \tau^{-\alpha q}\left[(t_1-\tau)^{\alpha q-1}-(t_2-\tau)^{\alpha q-1}\right] d\tau \\ \leq & \frac{\widehat{c}_\alpha M_{36} K_* \widetilde{C}_0}{(1-\alpha q)\alpha q} \|\varphi\|_{\mathbf{V}_{\beta(p+\widehat{q})}} \left(\frac{t_2}{t_1}-1\right)^{\alpha q}.\end{aligned}$$

In addition, by

$$\int_{t_1}^{t_2} \tau^{-\alpha q}(t_2-\tau)^{\alpha q-1} d\tau = \int_{t_1/t_2}^{1} \mu^{-\alpha q}(1-\mu)^{\alpha q-1} d\mu \leq \frac{1}{\alpha q}\left(\frac{t_2}{t_1}-1\right)^{\alpha q}$$

and (3.106), we can obtain the following chain of the inequalities:

$$\begin{aligned}\left\|\mathscr{I}_2^N\right\|_{\mathbf{V}_{-\beta q}} \leq & \widehat{c}_\alpha \int_{t_1}^{t_2} \|F(\tau,\cdot,u(\tau,\cdot))\| (t_2-\tau)^{\alpha q-1} d\tau \\ \leq & \widehat{c}_\alpha M_{36} K_* \widetilde{C}_0 \|\varphi\|_{\mathbf{V}_{\beta(p+\widehat{q})}} \int_{t_1}^{t_2} \tau^{-\alpha q}(t_2-\tau)^{\alpha q-1} d\tau \\ \leq & \widehat{c}_\alpha M_{36} K_* \widetilde{C}_0 \|\varphi\|_{\mathbf{V}_{\beta(p+\widehat{q})}} \frac{1}{\alpha q} t_1^{-\alpha q}(t_2-t_1)^{\alpha q}.\end{aligned}$$

Finally, the norm $\|\mathscr{I}_3^N\|_{V_{-\beta q}}$ has been estimated by (3.86), and the norm $\|\mathscr{I}_4^N\|_{V_{-\beta q}}$ can be estimated as follows:

$$\begin{aligned}\left\|\mathscr{I}_4^N\right\|_{\mathbf{V}_{-\beta q}} \leq & \frac{M_{24}}{\alpha} t_1^{-2\alpha}(t_2^\alpha - t_1^\alpha) \int_0^T \|F(\tau,\cdot,u(\tau,\cdot))\| (T-\tau)^{\alpha q-1} d\tau \\ \leq & \frac{M_{24}}{\alpha} t_1^{-2\alpha}(t_2^\alpha - t_1^\alpha) M_{36} K_* \widetilde{C}_0 \|\varphi\|_{\mathbf{V}_{\beta(p+\widehat{q})}} \int_0^T \tau^{-\alpha q}(T-\tau)^{\alpha q-1} d\tau \\ \leq & \frac{M_{24}}{\alpha} t_1^{-2\alpha}(t_2^\alpha - t_1^\alpha) M_{36} K_* \widetilde{C}_0 \|\varphi\|_{\mathbf{V}_{\beta(p+\widehat{q})}} B(\alpha q, 1-\alpha q).\end{aligned}$$

It follows from the above arguments that ${}_0^C D_t^\alpha u$ belongs to $C((0,T]; \mathbf{V}_{-\beta q})$. On the other hand, the estimate (3.107) also holds for $\widehat{p}=1$ and $\widehat{q}=0$, i.e., we have

$$t^\alpha \left\|{}_0^C D_t^\alpha u(t,\cdot)\right\|_{\mathbf{V}_{-\beta q}} \leq M_{38} \|\varphi\|_{\mathbf{V}_{\beta p}},$$

for $0 < t \le T$. Therefore, ${}_0^C D_t^\alpha u \in C_w^\alpha((0,T];\mathbf{V}_{-\beta q})$, and there exists a constant $M_{40} > 0$ such that

$$\left\| {}_0^C D_t^\alpha u(t,\cdot) \right\|_{C_w^\alpha((0,T];\mathbf{V}_{-\beta q})} \le M_{40} \left\| \varphi \right\|_{\mathbf{V}_{\beta(p+\widehat{q})}}, \tag{3.109}$$

by the Sobolev embedding $\mathbf{V}_{\beta(p+\widehat{q})} \hookrightarrow \mathbf{V}_{\beta p}$. The inequality (3.105) is derived by taking the inequalities (3.108) and (3.109) together. The proof is completed.

Remark 3.4 At the beginning part of step 1 of the above proof, we recall that $\psi_j^{N,2} = \psi_j^{(2)}$. This means that $\psi_j^{N,2}$ (with respect to the nonlinear case) can be similarly estimated in the same way as in the linear case. More precisely, this term can be estimated as (3.77). Here, the formula of $\psi_j^{(2)}$ is given by

$$\psi_j^{(2)} = -\varphi_j \frac{m_j^\beta E_{\alpha,1}(-m_j^\beta t^\alpha)}{E_{\alpha,1}(-m_j^\beta T^\alpha)}, \tag{3.110}$$

see the proof of Theorem 3.6. The appearance of the factor m_j^β in (3.110) tells that we need $\varphi \in \mathbf{V}_{\beta(p+\hat{q})}$ to obtain (3.77). In summary, in order to bound the Caputo fractional derivative ${}_0^C D_t^\alpha u$, we need the stronger assumption $\varphi \in \mathbf{V}_{\beta(p+\hat{q})}$ rather than $\varphi \in \mathbf{V}_{\beta p}$.

3.2.5 Existence

In the previous subsection, we found a solution u of FVP (3.32)–(3.34) in the set $\mathbf{W}_{\alpha q}^{\widehat{C}_0}(J \times D)$. This allows that $\|u(t,\cdot)\| \le \widehat{C}_0 t^{-\alpha q}$ for all $0 < t \le T$. Then, we obtain $u \in C_w^{\alpha q}((0,T];L^2(D))$ by establishing the time continuity of u, which corresponds to the boundedness

$$\sup_{0<t\le T} t^{\alpha q} \|u(t,\cdot)\| < +\infty. \tag{3.111}$$

However, the existence given in Theorem 3.7 requires the assumption $k_0(T) < 1$, which is equivalent to $KT^{\alpha q} < M_{41}$, where M_{41} is a constant. This can occur if K or T is small enough.

"Under what conditions is the contractivity condition $k_0(T) < 1$ satisfied?" This motivates the result in this subsection.

The purpose of this subsection is to discuss global existence of solutions, namely, the existence of solutions without any assumptions on K and T. To overcome the difficulties of finding solutions in $C_w^{\alpha q}((0,T];L^2(D))$, we shall seek solutions in a wider/weaker space than $C_w^{\alpha q}((0,T];L^2(D))$. The alternative solution space we are going to find is to take inspiration from replacing the supremum (3.111) by the following integral:

$$\int_0^T \left(t^b e^{-\rho t} \|u(t,\cdot)\|\right)^{\mu} dt < +\infty, \tag{3.112}$$

with suitable parameters b, ρ, and μ. We expect that the mapping $\mathscr{O}$ (formulated by Lemma 3.2) on the alternative solution space is contracted as ρ tends to positive infinity.

The above arguments motivate us to denote by $L^{\mu}_{\rho,b}(0,T;L^2(D))$, $\mu \geq 1$, $\rho > 0$, $b > 0$, the weighted Lebesgue space of all functions $v : (0,T) \to L^2(D)$ such that

$$\|v\|_{L^{\mu}_{\rho,b}(0,T;L^2(D))} := \left(\int_0^T \left(t^b e^{-\rho t} \|v(t,\cdot)\|\right)^{\mu} dt\right)^{\frac{1}{\mu}} < \infty.$$

In the next theorem, we present global existence for FVP (3.32) and (3.34) in $L^{\mu}_{\rho,b}(0,T;L^2(D))$. It is helpful to introduce the following special function:

$${}_1\mathscr{F}_1(a,b,z) := \frac{\Gamma(b)}{\Gamma(b-a)\Gamma(a)} \int_0^1 (1-\tau)^{b-a-1}\tau^{a-1}e^{z\tau}d\tau, \quad b > a > 0, z \in \mathbb{C},$$

which is called the Kummer function or hypergeometric function. We recall the following asymptotic behavior of this function:

$${}_1\mathscr{F}_1(a,b,z) := \Gamma(b)(\Gamma(a))^{-1}e^z z^{-(b-a)}\left(1 + O|z|^{-1}\right),$$

see [37] (Lemma 8) or [2] (chapter 13). Due to a simple integration by substitution, for all $t \in (0,T)$ we observe that

$$\begin{aligned}\int_0^t (t-\tau)^{a-1}\tau^{b-1}e^{z\tau}d\tau &= t^{a+b-1}\int_0^t (1-\tau)^{a-1}\tau^{b-1}e^{zt\tau}d\tau \\ &= t^{a+b-1}\frac{\Gamma(a)\Gamma(b)}{\Gamma(a+b)}{}_1\mathscr{F}_1(b,a+b,zt) \\ &= t^{a+b-1}\frac{e^{zt}\Gamma(a)}{(zt)^a}\left(1 + O\left(|zt|^{-1}\right)\right).\end{aligned} \tag{3.113}$$

Theorem 3.9 *Assume that* $\frac{1}{2} < \alpha < 1$. *Let* p *and* q *be defined by (R1) such that* $\frac{1}{2\alpha} < q < 1$. *Let* μ *and* b *be such that* $\frac{1}{\alpha q} < \mu < 2$, $\alpha q - \frac{1}{\mu} < b < 1 - \frac{1}{\mu}$. *If* φ *belongs to* $\mathbf{V}_{\beta p}$ *and* F *satisfies (A1), then there exists* $\widehat{\rho} > 0$ *such that FVP* (3.32)–(3.34) *has a unique solution* $u_G \in L^{\mu}_{\rho,b}(0,T;L^2(D))$ *with* $\rho \geq \widehat{\rho}$, *and furthermore*

$$\int_0^T \left(t^b e^{-\rho t} \|u_G(t,\cdot)\|\right)^{\mu} dt < +\infty.$$

Proof For $v_1, v_2 \in L^{\mu}_{b,\rho}(0,T;L^2(D))$, we first estimate the $L^{\mu}_{b,\rho}(0,T;L^2(D))$-norm of $\mathscr{O}_1 v_1 - \mathscr{O}_1 v_2$. The main idea is splitting the quantity $\|F(\tau,\cdot,v_1(\tau,\cdot)) - F(\tau,\cdot, v_2(\tau,\cdot))\|(t-\tau)^{\alpha q-1}$ into the product of $(t-\tau)^{\alpha q-1}\tau^{-b}e^{\rho\tau}$ and $\tau^b e^{-\rho\tau}\|F(\tau,\cdot, v_1(\tau,\cdot)) - F(\tau,\cdot,v_2(\tau,\cdot))\|$. Then the $L^{\mu}_{b,\rho}(0,T;L^2(D))$-norm of $\|v_1(\tau,\cdot)-v_2(\tau,\cdot)\|$ can be obtained by applying the Hölder inequality and the Lipschitz assumption (A1). Indeed, one can see that

$$
\begin{aligned}
&\int_0^T \left(t^b e^{-\rho t}\int_0^t \big\|v_1(\tau,\cdot)-v_2(\tau,\cdot)\big\|(t-\tau)^{\alpha q-1}d\tau\right)^{\mu} dt\\
\leq &\int_0^T \left(t^b e^{-\rho t}\right)^{\mu}\left(\int_0^t (t-\tau)^{\frac{(\alpha q-1)\mu}{\mu-1}}\tau^{-\frac{b\mu}{\mu-1}}e^{\frac{\rho\mu\tau}{\mu-1}}d\tau\right)^{\mu-1}\\
&\times\int_0^T \left(\tau^b e^{-\rho\tau}\big\|v_1(\tau,\cdot)-v_2(\tau,\cdot)\big\|\right)^{\mu} d\tau dt\\
\leq & C_1^{\mu-1}\big\|v_1-v_2\big\|^{\mu}_{L^{\mu}_{b,\rho}(0,T;L^2(D))}\int_0^T \left(t^b e^{-\rho t}\right)^{\mu}\\
&\times\left(t^{\frac{(\alpha q-b)\mu-1}{\mu-1}}e^{\frac{\rho\mu t}{\mu-1}}(\rho t)^{\frac{1-\alpha q\mu}{\mu-1}}\left(1+t^{-1}O(\rho^{-1})\right)\right)^{\mu-1} dt\\
\leq & C_1^{\mu-1}\big\|v_1-v_2\big\|^{\mu}_{L^{\mu}_{b,\rho}(0,T;L^2(D))}\left(T+O(\rho^{-1})\right)^{\mu-1}\rho^{1-\alpha q\mu}\int_0^T t^{1-\mu}dt,
\end{aligned}
$$

where we denote $C_1 := \Gamma\left((\alpha q\mu-1)/(\mu-1)\right)\left((\mu-1)/\mu\right)^{(\alpha q\mu-1)/(\mu-1)}$. Here, the asymptotic behavior (3.113) had been used in the second estimate, where we note that

$$
\frac{(\alpha q-1)\mu}{\mu-1}+1=\frac{\alpha q\mu-1}{\mu-1}>\frac{\alpha q(1/(\alpha q))-1}{\mu-1}=0,
$$

as $\mu > 1/(\alpha q)$, and $1-b\mu/(\mu-1)>0$ as $b<(\mu-1)/\mu$. Moreover, by taking ρ large enough, we can bound $\left(T+O(\rho^{-1})\right)^{\mu-1}$ by a constant independently of x,t. Since $\mu>1/(\alpha q)$, the factor $\rho^{1-\alpha q\mu}$ obviously tends to zero as ρ tends to infinity. Furthermore, the latter improper integral is convergent as $\mu<2$. Summarizing, we can find a constant $\rho_1>0$ such that

$$
\begin{aligned}
&\big\|\mathscr{O}_1 v_1-\mathscr{O}_1 v_2\big\|^{\mu}_{L^{\mu}_{b,\rho}(0,T;L^2(D))}\\
\leq &\left(\widehat{c}_\alpha m_1^{-\beta p}\right)^{\mu}\int_0^T\left(t^b e^{-\rho t}\int_0^t \big\|F(\tau,\cdot,v_1(\tau,\cdot))-F(\tau,\cdot,v_2(\tau,\cdot))\big\|\right.\\
&\qquad\left.(t-\tau)^{\alpha q-1}d\tau\right)^{\mu} dt
\end{aligned}
$$

$$\leq \left(K\widehat{c}_\alpha m_1^{-\beta p}\right)^\mu \int_0^T \left(t^b e^{-\rho t} \int_0^t \left\|v_1(\tau,\cdot) - v_2(\tau,\cdot)\right\| (t-\tau)^{\alpha q-1} d\tau\right)^\mu dt.$$

So we arrive at the following estimate:

$$\left\|\mathscr{O}_1 v_1 - \mathscr{O}_1 v_2\right\|_{L^\mu_{b,\rho}(0,T;L^2(D))} \leq \frac{1}{4}\left\|v_1 - v_2\right\|_{L^\mu_{b,\rho}(0,T;L^2(D))}, \tag{3.114}$$

for all $\rho \geq \rho_1$, where we have used (3.94) in the first estimate and the Lipschitz assumption (A1) in the second estimate.

Secondly, we will estimate the $L^\mu_{b,\rho}(0,T;L^2(D))$-norm of the difference $\mathscr{O}_3 v_1 - \mathscr{O}_3 v_2$. Based on estimating the $L^\mu_{b,\rho}(0,T;L^2(D))$-norm as above, this can be treated by combining the Hölder inequality, the Lipschitz assumption (A1), and the inequality (3.95). Indeed, we see that

$$\begin{aligned}
&\left\|\mathscr{O}_3(t,\cdot)v_1 - \mathscr{O}_3(t,\cdot)v_2\right\|^\mu_{L^\mu_{b,\rho}(0,T;L^2(D))}\\
&\leq M_3^\mu \int_0^T \left(t^b e^{-\rho t} t^{-\alpha q} \int_0^T \left\|F(\tau,\cdot,v_1(\tau,\cdot)) - F(\tau,\cdot,v_2(\tau,\cdot))\right\| (T-\tau)^{\alpha q-1} d\tau\right)^\mu dt\\
&\leq (KM_3)^\mu \int_0^T \left(t^b e^{-\rho t} t^{-\alpha q} \int_0^T \left\|v_1(\tau,\cdot) - v_2(\tau,\cdot)\right\| (T-\tau)^{\alpha q-1} d\tau\right)^\mu dt\\
&\leq (KM_3)^\mu C_1^{\mu-1} \left\|v_1 - v_2\right\|^\mu_{L^\mu_{b,\rho}(0,T;L^2(D))} \int_0^T \left(t^{b-\alpha q} e^{-\rho t}\right)^\mu \sigma_{b,\rho}(T) dt,
\end{aligned}$$

where

$$\sigma_{b,\rho}(T) := \left(T^{\frac{(\alpha q-b)\mu-1}{\mu-1}} e^{\frac{\rho\mu T}{\mu-1}} (\rho T)^{\frac{1-\alpha q\mu}{\mu-1}} \left(1 + T^{-1} O(\rho^{-1})\right)\right)^{\mu-1},$$

and the asymptotic behavior (3.113) was employed. Let us denote by $\mathscr{I}_{b,\rho}$ the latter integral; then

$$\mathscr{I}_{b,\rho} = \sigma_{b,\rho}(T) \int_0^T t^{(b-\alpha q)\mu} e^{-\rho\mu t} dt.$$

Note that $(b-\alpha q)\mu > ((\alpha q - 1/\mu) - \alpha q)\mu = -1$ as $b > \alpha q - 1/\mu$. Hence, using the asymptotic behavior (3.113), we have

$$\begin{aligned}
\mathscr{I}_{b,\rho} &\leq \sigma_{b,\rho}(T) T^{(b-\alpha q)\mu+1} e^{-\rho\mu T} (\rho\mu T)^{-1} \left(1 + O\left((\rho\mu T)^{-1}\right)\right)\\
&= \left(1 + T^{-1} O(\rho^{-1})\right)^{\mu-1} \left(1 + O\left((\rho\mu T)^{-1}\right)\right) \frac{\mu^{-1} T^{-\alpha q\mu}}{\rho^{\alpha q\mu}},
\end{aligned}$$

where the latter right-hand side tends to zero as ρ tends to infinity. Thus, there exists a $\rho_2 > 0$ such that

$$\left\|\mathscr{O}_3 v_1 - \mathscr{O}_3 v_2\right\|_{L^{\mu}_{b,\rho}(0,T;L^2(D))} \leq \frac{1}{4}\left\|v_1 - v_2\right\|_{L^{\mu}_{b,\rho}(0,T;L^2(D))}, \tag{3.115}$$

for all $\rho \geq \rho_2$. Taking the estimates (3.114) and (3.115) together gives

$$\begin{aligned}\left\|\mathscr{O} v_1 - \mathscr{O} v_2\right\|_{L^{\mu}_{b,\rho}(0,T;L^2(D))} &\leq \sum_{j\in\{1;3\}} \left\|\mathscr{O}_j v_1 - \mathscr{O}_j v_2\right\|_{L^{\mu}_{b,\rho}(0,T;L^2(D))} \\ &\leq \frac{1}{2}\left\|v_1 - v_2\right\|_{L^{\mu}_{b,\rho}(0,T;L^2(D))},\end{aligned}$$

for all $\rho \geq \max\{\rho_1, \rho_2\}$, namely, $\mathscr{O} : L^{\mu}_{b,\rho}(0, T; L^2(D)) \to L^{\mu}_{b,\rho}(0, T; L^2(D))$ is a contraction mapping. This means that $\mathscr{O}$ has only one fixed point in $L^{\mu}_{b,\rho}(0, T; L^2(D))$, and so FVP (3.32) and (3.34) has a unique solution u_G in the weighted Lebesgue space $L^{\mu}_{b,\rho}(0, T; L^2(D))$. The desiblack inequality is obvious.

Remark 3.5 In fact, we tried to find solutions in the space

$$\begin{aligned}&C^{b,\rho}_{w}((0, T]; L^2(D)) \\ &:= \left\{ v \in C((0, T]; L^2(D)) \,\middle|\, \left\|v\right\|_{C^{b,\rho}_{w}((0,T];L^2(D))} := \sup_{0<t\leq T} t^b e^{-\rho t}\left\|v(t, \cdot)\right\| < \infty \right\}.\end{aligned}$$

Note that the following inclusion holds:

$$C^{b,\rho}_{w}((0, T]; L^2(D)) \subset L^{\mu}_{b,\rho}(0, T; L^2(D)).$$

Indeed, if $v \in C^{b,\rho}_{w}((0, T]; L^2(D))$, then

$$\begin{aligned}\left\|v\right\|_{L^{\mu}_{\rho,b}(0,T;L^2(D))} &\leq \left(\int_0^T dt\right)^{\frac{1}{\mu}} \left(\sup_{0<t\leq T} t^b e^{-\rho t}\left\|v(t, \cdot)\right\|\right) \\ &= T^{\frac{1}{\mu}}\left\|v\right\|_{C^{b,\rho}_{w}((0,T];L^2(D))} < \infty.\end{aligned}$$

In order to find solutions in this space, for all $v_1, v_2 \in C^{b,\rho}_{w}((0, T]; L^2(D))$, it requires to bound the following quantities:

$$Q_1(b, \rho) := \sup_{0<t\leq T} t^b e^{-\rho t} \int_0^t \left\|v_1(\tau, \cdot) - v_2(\tau, \cdot)\right\|(t - \tau)^{\alpha q-1} d\tau,$$

$$Q_2(b,\rho) := \sup_{0<t\le T} t^b e^{-\rho t} t^{-\alpha q} \int_0^T \big\| v_1(\tau,\cdot) - v_2(\tau,\cdot) \big\| (T-\tau)^{\alpha q - 1} d\tau,$$

by $k_j(\rho)\|v_1 - v_2\|_{C_w^{b,\rho}((0,T];L^2(D))}$ with $k_j(\rho)$, $j = 1, 2$, tend to zero as ρ tends to infinity. Unfortunately, it does not occur with the term $Q_2(b,\rho)$. Indeed, since $v_1, v_2 \in C_w^{b,\rho}((0,T];L^2(D))$, we have

$$\big\| v_1(t,\cdot) - v_2(t,\cdot) \big\| \le t^{-b} e^{\rho t} \|v_1 - v_2\|_{C_w^{b,\rho}((0,T];L^2(D))},$$

which gives

$$Q_2(b,\rho) \le \sup_{0<t\le T} \widehat{\sigma}_{b,\rho}(t,T) \big\| v_1 - v_2 \big\|_{C_w^{b,\rho}((0,T];L^2(D))},$$

where

$$\widehat{\sigma}_{b,\rho}(t,T) := t^b e^{-\rho t} t^{-\alpha q} \int_0^T \tau^{-b} e^{\rho\tau} (T-\tau)^{\alpha q - 1} d\tau.$$

The following conclusions are obvious.

Due to the asymptotic behavior (3.113), the supremum of $\widehat{\sigma}_{b,\rho}(t,T)$ on $(0,T]$ tends to infinity as ρ tends to infinity. Hence, $\mathscr{O} : L^{\mu}_{b,\rho}(0,T;L^2(D)) \to L^{\mu}_{b,\rho}(0,T;L^2(D))$ cannot be a contraction mapping for arbitrary T. This is the main reason why we did not find solutions in $C_w^{b,\rho}((0,T];L^2(D))$.

The idea of using the space $L^{\mu}_{\rho,b}(0,T;L^2(D))$ fortuitously came when we realized that

$$\int_0^T \Big(\widehat{\sigma}_{b,\rho}(t,T) \Big)^{\mu} dt \to 0, \quad \text{as } \rho \to +\infty$$

with a suitable number $\mu \ge 1$. Here, we replaced the supremum (3.111) by the integral (3.112).

One can show that $Q_1(b,\rho)$ is bounded by $k_1(\rho)\|v_1 - v_2\|_{C_w^{b,\rho}((0,T];L^2(D))}$, where $k_1(\rho)$ tends to zero as ρ tends to infinity. This means that we can establish the existence of a mild solution to the initial value problem (3.32), (3.33), (3.35) in $C_w^{b,\rho}((0,T];L^2(D))$ without any assumptions on K and T.

Chapter 4
Well-Posedness and Regularity of Fractional Wave Equations

4.1 Damped Wave Equations

4.1.1 *Introduction*

The main purpose of this section is to investigate the initial/boundary value problems for time fractional damped wave equation

$$\partial_t^\beta u + \partial_t^\alpha u = \Delta u + f(u), \quad t > 0, \tag{4.1}$$

subject to Dirichlet's boundary condition

$$u(t, x) = 0, \quad x \in \partial\Omega, \quad t > 0 \tag{4.2}$$

and initial value conditions

$$u(0, x) = \phi(x), \quad \partial_t u(0, x) = \psi(x), \quad x \in \Omega, \tag{4.3}$$

where $\Omega \subset \mathbb{R}^d$ $(d \geq 1)$ is a bounded domain with the sufficiently smooth boundary $\partial\Omega$, and ∂_t^β, ∂_t^α are standard fractional derivatives in the sense of Caputo type of order $\beta \in (1, 2]$ and $\alpha \in (0, 1]$, respectively. f is an appropriate force function which will be special later. Taking the case of $\beta = 2$ and $\alpha = 1$ in (4.1), it becomes the standard damped wave equation, which is an important mathematical model in studying many physic problems. Readers can easily find a large number of related researches that are focused on the well-posedness of some linear or nonlinear Cauchy problems. In addition, various papers have considered to establish the asymptotic behavior and regularity estimates of the solutions; we refer to [8, 63, 93] and the references therein. Observe that if $\beta = 2\alpha$ for $\alpha \in (1/2, 1]$ associated with (4.1), this equation contains a typical time fractional telegraph equation, which

Y. Zhou, *Fractional Diffusion and Wave Equations*,
https://doi.org/10.1007/978-3-031-74031-2_4

is derived from the law of the iterated Brownian motion and Brownian time for the telegraph process, see, e.g., [173].

A strong motivation for investigating Eq. (4.1) comes from physical phenomena. The time fractional diffusion equation $\partial_t^\beta u = \Delta u$ of order $\beta \in (0, 1)$ can be used to model anomalous diffusion phenomena, which was driven by fractional Brownian motion, and it represents the subdiffusion behavior [234], while the time fractional wave equation of order $\beta \in (1, 2)$ will interpolate between the heat equation ($\beta = 1$) and the wave equation ($\beta = 2$) that govern intermediate processes between diffusion and wave propagation, and it is further interpreted as the superdiffusion behavior. Moreover, fractional wave equation can also model a cable made with special smart materials or a vibrating string in presence of a fractional friction with power-law memory kernel. From these physical points of view, some partial differential equations with fractional derivative will be more suitable to describe practical problems.

As for the current problem, in fact, without the term $\partial_t^\beta u$ associated with (4.1), there are more researches concerning with this fractional diffusion equation, the analysis of well-posedness, asymptotic analysis, decay estimates, and blow-up solutions have been studied in [1, 34, 116, 128]. Without the forcing term f and damped term $\partial_t^\alpha u$ associated with (4.1), the analysis theories of fractional wave equation have been studied by Luchko [143], Mainardi [144, 151], Sakamoto and Yamamoto [186], Schneider and Wyss [190], etc. Recently, concerning with fractional wave equation, Kian and Yamamoto [112] have investigated the existence of weak solution and some Strichartz estimates under the case of semilinear force function on bounded domain. The well-posedness results associated with a Dirichlet space have been considered by Alvarez et al. [10]. In addition, Otarola and Salgado [174] have studied the regularity of weak solutions and also discussed the spatial-time regularities of the solution for an extended problem. Djida et al. [54] have concerned with the well-posedness results on whole space $\mathbb{R}^N$, and they derived some L^p-L^r estimates of solution. Associated with an extra damping term in fractional wave equation, which can describe the interaction between the vector electric field and the electric and magnetic properties of the material (see, e.g., [73]), we observe that there are still few researches addressing the following wave equation with damping:

$$\partial_t^\beta u + \partial_t^\alpha u - \Delta u = 0.$$

Alaimia and Tatar [6] and Tatar [204] investigated the blow-up for the wave equation with a fractional damping. In one dimensional unbounded domain, Stojanovic and Gorenflo [200] obtained an upper viscosity solution for the case $\beta \in (1, 2)$ and $\alpha \in (0, 1)$, while on a bounded domain, Lin and Nakamura [137] investigated the Carleman estimate that gives the unique continuation property of solutions for an anomalous diffusion equation with multi-term time fractional Caputo derivative, as well as the case for fractional diffusion equation [136]. Consequently, it is natural to discuss the more general fractional wave equation with damping term.

Motivated by the abovementioned works, in this section, we will focus on the well-posedness and regularity of linear fractional damped wave equations; one reason to consider these properties is that there are few papers to establish the qualitative theory of damped wave equations in the sense of fractional versions. Especially in nonlinear problem, there is an urgent need for existence results to extend some known conclusions. The second reason is that the Laplacian operator associated with Dirichlet's boundary condition on a bounded domain with the sufficiently smooth boundary on $L^2(\Omega)$ can be expressed as a spectrum problem. It will lead to that relative solution operators are compact and uniformly continuous on their domains. According to these properties, we can get a general existence result without the Lipschitz condition or the smoothness assumption on nonlinear function.

This section is organized as follows. Section 4.1.2 recalls some concepts and known results. In Sect. 4.1.3, we first introduce a suitable definition of mild solutions for the linear problem, and then we obtain some existence and regularity of mild solutions. In Sect. 4.1.4, some exact upper bounds of several Mittag-Leffler functions are obtained. Under the local Lipschitz condition of nonlinear force function, a well-posedness result of the problem (4.1)–(4.3) is established. Next, we show the continuation and blow-up alternative of the solutions. In addition, we also prove the compactness of solution operators, which allows us to study the existence of mild solutions by removing the Lipschitz condition or higher regularity hypothesis of force function. Finally, an application is introduced to verify our main results.

4.1.2 Preliminary Results

Let us first recall the Riemann-Liouville fractional integral of order $\beta \in \mathbb{R}_+$ with the lower limit zero for a function $v \in L^1(0, T; X)$ and a Banach space X

$$J_t^\beta v(t) = \frac{1}{\Gamma(\beta)} \int_0^t (t-s)^{\beta-1} v(s) ds, \quad t \geq 0,$$

where $\Gamma(\beta)$ is the Gamma function.

Let $\alpha \in (0, 1)$, $\beta \in (1, 2)$, and $T > 0$. Consider a function $v \in L^1(0, T; X)$ such that $J_t^{1-\alpha} v \in W^{1,1}(0, T; X)$ or $J_t^{2-\beta} v \in W^{2,1}(0, T; X)$. The representations

$$\partial_t^\alpha v(t) = \partial_t \left(J_t^{1-\alpha} (v(t) - v(0)) \right)$$

and

$$\partial_t^\beta v(t) = \partial_{tt}^2 \left(J_t^{2-\beta} (v(t) - v(0) - t\, \partial_t v(0)) \right)$$

are called the Caputo fractional derivative of order α and β, respectively. In particular, when $\rho = 0$, one finds that $J_t^{\rho} v(t) = v(t)$. Hence, if $\alpha = 1$ or $\beta = 2$, then the Caputo fractional derivatives commute with integer order derivatives, respectively.

Let $L^2(\Omega)$ be the standard real Hilbert space with the norm $\|\cdot\|$ and scalar product $(\cdot,\cdot)$. $H^l(\Omega)$ and $H_0^m(\Omega)$ denote the usual Sobolev spaces for $l, m \geq 0$. Let X be a Banach space equipped with norm $|\cdot|$, and $\mathscr{B}(X)$ stands for the spaces of all bounded linear operators from X into itself. Let $C([0,T]; X)$ be the Banach space of all continuous functions from $[0,T]$ into X equipped with the supremum norm $\|u\|_{\mathscr{C}} = \sup_{t\in[0,T]} |u(t)|$. The symbol $L^p(0,T;X)$ denotes the Banach space of all p-integrable measurable functions u such that

$$\|u\|_{L^p(0,T;X)} = \begin{cases} \left(\int_0^T |u(t)|^p dt\right)^p < \infty, & \text{if } 1 \leq p < \infty, \\ \operatorname{ess\,sup}_{0\leq t\leq T} |u(t)| < \infty, & \text{if } p = \infty. \end{cases}$$

The symbol $W^{k,p}(0,T;X)$ $(k \geq 1)$ stands for the Banach space of all k-times differentiable functions u such that

$$\|u\|_{W^{k,p}(0,T;X)} = \sum_{n=0}^{k} \|\partial_t^n u\|_{L^p(0,T;X)} < \infty.$$

It is well known that the Laplacian operator $A = -\Delta$ is nonnegative and self-adjoint in Sobolev space $H_0^1(\Omega)$, and there exists an orthonormal basis of $L^2(\Omega)$ consisting of eigenfunctions $\{e_n\}_{n=1}^{\infty} \subset H_0^1(\Omega)$, which are corresponding to the discrete positive eigenvalues $\{\lambda_n\}_{n=1}^{\infty}$ for every $n \in \mathbb{N}^+$, here $0 < \lambda_1 \leq \lambda_2 \leq \ldots$ with $\lim_{n\to\infty} \lambda_n = \infty$ satisfying

$$Ae_n = \lambda_n e_n, \quad \text{in } \Omega\,; \qquad e_n = 0, \quad \text{on } \partial\Omega.$$

For any $\gamma \geq 0$, let fractional power operator A^{γ} possess the following representation:

$$A^{\gamma} u = \sum_{n=1}^{\infty} \lambda_n^{\gamma} (u, e_n) e_n, \quad u \in D(A^{\gamma}),$$

where

$$D(A^{\gamma}) = \left\{ u \in L^2(\Omega) : \sum_{n=1}^{\infty} \lambda_n^{2\gamma} |(u, e_n)|^2 < \infty \right\},$$

as a Hilbert space of functions

$$u := \sum_{n=1}^{\infty} u_n e_n = \sum_{n=1}^{\infty} (u, e_n) e_n \in L^2(\Omega),$$

equipped with the norm

$$\|u\|_{\gamma} := \|u\|_{D(A^{\gamma})} = \left(\sum_{n=1}^{\infty} \lambda_n^{2\gamma} |(u, e_n)|^2 \right)^{\frac{1}{2}}, \quad u \in D(A^{\gamma}).$$

By using the so-called Gelfand triple, we denote the duality space of $D(A^{\gamma})$ by $D(A^{-\gamma})$. It can be seen that $D(A^{-\gamma})$ is a Hilbert space endowed with the norm

$$\|u\|_{\gamma^*} := \|u\|_{D(A^{-\gamma})} = \left(\sum_{n=1}^{\infty} \lambda_n^{-2\gamma} |\langle u, e_n \rangle_{-\gamma,\gamma}|^2 \right)^{\frac{1}{2}}, \quad u \in D(A^{-\gamma}),$$

under the duality bracket $\langle \cdot, \cdot \rangle_{-\gamma,\gamma}$. Furthermore, we notice that

$$\langle u, v \rangle_{-\gamma,\gamma} = (u, v), \quad \text{for } u \in L^2(\Omega), \ v \in D(A^{\gamma}).$$

Specially, one has $D(A^{\gamma}) \subset H^{2\gamma}(\Omega)$ for $\gamma > 0$, $D(A^0) = L^2(\Omega)$, $D(A^{\frac{1}{2}}) = H_0^1(\Omega)$, see, e.g., [186].

4.1.3 Linear Problems

Consider the following linear problem:

$$\partial_t^{\beta} u(t, x) + \partial_t^{\alpha} u(t, x) = \Delta u(t, x) + f(t, x), \quad x \in \Omega, \ t \in (0, T), \tag{4.4}$$

$$u(t, x) = 0, \quad x \in \partial\Omega, \ t \in (0, T), \tag{4.5}$$

$$u(0, x) = \phi(x), \ \partial_t u(0, x) = \psi(x), \quad x \in \Omega. \tag{4.6}$$

Next, a suitable definition of mild solutions will be introduced to study the above linear problem; furthermore, the existence and regularity of solutions are discussed.

4.1.3.1 Solution Representation Formula

Let $\phi \in D(A^{\gamma})$, $\psi \in L^2(\Omega)$, with the aid of the spectrum property of operator A, and observe that the equation

$$\partial_t^{\beta} u(t, x) = -Au(t, x) + f(t, x), \quad t > 0, \ x \in \Omega \tag{4.7}$$

associated with initial/boundary value conditions (4.5)–(4.6) can be converted into

$$\begin{cases} {}_0^C D_t^\beta u_n(t) = -\lambda_n u_n(t) + f_n(t), \\ u_n(0) = \phi_n, \ u_n'(0) = \psi_n, \end{cases}$$

where $\phi_n = (\phi, e_n)$, $\psi_n = (\psi, e_n)$, $f_n(t) = (f(t, \cdot), e_n)$, and the solutions $u_n(t)$ are explicitly expressed as follows (see, e.g., [10, 112]):

$$\begin{aligned} u_n(t) =& E_{\beta,1}(-\lambda_n t^\beta)\phi_n + t E_{\beta,2}(-\lambda_n t^\beta)\psi_n \\ &+ \int_0^t (t-s)^{\beta-1} E_{\beta,\beta}(-\lambda_n (t-s)^\beta) f_n(s) ds, \end{aligned}$$

for all $t \geq 0$. With the help of the identities in Proposition 1.15, one can derive that the formula of first-order derivative with respect to t of $u_n(t)$ is equal to

$$\begin{aligned} &- \lambda_n t^{\beta-1} E_{\beta,\beta}(-\lambda_n t^\beta)\phi_n + E_{\beta,1}(-\lambda_n t^\beta)\psi_n \\ &+ \int_0^t (t-s)^{\beta-2} E_{\beta,\beta-1}(-\lambda_n (t-s)^\beta) f_n(s) ds. \end{aligned}$$

In view of the definition of the Caputo fractional derivative and (1.13), by changing the order of integration, one has

$$\begin{aligned} &\int_0^t (t-s)^{\beta-1} E_{\beta,\beta}(-\lambda_n (t-s)^\beta) {}_0^C D_s^\alpha u_n(s) ds \\ =& \int_0^t (t-s)^{\beta-\alpha} E_{\beta,\beta+1-\alpha}(-\lambda_n (t-s)^\beta) u_n'(s) ds. \end{aligned}$$

Therefore, after integration by parts in s, it follows that

$$\begin{aligned} &\int_0^t (t-s)^{\beta-\alpha} E_{\beta,\beta+1-\alpha}(-\lambda_n (t-s)^\beta) u_n'(s) ds \\ =& \int_0^t (t-s)^{\beta-1-\alpha} E_{\beta,\beta-\alpha}(-\lambda_n (t-s)^\beta) u_n(s) ds \\ &- t^{\beta-\alpha} E_{\beta,\beta+1-\alpha}(-\lambda_n t^\beta) u_n(0). \end{aligned} \tag{4.8}$$

By using the eigenfunction expansions, we set the operators

$$\mathscr{S}_\beta(t)v = \sum_{n=1}^\infty E_{\beta,1}(-\lambda_n t^\beta)(v, e_n)e_n, \quad \mathscr{P}_\beta(t)v = \sum_{n=1}^\infty t E_{\beta,2}(-\lambda_n t^\beta)(v, e_n)e_n$$

and

$$\mathcal{T}_\beta(t)v = \sum_{n=1}^{\infty} t^{\beta-1} E_{\beta,\beta}(-\lambda_n t^\beta)(v, e_n)e_n,$$

for all $v \in L^2(\Omega)$ and $t \geq 0$. In order to simplify the representation of solution, we just focus on the dependence of time variable t and sometimes omit the space variable x, writing $u(t) = u(t, \cdot)$, $f(t) = f(t, \cdot)$, and so on. Following the above arguments, (4.7) associated with (4.5)–(4.6) has an equivalent integral form as follows:

$$u(t) = \mathcal{S}_\beta(t)\phi + \mathcal{P}_\beta(t)\psi + \int_0^t \mathcal{T}_\beta(t-s)f(s)ds.$$

Hence, the formal solution of linear problem (4.4)–(4.6) can be represented as follows:

$$u(t) = \mathcal{S}_\beta(t)\phi + \mathcal{R}_\beta(t)\phi + \mathcal{P}_\beta(t)\psi - \int_0^t \mathcal{R}'_\beta(t-s)u(s)ds + \int_0^t \mathcal{T}_\beta(t-s)f(s)ds, \tag{4.9}$$

for $t \geq 0$, where

$$\mathcal{R}_\beta(t)\phi = \sum_{n=1}^{\infty} t^{\beta-\alpha} E_{\beta,\beta+1-\alpha}(-\lambda_n t^\beta)(\phi, e_n)e_n,$$

$$\mathcal{R}'_\beta(t)u = \sum_{n=1}^{\infty} t^{\beta-\alpha-1} E_{\beta,\beta-\alpha}(-\lambda_n t^\beta)(u, e_n)e_n.$$

As we can see, the damped term in linear problem can be regarded as a nonlinear term in nonlinear problem, which will avoid a lot of computations to check the properties of solution, for instance, when it is converted into a fundamental solution, see an application below, it is not easy to discuss the existence and regularities of solution and especially for considering nonlinear problem (4.1)–(4.3). For this reason, it is worthwhile to consider that such damped term converts into an integral representation at nonlinear term. Next, we shall introduce a suitable definition of mild solutions to the problem (4.4)–(4.6) which involves the Mittag-Leffler functions from the above arguments.

Definition 4.1 Let $T > 0$. If a function $u \in C([0, T]; L^2(\Omega))$ satisfies (4.9), then we say u is a mild solution of the problem (4.4)–(4.6).

4.1.3.2 Existence and Regularity

In the sequel, we will prove the existence and regularity of mild solutions of linear problem (4.4)–(4.6). The first result is concerned with the existence of the mild solution, and the regularities of the solution are given in the rest of results.

Theorem 4.1 *Let $(\phi, \psi) \in D(A^\gamma) \times L^2(\Omega)$ for $\gamma \in (0, 1)$, and let $f \in L^1(0, T; L^2(\Omega))$. Then there exists a unique mild solution u to the problem (4.4)–(4.6). Moreover,*

$$\|u(t)\| \lesssim \|\phi\|_\gamma + \|\psi\| + \|f\|_{L^1(0,T;L^2(\Omega))}. \tag{4.10}$$

The hidden constant, in the above inequality, is independent of t and γ but may be dependent on T.

Proof Let us first denote an operator $\mathscr{Q}$ on $C([0, T]; L^2(\Omega))$ as follows:

$$\begin{aligned}(\mathscr{Q}u)(t) =&\mathscr{S}_\beta(t)\phi + \mathscr{R}_\beta(t)\phi + \mathscr{P}_\beta(t)\psi\\ &- \int_0^t \mathscr{R}'_\beta(t-s)u(s)ds + \int_0^t \mathscr{T}_\beta(t-s)f(s)ds.\end{aligned}$$

Clearly, there exists a mild solution of the problem (4.4)–(4.6) if and only if operator $\mathscr{Q}$ has a fixed point in $C([0, T]; L^2(\Omega))$. In what follows, we shall show that operator $\mathscr{Q}$ is well defined on $C([0, T]; L^2(\Omega))$. Firstly, for any $\epsilon > 0$, let $0 \le t < t + \epsilon \le T$, we have

$$\begin{aligned}&(\mathscr{Q}u)(t+\epsilon) - (\mathscr{Q}u)(t)\\ =&\mathscr{S}_\beta(t+\epsilon)\phi - \mathscr{S}_\beta(t)\phi\\ &+ \mathscr{R}_\beta(t+\epsilon)\phi - \mathscr{R}_\beta(t)\phi + \mathscr{P}_\beta(t+\epsilon)\psi - \mathscr{P}_\beta(t)\psi\\ &- \int_0^{t+\epsilon} \mathscr{R}'_\beta(t+\epsilon-s)u(s)ds + \int_0^t \mathscr{R}'_\beta(t-s)u(s)ds\\ &+ \int_0^{t+\epsilon} \mathscr{T}_\beta(t+\epsilon-s)f(s)ds - \int_0^t \mathscr{T}_\beta(t-s)f(s)ds.\end{aligned} \tag{4.11}$$

In view of Proposition 1.15 and (i) in Proposition 1.18, we know that

$$\begin{aligned}\|\mathscr{S}_\beta(t+\epsilon)\phi - \mathscr{S}_\beta(t)\phi\| &= \left(\sum_{n=1}^{\infty}\left|\int_t^{t+\epsilon} -\lambda_n s^{\beta-1}E_{\beta,\beta}(-\lambda_n s^\beta)ds\right|^2 |\phi_n|^2\right)^{1/2}\\ &= \left(\sum_{n=1}^{\infty}\left(\int_t^{t+\epsilon} \lambda_n^{1-\gamma} s^{\beta-1}E_{\beta,\beta}(-\lambda_n s^\beta)ds\right)^2 \lambda_n^{2\gamma}|\phi_n|^2\right)^{1/2}\\ &\lesssim \left((t+\epsilon)^{\beta\gamma} - t^{\beta\gamma}\right)\|\phi\|_\gamma.\end{aligned}$$

From the definition of the fractional power space $D(A^\gamma)$ for $\gamma > 0$, in view of the Sobolev embedding $D(A^\gamma) \subset L^2(\Omega)$, it follows that $\|\phi\| \lesssim \|\phi\|_\gamma$. Moreover,

$$\begin{aligned}\|\mathscr{R}_\beta(t+\epsilon)\phi - \mathscr{R}_\beta(t)\phi\| &= \left(\sum_{n=1}^{\infty}\left|\int_t^{t+\epsilon} s^{\beta-\alpha-1}E_{\beta,\beta-\alpha}(-\lambda_n s^\beta)ds\right|^2 |\phi_n|^2\right)^{1/2}\\ &\lesssim \left((t+\epsilon)^{\beta-\alpha} - t^{\beta-\alpha}\right)\|\phi\|_\gamma.\end{aligned}$$

Propositions 1.15–1.16 imply

$$\|\mathscr{P}_\beta(t+\epsilon)\psi - \mathscr{P}_\beta(t)\psi\| = \left(\sum_{n=1}^{\infty}\left|\int_t^{t+\epsilon} E_{\beta,1}(-\lambda_n s^\beta)ds\right|^2 |\psi_n|^2\right)^{1/2} \lesssim \epsilon\,\|\psi\|.$$

By virtue of Proposition 1.16 again, one obtains

$$\|\mathscr{R}_\beta(t)v\| \lesssim t^{\beta-\alpha}\|v\|,\quad \|\mathscr{R}'_\beta(t)v\| \lesssim t^{\beta-\alpha-1}\|v\|,\quad v \in L^2(\Omega). \tag{4.12}$$

Therefore, it yields

$$\int_t^{t+\epsilon} \|\mathscr{R}'_\beta(t+\epsilon-s)u(s)\|ds \lesssim \epsilon^{\beta-\alpha}\,\|u\|_{\mathscr{C}}.$$

Moreover, by (i) in Proposition 1.18 with respect to $\mu = 1 - \frac{1}{\beta}$, we have

$$\begin{aligned}&\int_0^t \|(\mathscr{R}'_\beta(t+\epsilon-s) - \mathscr{R}'_\beta(t-s))u(s)\|ds\\ &= \int_0^t \left(\sum_{n=1}^{\infty}\left|\int_{t-s}^{t+\epsilon-s} \tau^{\beta-\alpha-2}E_{\beta,\beta-\alpha-1}(-\lambda_n\tau^\beta)d\tau\right|^2 (u(s),e_n)^2\right)^{1/2} ds\\ &\lesssim \int_0^t \left|\int_{t-s}^{t+\epsilon-s} \tau^{-\alpha-1}d\tau\right| ds\,\|u\|_{\mathscr{C}}\\ &\lesssim \left(\epsilon^{1-\alpha} + t^{1-\alpha} - (t+\epsilon)^{1-\alpha}\right)\|u\|_{\mathscr{C}}.\end{aligned}$$

Noting that

$$\|\mathscr{S}_\beta(t)v\| \lesssim \|v\|,\quad \|\mathscr{P}_\beta(t)v\| \lesssim t\,\|v\|,\quad \|\mathscr{T}_\beta(t)v\| \lesssim t^{\beta-1}\|v\|, \tag{4.13}$$

for all $v \in L^2(\Omega)$, we have

$$\begin{aligned}\int_t^{t+\epsilon} \|\mathscr{T}_\beta(t+\epsilon-s)f(s)\|ds &\lesssim \int_t^{t+\epsilon} (t+\epsilon-s)^{\beta-1}\|f(s)\|ds\\ &\lesssim \epsilon^{\beta-1}\,\|f\|_{L^1(0,T;L^2(\Omega))}.\end{aligned}$$

With the aid of Proposition 1.15, we deduce that

$$\begin{aligned}&\int_0^t \|(\mathscr{T}_\beta(t+\epsilon-s)-\mathscr{T}_\beta(t-s))f(s)\|ds\\&\lesssim\int_0^t\left((t+\epsilon-s)^{\beta-1}-(t-s)^{\beta-1}\right)\|f(s)\|ds\\&\lesssim\epsilon^{\beta-1}\|f\|_{L^1(0,T;L^2(\Omega))},\end{aligned}$$

where we use the following inequality:

$$\xi_1^\mu-\xi_2^\mu\le(\xi_1-\xi_2)^\mu,\quad \mu\in(0,1],\ \text{and}\ 0\le\xi_2\le\xi_1. \tag{4.14}$$

Therefore, together with the triangle inequality and the above estimates, we conclude that $\|(\mathscr{Q}u)(t+\epsilon)-(\mathscr{Q}u)(t)\|\to 0$ as ϵ tends to zero. An analogous argument can show that $\|(\mathscr{Q}u)(t)-(\mathscr{Q}u)(t-\epsilon)\|\to 0$ as ϵ tends to zero for $0\le t-\epsilon<t\le T$. Consequently, we obtain that $\mathscr{Q}u\in C([0,T];L^2(\Omega))$ for any $u\in C([0,T];L^2(\Omega))$.

We claim that $\mathscr{Q}$ has a unique fixed point. Indeed, for any $u_1,u_2\in C([0,T];L^2(\Omega))$, by (4.12), we have

$$\begin{aligned}\|(\mathscr{Q}u_1)(t)-(\mathscr{Q}u_2)(t)\|&\lesssim\int_0^t\|\mathscr{R}'_\beta(t-s)(u_1(s)-u_2(s))\|ds\\&\lesssim\int_0^t(t-s)^{\beta-\alpha-1}\|u_1(s)-u_2(s)\|ds\\&\le\frac{\Gamma(\beta-\alpha)}{\Gamma(\beta-\alpha+1)}t^{\beta-\alpha}\|u_1-u_2\|_{\mathscr{C}}.\end{aligned}$$

By mathematical induction, it follows that

$$\|(\mathscr{Q}^ju_1)(t)-(\mathscr{Q}^ju_2)(t)\|\lesssim\frac{(\Gamma(\beta-\alpha))^j}{\Gamma(j(\beta-\alpha)+1)}t^{j(\beta-\alpha)}\|u_1-u_2\|_{\mathscr{C}}. \tag{4.15}$$

If $j=1$, it has been proved. Assume that (4.15) holds for any $j>1$. We will show that (4.15) also holds for $j+1$. For this purpose, by using (4.12) again, one has

$$\begin{aligned}&\|(\mathscr{Q}^{j+1}u_1)(t)-(\mathscr{Q}^{j+1}u_2)(t)\|\\&\lesssim\int_0^t\|\mathscr{R}'_\beta(t-\tau)((\mathscr{Q}^ju_1)(\tau)-(\mathscr{Q}^ju_2)(\tau))\|d\tau\\&\lesssim\int_0^t(t-\tau)^{\beta-\alpha-1}\|(\mathscr{Q}^ju_1)(\tau)-(\mathscr{Q}^ju_2)(\tau)\|d\tau\end{aligned}$$

$$\lesssim \frac{(\Gamma(\beta-\alpha))^j}{\Gamma(j(\beta-\alpha)+1)} \int_0^t (t-\tau)^{\beta-\alpha-1} \tau^{j(\beta-\alpha)} d\tau \|u_1-u_2\|_{\mathscr{C}}$$
$$= \frac{(\Gamma(\beta-\alpha))^{j+1}}{\Gamma((j+1)(\beta-\alpha)+1)} t^{(j+1)(\beta-\alpha)} \|u_1-u_2\|_{\mathscr{C}}.$$

Thus, the inequality (4.15) follows for any $j+1$, and there exists a constant $C>0$ such that

$$\|(\mathscr{Q}^{j+1}u_1)(t)-(\mathscr{Q}^{j+1}u_2)(t)\| \le \frac{C(\Gamma(\beta-\alpha))^{j+1}}{\Gamma((j+1)(\beta-\alpha)+1)} t^{(j+1)(\beta-\alpha)} \|u_1-u_2\|_{\mathscr{C}}.$$

Let us choose $j=\hat{j}$ large enough so that

$$\varsigma := \frac{C(\Gamma(\beta-\alpha))^{\hat{j}}}{\Gamma(\hat{j}(\beta-\alpha)+1)} T^{\hat{j}(\beta-\alpha)} < 1.$$

Therefore, one has

$$\|\mathscr{Q}^{\hat{j}}u_1-\mathscr{Q}^{\hat{j}}u_2\|_{\mathscr{C}} \le \varsigma \|u_1-u_2\|_{\mathscr{C}}.$$

The contractility of $\mathscr{Q}^{\hat{j}}$ follows, and then $\mathscr{Q}^{\hat{j}}$ has a unique fixed point u^* on $C([0,T];L^2(\Omega))$. Since $\mathscr{Q}\mathscr{Q}^{\hat{j}}=\mathscr{Q}^{\hat{j}+1}=\mathscr{Q}^{\hat{j}}\mathscr{Q}$, one can see that $\mathscr{Q}^{\hat{j}}(\mathscr{Q}u^*)=\mathscr{Q}(\mathscr{Q}^{\hat{j}}u^*)=\mathscr{Q}u^*$ which deduce that $\mathscr{Q}u^*$ is the fixed point of $\mathscr{Q}^{\hat{j}}$. By virtue of the uniqueness, we conclude that $\mathscr{Q}u^*=u^*$. Consequently, there exists a unique mild solution.

Let us check (4.10). It follows from (4.12) and (4.13) that $\|\mathscr{S}_\beta(t)\phi\| \lesssim \|\phi\|_\gamma$, $\|\mathscr{P}_\beta(t)\psi\| \lesssim t\,\|\psi\|$, and $\|\mathscr{R}_\beta(t)\phi\| \lesssim t^{\beta-\alpha}\,\|\phi\|_\gamma$. Hence, one obtains

$$\|u(t)\| \lesssim \|\phi\|_\gamma + t^{\beta-\alpha}\,\|\phi\|_\gamma + t\|\psi\|$$
$$+ \int_0^t (t-\tau)^{\beta-\alpha-1}\|u(\tau)\|d\tau + \int_0^t (t-s)^{\beta-1}\|f(s)\|ds.$$

From the generalized Gronwall inequality (see, e.g., [233, Corollary 2]) and Proposition 1.17, there exists a positive constant C such that

$$\|u(t)\| \lesssim \varrho(t)\exp\left(\left(C\Gamma(\beta-\alpha)\right)^{\frac{1}{\beta-\alpha}} t\right),$$

where

$$\varrho(t) = \|\phi\|_\gamma + t^{\beta-\alpha}\|\phi\|_\gamma + t\|\psi\| + t^{\beta-1}\|f\|_{L^1(0,T;L^2(\Omega))}.$$

Thus, the main conclusion is obtained. We have completed this proof.

In what follows, we are in position to show the regularity of solution.

Theorem 4.2 *Let $(\phi, \psi) \in D(A^\gamma) \times L^2(\Omega)$ for $\gamma \in (0, 1)$, and let $\beta - 1 > \alpha$, $f \in L^p(0, T; L^2(\Omega))$ for $p > \frac{1}{\beta-1}$. Then, the solution u of the problem (4.4)–(4.6) satisfies*

$$\|\partial_t u(t)\| \lesssim \begin{cases} \|\phi\|_\gamma + \|\psi\| + \|f\|_{L^p(0,T;L^2(\Omega))}, & \beta\gamma \geq 1, \\ t^{\beta\gamma-1}\left(\|\phi\|_\gamma + \|\psi\| + \|f\|_{L^p(0,T;L^2(\Omega))}\right), & \beta\gamma < 1. \end{cases} \tag{4.16}$$

Proof Theorem 4.1 ensures a mild solution of the problem (4.4)–(4.6). Hence, it remains to check (4.16). For any $v \in L^2(\Omega)$, let

$$\mathscr{S}'_\beta(t)v = \sum_{n=1}^{\infty} -\lambda_n t^{\beta-1} E_{\beta,\beta}(-\lambda_n t^\beta)(v, e_n)e_n,$$

$$\mathscr{T}'_\beta(t)v = \sum_{n=1}^{\infty} t^{\beta-2} E_{\beta,\beta-1}(-\lambda_n t^\beta)(v, e_n)e_n,$$

and

$$\mathscr{R}''_\beta(t)v = \sum_{n=1}^{\infty} t^{\beta-\alpha-2} E_{\beta,\beta-\alpha-1}(-\lambda_n t^\beta)(v, e_n)e_n.$$

It is not difficult to check that $\|\mathscr{S}'_\beta(t)\phi\| \lesssim t^{\beta\gamma-1}\|\phi\|_\gamma$ and $\|\mathscr{T}'_\beta(t)f(\cdot)\| \lesssim t^{\beta-2}\|f(\cdot)\|$, respectively. It follows from (4.13) that $\|\mathscr{S}_\beta(t)\psi\| \lesssim \|\psi\|$. As the same argument, Proposition 1.16 shows that $\|\mathscr{R}''_\beta(t)u(\cdot)\| \lesssim t^{\beta-\alpha-2}\|u(\cdot)\|$. In view of Proposition 1.15, we have

$$\partial_t u(t) = \mathscr{S}'_\beta(t)\phi + \mathscr{R}'_\beta(t)\phi + \mathscr{S}_\beta(t)\psi - \int_0^t \mathscr{R}''_\beta(t-s)u(s)ds + \int_0^t \mathscr{T}'_\beta(t-s)f(s)ds. \tag{4.17}$$

Therefore, it follows that

$$\begin{aligned} \|\partial_t u(t)\| \lesssim & t^{\beta\gamma-1}\|\phi\|_\gamma + t^{\beta-\alpha-1}\|\phi\|_\gamma + \|\psi\| \\ & + \int_0^t (t-s)^{\beta-\alpha-2}\|u(s)\|ds + t^{\beta-1-\frac{1}{p}}\|f\|_{L^p(0,T;L^2(\Omega))}. \end{aligned}$$

Substituting (4.10) into the above inequality, we thus obtain the desired result.

Theorem 4.3 *Let $(\phi, \psi) \in D(A^\gamma) \times L^2(\Omega)$ for $\gamma \in (0, 1)$ satisfying $\frac{1}{\beta} \leq \gamma < \frac{\beta-\alpha}{\beta}$, where $\beta - 1 > \alpha$, and let $f \in L^p(0, T; L^2(\Omega))$ for $p > \frac{1}{\beta-\beta\gamma}$. Then the solution u belongs to $C((0, T]; H^{2\gamma}(\Omega))$ and satisfies*

$$\|u(t)\|_{H^{2\gamma}(\Omega)} \lesssim t^{-(\beta\gamma-1)} \left(\|\phi\|_\gamma + \|\psi\| + \|f\|_{L^p(0,T;L^2(\Omega))}\right). \tag{4.18}$$

Proof From the assumption of $1 \le \beta\gamma < \beta - \alpha$, one can see that there exists a mild solution u such that $u(t), \partial_t u(t) \in L^2(\Omega)$ for $t \in [0, T]$ by Theorems 4.1 and 4.2. Consequently, by virtue of (4.8) and transposition of term, we have

$$\int_0^t \mathscr{R}'_\beta(t-s)u(s)ds = \mathscr{R}_\beta(t)\phi + \int_0^t \mathscr{R}_\beta(t-s)\partial_s u(s)ds. \tag{4.19}$$

Hence, we only need to consider $u \in C((0, T]; H^{2\gamma}(\Omega))$, and further it satisfies (4.18). Initially, the Sobolev embedding $D(A^\gamma) \subset H^{2\gamma}(\Omega)$ for $\gamma > 0$ implies that if u belongs to $D(A^\gamma)$, then one has u belonging to $H^{2\gamma}(\Omega)$. Repeating the existence proof process of Theorem 4.1, we can verify $u \in C((0, T]; D(A^\gamma))$. It means $u \in C((0, T]; H^{2\gamma}(\Omega))$. Thus, it is sufficient to check the estimate (4.18). Now, from the definition of fractional power operators, it follows that

$$A^\gamma \mathscr{T}_\beta(t)v = t^{\beta-1}\sum_{n=1}^{\infty} \lambda_n^\gamma E_{\beta,\beta}(-\lambda_n t^\beta)(v, e_n)e_n, \quad t \ge 0,$$

for any $v \in L^2(\Omega)$ and $\gamma \in (0, 1)$. By Proposition 1.18, we have

$$\|A^\gamma \mathscr{T}_\beta(t)v\| \lesssim t^{\beta-\beta\gamma-1}\|v\|, \quad t > 0. \tag{4.20}$$

For $t \in (0, T]$, we set

$$\chi(t) = \int_0^t A^\gamma \mathscr{R}_\beta(t-s)\partial_s u(s)ds, \ \varphi(t) = \int_0^t A^\gamma \mathscr{T}_\beta(t-s)f(s)ds.$$

Applying (4.9) and (4.19), we conclude that

$$A^\gamma u(t) = A^\gamma \mathscr{S}_\beta(t)\phi + A^\gamma \mathscr{P}_\beta(t)\psi - \chi(t) + \varphi(t), \quad t \in (0, T].$$

On the other hand, Sobolev embedding theorem shows that $\|u(t)\|_{H^{2\gamma}(\Omega)} \lesssim \|u(t)\|_\gamma$. Thus, it is sufficient to estimate these terms $\|\mathscr{S}_\beta(t)\phi\|_\gamma$, $\|\mathscr{P}_\beta(t)\psi\|_\gamma$, $\|\chi(t)\|$, and $\|\varphi(t)\|$.

Proposition 1.16 implies

$$\|\mathscr{S}_\beta(t)\phi\|_\gamma = \left(\sum_{n=1}^{\infty} |E_{\beta,1}(-\lambda_n t^\beta)|^2 \lambda_n^{2\gamma} |\phi_n|^2\right)^{1/2} \lesssim \|\phi\|_\gamma. \tag{4.21}$$

With the help of (i) in Proposition 1.18, we get

$$\|\mathscr{P}_\beta(t)\psi\|_\gamma \le \left(\sum_{n=1}^{\infty} \left|\lambda_n^\gamma t E_{\beta,2}(-\lambda_n t^\beta)\right|^2 |\psi_n|^2\right)^{1/2} \lesssim t^{-(\beta\gamma-1)}\|\psi\|. \tag{4.22}$$

For the fourth term containing φ, by applying (4.20), we have the estimate

$$\|\varphi(t)\| \lesssim \int_0^t (t-s)^{\beta-\beta\gamma-1}\|f(s)\|ds \lesssim t^{\beta-\beta\gamma-\frac{1}{p}}\|f\|_{L^p(0,T;L^2(\Omega))}. \tag{4.23}$$

Therefore, it remains to verify the third term containing χ. Obviously, we get the inequality

$$\|A^\gamma \mathscr{R}_\beta(t)v\| \lesssim t^{\beta-\beta\gamma-\alpha-1}\|v\|, \quad v \in L^2(\Omega).$$

Therefore, one can see

$$\|\chi(t)\| \lesssim \int_0^t (t-s)^{\beta-\beta\gamma-\alpha-1}\|\partial_s u(s)\|ds.$$

By virtue of (4.16) and $\beta\gamma \ge 1$, the following estimate is established:

$$\|\chi(t)\| \lesssim t^{\beta-\beta\gamma-\alpha}\left(\|\phi\|_\gamma + \|\psi\| + \|f\|_{L^p(0,T;L^2(\Omega))}\right). \tag{4.24}$$

Together with (4.21) and (4.24), the proof is completed.

Using a similar argument as in Theorems 4.1 and 4.2, we can deduce the following conclusion.

Theorem 4.4 *Let $(\phi, \psi) \in D(A^\gamma) \times L^2(\Omega)$ for $\gamma \in (0,1)$ satisfying $\gamma \le \frac{\beta-1}{2\beta}$, and let $f \in L^\infty(0,T; D(A^{-\gamma}))$. Then the solution u of the problem (4.4)–(4.6) belongs to $L^\infty(0,T; H^{2\gamma}(\Omega))$. Moreover,*

$$\|u\|_{L^\infty(0,T;H^{2\gamma}(\Omega))} \lesssim \|\phi\|_\gamma + \|\psi\| + \|f\|_{L^\infty(0,T;D(A^{-\gamma}))}. \tag{4.25}$$

Proof Indeed, noting that $2\beta\gamma \le \beta - 1$, we get $0 < \beta\gamma < 1/2$, $\beta - \beta\gamma - 1 > 0$ and

$$\begin{aligned}\|\mathscr{T}_\beta(t)v\| &\le \left(\sum_{n=1}^{\infty} \left|\lambda_n^\gamma t^{\beta-1} E_{\beta,\beta}(-\lambda_n t^\beta)\right|^2 \lambda_n^{-2\gamma}|(v,e_n)|^2\right)^{1/2} \\ &\lesssim t^{\beta-\beta\gamma-1}\|v\|_{\gamma*}, \quad v \in D(A^{-\gamma}).\end{aligned}$$

From the assumption of f, by applying an analogous method of existence proof in Theorem 4.1, one can easily check that there exists a unique mild solution u, which satisfies

$$\|u(t)\| \lesssim \|\phi\|_\gamma + \|\psi\| + \|f\|_{L^\infty(0,T;D(A^{-\gamma}))}. \tag{4.26}$$

On the other hand, by virtue of Proposition 1.16, we see that

$$\|A^\gamma \mathscr{T}_\beta(t) f(\cdot)\| \lesssim t^{\beta-2\beta\gamma-1}\|f(\cdot)\|_{\gamma*}.$$

Therefore, one has

$$\begin{aligned}\left\|A^\gamma \mathscr{S}_\beta(t)\phi + A^\gamma \mathscr{P}_\beta(t)\psi + \varphi(t)\right\| \lesssim \|\phi\|_\gamma + t^{1-\beta\gamma}\|\psi\| \\ + t^{\beta-2\beta\gamma}\|f\|_{L^\infty(0,T;D(A^{-\gamma}))},\end{aligned}$$

where φ is defined in Theorem 4.3. Hence, it remains to estimate

$$A^\gamma \mathscr{R}_\beta(t)\phi - \int_0^t A^\gamma \mathscr{R}'_\beta(t-s)u(s)ds. \tag{4.27}$$

It is easy to estimate $\|A^\gamma \mathscr{R}_\beta(t)\phi\| \lesssim t^{\beta-\alpha}\|\phi\|_\gamma$. Now, we estimate another term of (4.27). Indeed, by (ii) in Proposition 1.18, we have

$$\begin{aligned}&\int_0^t \|A^\gamma \mathscr{R}'_\beta(t-s)u(s)\|ds \\ &= \int_0^t \left(\sum_{n=1}^\infty |\lambda_n^\gamma (t-s)^{\beta-\alpha-1} E_{\beta,\beta-\alpha}(-\lambda_n (t-s)^\beta)|^2 |u_n(s)|^2\right)^{\frac{1}{2}} ds \\ &\lesssim \int_0^t (t-s)^{\beta-\beta\gamma-\alpha-1}\|u(s)\|ds.\end{aligned}$$

Noting the assumption $2\beta\gamma \le \beta - 1$ and $\alpha \in (0, 1]$, by substituting (4.26) to the above inequality, we thus immediately conclude that the assertion of (4.25) is satisfied. This completes the proof.

In [186], the authors considered a fractional diffusion-wave problem, and further they obtained that the regularity property in time is of infinity order which means that $u \in C^\infty$ for $t > 0$. In [174], the authors derived some time regularity estimates for a weak solution of fractional wave equation, and they also corrected some papers including numerical technique, which ignores the situation that the solution will blow up at point $t = 0$ for the time regularity $u \in C^3$. Inspired by these works, we establish the following regularity results for time fractional damped wave equations.

Theorem 4.5 *Let $(\phi, \psi) \in D(A^\gamma) \times L^2(\Omega)$ for $\gamma \in (0, 1)$ satisfying $\frac{1}{\beta} \le \gamma$, and let $\beta - 1 > \alpha$. Assume that $f(0) \in L^2(\Omega)$ is finite and $f \in W^{1,p}(0, T; L^2(\Omega))$ for $p > \frac{1}{\beta-1}$. Then the mild solution u of the problem (4.4)–(4.6) satisfies*

$$\left\|\partial_t^2 u(t)\right\| \lesssim t^{-1}\left(\|\phi\|_\gamma + \|\psi\| + \|f\|_{W^{1,p}(0,T;L^2(\Omega))} + \|f(0)\|\right).$$

Proof By Theorems 4.1 and 4.2, the mild solution belongs to $C^1([0,T];L^2(\Omega))$. Hence, we can find a u satisfying (4.17). Invoking the initial value conditions $u(0) = \phi$ and $\partial_t u(0) = \psi$, by changing of variable and taking the derivative with respect to t in (4.17), we conclude that for $t > 0$,

$$\begin{aligned}\partial_t^2 u(t) =& \mathscr{S}_\beta''(t)\phi + \mathscr{S}_\beta'(t)\psi - \int_0^t \mathscr{R}_\beta''(s)\partial_t u(t-s)ds \\ &+ \mathscr{T}_\beta'(t)f(0) + \int_0^t \mathscr{T}_\beta'(s)\partial_t f(t-s)ds.\end{aligned}$$

On the other hand, Proposition 1.16 shows that

$$\|\mathscr{S}_\beta''(t)\phi\| \le t^{\beta-2}\left(\sum_{n=1}^\infty \left(\lambda_n^{1-\gamma}E_{\beta,\beta-1}(-\lambda_n t^\beta)\right)^2 \lambda_n^{2\gamma}|\phi_n|^2\right)^{1/2} \lesssim t^{\beta-2}\|\phi\|_\gamma.$$

For $t > 0$, we have the following inequalities:

$$\|\mathscr{S}_\beta''(t)\psi\| \lesssim t^{-1}\|\psi\|,\quad \|\mathscr{R}_\beta''(t)v\| \lesssim t^{\beta-\alpha-2}\|v\|,\quad \|\mathscr{T}_\beta'(t)v\| \lesssim t^{\beta-2}\|v\|,$$

for $v \in L^2(\Omega)$. Hence, we obtain the estimate

$$\begin{aligned}\left\|\partial_t^2 u(t)\right\| \lesssim& t^{\beta-2}\|\phi\|_\gamma + t^{-1}\|\psi\| + \int_0^t s^{\beta-\alpha-2}\|\partial_t u(t-s)\|ds \\ &+ t^{\beta-2}\|f(0)\| + \int_0^t s^{\beta-2}\|\partial_t f(t-s)\|ds,\end{aligned}$$

which implies from (4.16) the desired result. The proof is completed.

Remark 4.1 Let us mention that the time regularity of mild solutions in the present problem just achieves the second time derivative under the assumptions of Theorem 4.5. This is the difference between our results and the previous papers [174, 186] where they could establish more higher time regularity of solutions. Nevertheless, if we alter the initial value of ψ belonging to $D(A^\gamma)$, $f \in W^{2,p}(0,T;L^2(\Omega))$ such that $\partial_t f(0) \in L^2(\Omega)$, by the Sobolev embedding relationship $D(A^\gamma) \subset L^2(\Omega)$ for $\gamma \in (0,1)$, based on the existing assumptions in Theorem 4.5, we also establish a unique mild solution on $C^1([0,T];L^2(\Omega))$, and the solution will possess the third time derivative

$$\left\|\partial_t^3 u(t)\right\| \lesssim t^{\beta-\alpha-3}\left(\|\phi\|_\gamma + \|\psi\|_\gamma + \|f\|_{W^{2,p}(0,T;L^2(\Omega))} + \|f(0)\| + \|\partial_t f(0)\|\right).$$

4.1.4 Nonlinear Problems

In this subsection, we will take into account of the nonlinear problem for fractional wave equation with damping. Initially, as before, we introduce a suitable definition of mild solutions to the nonlinear problem.

Definition 4.2 Let $T > 0$. A function $u \in C([0, T]; L^2(\Omega))$ is said to be a mild solution of the problem (4.1)–(4.3), if u satisfies the following equation:

$$\begin{aligned} u(t) =& \mathscr{S}_\beta(t)\phi + \mathscr{R}_\beta(t)\phi + \mathscr{P}_\beta(t)\psi \\ &- \int_0^t \mathscr{R}'_\beta(t-s)u(s)ds + \int_0^t \mathscr{T}_\beta(t-s)f(u(s))ds. \end{aligned}$$

4.1.4.1 Well-Posedness

In the following part, we shall infer that the present problem is well-posed.

Lemma 4.1 *Let $\beta \in (1, 2)$. Then for $z \in \mathbb{C}$, there are important formulas between Mittag-Leffler functions, the Wright-type function, and sine/cosine functions given by*

$$\begin{aligned} E_{\beta,1}(-z^2) &= \int_0^\infty \mathscr{M}_{\beta/2}(\theta)\cos(z\theta)d\theta, \quad E_{\beta,\beta}(-z^2) \\ &= \frac{\beta}{2z}\int_0^\infty \theta\mathscr{M}_{\beta/2}(\theta)\sin(z\theta)d\theta. \end{aligned}$$

Proof The first identity was proved in [151, pp. 252]. Hence, it is sufficient to verify the second identity. Indeed, by developing the sine function in series, we get

$$\frac{1}{z}\int_0^\infty \theta\mathscr{M}_{\beta/2}(\theta)\sin(z\theta)d\theta = \sum_{k=0}^\infty \frac{(-1)^k z^{2k}}{(2k+1)!}\int_0^\infty \theta^{2k+2}\mathscr{M}_{\beta/2}(\theta)d\theta, \quad \text{for } z \in \mathbb{C}.$$

Applying the formula in (W3) of Proposition 1.19, it is easily seen that

$$\frac{1}{z}\int_0^\infty \theta\mathscr{M}_{\beta/2}(\theta)\sin(z\theta)d\theta = \sum_{k=0}^\infty \frac{(-1)^k z^{2k}(2k+2)}{\Gamma(1+(k+1)\beta)} = \frac{2}{\beta}E_{\beta,\beta}(-z^2).$$

Consequently, we get the desired formulas.

It is interesting to notice that the Mittag-Leffer function has a strong connection with the sine/cosine functions and the exponential function $\exp(z)$ (see, e.g., [128, 240]) such as

$$E_{\alpha,1}(z) = \int_0^\infty \mathscr{M}_\alpha(\theta)\exp(z\theta)d\theta,$$

$$E_{\alpha,\alpha}(z) = \alpha\int_0^\infty \theta\mathscr{M}_\alpha(\theta)\exp(z\theta)d\theta, \quad z \in \mathbb{C},\ \alpha \in (0,1).$$

This means that the Wright-type function acts as a bridge between the classical and fractional differential equations.

Lemma 4.2 *Let $\beta \in (1,2]$, $\alpha \in (0,1]$, and $\lambda > 0$. Then the following estimates hold for $t \geq 0$:*

$$|E_{\beta,\beta'}(-\lambda t^\beta)| \leq \frac{1}{\Gamma(\beta')},\ \text{for } \beta' = 1, 2, \beta;\ |E_{\beta,\beta-\alpha}(-\lambda t^\beta)| \leq \frac{1}{(\beta-1)\Gamma(\beta-\alpha)}.$$

Proof By properties of the Mittag-Leffler function in series, the case of $t = 0$ is obvious. Hence, for any $t > 0$, $z \in \mathbb{R}^+$, from the fact $E_{2,1}(-z^2) = \cos(z)$ and $zE_{2,2}(-z^2) = \sin(z)$, by using the inequalities $|\cos(z)| \leq 1$ and $\sin(z) \leq z$, it is easy to check the first inequality for $\beta = 2$, $\beta' = 1, 2$. By Lemma 4.1, it yields from (W3) of Proposition 1.19 and $|\cos(z)| \leq 1$ that $|E_{\beta,1}(-\lambda t^\beta)| \leq 1$. Proposition 1.15 checks $\frac{d}{dt}(tE_{\beta,2}(-\lambda t^\beta)) = E_{\beta,1}(-\lambda t^\beta)$, and hence $|E_{\beta,2}(-\lambda t^\beta)| \leq 1$ follows. Lemma 4.1 implies

$$t^{\beta-1}E_{\beta,\beta}(-\lambda t^\beta) = \frac{1}{\sqrt{\lambda}}\frac{\beta}{2}t^{\frac{\beta}{2}-1}\int_0^\infty \theta\mathscr{M}_{\beta/2}(\theta)\sin\left(\sqrt{\lambda}t^{\frac{\beta}{2}}\theta\right)d\theta. \tag{4.28}$$

We notice that the left side of the above equation tends to zero when $t \to 0$ and the right-hand side of (4.28), because of the fact $\lim_{z\to 0}\frac{\sin(z)}{z} = 1$. Consequently, by using $\sin(z) \leq z$ for $z \in \mathbb{R}^+$ and (W3) of Proposition 1.19, we get

$$|E_{\beta,\beta}(-\lambda t^\beta)| \leq \frac{\beta}{2}\int_0^\infty \theta^2\mathscr{M}_{\beta/2}(\theta)d\theta = \frac{1}{\Gamma(\beta)}.$$

Taking the derivative with respect to t in (4.28), noting that

$$\begin{aligned}
&\frac{d}{dt}\left(t^{\beta-1}E_{\beta,\beta}(-\lambda t^\beta)\right)\\
=&t^{\beta-2}\frac{\beta^2}{4}\int_0^\infty \theta^2\mathscr{M}_{\beta/2}(\theta)\cos\left(\sqrt{\lambda}t^{\frac{\beta}{2}}\theta\right)d\theta\\
&-t^{\frac{\beta}{2}-2}\frac{1}{\sqrt{\lambda}}\frac{\beta}{2}\left(1-\frac{\beta}{2}\right)\int_0^\infty \theta\mathscr{M}_{\beta/2}(\theta)\sin\left(\sqrt{\lambda}t^{\frac{\beta}{2}}\theta\right)d\theta,
\end{aligned} \tag{4.29}$$

in view of (1.13) and Proposition 1.15, we find

$$t^{\beta-\alpha-1}E_{\beta,\beta-\alpha}(-\lambda t^{\beta}) = \frac{1}{\Gamma(1-\alpha)}\int_0^t (t-s)^{-\alpha}\frac{d}{ds}\left(s^{\beta-1}E_{\beta,\beta}(-\lambda s^{\beta})\right)ds.$$

Substituting (4.29) into (4.28) implies

$$|t^{\beta-\alpha-1}E_{\beta,\beta-\alpha}(-\lambda t^{\beta})| \leq \frac{1}{(\beta-1)\Gamma(\beta-\alpha)}t^{\beta-\alpha-1}, \quad t>0.$$

Hence, it remains to check the case of $\beta = 2$, $\alpha \in (0, 1)$. Indeed, note that

$$t^{1-\alpha}E_{2,2-\alpha}(-\lambda t^2) = \frac{1}{\Gamma(1-\alpha)}\int_0^t (t-s)^{-\alpha}E_{2,1}(-\lambda s^2)ds,$$

which is easy to deduce that $|E_{2,2-\alpha}(-\lambda t^2)| \leq 1/\Gamma(2-\alpha)$. Thus, we obtain the desired results.

On the basis of the above arguments, we now show that the problem (4.1)–(4.3) is well-posed.

Theorem 4.6 *Let $\gamma \in (0, 1)$. Assume that there exist two positive constants a, b such that the nonlinear function $f \in L^1(\mathbb{R}, L^2(\Omega))$ satisfies the following conditions:*

$$\|f(u) - f(v)\| \leq a\left(\|u\|^{\vartheta-1} + \|v\|^{\vartheta-1}\right)\|u - v\|,$$

$$\|f(u)\| \leq b\left(1 + \|u\|^{\vartheta}\right),$$

for each $u, v \in L^2(\Omega)$, where $\vartheta \geq 1$ is a constant. Then for $\phi \in D(A^{\gamma})$, $\psi \in L^2(\Omega)$, the problem (4.1)–(4.3) possesses a unique mild solution on $C([0, T_0]; L^2(\Omega))$ for some $T_0 \in (0, T]$. Moreover, let $(u, \widetilde{u})$ be two mild solutions of the problem (4.1)–(4.3) associated with the initial conditions $(\phi, \widetilde{\phi})$ and $(\psi, \widetilde{\psi})$; then,

$$\|u(t) - \widetilde{u}(t)\| \lesssim \|\phi - \widetilde{\phi}\|_{\gamma} + \|\psi - \widetilde{\psi}\|. \tag{4.30}$$

Proof For fixed $r > 0$, let us introduce a metric space:

$$B_r(\phi, \psi) = \left\{u \in C([0, T]; L^2(\Omega)) : \rho_T\left(u, \mathscr{S}_{\beta}(t)\phi + \mathscr{R}_{\beta}(t)\phi + \mathscr{P}_{\beta}(t)\psi\right) \leq r\right\},$$

where

$$\rho_T(u_1, u_2) = \sup_{t\in[0,T]} \|u_1(t) - u_2(t)\|.$$

It is not difficult to check that $B_r(\phi, \psi)$ is a complete metric space with the above metric.

Let us consider an operator $\mathscr{Q}$ given by

$$\begin{aligned}(\mathscr{Q}u)(t) =&\mathscr{S}_\beta(t)\phi + \mathscr{P}_\beta(t)\psi + \mathscr{R}_\beta(t)\phi \\ &- \int_0^t \mathscr{R}'_\beta(t-s)u(s)ds + \int_0^t \mathscr{T}_\beta(t-s)f(u(s))ds,\end{aligned} \tag{4.31}$$

for any $u \in B_r(\phi, \psi)$. Clearly, $\mathscr{Q}$ is well defined in $C([0, T]; L^2(\Omega))$, as it follows from the assumptions of f. Next, we are planning to show the existence and uniqueness. It is sufficient to verify that $\mathscr{Q}$ has a unique fixed point in $B_r(\phi, \psi)$.

In view of Lemma 4.2, for $t \in [0, T]$, we get some exact upper bounds

$$\|\mathscr{S}_\beta(t)v\| \le \|v\|, \ \|\mathscr{P}_\beta(t)v\| \le t\,\|v\|, \ \|\mathscr{T}_\beta(t)v\| \le \frac{t^{\beta-1}}{\Gamma(\beta)}\|v\| \tag{4.32}$$

and $\|\mathscr{R}'_\beta(t)v\| \le \sigma\, t^{\beta-\alpha-1}\|v\|$, $\|\mathscr{R}_\beta(t)v\| \le \varrho\, t^{\beta-\alpha}\|v\|$ for any $v \in L^2(\Omega)$, where $\sigma := 1/((\beta-1)\Gamma(\beta-\alpha))$, $\varrho = \sigma/(\beta-\alpha)$. In view of the Sobolev embedding $D(A^\gamma) \subset L^2(\Omega)$, $\gamma \in (0, 1)$, one finds that $\|\phi\| \le \lambda_1^{-\gamma}\|\phi\|_\gamma$, where λ_1 is the first eigenvalue of operator A. Hence, taking

$$L_r := r + (1 + \varrho T^{\beta-\alpha})\lambda_1^{-\gamma}\|\phi\|_\gamma + T\|\psi\|, \tag{4.33}$$

the following estimate is established:

$$\begin{aligned}\|u(t)\| \le& \left\|u(t) - \mathscr{S}_\beta(t)\phi - \mathscr{R}_\beta(t)\phi - \mathscr{P}_\beta(t)\psi\right\| \\ &+ \|\mathscr{S}_\beta(t)\phi\| + \|\mathscr{R}_\beta(t)\phi\| + \|\mathscr{P}_\beta(t)\psi\| \\ \le& L_r.\end{aligned}$$

Choose $T_0 \in (0, T]$ such that

$$\varrho\, L_r T_0^{\beta-\alpha} + \frac{b}{\Gamma(\beta+1)}\left(1 + L_r^{\vartheta-1}\right)T_0^\beta \le r \tag{4.34}$$

and

$$\varrho T_0^{\beta-\alpha} + \frac{2a}{\Gamma(\beta+1)}L_r^{\vartheta-1}T_0^\beta \le \frac{1}{2}. \tag{4.35}$$

Therefore, from (4.34), we get

$$\begin{aligned}&\|(\mathscr{Q}u)(t) - \mathscr{S}_\beta(t)\phi - \mathscr{R}_\beta(t)\phi - \mathscr{P}_\beta(t)\psi\| \\ &\le \int_0^t \|\mathscr{R}'_\beta(t-s)u(s)\|ds + \int_0^t \|\mathscr{T}_\beta(t-s)f(u(s))\|ds\end{aligned}$$

$$\leq \sigma \int_0^t (t-s)^{\beta-\alpha-1} \|u(s)\| ds + \frac{b}{\Gamma(\beta)} \int_0^t (t-s)^{\beta-1}(1+\|u(s)\|^{\vartheta}) ds$$
$$\leq \varrho L_r T_0^{\beta-\alpha} + \frac{b}{\Gamma(\beta+1)} \left(1+L_r^{\vartheta-1}\right) T_0^{\beta} \leq r.$$

This implies that $\mathscr{Q}$ maps $B_r(\phi, \psi)$ into itself. In addition, for any $u, v \in B_r(\phi, \psi)$, by the assumption of f, we have

$$\|(\mathscr{Q}u)(t) - (\mathscr{Q}v)(t)\|$$
$$\leq \left\| \int_0^t \mathscr{R}'_\beta(t-s)\big(u(s)-v(s)\big) ds \right\|$$
$$+ \left\| \int_0^t \mathscr{T}_\beta(t-s)(f(u(s)) - f(v(s))) ds \right\|$$
$$\leq \sigma \int_0^t (t-s)^{\beta-\alpha-1} \|u(s)-v(s)\| \, ds$$
$$+ \frac{a}{\Gamma(\beta)} \int_0^t (t-s)^{\beta-1} \left(\|u(s)\|^{\vartheta-1} + \|v(s)\|^{\vartheta-1}\right) \|u(s)-v(s)\| ds$$
$$\leq \varrho t^{\beta-\alpha} \rho_t(u,v) + \frac{2a}{\Gamma(\beta+1)} L_r^{\vartheta-1} t^{\beta} \rho_t(u,v)$$
$$\leq \left(\varrho T_0^{\beta-\alpha} + \frac{2a}{\Gamma(\beta+1)} L_r^{\vartheta-1} T_0^{\beta} \right) \rho_{T_0}(u,v).$$

Hence, in view of (4.35), we conclude that $\mathscr{Q}$ is a contraction on $B_r(\phi, \psi)$. Thus, according to Banach fixed point theorem, the operator $\mathscr{Q}$ has a unique fixed point that is the mild solution of the problem (4.1)–(4.3) on $[0, T_0]$.

We next show the continuous dependence of the mild solution on the initial data.

$$\|u(t) - \widetilde{u}(t)\|$$
$$\leq \|\mathscr{S}_\beta(t)\phi - \mathscr{S}_\beta(t)\widetilde{\phi}\| + \|\mathscr{R}_\beta(t)\phi - \mathscr{R}_\beta(t)\widetilde{\phi}\| + \|\mathscr{P}_\beta(t)\psi - \mathscr{P}_\beta(t)\widetilde{\psi}\|$$
$$+ \int_0^t \left\| \mathscr{R}'_\beta(t-s)\big(u(s) - \widetilde{u}(s)\big) \right\| ds$$
$$+ \int_0^t \left\| \mathscr{T}_\beta(t-s)(f(u(s)) - f(\widetilde{u}(s))) \right\| ds$$
$$\leq \zeta(t) \|\phi - \widetilde{\phi}\|_\gamma + t \|\psi - \widetilde{\psi}\| + \sigma \int_0^t (t-s)^{\beta-\alpha-1} \|u(s) - \widetilde{u}(s)\| \, ds$$
$$+ \frac{a}{\Gamma(\beta)} \int_0^t (t-s)^{\beta-1} \left(\|u(s)\|^{\vartheta-1} + \|\widetilde{u}(s)\|^{\vartheta-1}\right) \|u(s) - \widetilde{u}(s)\| ds$$
$$\leq \Phi(t) + \Phi_r(t) \int_0^t \|u(s) - \widetilde{u}(s)\| ds,$$

where $\zeta(t) = \lambda_1^{-\gamma}\left(1 + \varrho t^{\beta-\alpha}\right)$ and

$$\Phi(t) = \zeta(t)\,\|\phi - \widetilde{\phi}\|_\gamma + t\,\|\psi - \widetilde{\psi}\|, \quad \Phi_r(t) = \left(\sigma + \frac{2a}{\Gamma(\beta)} L_r^{\vartheta-1} t^\alpha\right).$$

Thus, the generalized Gronwall inequality implies that

$$\|u(t) - \widetilde{u}(t)\| \lesssim \Phi(t) \exp\left(\left(\Phi_r(t)\Gamma(\beta-\alpha)\right)^{\frac{1}{\beta-\alpha}} t\right),$$

which means that the solution is continuous dependence on the initial conditions for any positive real number r. We thus have proved this theorem.

Remark 4.2 Noting that if the assumption of nonlinear function f is replaced by another local Lipschitz condition: There exists a nondecreasing function $L_f(\cdot) \in L^\infty(\mathbb{R}_+)$ such that the nonlinear mapping f is continuous with respect to t and satisfies the condition

$$\|f(u) - f(v)\| \le L_f(r)\|u - v\|, \quad r > 0,$$

for each $u, v \in L^2(\Omega)$ satisfying $\|u\|, \|v\| \le r$. Then, for some $T_0 \in (0, T)$, we get an analogous result of Theorem 4.6 on the following Banach space:

$$B_r(T_0, \phi) = \left\{u \in C([0, T_0]; L^2(\Omega)) : \sup_{t\in[0,T_0]} \|u(t)\| \le r\right\},$$

that is, for $\phi \in D(A^\gamma)$, $\psi \in L^2(\Omega)$, $\gamma \in (0, 1)$, the problem (4.1)–(4.3) possesses a unique mild solution on $C([0, T_0]; L^2(\Omega))$. Moreover, the solution depends continuously on the initial conditions.

4.1.4.2 Continuation and Blow-Up Alternative

Given a mild solution $u \in C([0, T_0]; L^2(\Omega))$ of the problem (4.1)–(4.3), we say that $\bar{u} : [0, T_1] \to L^2(\Omega)$ is a continuation of u with $T_1 > T_0$ if $\bar{u}$ is a mild solution, and $u(t) = \bar{u}(t)$ whenever $t \in [0, T_0]$.

Theorem 4.7 *Let the assumptions of Theorem 4.6 hold and u be a mild solution of the problem* (4.1)–(4.3) *on* $[0, T_0]$. *Then u can be uniquely continued up a time* T_1.

Proof Fix $R > 0$. Taking $T_1 > T_0$, we denote a metric space:

$$\mathscr{B}_R = \left\{v \in C([0, T_1]; L^2(\Omega)) : \rho_{T_1}(v, u(T_0)) \le R, \text{ and } v(t) = u(t),\ t \in [0, T_0]\right\},$$

equipped with the metric

$$\rho_{T_1}(v, u) = \sup_{t \in [0, T_1]} \|v(t) - u(t)\|.$$

It is not difficult to check that $\mathscr{B}_R$ is a complete metric space. Let us define $\mathscr{G} : \mathscr{B}_R \to \mathscr{B}_R$ by

$$(\mathscr{G}v)(t) = (\mathscr{G}_1 v)(t) + (\mathscr{G}_2 v)(t),$$

where

$$(\mathscr{G}_1 v)(t) = \mathscr{S}_\beta(t)\phi + \mathscr{R}_\beta(t)\phi + \mathscr{P}_\beta(t)\psi,$$
$$(\mathscr{G}_2 v)(t) = -\int_0^t \mathscr{R}'_\beta(t-s)v(s)ds + \int_0^t \mathscr{T}_\beta(t-s)f(v(s))ds.$$

If $v \in \mathscr{B}_R$, it is clear to obtain that $\mathscr{G}v(t) = u(t)$ for any $t \in [0, T_0]$. Let $t \in [T_0, T_1]$. For any $v \in \mathscr{B}_R$, by some simple computations, we get

$$\begin{aligned}\|(\mathscr{G}v)(t) - u(T_0)\| \le& \|(\mathscr{G}_1 v)(t) - (\mathscr{G}_1 v)(T_0)\| \\ &+ \int_0^{T_0} \|(\mathscr{R}'_\beta(t-s) - \mathscr{R}'_\beta(T_0-s))u(s)\|ds \\ &+ \int_0^{T_0} \|(\mathscr{T}_\beta(t-s) - \mathscr{T}_\beta(T_0-s))f(u(s))\|ds \\ &+ \int_{T_0}^t \|\mathscr{R}'_\beta(t-s)v(s)\|ds + \int_{T_0}^t \|\mathscr{T}_\beta(t-s)f(v(s))\|ds.\end{aligned}$$

Since the mappings $t \mapsto \mathscr{S}_\beta(t)\phi$, $t \mapsto \mathscr{R}_\beta(t)\phi$, and $t \mapsto \mathscr{P}_\beta(t)\psi$ belong to $C([0, T]; L^2(\Omega))$ for every $t \in [0, T]$ with $T > T_0$, it means that we can pick $T_a \in [T_0, T)$ such that for $t \in [T_0, T_a]$,

$$\|(\mathscr{G}_1 v)(t) - (\mathscr{G}_1 v)(T_0)\| \le \frac{R}{3}.$$

Processing as the proof of Theorem 4.1, one can see that for $t \in [T_0, T)$, Proposition 1.16, and by (i) in Proposition 1.18 with respect to $\mu = 1 - \frac{1}{\beta}$, we have

$$\begin{aligned}&\int_0^{T_0} \|(\mathscr{R}'_\beta(t-s) - \mathscr{R}'_\beta(T_0-s))u(s)\|ds \\ &+ \int_0^{T_0} \|(\mathscr{T}_\beta(t-s) - \mathscr{T}_\beta(T_0-s))f(u(s))\|ds \\ \lesssim& \int_0^{T_0} \left|\int_{T_0-s}^{t-s} \tau^{-\alpha-1}d\tau\right| \|u(s)\|ds + \int_0^{T_0} \left|\int_{T_0-s}^{t-s} \tau^{\beta-2}d\tau\right| \|f(u(s))\|ds\end{aligned}$$

$$\lesssim r\left(T_0^{1-\alpha} + (t-T_0)^{1-\alpha} - t^{1-\alpha}\right) + (t-T_0)^{\beta} L_r^{\vartheta}$$
$$\to 0, \quad \text{as } t \to T_0,$$

where r is picked as in Theorem 4.6 and L_r is defined in (4.33). Therefore, we can choose $T_b \in [T_0, T)$ such that for $t \in [T_0, T_b]$,

$$\int_0^{T_0} \|(\mathscr{R}'_\beta(t-s) - \mathscr{R}'_\beta(T_0-s))u(s)\|ds$$
$$+ \int_0^{T_0} \|(\mathscr{T}_\beta(t-s) - \mathscr{T}_\beta(T_0-s))f(u(s))\|ds \le \frac{R}{3}.$$

On the other hand, it is easy to see that

$$\begin{aligned}
&\int_{T_0}^t \|\mathscr{R}'_\beta(t-s)v(s)\|ds + \int_{T_0}^t \|\mathscr{T}_\beta(t-s)f(v(s))\|ds \\
\le& \sigma \int_{T_0}^t (t-s)^{\beta-\alpha-1}\|v(s)\|ds + \frac{1}{\Gamma(\beta)}\int_{T_0}^t (t-s)^{\beta-1}\|f(v(s))\|ds \\
\le& \varrho(t-T_0)^{\beta-\alpha}(R+L_r) + \frac{b}{\Gamma(\beta)}(t-T_0)^{\beta}(1+L_r^{\vartheta}),
\end{aligned}$$

where we use the fact $\|u(T_0)\| \le L_r$. With the same argument, one can choose $T_c \in [T_0, T)$ such that for $t \in [T_0, T_c]$,

$$\int_{T_0}^t \|\mathscr{R}'_\beta(t-s)v(s)\|ds + \int_{T_0}^t \|\mathscr{T}_\beta(t-s)f(v(s))\|ds \le \frac{R}{3}.$$

Consequently, let $T_1 := \min\{T_a, T_b, T_c\}$ and then

$$\|(\mathscr{G}v)(t) - u(T_0)\| \le R.$$

We thus prove that $\mathscr{G}$ maps $\mathscr{B}_R$ into itself. Now, for any $v, w \in \mathscr{B}_R$, one has

$$\begin{aligned}
\|(\mathscr{G}v)(t) - (\mathscr{G}w)(t)\| =& \|(\mathscr{G}_2v)(t) - (\mathscr{G}_2w)(t)\| \\
=& \int_0^t \|\mathscr{R}'_\beta(t-s)(v(s)-w(s))\|ds \\
&+ \int_0^t \|\mathscr{T}_\beta(t-s)(f(v(s)) - f(w(s))\|ds.
\end{aligned}$$

By the uniqueness, clearly for $t \in [0, T_0]$, $\mathscr{G}$ is contractive on $\mathscr{B}_R$. Let $t \in [T_0, T_1]$, and from the assumption of f, we have

$$
\begin{aligned}
&\|(\mathscr{G}v)(t) - (\mathscr{G}w)(t)\| \\
&\quad \le \sigma \int_{T_0}^{t} (t-s)^{\beta-\alpha-1} \|v(s) - w(s)\| ds \\
&\quad + \frac{a}{\Gamma(\beta)} \int_{T_0}^{t} (t-s)^{\beta-1} (\|v(s)\|^{\vartheta} + \|w(s)\|^{\vartheta}) \|v(s) - w(s)\| ds \\
&\quad \le \varrho (t - T_0)^{\beta-\alpha} \rho_{T_1}(v, w) + \frac{2a}{\Gamma(\beta)} (t - T_0)^{\beta} (R + L_r)^{\vartheta} \rho_{T_1}(v, w).
\end{aligned}
$$

Therefore, choosing T_1 such that for $t \in [T_0, T_1]$,

$$
\varrho (t - T_0)^{\beta-\alpha} + \frac{2a}{\Gamma(\beta)} (t - T_0)^{\beta} (R + L_r)^{\vartheta} < 1,
$$

we thus conclude that $\mathscr{G}$ is a contraction map on $\mathscr{B}_R$. This implies that $\mathscr{G}$ has a unique fixed point v on $\mathscr{B}_R$. We have finished this proof.

Theorem 4.8 *Let the assumptions of Theorem 4.6 hold and* $u \in C([0, T_{max}); L^2(\Omega))$ *be a mild solution of the problem* (4.1)–(4.3) *defined on the maximal interval* $[0, T_{max})$ *of existence. Then* $T_{max} = +\infty$ *or* $\lim_{t \to T_{max}^-} \|u(t)\| = \infty$ *if* $T_{max} < +\infty$.

Proof Let $T_{max} = \sup\{T \in [0, \infty) : \exists$ unique local solution u to (4.1)–(4.3) in $(0, T)\}$. Suppose that $T_{max} < \infty$, and there exists a positive constant $M < \infty$ such that $\|u(t)\| \le M$ for any $t \in [0, T_{max})$. Let $\{t_i\}_{i \in \mathbb{N}^+}$ be a sequence of $[0, T_{max})$ such that $t_i \to T_{max}^-$ as $i \to \infty$, we now consider the sequence $\{u(t_i)\}_{i \in \mathbb{N}^+} \in L^2(\Omega)$, and we will check that it is a Cauchy sequence in the space $L^2(\Omega)$. Setting $t_i > t_j$, we get

$$
\begin{aligned}
u(t_i) - u(t_j) =& \mathscr{S}_\beta(t_i)\phi - \mathscr{S}_\beta(t_j)\phi + \mathscr{R}_\beta(t_i)\phi - \mathscr{R}_\beta(t_j)\phi + \mathscr{P}_\beta(t_i)\psi - \mathscr{P}_\beta(t_j)\psi \\
&- \int_{t_j}^{t_i} \mathscr{R}'_\beta(t_i - s) u(s) ds - \int_0^{t_j} (\mathscr{R}'_\beta(t_i - s) - \mathscr{R}'_\beta(t_j - s)) u(s) ds \\
&+ \int_{t_j}^{t_i} \mathscr{T}_\beta(t_i - s) f(u(s)) ds + \int_0^{t_j} (\mathscr{T}_\beta(t_i - s) \\
&- \mathscr{T}_\beta(t_j - s)) f(u(s)) ds.
\end{aligned}
$$

Therefore, the same reasoning used as (4.11) in Theorem 4.1 and the similar process in Theorem 4.7 ensure that

$$
\|u(t_i) - u(t_j)\| \to 0, \quad \text{as } i, j \to \infty.
$$

Hence, $\{u(t_i)\}_{i \in \mathbb{N}^+}$ is a Cauchy sequence, and then there exists the limit

$$\lim_{i\to\infty} u(t_i) =: u(T_{max}) \in L^2(\Omega).$$

For the above reasons, we may extend u over a large interval $[0, T_{max}]$. This shows a contradiction with the maximality of T_{max}. The proof is completed.

4.1.4.3 Compactness Method

In the sequel, we remove the Lipschitz condition or higher smoothness assumption of $f \in C^1(\mathbb{R})$, and we also consider a more general condition. For this purpose, we need the following lemma.

Lemma 4.3 ([218]) *Let X be a Banach space, and let $R(Q)$ be the range of operator Q. Assume that $Q : X \to X$ is linear.*

(i) *If the dimension of $R(Q)$ is finite, then Q is compact.*
(ii) *If $\{Q_n\}_{n\in\mathbb{N}^+}$ is a sequence of compact operators in $\mathscr{B}(X)$ that converge uniformly to Q, then Q is compact.*

Definition 4.3 Let X be a Banach space. An operator-valued function $T(\cdot)$ defined on $\mathbb{R}_+$ is said to be:

(i) *Uniformly continuous*, if the map $t \mapsto T(t)x$ from $\mathbb{R}_+$ to $\mathscr{B}(X)$ is continuous with respect to the operator topology.
(ii) *Strongly continuous*, if the map $t \mapsto T(t)x$ from $\mathbb{R}_+$ to X is continuous for every $x \in X$.

Lemma 4.4 *Operator $\mathscr{T}_\beta(t)$ is compact for every $t \geq 0$ and is uniformly continuous on $L^2(\Omega)$ for all $t \geq 0$.*

Proof It is clear that $\mathscr{T}_\beta(0)$ is a zero operator, which is trivial involving with the compactness result. Let $t > 0$ be fixed, and let $\Omega_N = \text{span}\{e_1(x), \ldots, e_N(x)\}$, for every $N \in \mathbb{N}^+$. It is easy to see that $L^2(\Omega)$ can be expressed by $\text{span}\{e_1(x), \ldots, e_N(x), \ldots\}$. Obviously, Ω_N is a finite dimensional subspace of $L^2(\Omega)$. For all $N \in \mathbb{N}^+$, we denote operators $\mathscr{T}_\beta^N(t)\cdot : L^2(\Omega) \to \Omega_N$ by

$$\mathscr{T}_\beta^N(t)v = \sum_{n=1}^{N} t^{\beta-1} E_{\beta,\beta}(-\lambda_n t^\beta)(v, e_n)e_n(x).$$

Clearly, $\mathscr{T}_\beta^N(t)$ is also a linear finite dimensional operator. Applying (i) in Proposition 1.18 with respect to $\mu = 1 - \frac{1}{\beta}$, we have

$$\|\mathscr{T}_\beta^N(t)v\| = \left(\sum_{n=1}^{N} \lambda_n^{-\mu} \left(\lambda_n^\mu t^{\beta-1} E_{\beta,\beta}(-\lambda_n t^\beta)\right)^2 (v, e_n)^2\right)^{1/2} \lesssim \|v\|.$$

This yields that $\mathscr{T}_\beta^N(t)$ is well defined in $\mathscr{B}(L^2(\Omega))$. Thus, $R(\mathscr{T}_\beta^N(t))$ is finite, we conclude from (i) in Lemma 4.3 that the operator $\mathscr{T}_\beta^N(t)$ is a compact operator for every $N \in \mathbb{N}^+$.

Now, we shall prove that $\mathscr{T}_\beta^N(t)$ converges uniformly to $\mathscr{T}_\beta(t)$ whenever N tends to infinite. By applying the above argument, it is noticed that from the asymptotic property of the eigenvalues $\lambda_n \to \infty$ as $n \to \infty$, we have when $N \to \infty$,

$$\begin{aligned}\|\mathscr{T}_\beta(t)v - \mathscr{T}_\beta^N(t)v\| &\le \left(\sum_{n=N+1}^{\infty} \left(t^{\beta-1}E_{\beta,\beta}(-\lambda_n t^\beta)\right)^2 (v, e_n)^2\right)^{1/2} \\ &\lesssim \lambda_{N+1}^{-\mu}\|v\| \to 0.\end{aligned}$$

It means from (ii) in Lemma 4.3 that the operator $\mathscr{T}_\beta(t)$ is a compact operator on $\mathscr{B}(L^2(\Omega))$ for every $t \ge 0$.

In addition, in view of Proposition 1.15 and (4.14), for any $v \in L^2(\Omega)$, for $t_1, t_2 \in \mathbb{R}_+$ with $t_1 < t_2$, we find

$$\|\mathscr{T}_\beta(t_2)v - \mathscr{T}_\beta(t_1)v\| \lesssim (t_2 - t_1)^{\beta-1}\|v\| \to 0, \quad \text{as } t_2 \to t_1.$$

Therefore, we conclude that $\mathscr{T}_\beta(t)v$ is strong continuous for all $t \ge 0$. Combined with the compactness of $\mathscr{T}_\beta(t)$, this implies the desired result. The proof is completed.

Remark 4.3 It is notice that, by using the method of finite dimensional approximation and the compact results, Lemma 4.3, one can easily prove that operators $\mathscr{S}_\beta(t)$, $\mathscr{P}_\beta(t)$, and $\mathscr{R}_\beta(t)$ are compact for every $t \ge 0$ and are uniformly continuous for all $t \ge 0$. Additionally, $\mathscr{R}'_\beta(t)$ is also compact for every $t > 0$ since it may be unbounded at time $t = 0$ on $\mathscr{B}(L^2(\Omega))$ for $\beta - \alpha < 1$ and is uniformly continuous for all $t > 0$.

Theorem 4.9 *Let $(\phi, \psi) \in D(A^\gamma) \times L^2(\Omega)$ for $\gamma \in (0, 1)$. Assume that f is Lebesgue measurable with respect to t and is continuous with respect to u, and there exists a nonnegative nondecreasing function $W(\cdot) : \mathbb{R}^+ \to \mathbb{R}^+$ such that*

$$\|f(u(t))\| \le W(\|u(t)\|).$$

Assume further that

$$\liminf_{r\to\infty} \frac{W(r)}{r} = \varpi < \infty$$

and the following inequality

$$\frac{T^{\beta-\alpha}}{(\beta-1)\Gamma(\beta-\alpha+1)} + \frac{\varpi T^\beta}{\Gamma(\beta+1)} \le 1. \tag{4.36}$$

Then the problem (4.1)–(4.3) possesses at least one mild solution on $C([0,T]; L^2(\Omega))$.

Proof For each $r > 0$, let us set

$$B_r = \{u \in C([0,T]; L^2(\Omega)) : \|u\|_{\mathscr{C}} \le r\}.$$

Then B_r is a bounded closed and convex subset of $C([0,T]; L^2(\Omega))$. Consequently, we need to show that the operator equation $u = \mathscr{Q}u$ has a solution where $\mathscr{Q}$ is defined in Theorem 4.6.

Let us first check that operator $\mathscr{Q}$ maps B_r into itself. In fact, if this is not true, then for each $r > 0$, there exists $u^r \in B_r$ such that $\|(\mathscr{Q}u^r)(t_*)\| > r$ for some $t_* \in [0,T]$. In view of Lemma 4.2 and (4.32), one finds

$$\begin{aligned}
r <& \|(\mathscr{Q}u^r)(t_*)\| \\
\le& \|\mathscr{S}_\beta(t_*)\phi\| + \|\mathscr{R}_\beta(t_*)\phi\| + \|\mathscr{P}_\beta(t_*)\psi\| \\
&+ \int_0^{t_*} \|\mathscr{R}'_\beta(t_* - s)u^r(s)\|ds + \int_0^{t_*} \|\mathscr{T}_\beta(t_* - s)f(u^r(s))\|ds \\
\le& \lambda_1^{-\gamma}\|\phi\|_\gamma + t_*\|\psi\| + \sigma\left(\|\phi\| + r\right)\frac{t_*^{\beta-\alpha}}{\beta-\alpha} + \frac{1}{\Gamma(\beta)} \\
&\times \int_0^{t_*} (t_* - s)^{\beta-1} W(\|u^r(s)\|)ds \\
\le& \lambda_1^{-\gamma}\|\phi\|_\gamma + T\|\psi\| + \sigma\left(\lambda_1^{-\gamma}\|\phi\|_\gamma + r\right)\frac{T^{\beta-\alpha}}{\beta-\alpha} + \frac{T^\beta}{\Gamma(\beta+1)}W(r),
\end{aligned}$$

where $\sigma = 1/((\beta-1)\Gamma(\beta-\alpha))$. Dividing both sides by r and taking the lower limit as $r \to \infty$, we obtain that

$$1 < \frac{\sigma}{\beta-\alpha}T^{\beta-\alpha} + \frac{\varpi}{\Gamma(\beta+1)}T^\beta,$$

which contradicts (4.36). Therefore, one can select r such that $\|\mathscr{Q}u\|_{\mathscr{C}} \le r$. This implies that $\mathscr{Q}(B_r) \subseteq B_r$.

We claim that operator $\mathscr{Q}$ is completely continuous. To prove this property, we will divide the proof into three steps. First, we show that the set $\Theta = \{\mathscr{Q}u,\ u \in B_r\}$ is equicontinuous. Indeed, for $0 \le t_1 < t_2 \le T$, and we have

$$\begin{aligned}
&\|(\mathscr{Q}u)(t_2) - (\mathscr{Q}u)(t_1)\| \\
\le& \|\mathscr{S}_\beta(t_2)\phi - \mathscr{S}_\beta(t_1)\phi\| + \|\mathscr{R}_\beta(t_2)\phi - \mathscr{R}_\beta(t_1)\phi\| + \|\mathscr{P}_\beta(t_2)\psi - \mathscr{P}_\beta(t_1)\psi\| \\
&+ \left\|\int_0^{t_2} \mathscr{R}'_\beta(t_2 - s)u(s)ds - \int_0^{t_1} \mathscr{R}'_\beta(t_1 - s)u(s)ds\right\|
\end{aligned}$$

$$+\left\|\int_0^{t_2}\mathscr{T}_\beta(t_2-s)f(u(s))ds-\int_0^{t_1}\mathscr{T}_\beta(t_1-s)f(u(s))ds\right\|$$
$$=:J_1+J_2+J_3.$$

From Remark 4.3, we get

$$J_1=\|\mathscr{S}_\beta(t_2)\phi-\mathscr{S}_\beta(t_1)\phi\|+\|\mathscr{R}_\beta(t_2)\phi-\mathscr{R}_\beta(t_1)\phi\|+\|\mathscr{P}_\beta(t_2)\psi-\mathscr{P}_\beta(t_1)\psi\|$$
$$\to 0,\quad \text{as } t_2\to t_1.$$

As for J_2, for any $\varepsilon>0$, we estimate

$$\begin{aligned}
J_2\le&\int_0^{t_1-\varepsilon}\left\|\left(\mathscr{R}'_\beta(t_2-s)-\mathscr{R}'_\beta(t_1-s)\right)u(s)\right\|ds\\
&+\int_{t_1-\varepsilon}^{t_1}\left\|\left(\mathscr{R}'_\beta(t_2-s)-\mathscr{R}'_\beta(t_1-s)\right)u(s)\right\|ds+\int_{t_1}^{t_2}\left\|\mathscr{R}'_\beta(t_2-s)u(s)\right\|ds\\
\le&\int_0^{t_1-\varepsilon}\|u(s)\|\,ds\sup_{s\in[0,t_1-\varepsilon]}\left\|\mathscr{R}'_\beta(t_2-s)-\mathscr{R}'_\beta(t_1-s)\right\|_{\mathscr{B}(L^2(\Omega))}\\
&+\sigma\int_{t_1-\varepsilon}^{t_2}(t_2-s)^{\beta-\alpha-1}\|u(s)\|ds+\sigma\int_{t_1-\varepsilon}^{t_1}(t_1-s)^{\beta-\alpha-1}\|u(s)\|ds\\
\lesssim& r\sup_{s\in[0,t_1-\varepsilon]}\left\|\mathscr{R}'_\beta(t_2-s)-\mathscr{R}'_\beta(t_1-s)\right\|_{\mathscr{B}(L^2(\Omega))}\\
&+(\|\phi\|+r)((t_2-t_1+\varepsilon)^{\beta-\alpha}+\varepsilon^{\beta-\alpha})\\
\to& 0,\quad \text{as } t_2\to t_1,\ \varepsilon\to 0.
\end{aligned}$$

Next, we estimate J_3. In fact, Lemma 4.2 shows that

$$\begin{aligned}
J_3\le&\int_0^{t_1}W(\|u(s)\|)ds\sup_{s\in[0,t_1]}\left\|\mathscr{T}_\beta(t_2-s)-\mathscr{T}_\beta(t_1-s)\right\|_{\mathscr{B}(L^2(\Omega))}\\
&+\frac{1}{\Gamma(\beta)}\int_{t_1}^{t_2}(t_2-s)^{\beta-1}W(\|u(s)\|)ds\\
\lesssim& W(r)\sup_{s\in[0,t_1]}\left\|\mathscr{T}_\beta(t_2-s)-\mathscr{T}_\beta(t_1-s)\right\|_{\mathscr{B}(L^2(\Omega))}+W(r)(t_2-t_1)^\beta\\
\to& 0,\quad \text{as } t_2\to t_1.
\end{aligned}$$

Hence, it follows that $\|(\mathscr{Q}u)(t_2)-(\mathscr{Q}u)(t_1)\|$ tends to zero as $t_2-t_1\to 0$ independent of $u\in B_r$. Thus, we conclude that the set Θ is equicontinuous.

Secondly, we show that $\mathscr{Q}$ is continuous. For any $\{u_m\}_{m=1}^\infty\subset B_r$, $u\in B_r$ with $u_m\to u$ as $m\to\infty$. In view of the assumptions of f, one has

$$\lim_{m\to\infty} f(u_m(t)) = f(u(t)).$$

On the other hand, one has the inequalities

$$(t-s)^{\beta-\alpha-1}\|u_m(s)-u(s)\| \le 2(t-s)^{\beta-\alpha-1}r$$

and

$$(t-s)^{\beta-1}\|f(u_m(s))-f(u(s))\| \le 2(t-s)^{\beta-1}W(r)$$

which are integrable with respect to a.e. $s \in [0,t]$ and $t \in [0,T]$. Therefore, Lebesgue's dominated convergence theorem implies

$$\int_0^t (t-s)^{\beta-1}\|u_m(s)-u(s)\|ds \to 0, \quad \int_0^t (t-s)^{\beta-1}\|f(u_m(s))-f(u(s))\|ds \to 0,$$

as $m \to \infty$. Consequently, we get

$$\begin{aligned}
&\|(\mathscr{Q}u_m)(t)-(\mathscr{Q}u)(t)\| \\
\le&\sigma \int_0^t (t-s)^{\beta-\alpha-1}\|u_m(s)-u(s)\|ds \\
&+\frac{1}{\Gamma(\beta)}\int_0^t (t-s)^{\beta-1}\|f(u_m(s))-f(u(s))\|ds \to 0, \quad \text{as } m\to\infty.
\end{aligned}$$

This proves that $\mathscr{Q}u_m \to \mathscr{Q}u$ pointwise on $[0,T]$ as $m\to\infty$, which follows from the equicontinuity of Θ that $\mathscr{Q}u_m \to \mathscr{Q}u$ uniformly on $[0,T]$ as $m\to\infty$. Thus, $\mathscr{Q}$ is continuous.

Finally, we show that operator $\mathscr{Q}$ is compact. It is sufficient to prove that for any $t\in[0,T]$, $\Theta(t)$ is relatively compact in $L^2(\Omega)$. Obviously, for the case $t=0$, it is easy to see that $\Theta(0)$ is relatively compact. Let $t\in(0,T]$ be fixed, and since $\mathscr{R}'_\beta(t)$ and $\mathscr{T}_\beta(t)$ are compact for every $t>0$ in view of Lemma 4.2 and Remark 4.3, we can structure a family of finite dimensional compact operators as the same way in Lemma 4.2 by

$$\begin{aligned}
(\mathscr{Q}^N u)(t) =&\mathscr{S}_\beta(t)\phi+\mathscr{R}_\beta(t)\phi+\mathscr{P}_\beta(t)\psi \\
&-\int_0^t \mathscr{R}'^N_\beta(t-s)u(s)ds+\int_0^t \mathscr{T}^N_\beta(t-s)f(u(s))ds,
\end{aligned}$$

for every $N\in\mathbb{N}^+$, in which $\mathscr{T}^N_\beta(\cdot)$ is defined as in Lemma 4.2 and

$$\mathscr{R}'^N_\beta(t)u = \sum_{n=1}^N t^{\beta-\alpha-1}E_{\beta,\beta-\alpha}(-\lambda_n t^\beta)(u,e_n)e_n.$$

Obviously, one can repeat the above proof process, and then the relatively compactness of set $\Theta^N = \{\mathscr{Q}^N u : u \in B_r\}$ follows. On the other hand, by virtue of Proposition 1.16 and (i) in Proposition 1.18 with respect to $\mu \in \left(0, \frac{\beta-\alpha}{\beta}\right)$, it yields

$$\begin{aligned}\|(\mathscr{Q}u)(t) - (\mathscr{Q}^N u)(t)\| \leq & \int_0^t \|(\mathscr{R}'_\beta(t-s) - \mathscr{R}'^N_\beta(t-s))u(s)\|ds \\ & + \int_0^t \|(\mathscr{T}_\beta(t-s) - \mathscr{T}^N_\beta(t-s))f(u(s))\|ds \\ \lesssim & \lambda_{N+1}^{-\mu} \int_0^t (t-s)^{\beta(1-\mu)-\alpha-1}\|u(s)\|ds \\ & + \lambda_{N+1}^{-\mu} \int_0^t (t-s)^{\beta(1-\mu)-1}\|f(u(s))\|ds.\end{aligned}$$

Hence, it is easy to show that

$$\|(\mathscr{Q}u)(t) - (\mathscr{Q}^N u)(t)\| \to 0, \quad \text{as } N \to \infty.$$

This means that there are relatively compact sets arbitrarily close to the set $\Theta(t)$. Therefore, $\Theta(t)$ is relatively compact in $L^2(\Omega)$, and we derive that $\mathscr{Q}$ is a compact operator.

Now, let us finish this proof. By the above arguments and the Arzelà-Ascoli theorem, we know that $\mathscr{Q}$ is completely continuous. Therefore, the Schauder fixed point theorem implies that $\mathscr{Q}$ has at least one fixed point, which means that there exists at least one mild solution to the problem (4.1)–(4.3). The proof is completed.

Remark 4.4 Concerning the well-posedness in Theorem 4.6, it is indeed a local result corresponding existence interval $(0, T_0)$ sufficiently small such that (4.34) and (4.35) must be satisfied. Despite all of this, it is a new result to some special nonlinear functions, such as $f(u) = |u|^{\vartheta-1}u$, $\vartheta \geq 1$. Besides, the existence interval of Theorem 4.9 is not needed to make sufficiently small since we can get the exact interval of time from the exact upper bounds of Mittag-Leffler functions by (4.36) and Lemma 4.2, and it means that there may appear multiple solutions. Thus this conclusion extends certain results in literatures.

4.1.5 An Application

Let us take into account of the following time fractional telegraph equation:

$$\partial_t^{2\alpha} u(t,x) + \partial_t^\alpha u(t,x) = u_{xx}(t,x) + f(u(t,x)), \quad 0 < x < 1, \ t > 0,$$

where $\partial_t^{2\alpha}$ and ∂_t^{α} are fractional derivatives in the sense of Caputo type with respect to t of order $1/2 < \alpha \leq 1$. Specially, the case $\alpha = 1$ is related to the well-known telegraph process, which describes the propagation process of electron in telegraph cable, and it can be regarded as an integral order wave equation with damped term $\partial_t u$.

In the sequel, let us consider the boundary conditions $u(0,t) = u(1,t) = 0$, and let $\lambda_n = n^2\pi^2$ and $e_n = \sin(n\pi x)$, $n \in \mathbb{N}^+$, $\Omega = [0, 1]$; obviously, $\{-\lambda_n, e_n\}_{n=1}^{\infty}$ is the eigensystem in $L^2(\Omega)$ associated with operator $A = \frac{\partial^2}{\partial x^2}$. If $\phi(x) = \cos(x\pi/2)$, $\psi(x) = x$, and linear function $f(t,x) = t^2 \sin(x\pi/2)$, then it is easy to check that $\phi \in D(A^{\gamma})$ for $0 < \gamma < 1$, $\psi \in L^2(\Omega)$, and $f \in L^1(0, T; L^2(\Omega))$; then from [36], one can find a solution given by

$$
\begin{aligned}
u(t,x) = &\sum_{n=1}^{\infty} \Bigg(\int_0^t \tau^{2\alpha-1} E_{(\alpha,2\alpha),2\alpha}\left(-\tau^{\alpha}, -\lambda_n \tau^{2\alpha}\right) f_n(t-\tau) d\tau \\
&+ A_{1n}(0) B_1(t) + A_{2n}(0) B_2(t) \Bigg) e_n(x),
\end{aligned}
$$

where

$$
\begin{aligned}
&f_n(t) = 2 \int_0^1 f(t,x) e_n(x) dx, \\
&A_{1n}(0) = 2 \int_0^1 \phi(x) e_n(x) dx, \quad A_{2n}(0) = 2 \int_0^1 \psi(x) e_n(x) dx, \\
&B_1(t) = 1 - \lambda_n t^{2\alpha} E_{(\alpha,2\alpha),2\alpha+1}\left(-t^{\alpha}, -\lambda_n t^{2\alpha}\right), \\
&B_2(t) = t - t^{\alpha+1} E_{(\alpha,2\alpha),2\alpha+2}\left(-\tau^{\alpha}, -\lambda_n \tau^{2\alpha}\right) \\
&\quad - \lambda_n t^{2\alpha+1} E_{(\alpha,2\alpha),2\alpha+2}\left(-t^{\alpha}, -\lambda_n t^{2\alpha}\right),
\end{aligned}
$$

and for $b > 0$, $a_i > 0$, $|z_i| < \infty$, $i = 1, 2$, the multivariate Mittag-Leffler function is defined as

$$
E_{(\cdot),b}(\cdot) = E_{(a_1,a_2),b}(z_1, z_2) = \sum_{k=0}^{\infty} \sum_{\substack{l_1+l_2=k \\ l_1 \geq 0, l_2 \geq 0}} \frac{k!}{l_1! \times l_2!} \frac{z_1^{l_1} z_2^{l_2}}{\Gamma(b + a_1 l_1 + a_2 l_2)}.
$$

This ensures that the above solution also belongs to $C([0, T]; L^2(\Omega))$ according to Theorem 4.1, and further the solution will satisfy (4.10). Nevertheless, if nonlinear function $f(u) = |u|^{\vartheta-1}u$ for $\vartheta \geq 1$ or $f(u) = \sin(u)$, then we cannot use the method of [36] to establish the existence of solution, while there will exist a solution to the current problem by Theorem 4.6 and Theorem 4.9. Consequently, this extends the results in [36], and we will get a more general existence result for the case from the exponent $\beta = 2\alpha$ to $\beta \in (1, 2]$, $\alpha \in (0, 1]$.

4.2 Wave Equations on $\mathbb{R}^N$

4.2.1 Introduction

In this section, we focus on the following time fractional wave equation:

$$\partial_t^\beta u - \Delta u = f(u), \quad t > 0, \tag{4.37}$$

supplemented with the initial conditions

$$u(0, x) = \phi(x), \ \partial_t u(0, x) = \psi(x), \quad x \in \mathbb{R}^N, \tag{4.38}$$

where ∂_t^β stands for the Caputo fractional derivative operator of order $\beta \in (1, 2)$, Δ is the Laplacian operator, f is the semilinear data to be specified later, and ϕ and ψ are given initial data on some fractional Sobolev spaces, likely $(\phi, \psi) \in H^s(\mathbb{R}^N) \times H^{s-1}(\mathbb{R}^N)$ for $s \in \mathbb{R}$.

The time fractional partial differential equation $\partial_t^\beta u = \Delta u$ of order $\beta \in (0, 1)$ models anomalous diffusion phenomena ensuring the behavior of a subdiffusion process driven by a fractional Brownian motion [161]. The case of order $\beta \in (1, 2)$ will govern intermediate processes between diffusion and wave propagation, for example, see [151], and it also ensures the behavior of superdiffusion process, for instance, see [14]. For the cases $\beta \to 1^+$ and $\beta \to 2^-$, the equation corresponds to the diffusion equation (heat equation) and ballistic diffusion (wave equation), respectively. Besides, in anomalous diffusion equations of order $\beta \in (0, 1)$ or $\beta \in (1, 2)$, the mean squared displacement of a diffusive particle behaves likes $\langle x^2(t) \rangle \sim t^\beta$, in contrast to normal diffusion behavior (Brownian motion) of the form $\langle x^2(t) \rangle \sim t$.

There are many interesting works about time fractional wave equation. One of the most favorable reasons is that the integral kernel in time fractional derivative represents memory of a long-time tail of the power order. The investigation of existence and uniqueness of solutions for a low regularity initial data is a matter of interest in the mathematical analysis. For instance, Kian and Yamamoto [112] investigated a weak solution for semilinear case of (4.37) in bounded domain Ω for dimension $n = 2, 3$. By using the technique of eigenvalue expansion together with the properties of Mittag-Leffler functions, they established the existence and uniqueness results, which shall lie in $L^p(0, T; L^q(\Omega)) \cap C([0, T]; H^{2r}(\Omega))$ for $r = \min\{1 - 1/\beta, \gamma\}$ with some $1 \le p, q \le \infty, 0 < \gamma < 1$. Following this technique, Alvarez et al. [10] considered the well-posedness for an abstract Cauchy problem in a Hilbert space, where the solutions will lie in $L^{rq}(0, T; L^{2r}(X))$ associated with initial data $(\phi, \psi) \in D(A^{\gamma}) \times L^2(X)$, $D(A^{\gamma})$ is the fractional power spaces with spectrum form of A (nonnegative self-adjoint operator) for some $\beta \in (1, 2)$, $\gamma = 1/\beta$, X is a (relatively) compact metric space, and $q \in (1/(\beta - 1), \infty]$, $r > 1$. Moreover, Otarola and Salgado [174] studied the time and space regularities of weak solutions for the space-time fractional wave equation, Zhou and He [242] discussed well-posedness and time regularity of mild solutions for time fractional damped wave equations, etc.

However, few results discuss the unbounded domain case, likely the whole Euclidean space $\mathbb{R}^N$. This is due to the difficulty to establish the relevant estimates on solution operators, and also the method of eigenvalue expansion is not appropriate to such a problem. As we know, Alemida and Precioso [44] investigated the global existence with large initial data in the framework of Besov-Morrey spaces, and Alemida and Viana [45] studied existence, stability, self-similarity, and symmetries of solutions with initial data in Sobolev-Morrey space. Zhang and Li [238] studied the local existence on $C([0, T]; L^q(\mathbb{R}^N))$ for $\alpha N(p-1)/2 < q$ for a special semilinear function $f \sim |u|^p$, and also the critical exponents of the global existence and blow-up solutions are determined when $\psi \equiv 0$ and $\psi \not\equiv 0$. Djida et al. [54] worked on a well-posedness result for semilinear space-time fractional wave equation, by adopting the method of the Laplace-Fourier transforms, the properties of the Mittag-Leffler functions, and Fox H-functions.

Our goal in this section is to establish the well-posedness results for time fractional wave equation with initial conditions in certain function spaces. More precisely, we consider linear and semilinear problems on $\mathbb{R}^N$ and present a general assumption in semilinear function and a special case $\lambda|u|^{\alpha}u$ to deal with the current problem. In addition, this section is devoted to studying the well-posedness of mild solutions to the problems. In order to obtain the solution operators for time fractional wave equation and their properties, we establish some useful estimates about solution operators that should be required for the estimates of wave operators. We remark that our proof and results are completely different from the previous mentioned works. We will concern about the local/global well-posedness of solutions. Let us now enlist the main results presented in this section.

I. The solution operators of Eq. (4.37). Let $\varpi = (-\Delta)^{1/2}$, and we know that the wave operators can be given by $\cos(\varpi t)$ and $\varpi^{-1}\sin(\varpi t)$. Concerning the principle of subordination, two inherent relationships between probability density function and wave operators are presented. More precisely, we get

$$\mathscr{C}_{\sigma}(t) = \int_0^{\infty} \mathscr{M}_{\sigma}(\theta)\cos(\varpi t^{\sigma}\theta)d\theta,$$

$$\mathscr{S}_{\sigma}(t) = \sigma t^{\sigma-1}\int_0^{\infty} \theta\mathscr{M}_{\sigma}(\theta)\varpi^{-1}\sin(\varpi t^{\sigma}\theta)d\theta.$$

These forms of solution operators relied upon the estimates of wave operators, which are completely different from previous literatures. Based on this, some useful estimates of solution operators can be derived, which provide very helpful tools for the proof. Observe that the probability density function builds a bridge between integer wave equations and the fractional one.

II. The well-posedness results on $\mathbb{R}^N$. We first establish some estimates on several spaces for the linear problem with initial data $(\phi, \psi) \in H^s(\mathbb{R}^N) \times H^{s-1}(\mathbb{R}^N)$, and then an existence of L^2-solution is established on a space of continuous functions. Next, under the case that the semilinear function $f : L^2(\mathbb{R}^N) \cap L^r(\mathbb{R}^N) \to L^{r'}(\mathbb{R}^N)$ and satisfies

$$\|f(u) - f(v)\|_{L^{r'}(\mathbb{R}^N)} \le C_\alpha(R)\left(\|u\|^\alpha_{L^r(\mathbb{R}^N)} + \|v\|^\alpha_{L^r(\mathbb{R}^N)}\right)\|u - v\|_{L^r(\mathbb{R}^N)},$$

where constants r and α satisfy some restriction requirements, the local well-posedness results of mild solutions are established in the framework of $L^\gamma(L^{\gamma'})$ spaces; here γ and γ' are the conjugate indices. By concerning a special semilinear data $f \sim \lambda|u|^\alpha u$, we show the local well-posedness on Besov spaces $B^s_{r,2}(\mathbb{R}^N)$. Furthermore, when $\psi \equiv 0$, we also show the global existence to the Cauchy problem (4.37)–(4.38).

The rest of this section is divided into three subsections. In Sect. 4.2.2, some basic notations and useful preliminaries are introduced. In Sect. 4.2.3, for the linear problem, we derive the solution operators and establish their some properties. In addition, an existence of L^2-solutions is given. In Sect. 4.2.4, we prove some local/global well-posedness results on Lebesgue and Besov spaces for the semilinear problems.

4.2.2 Preliminaries

In this subsection, some notations and preliminaries related to our work will be introduced.

Denote by $L^q(\mathbb{R}^N)$ $(q \ge 1)$ the Lebesgue space of q-integrable functions with the norm $\|\cdot\|_{L^q(\mathbb{R}^N)}$. Let $\mathscr{S}(\mathbb{R}^N)$ be the Schwartz space, and $\mathscr{S}'(\mathbb{R}^N)$ be its topological dual; for any $v \in \mathscr{S}(\mathbb{R}^N)$, $\mathscr{F}$ represents the Fourier transform

$$\hat{v}(\xi) = \mathscr{F}(v)(\xi) = (2\pi)^{-N/2}\int_{\mathbb{R}^N} e^{-ix\cdot\xi}v(x)dx,$$

with its inverse

$$\check{v}(x) = \mathscr{F}^{-1}(v)(x) = (2\pi)^{-N/2}\int_{\mathbb{R}^N} e^{i\xi\cdot x}v(\xi)d\xi.$$

Define the Sobolev space by

$$H^{s,q}(\mathbb{R}^N) = \{u \in \mathscr{S}'(\mathbb{R}^N) : \mathscr{F}^{-1}[(1+|\xi|^2)^{\frac{s}{2}}\mathscr{F}(u)] \in L^q(\mathbb{R}^N)\},$$

which is equipped with the norm

$$\|u\|_{H^{s,q}(\mathbb{R}^N)} := \|\mathscr{F}^{-1}[(1+|\xi|^2)^{\frac{s}{2}}\mathscr{F}(u)]\|_{L^q(\mathbb{R}^N)},$$

for $s \in \mathbb{R}$, $1 \le q \le \infty$. We also use the Besov space $B^s_{p,q} := B^s_{p,q}(\mathbb{R}^N)$ and homogeneous Besov space $\dot{B}^s_{p,q} := \dot{B}^s_{p,q}(\mathbb{R}^N)$, for $s \in \mathbb{R}$, $1 \le p, q \le \infty$. For the definitions and properties of Besov spaces, we refer to see [18, 32]. In particular, it yields $H^{0,p}(\mathbb{R}^N) = L^p(\mathbb{R}^N)$ and $H^s(\mathbb{R}^N) = H^{s,2}(\mathbb{R}^N) = B^s_{2,2}(\mathbb{R}^N)$ for $p \ge 1$ and $s \in \mathbb{R}$. Throughout this section, we denote the notation $a \lesssim b$ that stands for $a \le Cb$, with a positive generic constant C that does not depend on a, b, and the notations $\vee$ and $\wedge$ stand for $a \vee b = \max\{a, b\}$ and $a \wedge b = \min\{a, b\}$, respectively. Let p and p' be the conjugate indices such that $1/p + 1/p' = 1$.

Let $T > 0$, and let X be a usual Banach space. For any $u \in L^1(0, T; X)$ and $v \in L^1(0, T; X)$, denote $*$ the convolution by

$$(u * v)(t) = \int_0^t u(t-s)v(s)ds, \quad t \ge 0$$

and for $\beta \ge 0$. Let the weak singular kernel $g_\beta(\cdot)$ be defined by

$$g_\beta(t) = t^{\beta-1}/\Gamma(\beta), \quad t > 0,$$

where $\Gamma(\cdot)$ is the Gamma function. Next, let us recall the concepts of fractional calculus and Mittag-Leffler functions. The Riemann-Liouville fractional integral of order $\beta \ge 0$ for a function $v \in L^1(0, T; X)$ is defined as

$$J_t^\beta v(t) = \frac{1}{\Gamma(\beta)} \int_0^t (t-s)^{\beta-1} v(s)ds = (g_\beta * v)(t), \; t > 0.$$

Definition 4.4 Let $\beta \in (1, 2)$ and $T > 0$. Consider a function $v \in L^1(0, T; X)$ such that convolution $g_{2-\beta} * v \in W^{2,1}(0, T; X)$. The representation

$$\partial_t^\beta v(t) = \partial_{tt}^2 \big(g_{2-\beta} * [v(t) - v(0) - t\, \partial_t v(0)]\big)$$

is called the Caputo fractional derivative of order β.

An important special function for the fractional differential equations involving the Caputo fractional derivative is the Mittag-Leffler function, which is defined by

$$E_{\nu,\mu}(z) = \sum_{k=0}^{\infty} \frac{z^k}{\Gamma(\nu k + \mu)}, \quad z \in \mathbb{C}, \; \nu > 0, \; \mu \in \mathbb{R}.$$

Note that if $w(t) := E_{\nu,1}(at^\nu)$, $\nu \in (0, 2)$, $a \in \mathbb{R}$, then one can check that w is the solution of the equation ${}_0^C D_t^\nu w(t) = aw(t)$.

Let $\mathscr{M}_\upsilon(\cdot)$ be the Wright-type function as in Definition 1.8. For $\theta > 0$, $\mathscr{M}_\upsilon(\cdot)$ has the properties

$$\mathscr{M}_\upsilon(\theta) \geq 0, \quad \int_0^\infty \theta^\delta \mathscr{M}_\upsilon(\theta) d\theta = \frac{\Gamma(1+\delta)}{\Gamma(1+\upsilon\delta)}, \quad \text{for } -1 < \delta < \infty. \tag{4.39}$$

Next, we see that the Wright-type function can be viewed as a bridge between the classical and fractional wave equations.

Lemma 4.5 ([242]) *Let $\beta \in (1,2)$. Then for $z \in \mathbb{C}$, the following formulas expressing the Mittag-Leffler function in terms of probability density function hold:*

$$E_{\beta,1}(-z^2) = \int_0^\infty \mathscr{M}_{\beta/2}(\theta)\cos(z\theta)d\theta, \quad E_{\beta,\beta}(-z^2) = \frac{\beta}{2z}\int_0^\infty \theta \mathscr{M}_{\beta/2}(\theta)\sin(z\theta)d\theta.$$

Lemma 4.6 ([115]) *Let $f \in L^1(0,T;\mathbb{R})$. The unique solution of the fractional order problem*

$$\begin{cases} {}_0^C D_t^\beta u(t) + au(t) = f(t), & a \geq 0,\ t \geq 0, \\ u(0) = u_0, \ u'(0) = u_1 \end{cases}$$

is given by

$$u(t) = E_{\beta,1}(-at^\beta)u_0 + tE_{\beta,2}(-at^\beta)u_1 + \int_0^t (t-s)^{\beta-1}E_{\beta,\beta}(-a(t-s)^\beta)f(s)ds.$$

In particular,

$$u(t) = \cos(\sqrt{a}\,t)u_0 + \frac{1}{\sqrt{a}}\sin(\sqrt{a}\,t)u_1 + \frac{1}{\sqrt{a}}\int_0^t \sin(\sqrt{a}\,(t-s))f(s)ds,$$

which is the unique solution to the corresponding classical wave equation, i.e., $\beta = 2$.

4.2.3 Local/Global Solutions of Linear Problems

In this subsection, we are concerned with the following linear Cauchy problem:

$$\begin{cases} \partial_t^\beta u(t,x) - \Delta u(t,x) = f(t,x), & t > 0,\ x \in \mathbb{R}^N, \\ u(0,x) = \phi(x),\ \partial_t u(0,x) = \psi(x). \end{cases} \tag{4.40}$$

Without loss of generality, the solutions in this subsection are defined as mild solutions associated with the corresponding initial data.

4.2.3.1 Solution Representation

We first establish the solution representation of linear problem (4.40). Let u satisfy (4.40); taking the Fourier transform of both sides in (4.40) with respect to $x \in \mathbb{R}^N$, we obtain

$$\begin{cases} \partial_t^\beta \hat{u}(t,\xi) + |\xi|^2 \hat{u}(t,\xi) = \hat{f}(t,\xi), & t > 0, \\ \hat{u}(0,\xi) = \hat{\phi}(\xi), \ \partial_t \hat{u}(0,\xi) = \hat{\psi}(\xi). \end{cases}$$

It follows from [180, (1.100)] that

$$tE_{\beta,2}(-t^\beta|\xi|^2) = \frac{1}{\Gamma(2-\beta)} \int_0^t (t-s)^{1-\beta} E_{\beta,\beta}(-s^\beta|\xi|^2) s^{\beta-1} ds.$$

Therefore, by virtue of Lemma 4.6, we get

$$\hat{u}(t,\xi) = \hat{\varphi}(t,\xi)\hat{\phi}(\xi) + (\hat{\vartheta}(\cdot,\xi) * g_{2-\beta})(t)\hat{\psi}(\xi) + (\hat{\vartheta}(\cdot,\xi) * \hat{f})(t),$$

where

$$\hat{\varphi}(t,\xi) = E_{\beta,1}(-t^\beta|\xi|^2), \quad \hat{\vartheta}(t,\xi) = t^{\beta-1} E_{\beta,\beta}(-t^\beta|\xi|^2).$$

By using the inverse Fourier transform, we get

$$\begin{aligned} u(t,x) = &\int_{\mathbb{R}^N} \varphi(t,x-y)\phi(y)dy + \int_0^t \int_{\mathbb{R}^N} g_{2-\beta}(t-s)\vartheta(s,x-y)\psi(y)dyds \\ &+ \int_0^t \int_{\mathbb{R}^N} \vartheta(t-s,x-y)f(s,y)dyds, \end{aligned}$$

where

$$\begin{aligned} \varphi(t,x) &= (2\pi)^{-N/2} \int_{\mathbb{R}^N} e^{ix\cdot\xi} E_{\beta,1}(-t^\beta|\xi|^2)d\xi, \\ \vartheta(t,x) &= (2\pi)^{-N/2} \int_{\mathbb{R}^N} e^{ix\cdot\xi} t^{\beta-1} E_{\beta,\beta}(-t^\beta|\xi|^2)d\xi. \end{aligned}$$

On the other hand, set $\sigma = \beta/2 \in (1/2, 1)$, and it follows from Lemma 4.5 that

$$\varphi(t,x) = (2\pi)^{-N/2} \int_0^\infty \int_{\mathbb{R}^N} e^{ix\cdot\xi} \mathscr{M}_\sigma(\theta) \cos(t^\sigma \theta|\xi|) d\xi d\theta$$

and

$$\vartheta(t,x) = t^{\sigma-1} (2\pi)^{-N/2} \int_0^\infty \int_{\mathbb{R}^N} e^{ix\cdot\xi} \sigma\theta \mathscr{M}_\sigma(\theta) \frac{\sin(t^\sigma\theta|\xi|)}{|\xi|} d\xi d\theta.$$

Let

$$\dot{K}(t)v = \mathscr{F}^{-1}[\cos(t|\xi|)\hat{v}(\xi)], \quad K(t)v = \mathscr{F}^{-1}\left[\frac{\sin(t|\xi|)}{|\xi|}\hat{v}(\xi)\right]$$

and

$$\mathscr{S}_\sigma(t)v = \sigma t^{\sigma-1} \int_0^\infty \theta \mathscr{M}_\sigma(\theta) K(t^\sigma\theta) v d\theta.$$

Rewriting $u(t)$ for the function $u(t,\cdot)$, we get an equivalent integral representation for the problem (4.40) by

$$u(t) = \mathscr{C}_\sigma(t)\phi + \mathscr{P}_\sigma(t)\psi + \int_0^t \mathscr{S}_\sigma(t-s) f(s) ds, \tag{4.41}$$

where solution operators $\mathscr{C}_\sigma(\cdot)$ and $\mathscr{P}_\sigma(\cdot)$ are defined by

$$\mathscr{C}_\sigma(t)\phi = \int_0^\infty \mathscr{M}_\sigma(\theta) \dot{K}(t^\sigma\theta)\phi d\theta, \quad \mathscr{P}_\sigma(t)\psi = (g_{2-2\sigma} * \mathscr{S}_\sigma)(t)\psi.$$

4.2.3.2 Some Properties of Solution Operators

Let $\alpha(r) = \frac{1}{2} - \frac{1}{r}$ for $r \in [2,\infty]$, and

$$\beta(r) = \frac{N+1}{2}\alpha(r), \quad \gamma(r) = (N-1)\alpha(r), \quad \delta(r) = N\alpha(r).$$

Let us recall the following two results in [176], which play a key role in proving the general results on the solution operators.

Lemma 4.7 *Let $2 \le p < \infty$ and $2\beta(p) \le \nu \le 2\delta(p)$. Then for $t \neq 0$, it follows that*

$$\left\| \mathscr{F}^{-1}\left[|\xi|^{-\nu} e^{it|\xi|} \hat{v}(\xi)\right] \right\|_{L^p(\mathbb{R}^N)} \lesssim |t|^{\nu - 2\delta(p)} \|v\|_{L^{p'}(\mathbb{R}^N)}.$$

Lemma 4.8 *Let $2 \le p < \infty$ and $2\beta(p) \le \nu \le 2\delta(p)$, $0 \le \mu \le s + \nu$. Then for $1 \le q \le \infty$, $t \neq 0$, it follows that*

$$\left\| \mathscr{F}^{-1}\left[|\xi|^{-\mu}e^{it|\xi|}\hat{v}(\xi)\right]\right\|_{B^s_{p,q}(\mathbb{R}^N)} \lesssim |t|^{\nu-2\delta(p)}\|v\|_{B^{s+\nu-\mu}_{p',q}(\mathbb{R}^N)}.$$

In the sequel, we set $\varpi = (-\Delta)^{\frac{1}{2}}$ and $U(t) = \exp(i\varpi t) = \mathscr{F}^{-1}[\exp(i|\xi|t)\mathscr{F}]$, so that $K(t) = \varpi^{-1}\sin(\varpi t)$, $\dot{K}(t) = \cos(\varpi t)$. Hence, it follows that

$$K(t) = \varpi^{-1}\frac{U(t) - U(-t)}{2i}, \quad \dot{K}(t) = \frac{U(t) + U(-t)}{2}.$$

Lemma 4.9 *Let $N \geq 2$, $2N/(N-1) \leq p \leq 2(N+1)/(N-1)$, then $\mathscr{S}_\sigma(t) : L^{p'}(\mathbb{R}^N) \to L^p(\mathbb{R}^N)$, and moreover*

$$\|\mathscr{S}_\sigma(t)v\|_{L^p(\mathbb{R}^N)} \lesssim t^{2\sigma-2\sigma\delta(p)-1}\|v\|_{L^{p'}(\mathbb{R}^N)}, \quad v \in L^{p'}(\mathbb{R}^N), \ t > 0.$$

Proof Obviously, the condition $\frac{2N}{N-1} \leq p \leq \frac{2(N+1)}{N-1}$ for $N \geq 2$ implies that $1/2 \leq \delta(p) \leq N/(N+1)$. Hence, it follows from Lemma 4.7 that

$$\|K(t)v\|_{L^p(\mathbb{R}^N)} \lesssim t^{1-2\delta(p)}\|v\|_{L^{p'}(\mathbb{R}^N)}.$$

By the definition of $\mathscr{S}_\sigma(t)$ and (4.39), we obtain

$$\begin{aligned}
\|\mathscr{S}_\sigma(t)v\|_{L^p(\mathbb{R}^N)} &\leq \sigma t^{\sigma-1}\int_0^\infty \theta\mathscr{M}_\sigma(\theta)\|K(t^\sigma\theta)v\|_{L^p(\mathbb{R}^N)}d\theta \\
&\lesssim \sigma t^{\sigma-1+\sigma(1-2\delta(p))}\int_0^\infty \theta^{2-2\delta(p)}\mathscr{M}_\sigma(\theta)d\theta\|v\|_{L^{p'}(\mathbb{R}^N)} \\
&\lesssim t^{2\sigma-2\sigma\delta(p)-1}\|v\|_{L^{p'}(\mathbb{R}^N)}.
\end{aligned}$$

Consequently, we get the desired inequality.

Lemma 4.10 *Let $2 \leq p < \infty$, $1 \leq q \leq \infty$, $s \in \mathbb{R}$, $t > 0$.*
If $(-s) \vee (2\delta(p) - 1) \vee (2\beta(p)) \leq \nu \leq 2\delta(p)$, then

$$\|\mathscr{C}_\sigma(t)v\|_{B^s_{p,q}(\mathbb{R}^N)} \lesssim t^{\sigma\left(\nu-2\delta(p)\right)}\|v\|_{B^{s+\nu}_{p',q}(\mathbb{R}^N)}.$$

If $(1-s) \vee (2\delta(p) - 2) \vee (2\beta(p)) \leq \nu \leq 2\delta(p)$, then

$$\|\mathscr{S}_\sigma(t)v\|_{B^s_{p,q}(\mathbb{R}^N)} \lesssim t^{\sigma\left(\nu-2\delta(p)+1\right)-1}\|v\|_{B^{s+\nu-1}_{p',q}(\mathbb{R}^N)}.$$

Proof From Lemma 4.8, for $t > 0$ we have the following estimates:

$$\left\|\dot{K}(t)v\right\|_{B^s_{p,q}(\mathbb{R}^N)} \lesssim t^{\nu-2\delta(p)}\|v\|_{B^{s+\nu}_{p',q}(\mathbb{R}^N)}, \quad 0 \leq s+\nu,$$

$$\|K(t)v\|_{B^s_{p,q}(\mathbb{R}^N)} \lesssim t^{\nu-2\delta(p)}\|v\|_{B^{s+\nu-1}_{p',q}(\mathbb{R}^N)}, \quad 1 \le s+\nu.$$

Therefore, using the same argument as employed in Lemma 4.9, we obtain the desired results.

4.2.3.3 The Existence Results

Let $\sigma = \beta/2$ for $\beta \in (1,2)$. We now introduce an operator $\mathcal{Q}_\sigma$ defined by

$$(\mathcal{Q}_\sigma f)(t) = \int_0^t \mathcal{S}_\sigma(t-s)f(s)ds.$$

In order to obtain the existence of L^2-solution, we need the following lemma.

Lemma 4.11 ([236]) *For each $h \in L^p(0,T;X)$ with $1 \le p < +\infty$, we have*

$$\lim_{\tau\to 0}\int_0^T |h(t+\tau)-h(t)|^p dt = 0,$$

where we suppose that $h(s) = 0$ for s not belonging to $[0,T]$.

Lemma 4.12 *For any q and μ with $1 < q \le 2$ and $\mu = N(1/q - 1/2 - 1/N)$, $N \ge 1$, let $f \in L^r(0,T;H^{\mu,q}(\mathbb{R}^N))$, and then*

$$\|\mathcal{Q}_\sigma f\|_{L^\infty(0,T;L^2(\mathbb{R}^N))} \lesssim \|f\|_{L^r(0,T;H^{\mu,q}(\mathbb{R}^N))},$$

for $r > 1/\sigma$. Furthermore, let $q = 2N/(N+2)$ for $N \ge 3$, and $f \in L^r(0,T;L^q(\mathbb{R}^N))$, then

$$\|\mathcal{Q}_\sigma f\|_{L^\infty(0,T;L^2(\mathbb{R}^N))} \lesssim \|f\|_{L^r(0,T;L^q(\mathbb{R}^N))}.$$

Moreover the operator $\mathcal{Q}_\sigma$ maps $L^r(0,T;H^{-1}(\mathbb{R}^N))$ into $C([0,T];L^2(\mathbb{R}^N))$.

Proof The inequality $|\sin\alpha| \le 1 \wedge |\alpha|$ for any $\alpha \in \mathbb{R}$ implies

$$\begin{aligned}|(\sin(t|\xi|)\langle\xi\rangle)/|\xi|| \le& (1\wedge(t|\xi|))\cdot\langle\xi\rangle/|\xi| \le (1+t)\cdot(1\wedge|\xi|)\cdot\langle\xi\rangle/|\xi| \\ \le& \sqrt{2}(1+t),\end{aligned} \tag{4.42}$$

where $\langle\xi\rangle = \sqrt{1+|\xi|^2}$, for $t \ge 0$, $\xi \in \mathbb{R}^N$. Let $g(t,s,\xi) = \sin((t-s)^\sigma\theta|\xi|)\hat{f}(s,\xi)/|\xi|$ and $y(t,x) = \langle x\rangle^{-1}\hat{f}(s,x)$, for $s \in (0,t)$, $\xi, x \in \mathbb{R}^N$. It is clear by (4.42) that

$$|g(t,s,\xi)| \le \sqrt{2}(1+(t-s)^\sigma\theta)|y(s,\xi)|, \quad \text{for } s \in (0,t),\ \xi \in \mathbb{R}^N.$$

Then it follows from the Plancherel theorem (see, e.g., [59]) that

$$\begin{aligned}
&\left\|\check{g}(t,s,\cdot)\right\|_{L^2(\mathbb{R}^N)}\\
=&\|g(t,s,\cdot)\|_{L^2(\mathbb{R}^N)}\\
\leq&\sqrt{2}(1+(t-s)^\sigma\theta)\|y(s,\cdot)\|_{L^2(\mathbb{R}^N)}=\sqrt{2}(1+(t-s)^\sigma\theta)\|\check{y}(s,\cdot)\|_{L^2(\mathbb{R}^N)}.
\end{aligned}$$

Therefore, we have

$$\begin{aligned}
&\|(\mathscr{Q}_\sigma f)(t)\|_{L^2(\mathbb{R}^N)}\\
\leq&\int_0^t\|\mathscr{S}_\sigma(t-s)f(s,\cdot)\|_{L^2(\mathbb{R}^N)}ds\\
\leq&\int_0^t\int_0^\infty\sigma(t-s)^{\sigma-1}\theta\mathscr{M}_\sigma(\theta)\left\|\check{g}(t,s,\cdot)\right\|_{L^2(\mathbb{R}^N)}d\theta ds\\
\leq&\sqrt{2}\int_0^t\int_0^\infty\sigma(t-s)^{\sigma-1}\theta\mathscr{M}_\sigma(\theta)(1+(t-s)^\sigma\theta)\|\check{y}(s,\cdot)\|_{L^2(\mathbb{R}^N)}d\theta ds.
\end{aligned}$$

Due to $\|\check{y}(s,\cdot)\|_{L^2(\mathbb{R}^N)}=\|\mathscr{F}^{-1}[\langle\xi\rangle^{-1}\hat{f}(s,\xi)]\|_{L^2(\mathbb{R}^N)}=\|f(s,\cdot)\|_{H^{-1}(\mathbb{R}^N)}$, in view of (4.39) and Hölder's inequality, for $r>1/\sigma$, we get

$$\begin{aligned}
\|(\mathscr{Q}_\sigma f)(t)\|_{L^2(\mathbb{R}^N)}\leq&\sqrt{2}\int_0^t\int_0^\infty\sigma(t-s)^{\sigma-1}\theta\mathscr{M}_\sigma(\theta)\|f(s,\cdot)\|_{H^{-1}(\mathbb{R}^N)}d\theta ds\\
&+\sqrt{2}\int_0^t\int_0^\infty\sigma(t-s)^{2\sigma-1}\theta^2\mathscr{M}_\sigma(\theta)\|f(s,\cdot)\|_{H^{-1}(\mathbb{R}^N)}d\theta ds\\
=&\sqrt{2}\int_0^t g_\sigma(t-s)\|f(s,\cdot)\|_{H^{-1}(\mathbb{R}^N)}ds\\
&+\sqrt{2}\int_0^t g_{2\sigma}(t-s)\|f(s,\cdot)\|_{H^{-1}(\mathbb{R}^N)}ds\\
\leq&\left(C_{r1}T^{\sigma-1/r}+C_{r2}T^{2\sigma-1/r}\right)\|f\|_{L^r(0,T;H^{-1}(\mathbb{R}^N))},
\end{aligned}\tag{4.43}$$

where constants $C_{r1}=\sqrt{2}[(r-1)/(\sigma r-1)]^{1-1/r}/\Gamma(\sigma)$ and $C_{r2}=\sqrt{2}[(r-1)/(2\sigma r-1)]^{1-1/r}/\Gamma(2\sigma)$.

Now, let us show the first estimate. Due to the embedding $H^{s,p}(\mathbb{R}^N)\hookrightarrow H^{s_1,p_1}(\mathbb{R}^N)$ for $1<p\leq p_1<\infty$, $s,s_1\in\mathbb{R}$, and $s-N/p=s_1-N/p_1$, (see, e.g., [18, 32]), we know that $H^{\mu,q}(\mathbb{R}^N)\hookrightarrow H^{-1}(\mathbb{R}^N)$ for $\mu=N(1/q-1/2-1/N)$. This means that

$$\|(\mathscr{Q}_\sigma f)(t)\|_{L^2(\mathbb{R}^N)}\lesssim\|f\|_{L^r(0,T;H^{\mu,q}(\mathbb{R}^N))},$$

where C depends on σ, r, N, and T. Hence, the first estimate holds. On the other hand, since $B^s_{2,2}(\mathbb{R}^N) = H^{s,2}(\mathbb{R}^N)$ for $s \in \mathbb{R}$, recall the embedding $H^{s,q}(\mathbb{R}^N) \hookrightarrow B^s_{q,2}(\mathbb{R}^N)$ for $1 < q \le 2$ and the embedding (see, e.g., [18, 32])

$$B^{s_0}_{p_0,q_0}(\mathbb{R}^N) \hookrightarrow B^{s_1}_{p_1,q_1}(\mathbb{R}^N), \quad \text{for } s_0 - N/p_0 = s_1 - N/p_1, \tag{4.44}$$

for any $s_0, s_1 \in \mathbb{R}$, $1 \le p_0 \le p_1 \le \infty$, $1 \le q_0 \le q_1 \le \infty$, we have

$$B^0_{q,2}(\mathbb{R}^N) \hookrightarrow B^{-1}_{2,2}(\mathbb{R}^N), \quad \text{for } q = \frac{2N}{N+2}.$$

By virtue of $H^{0,q}(\mathbb{R}^N) \hookrightarrow B^0_{q,2}(\mathbb{R}^N)$ and $H^{0,q}(\mathbb{R}^N) = L^q(\mathbb{R}^N)$ for $q \ge 1$, we obtain the embedding $L^q(\mathbb{R}^N) \hookrightarrow H^{-1}(\mathbb{R}^N)$ when $q = 2N/(N+2)$. This means that

$$\|(\mathscr{Q}_\sigma f)(t)\|_{L^2(\mathbb{R}^N)} \lesssim \|f\|_{L^r(0,T;L^q(\mathbb{R}^N))}.$$

Hence, the second estimate holds.

In order to obtain the conclusion that the operator $\mathscr{Q}_\sigma$ maps $L^r(0, T; H^{-1}(\mathbb{R}^N))$ into $C([0, T]; L^2(\mathbb{R}^N))$, by (4.43) it suffices to prove the continuity of operator $\mathscr{Q}_\sigma$ for any $f \in L^r(0, T; H^{-1}(\mathbb{R}^N))$. That means we need to check that for $0 \le t < t + h \le T$,

$$\|(\mathscr{Q}_\sigma f)(t+h) - (\mathscr{Q}_\sigma f)(t)\|_{L^2(\mathbb{R}^N)} \to 0, \quad \text{as } h \to 0.$$

In fact, we first have

$$\begin{aligned}
&\|(\mathscr{Q}_\sigma f)(t+h) - (\mathscr{Q}_\sigma f)(t)\|_{L^2(\mathbb{R}^N)} \\
\le& \left\| \int_t^{t+h} \mathscr{S}_\sigma(t+h-s) f(s,\cdot) ds \right\|_{L^2(\mathbb{R}^N)} \\
&+ \left\| \int_0^t \int_0^\infty \sigma((t+h-s)^{\sigma-1} - (t-s)^{\sigma-1}) \theta \mathscr{M}_\sigma(\theta) \breve{g}(t+h, s, \cdot) d\theta ds \right\|_{L^2(\mathbb{R}^N)} \\
&+ \left\| \int_0^t \int_0^\infty \sigma(t-s)^{\sigma-1} \theta \mathscr{M}_\sigma(\theta) \big(\breve{g}(t+h, s, \cdot) - \breve{g}(t, s, \cdot)\big) d\theta ds \right\|_{L^2(\mathbb{R}^N)} \\
=:& I_1 + I_2 + I_3.
\end{aligned}$$

Obviously, applying (4.39) and Hölder's inequality, it follows that

$$\begin{aligned}
I_1 \le& \sqrt{2} \int_t^{t+h} \int_0^\infty \sigma(t+h-s)^{\sigma-1} \theta \mathscr{M}_\sigma(\theta) \|f(s,\cdot)\|_{H^{-1}(\mathbb{R}^N)} d\theta ds \\
&+ \sqrt{2} \int_t^{t+h} \int_0^\infty \sigma(t+h-s)^{2\sigma-1} \theta^2 \mathscr{M}_\sigma(\theta) \|f(s,\cdot)\|_{H^{-1}(\mathbb{R}^N)} d\theta ds
\end{aligned}$$

$$
\begin{aligned}
&=\sqrt{2}\int_t^{t+h} g_\sigma(t+h-s)\|f(s,\cdot)\|_{H^{-1}(\mathbb{R}^N)}ds \\
&\quad+\sqrt{2}\int_t^{t+h} g_{2\sigma}(t+h-s)\|f(s,\cdot)\|_{H^{-1}(\mathbb{R}^N)}ds \\
&\lesssim h^{\sigma-1/r}\|f\|_{L^r(0,T;H^{-1}(\mathbb{R}^N))}+h^{2\sigma-1/r}\|f\|_{L^r(0,T;H^{-1}(\mathbb{R}^N))} \\
&\to 0, \quad \text{as } h\to 0.
\end{aligned}
$$

For the second term I_2, similarly to (4.43) we have

$$
\begin{aligned}
I_2 \leq&\sqrt{2}\int_0^t\int_0^\infty \sigma|(t+h-s)^{\sigma-1}-(t-s)^{\sigma-1}|\theta\mathscr{M}_\sigma(\theta)\|f(s,\cdot)\|_{H^{-1}(\mathbb{R}^N)}d\theta ds \\
&+\sqrt{2}\int_0^t\int_0^\infty \sigma|(t+h-s)^{\sigma-1}-(t-s)^{\sigma-1}| \\
&\times(t+h-s)^\sigma\theta^2\mathscr{M}_\sigma(\theta)\|f(s,\cdot)\|_{H^{-1}(\mathbb{R}^N)}d\theta ds \\
=&C_h\int_0^t |(t+h-s)^{\sigma-1}-(t-s)^{\sigma-1}|\|f(s,\cdot)\|_{H^{-1}(\mathbb{R}^N)}ds \\
\leq&C_h\left(\int_0^t |(t+h-s)^{\sigma-1}-(t-s)^{\sigma-1}|^{r/(r-1)}ds\right)^{1-1/r}\|f\|_{L^r(0,T;H^{-1}(\mathbb{R}^N))},
\end{aligned}
$$

where $C_h=\left(\sqrt{2}/\Gamma(\sigma)+\sqrt{2}(T+h)^\sigma/\Gamma(2\sigma)\right)$. Using Lebesgue's dominated convergence theorem and Lemma 4.11, we find that $I_2\to 0$ as $h\to 0$.

For estimating the third term I_3, by virtue of (4.42), we first have

$$
\left\|\breve{g}(t+h,s,\cdot)-\breve{g}(t,s,\cdot)\right\|_{L^2(\mathbb{R}^N)}\leq 2\sqrt{2}(1+(t+h-s)^\sigma\theta)\|f(s,\cdot)\|_{H^{-1}(\mathbb{R}^N)},
$$

which means that

$$
\begin{aligned}
I_3 \leq&\int_0^t\int_0^\infty \sigma(t-s)^{\sigma-1}\theta\mathscr{M}_\sigma(\theta)\left\|\breve{g}(t+h,s,\cdot)-\breve{g}(t,s,\cdot)\right\|_{L^2(\mathbb{R}^N)}d\theta ds \\
\leq&2\sqrt{2}\int_0^t\int_0^\infty \sigma(t-s)^{\sigma-1}\theta\mathscr{M}_\sigma(\theta)\|f(s,\cdot)\|_{H^{-1}(\mathbb{R}^N)}d\theta ds \\
&+2\sqrt{2}\int_0^t\int_0^\infty \sigma(t-s)^{\sigma-1}(t+h-s)^\sigma\theta^2\mathscr{M}_\sigma(\theta)\|f(s,\cdot)\|_{H^{-1}(\mathbb{R}^N)}d\theta ds \\
\lesssim&\|f\|_{L^r(0,T;H^{-1}(\mathbb{R}^N))}.
\end{aligned}
$$

Hence, by Lebesgue's dominated convergence theorem, we conclude that $I_3\to 0$ as $h\to 0$. Similarly, for any $0\leq t-h<t\leq T$, it is not difficult to verify that

$\|(\mathcal{Q}_\sigma f)(t) - (\mathcal{Q}_\sigma f)(t-h)\|_{L^2(\mathbb{R}^N)} \to 0$, as $h \to 0$. Thus, we obtain the desired result.

Lemma 4.13 *For any $s \in \mathbb{R}$, operators $\mathscr{C}_\sigma(\cdot)$ and $\mathscr{P}_\sigma(\cdot)$ satisfy*

$$\|\mathscr{C}_\sigma(\cdot)\phi\|_{C([0,T];H^s(\mathbb{R}^N))} \leq \|\phi\|_{H^s(\mathbb{R}^N)}, \quad \|\mathscr{P}_\sigma(\cdot)\psi\|_{C([0,T];H^s(\mathbb{R}^N))} \lesssim \|\psi\|_{H^{s-1}(\mathbb{R}^N)},$$

for any $(\phi, \psi) \in H^s(\mathbb{R}^N) \times H^{s-1}(\mathbb{R}^N)$.

Proof By virtue of $|\cos(t|x|)| \leq 1$ for all $t \geq 0$, $x \in \mathbb{R}^N$, we have $\|\dot{K}(t)\phi\|_{H^s(\mathbb{R}^N)} \leq \|\phi\|_{H^s(\mathbb{R}^N)}$. In fact, for any $\phi \in H^s(\mathbb{R}^N)$, we have

$$\begin{aligned}\|\dot{K}(t)\phi\|_{H^s(\mathbb{R}^N)} &= \|\mathscr{F}^{-1}[\cos(t|\cdot|)\hat{\phi}]\|_{H^s(\mathbb{R}^N)} \\ &= \|\mathscr{F}^{-1}[(1+|\cdot|^2)^{s/2}\cos(t|\cdot|)\hat{\phi}]\|_{L^2(\mathbb{R}^N)}.\end{aligned}$$

The Plancherel theorem implies

$$\begin{aligned}\|\mathscr{F}^{-1}[(1+|\cdot|^2)^{s/2}\cos(t|\cdot|)\hat{\phi}]\|_{L^2(\mathbb{R}^N)} &= \|(1+|\cdot|^2)^{s/2}\cos(t|\cdot|)\hat{\phi}\|_{L^2(\mathbb{R}^N)} \\ &\leq \|(1+|\cdot|^2)^{s/2}\hat{\phi}\|_{L^2(\mathbb{R}^N)} \\ &= \|\mathscr{F}^{-1}[(1+|\cdot|^2)^{s/2}\hat{\phi}]\|_{L^2(\mathbb{R}^N)},\end{aligned}$$

which means that $\|\dot{K}(t)\phi\|_{H^s(\mathbb{R}^N)} \leq \|\phi\|_{H^s(\mathbb{R}^N)}$. In addition, by virtue of the inequality $|\sin(t|x|)| \leq t|x|$, for all $t \geq 0$, $x \in \mathbb{R}^N$, as repeating the above processes, by (4.42) it is easy to check that $\|K(t)\psi\|_{H^s(\mathbb{R}^N)} \leq \sqrt{2}(1+t)\|\psi\|_{H^{s-1}(\mathbb{R}^N)}$ for any $\psi \in H^{s-1}(\mathbb{R}^N)$.

Let us show that $\|\mathscr{C}_\sigma(t)\phi\|_{H^s(\mathbb{R}^N)} \leq \|\phi\|_{H^s(\mathbb{R}^N)}$. In fact, from the definition of operator $\mathscr{C}_\sigma(\cdot)$, we have

$$\|\mathscr{C}_\sigma(t)\phi\|_{H^s(\mathbb{R}^N)} \leq \int_0^\infty \mathscr{M}_\sigma(\theta)\|\dot{K}(t^\sigma\theta)\phi\|_{H^s(\mathbb{R}^N)}d\theta \leq \|\phi\|_{H^s(\mathbb{R}^N)},$$

where we have used the identity (4.39). Moreover, from the definition of operator $\mathscr{S}_\sigma(\cdot)$ and the semigroup $(g_a * g_b)(t) = g_{a+b}(t)$ for $a, b > 0$, we have

$$\begin{aligned}&\|\mathscr{P}_\sigma(t)\psi\|_{H^s(\mathbb{R}^N)} \\ \leq &\int_0^t g_{2-2\sigma}(t-s)\|\mathscr{S}_\sigma(s)\psi\|_{H^s(\mathbb{R}^N)}ds \\ \leq &\int_0^t\int_0^\infty g_{2-2\sigma}(t-s)\sigma s^{\sigma-1}\theta\mathscr{M}_\sigma(\theta)\|K(s^\sigma\theta)\psi\|_{H^s(\mathbb{R}^N)}d\theta ds \\ \leq &\sqrt{2}\int_0^t\int_0^\infty g_{2-2\sigma}(t-s)\sigma s^{\sigma-1}(1+s^\sigma\theta)\theta\mathscr{M}_\sigma(\theta)\|\psi\|_{H^{s-1}(\mathbb{R}^N)}d\theta ds\end{aligned}$$

$$=\sqrt{2}((g_{2-2\sigma} * g_{\sigma})(t) + (g_{2-2\sigma} * g_{2\sigma})(t))\|\psi\|_{H^{s-1}(\mathbb{R}^N)}$$
$$\lesssim \|\psi\|_{H^{s-1}(\mathbb{R}^N)},$$

where $(g_{2-2\sigma} * g_{\sigma})(t) + (g_{2-2\sigma} * g_{2\sigma})(t) \leq g_{2-\sigma}(T) + g_2(T)$. Hence, it yields $\|\mathscr{P}_{\sigma}(t)\psi\|_{H^s(\mathbb{R}^N)} \lesssim \|\psi\|_{H^{s-1}(\mathbb{R}^N)}$. To end this proof, it suffices to check the continuity of $\mathscr{C}_{\sigma}(t)\phi$ and $\mathscr{P}_{\sigma}(t)\psi$.

By the continuity of $\cos(t|\xi|)$, for $h > 0$ and $0 \leq t < t + h \leq T$, by the Plancherel theorem we know that

$$\begin{aligned}
&\|\dot{K}((t+h)^{\sigma}\theta)\phi - \dot{K}(t^{\sigma}\theta)\phi\|_{H^s(\mathbb{R}^N)} \\
=&\|\mathscr{F}^{-1}[\langle\cdot\rangle^s(\cos((t+h)^{\sigma}\theta|\cdot|) - \cos(t^{\sigma}\theta|\cdot|))\hat{\phi}]\|_{L^2(\mathbb{R}^N)} \\
=&\|\langle\cdot\rangle^s(\cos((t+h)^{\sigma}\theta|\cdot|) - \cos(t^{\sigma}\theta|\cdot|))\hat{\phi}\|_{L^2(\mathbb{R}^N)} \\
\leq&2\|\langle\cdot\rangle^s\hat{\phi}\|_{L^2(\mathbb{R}^N)} = 2\|\phi\|_{H^s(\mathbb{R}^N)}.
\end{aligned}$$

Hence, passing to the Fourier representation and the Lebesgue dominated convergence theorem, we have the pointwise convergence

$$\|\dot{K}((t+h)^{\sigma}\theta)\phi - \dot{K}(t^{\sigma}\theta)\phi\|_{H^s(\mathbb{R}^N)} \to 0, \quad \text{as } h \to 0, \text{ a.e. } \theta \in (0, \infty).$$

On the other hand, by (4.39), we have

$$\mathscr{M}_{\sigma}(\theta)\|\dot{K}((t+h)^{\sigma}\theta)\phi - \dot{K}(t^{\sigma}\theta)\phi\|_{H^s(\mathbb{R}^N)} \leq \mathscr{M}_{\sigma}(\theta)\|\phi\|_{H^s(\mathbb{R}^N)},$$

which is integrable for a.e. $\theta \in (0, \infty)$. Hence, Lebesgue's dominated convergence theorem implies

$$\begin{aligned}
\|\mathscr{C}_{\sigma}(t+h)\phi - \mathscr{C}_{\sigma}(t)\phi\|_{H^s(\mathbb{R}^N)} =& \left\| \int_0^{\infty} \mathscr{M}_{\sigma}(\theta)(\dot{K}((t+h)^{\sigma}\theta)\phi \right. \\
&\left. - \dot{K}(t^{\sigma}\theta)\phi d\theta \right\|_{H^s(\mathbb{R}^N)} \\
\to& 0, \quad \text{as } h \to 0.
\end{aligned}$$

This means that $\mathscr{C}_{\sigma}(\cdot)\phi \in C([0, T]; H^s(\mathbb{R}^N))$. Furthermore, as repeating the above processes, we also get the continuity of $\mathscr{P}_{\sigma}(t)\psi$. Hence, $\mathscr{P}_{\sigma}(\cdot)\psi \in C([0, T]; H^s(\mathbb{R}^N))$. The proof is completed.

Remark 4.5 Obviously, in view of the inequality $|\sin(t|x|)| \leq t|x|$, for all $t \geq 0$, $x \in \mathbb{R}^N$, from the same way as in Lemma 4.13, we get

$$\|\mathscr{Q}_{\sigma} f\|_{C([0,T];L^2(\mathbb{R}^N))} \lesssim \|f\|_{L^{\infty}(0,T;L^2(\mathbb{R}^N))}.$$

By using Lemma 4.12 and Lemma 4.13, it is not difficult to obtain the existence theorem to the problem (4.40).

Theorem 4.10 *Given $(\phi, \psi) \in H^s(\mathbb{R}^N) \times H^{s-1}(\mathbb{R}^N)$ for $s \in \mathbb{R}$, let $r > 2/\beta$ and $f \in L^r(0, T; H^{\mu,q}(\mathbb{R}^N))$, for any q and μ satisfying $1 < q \le 2$ and $\mu = N(1/q - 1/2 - 1/N)$, $N \ge 1$, and then there exists a unique solution $u \in C([0, T]; L^2(\mathbb{R}^N))$ to the linear problem (4.40), and moreover*

$$\|u\|_{L^\infty(0,T;L^2(\mathbb{R}^N))} \lesssim \|\phi\|_{H^s(\mathbb{R}^N)} + \|\psi\|_{H^{s-1}(\mathbb{R}^N)} + \|f\|_{L^r(0,T;H^{\mu,q}(\mathbb{R}^N))}.$$

Let $q = 2N/(N+2)$ for $N \ge 3$; if $f \in L^r(0, T; L^q(\mathbb{R}^N))$, then

$$\|u\|_{L^\infty(0,T;L^2(\mathbb{R}^N))} \lesssim \|\phi\|_{H^s(\mathbb{R}^N)} + \|\psi\|_{H^{s-1}(\mathbb{R}^N)} + \|f\|_{L^r(0,T;L^q(\mathbb{R}^N))}.$$

In the sequel, we consider the global existence to the linear Cauchy problem (4.40).

Theorem 4.11 *Given $\phi \in H^s(\mathbb{R}^N)$, $\psi \equiv 0$ for $s \in \mathbb{R}$. Let $N \ge 2$, $\sigma = \beta/2$ for $\beta \in (1, 2)$ and $f \in C_\sigma([0, \infty); H^{s-1}(\mathbb{R}^N))$, where the Banach space*

$$\begin{aligned} &C_\sigma([0, \infty); H^{s-1}(\mathbb{R}^N)) \\ =&\{u \in C([0, \infty); H^{s-1}(\mathbb{R}^N)) : t^\sigma \|u(t)\|_{H^{s-1}(\mathbb{R}^N)} < \infty, \ t \ge 0\}, \end{aligned}$$

equipped with its natural norm $\|u\|_\sigma = \sup_{t\in[0,\infty)} t^\sigma \|u(t)\|_{H^{s-1}(\mathbb{R}^N)}$. Then there exists a unique solution $u \in C([0, \infty); H^s(\mathbb{R}^N))$ to the linear problem (4.40), and moreover

$$\|u\|_{L^\infty(0,\infty;H^s(\mathbb{R}^N))} \lesssim \|\phi\|_{H^s(\mathbb{R}^N)} + \|f\|_{C_\sigma([0,\infty);H^{s-1}(\mathbb{R}^N))}. \tag{4.45}$$

Proof Observe that, for $t \ge 0$, from Lemma 4.13, $\|\mathscr{C}_\sigma(t)\phi\|_{H^s(\mathbb{R}^N)} \le \|\phi\|_{H^s(\mathbb{R}^N)}$. Moreover, for any $v \in H^{s-1}(\mathbb{R}^N)$, Lemma 4.10 implies

$$\|\mathscr{S}_\sigma(t)v\|_{B^s_{2,2}(\mathbb{R}^N)} \lesssim t^{\sigma-1}\|v\|_{B^{s-1}_{2,2}(\mathbb{R}^N)},$$

which means that $\|\mathscr{S}_\sigma(t)v\|_{H^s(\mathbb{R}^N)} \lesssim t^{\sigma-1}\|v\|_{H^{s-1}(\mathbb{R}^N)}$. Hence, we have

$$\begin{aligned} &\|\mathscr{C}_\sigma(t)\phi\|_{H^s(\mathbb{R}^N)} + \|(\mathscr{Q}_\sigma f)(t)\|_{H^s(\mathbb{R}^N)} \\ &\lesssim \|\phi\|_{H^s(\mathbb{R}^N)} + \int_0^t \|\mathscr{S}_\sigma(t-s)f(s,\cdot)\|_{H^s(\mathbb{R}^N)}ds \\ &\lesssim \|\phi\|_{H^s(\mathbb{R}^N)} + \int_0^t (t-s)^{\sigma-1}\|f(s)\|_{H^{s-1}(\mathbb{R}^N)}ds \\ &\lesssim \|\phi\|_{H^s(\mathbb{R}^N)} + \|f\|_{C_\sigma([0,\infty);H^{s-1}(\mathbb{R}^N))}, \end{aligned}$$

where we have used the semigroup $(g_a * g_b)(t) = g_{a+b}(t)$ for $a, b > 0$. Similarly to Lemma 4.12 and Lemma 4.13, the continuity is easy to check, where we can use the Plancherel theorem and the Lebesgue dominated convergence theorem, and we need to require function $f \in C_\sigma([0,\infty); H^{s-1}(\mathbb{R}^N))$ such that the time T can tend to infinity. Consequently, there exists a solution satisfying (4.41), its values lie in $C([0,\infty); H^s(\mathbb{R}^N))$, and (4.45) holds. Moreover, the uniqueness follows (4.45). The proof is completed.

4.2.4 Results of Semilinear Problems

In this subsection, we focus on the well-posedness results of the semilinear problem, we first establish a local well-posedness result of L^2-solution that also belongs to the setting of $L^\gamma(0,T; L^r(\mathbb{R}^N))$, and furthermore, by a similar way, another conclusion will be given in the framework of $C([0,T]; H^s(\mathbb{R}^N))$. In the sequel, for a given semilinear function, we obtain the well-posedness results in the setting of Besov space $B^s_{r,2}$.

Theorem 4.12 *Let $N \geq 2$ and $(\phi, \psi) \in H^s(\mathbb{R}^N) \times H^{s-1}(\mathbb{R}^N)$ for any $s \in (1/2, N/2)$. Assume that*

$$f : L^2(\mathbb{R}^N) \cap L^r(\mathbb{R}^N) \to L^{r'}(\mathbb{R}^N),$$

where index r satisfies

$$\frac{2N}{N-1} \leq r \leq \frac{2(N+1)}{N-1}\bigg|_{N\geq 2} \wedge \frac{2N}{N-2s}\bigg|_{N>2s} \wedge \frac{2N}{N-2}\bigg|_{N\geq 3}.$$

For every $R > 0$, there exist $\alpha \geq 0$ and constant $C_\alpha(R) \in (0,\infty)$ such that

$$\|f(u) - f(v)\|_{L^{r'}(\mathbb{R}^N)} \leq C_\alpha(R)\left(\|u\|^\alpha_{L^r(\mathbb{R}^N)} + \|v\|^\alpha_{L^r(\mathbb{R}^N)}\right)\|u - v\|_{L^r(\mathbb{R}^N)},$$

for all $u, v \in L^2(\mathbb{R}^N) \cap L^r(\mathbb{R}^N)$ with $\|u\|_{L^2(\mathbb{R}^N)}, \|v\|_{L^2(\mathbb{R}^N)} \leq R$. Let $\gamma > 0$ be an element of the admissible set

$$\{\gamma \in \mathbb{R}^+ : \gamma > 2 + \alpha,\ \gamma(\beta - 2) + 2 > 0,\ \gamma(\sigma_N - 1) + 1 > 0\},$$

where $\sigma_N = \beta(1 - \delta(r))$. Then there exists a unique mild solution $u \in C([0,T]; L^2(\mathbb{R}^N)) \cap L^\gamma(0,T; L^r(\mathbb{R}^N))$ of the problem (4.37)–(4.38) for some $T > 0$. Moreover, u depends continuously on ϕ, ψ in the following sense. If $\phi_m \to \phi$ in $H^s(\mathbb{R}^N)$ and $\psi_m \to \psi$ in $H^{s-1}(\mathbb{R}^N)$ and if u_m denotes the solution of the problem (4.37)–(4.38) with the initial values ϕ_m, ψ_m, then for all sufficiently large m, u_m converges to u in $C([0,T]; L^2(\mathbb{R}^N)) \cap L^\gamma(0,T; L^r(\mathbb{R}^N))$.

Proof Let $\sigma = \beta/2$ for $\beta \in (1, 2)$. We want to construct a local (in t) solution to the integral equation

$$(\Phi u)(t) = \mathscr{C}_\sigma(t)\phi + \mathscr{P}_\sigma(t)\psi + (\mathscr{Q}_\sigma f)(u)(t).$$

By applying Lemma 4.12, Lemma 4.13, and Remark 4.5, it is clear that Φ is well defined. We shall next use a fixed point theorem to verify this proof. Fixed $T, R > 0$ and set a space by

$$U_T = \{u \in L^\infty(0, T; L^2(\mathbb{R}^N)) \cap L^\gamma(0, T; L^r(\mathbb{R}^N)) : d(u, 0) \le R\},$$

where $d(\cdot, \cdot)$ is the distance to the space U_T given by

$$d(u, v) = \|u - v\|_{L^\infty(0,T;L^2(\mathbb{R}^N))} + \|u - v\|_{L^\gamma(0,T;L^r(\mathbb{R}^N))}, \quad \text{for } u, v \in U_T.$$

Obviously, (U_T, d) is a complete metric space.

For $u \in U_T$ and every $R > 0$, from the assumption of nonlinearity f, by the trigonometric inequality we first have

$$\begin{aligned}\|f(u)\|_{L^{r'}(\mathbb{R}^N)} \le& \|f(u) - f(0)\|_{L^{r'}(\mathbb{R}^N)} + \|f(0)\|_{L^{r'}(\mathbb{R}^N)}\\ \le& C_\alpha(R)\|u\|^{\alpha+1}_{L^r(\mathbb{R}^N)} + \|f(0)\|_{L^{r'}(\mathbb{R}^N)}.\end{aligned}$$

Therefore, for $\gamma > \alpha + 2$, Hölder's inequality yields

$$\begin{aligned}\|f(u)\|_{L^{\gamma'}(0,T;L^{r'}(\mathbb{R}^N))} \le& C_\alpha(R)\|u\|^{\alpha+1}_{L^{(\alpha+1)\gamma'}(0,T;L^r(\mathbb{R}^N))} + T^{\frac{1}{\gamma'}}\|f(0)\|_{L^{r'}(\mathbb{R}^N)}\\ \le& C_\alpha(R)T^{\frac{\gamma-(\alpha+2)}{\gamma}}\|u\|^{\alpha+1}_{L^\gamma(0,T;L^r(\mathbb{R}^N))} + T^{\frac{\gamma-1}{\gamma}}\|f(0)\|_{L^{r'}(\mathbb{R}^N)}.\end{aligned}$$

Similarly, for $u, v \in U_T$, we have

$$\begin{aligned}&\|f(u) - f(v)\|_{L^{\gamma'}(0,T;L^{r'}(\mathbb{R}^N))}\\ \le& C_\alpha(R)\left(\int_0^T \left(\|u(t)\|^\alpha_{L^r(\mathbb{R}^N)} + \|v(t)\|^\alpha_{L^r(\mathbb{R}^N)}\right)^{\gamma'} \|u(t) - v(t)\|^{\gamma'}_{L^r(\mathbb{R}^N)}dt\right)^{1/\gamma'},\end{aligned}$$

which, by Hölder's inequality and Minkowski's inequality, leads to

$$\begin{aligned}&\|f(u) - f(v)\|_{L^{\gamma'}(0,T;L^{r'}(\mathbb{R}^N))}\\ \le& C_\alpha(R)\Big(\Big\| \|u\|^\alpha_{L^r(\mathbb{R}^N)} + \|v\|^\alpha_{L^r(\mathbb{R}^N)}\Big\|_{L^{\frac{\gamma}{\gamma-2}}(0,T)}\Big)\|u - v\|_{L^\gamma(0,T;L^r(\mathbb{R}^N))} \qquad (4.46)\\ \le& 2C_\alpha(R)T^{\frac{\gamma-(\alpha+2)}{\gamma}}R^\alpha\|u - v\|_{L^\gamma(0,T;L^r(\mathbb{R}^N))}.\end{aligned}$$

Next, we obtain the existence and uniqueness result for T small. By Lemma 4.9, for $2N/(N-1) \le r \le 2(N+1)/(N-1)$ and $(\sigma_N - 1)\gamma + 1 > 0$, we have

$$\begin{aligned}\|(\mathscr{Q}_\sigma f)(u)(t)\|_{L^r(\mathbb{R}^N)} &\lesssim \int_0^t (t-s)^{\sigma_N-1}\|f(u)(s)\|_{L^{r'}(\mathbb{R}^N)}ds\\ &\lesssim t^{\sigma_N-1+\frac{1}{\gamma}}\|f(u)\|_{L^{\gamma'}(0,T;L^{r'}(\mathbb{R}^N))}.\end{aligned}$$

Observe the embedding $H^s(\mathbb{R}^N) \hookrightarrow L^r(\mathbb{R}^N)$ for $s \in [0, N/2)$, $2 \le r < 2N/(N-2s)$, and we have

$$\|\Phi(u)(t)\|_{L^r(\mathbb{R}^N)} \lesssim \|\mathscr{C}_\sigma(t)\phi\|_{H^s(\mathbb{R}^N)} + \|\mathscr{P}_\sigma(t)\psi\|_{H^s(\mathbb{R}^N)} + \|(\mathscr{Q}_\sigma f)(u)(t)\|_{L^r(\mathbb{R}^N)}.$$

Therefore, we have

$$\begin{aligned}&\|\Phi(u)\|_{L^\gamma(0,T;L^r(\mathbb{R}^N))}\\ &\lesssim T^{\frac{1}{\gamma}}\|\phi\|_{H^s(\mathbb{R}^N)} + T^{\frac{1}{\gamma}}\|\psi\|_{H^{s-1}(\mathbb{R}^N)} + \|(\mathscr{Q}_\sigma f)(u)(t)\|_{L^\gamma(0,T;L^r(\mathbb{R}^N))}\\ &\lesssim T^{\frac{1}{\gamma}}\|\phi\|_{H^s(\mathbb{R}^N)} + T^{\frac{1}{\gamma}}\|\psi\|_{H^{s-1}(\mathbb{R}^N)} + T^{\sigma_N}\|f(0)\|_{L^{r'}(\mathbb{R}^N)}\\ &\quad + C_\alpha(R)T^{\sigma_N-\frac{\alpha+1}{\gamma}}R^{\alpha+1}.\end{aligned}$$

On the other hand, by Lemma 4.12, for any $u \in U_T$, we have

$$\begin{aligned}\|\Phi(u)\|_{L^\infty(0,T;L^2(\mathbb{R}^N))} &\lesssim \|\phi\|_{H^s(\mathbb{R}^N)} + \|\psi\|_{H^{s-1}(\mathbb{R}^N)} + \|(\mathscr{Q}_\sigma f)(u)\|_{L^\infty(0,T;L^2(\mathbb{R}^N))}\\ &\lesssim \|\phi\|_{H^s(\mathbb{R}^N)} + \|\psi\|_{H^{s-1}(\mathbb{R}^N)} + \|f(u)\|_{L^{\gamma'}(0,T;L^{r'}(\mathbb{R}^N))}\\ &\lesssim \|\phi\|_{H^s(\mathbb{R}^N)} + \|\psi\|_{H^{s-1}(\mathbb{R}^N)}\\ &\quad + T^{\frac{\gamma-1}{\gamma}}\|f(0)\|_{L^{r'}(\mathbb{R}^N)} + C_\alpha(R)T^{\frac{\gamma-(\alpha+2)}{\gamma}}R^{\alpha+1}.\end{aligned}$$

Therefore, there exists a constant $C \ge 1$ such that

$$\begin{aligned}&\|\Phi(u)\|_{L^\infty(0,T;L^2(\mathbb{R}^N))} + \|\Phi(u)\|_{L^\gamma(0,T;L^r(\mathbb{R}^N))}\\ &\le C(1+T^{\frac{1}{\gamma}})(\|\phi\|_{H^s(\mathbb{R}^N)} + \|\psi\|_{H^{s-1}(\mathbb{R}^N)}) + C(T^{\frac{\gamma-1}{\gamma}} + T^{\sigma_N})\|f(0)\|_{L^{r'}(\mathbb{R}^N)}\\ &\quad + CC_\alpha(R)(T^{\frac{\gamma-(\alpha+2)}{\gamma}} + T^{\sigma_N-\frac{\alpha+1}{\gamma}})R^{\alpha+1}.\end{aligned}$$

Fixed $R > 0$ satisfying $C(\|\phi\|_{H^s(\mathbb{R}^N)} + \|\psi\|_{H^{s-1}(\mathbb{R}^N)}) \le R/4$, let T be small enough such that

$$C_T := CC_\alpha(R)R^\alpha\left(T^{\frac{\gamma-(\alpha+2)}{\gamma}} + T^{\sigma_N-\frac{\alpha+1}{\gamma}}\right) < \frac{1}{2}, \tag{4.47}$$

$$C\left(\|\phi\|_{H^s(\mathbb{R}^N)} + \|\psi\|_{H^{s-1}(\mathbb{R}^N)}\right) T^{\frac{1}{\gamma}} + C\left(T^{\sigma_N} + T^{\frac{\gamma-1}{\gamma}}\right)\|f(0)\|_{L^{r'}(\mathbb{R}^N)} < \frac{R}{4}.$$

Hence, it follows that $\Phi(u) \in U_T$ for any $u \in U_T$. Moreover, from Lemma 4.12, we deduce

$$\begin{aligned}\|\Phi(u) - \Phi(v)\|_{L^\infty(0,T;L^2(\mathbb{R}^N))} &= \|(\mathcal{Q}_\sigma f)(u) - (\mathcal{Q}_\sigma f)(v)\|_{L^\infty(0,T;L^2(\mathbb{R}^N))} \\ &\lesssim \|f(u) - f(v)\|_{L^{\gamma'}(0,T;L^{r'}(\mathbb{R}^N))},\end{aligned}$$

where $\gamma' > 1/\sigma$, $1 < r' \le 2$ for $N = 1, 2$ and $2N/(N+2) \le r' \le 2$ for $N \ge 3$. Hence, by selecting T small enough so that (4.47) holds, by virtue of (4.46), we have

$$\|\Phi(u) - \Phi(v)\|_{L^\infty(0,T;L^2(\mathbb{R}^N))} \le \frac{1}{3} d(u, v),$$

for all $u, v \in U_T$. Since

$$\|(\mathcal{Q}_\sigma f)(u)(t) - (\mathcal{Q}_\sigma f)(v)(t)\|_{L^r(\mathbb{R}^N)} \lesssim t^{\sigma_N - 1 + \frac{1}{\gamma}} \|f(u) - f(v)\|_{L^{\gamma'}(0,T;L^{r'}(\mathbb{R}^N))},$$

for T small enough so that (4.47) holds, by virtue of (4.46), it follows that

$$\|\Phi(u) - \Phi(v)\|_{L^\gamma(0,T;L^r(\mathbb{R}^N))} = \|(\mathcal{Q}_\sigma f)(u) - (\mathcal{Q}_\sigma f)(v)\|_{L^\gamma(0,T;L^r(\mathbb{R}^N))} \le \frac{1}{3} d(u, v).$$

Consequently, Φ is a strict contraction on U_T. From the similar proof of continuity in Theorem 4.10 taking on $\Phi(u)$, it follows that Φ has a fixed point $u \in C([0, T]; L^2(\mathbb{R}^N)) \cap L^\gamma(0, T; L^r(\mathbb{R}^N))$, which is the unique mild solution of the problem (4.37)–(4.38).

For the choice of T (independent of $\|\phi\|_{H^s(\mathbb{R}^N)}$ and $\|\psi\|_{H^{s-1}(\mathbb{R}^N)}$), as before, R is determined only by the size of the norm of initial data. Hence T and R are independent of $u_m \in U_T$ for m sufficiently large. Suppose $\phi_m \to \phi$ in $H^s(\mathbb{R}^N)$ and $\psi_m \to \psi$ in $H^{s-1}(\mathbb{R}^N)$ when $m \to \infty$. Now, let m be large enough; then

$$\begin{aligned}&\|u - u_m\|_{L^\infty(0,T;L^2(\mathbb{R}^N))} \\ &\lesssim \|\phi - \phi_m\|_{H^s(\mathbb{R}^N)} + \|\psi - \psi_m\|_{H^{s-1}(\mathbb{R}^N)} + \|\mathcal{Q}_\sigma(f(u) - f(u_m))\|_{L^\infty(0,T;L^2(\mathbb{R}^N))} \\ &\lesssim \|\phi - \phi_m\|_{H^s(\mathbb{R}^N)} + \|\psi - \psi_m\|_{H^{s-1}(\mathbb{R}^N)} + \|f(u) - f(u_m)\|_{L^{\gamma'}(0,T;L^{r'}(\mathbb{R}^N))}.\end{aligned}$$

By the same argument, one can conclude that

$$\begin{aligned}\|u - u_m\|_{L^\gamma(0,T;L^r(\mathbb{R}^N))} \lesssim\, & T^{\frac{1}{\gamma}} \|\phi - \phi_m\|_{H^s(\mathbb{R}^N)} + T^{\frac{1}{\gamma}} \|\psi - \psi_m\|_{H^{s-1}(\mathbb{R}^N)} \\ & + T^{\sigma_N - 1 + \frac{1}{\gamma}} \|f(u) - f(u_m)\|_{L^{\gamma'}(0,T;L^{r'}(\mathbb{R}^N))}.\end{aligned}$$

Then, (4.46) implies that

$$d(u,u_m) \lesssim \left(1+T^{\frac{1}{\gamma}}\right)\|\phi-\phi_m\|_{H^s(\mathbb{R}^N)} + \left(1+T^{\frac{1}{\gamma}}\right)\|\psi-\psi_m\|_{H^{s-1}(\mathbb{R}^N)}$$
$$+ 2C_\alpha(R)T^{\frac{\gamma-(\alpha+2)}{\gamma}}R^\alpha\left(1+T^{\sigma_n-1+\frac{1}{\gamma}}\right)d(u,u_m).$$

Let T be chosen so small so that (4.47) holds; then

$$(1-C_T)d(u,u_m) \lesssim \left(1+T^{\frac{1}{\gamma}}\right)\left(\|\phi-\phi_m\|_{H^s(\mathbb{R}^N)} + \|\psi-\psi_m\|_{H^{s-1}(\mathbb{R}^N)}\right).$$

Consequently, we deduce that $d(u,u_m) \to 0$ as $m \to \infty$. The proof is completed.

Remark 4.6 From the above theorem, we have the following remarks:

(i) The admissible set of γ is not empty. Indeed, for $\alpha = 2$, $r = 3$, $N = 3$ and taking initial values $(\phi,\psi) \in H^1(\mathbb{R}^3) \times L^2(\mathbb{R}^3)$, for a suitable semilinear function $f(u)$ satisfying the assumptions in Theorem 4.12, likely $f(u) = \lambda|u|^2u$ for $\lambda > 0$, it follows that $4 < \gamma < 2/(2-\beta)$ for $\beta \in (3/2, 2)$.
(ii) Let α, r, N, and f satisfy the assumptions of Theorem 4.12, noting that if $\delta(r) \le 1/2$, then the restrictions in the admissible set of γ reduce to $2+\alpha < \gamma < 2/(2-\beta)$ or if $0 \le \alpha < 2(\beta-1)/(2-\beta)$ and $1/2 < \delta(r) < 1-(1+\alpha)/(\beta(2+\alpha))$, then the restrictions in the admissible set of γ reduce to $2+\alpha < \gamma < 1/(1-\sigma_N)$.
(iii) By the embedding $H^s(\mathbb{R}^N) \hookrightarrow L^r(\mathbb{R}^N)$ for $s \in [0, N/2)$, the requirement $2N/(N-1) \le r < 2N/(N-2s)$ implies that the conclusion fails for $0 \le s \le 1/2$, and thus the assumption $s > 1/2$ is needed in the initial value conditions.

Noting that, by virtue of the critical embedding $H^s(\mathbb{R}^N) \hookrightarrow L^r(\mathbb{R}^N)$, for $2 \le r < \infty$, $s = N/2$, we obtain a weaken requirement of index r in Theorem 4.12.

Corollary 4.1 *Let $N \ge 2$ and $(\phi,\psi) \in H^{N/2}(\mathbb{R}^N) \times H^{N/2-1}(\mathbb{R}^N)$. Assume that*

$$f : L^2(\mathbb{R}^N) \cap L^r(\mathbb{R}^N) \to L^{r'}(\mathbb{R}^N),$$

where index r satisfies

$$\frac{2N}{N-1} \le r \le \frac{2(N+1)}{N-1}\Big|_{N\ge 2} \wedge \frac{2N}{N-2}\Big|_{N\ge 3}.$$

For every $R > 0$, there exist $\alpha \ge 0$ and constant $C_\alpha(R) \in (0,\infty)$ such that

$$\|f(u)-f(v)\|_{L^{r'}(\mathbb{R}^N)} \le C_\alpha(R)\left(\|u\|^\alpha_{L^r(\mathbb{R}^N)} + \|v\|^\alpha_{L^r(\mathbb{R}^N)}\right)\|u-v\|_{L^r(\mathbb{R}^N)},$$

for all $u, v \in L^2(\mathbb{R}^N) \cap L^r(\mathbb{R}^N)$ with $\|u\|_{L^2(\mathbb{R}^N)}, \|v\|_{L^2(\mathbb{R}^N)} \le R$. Let $\gamma > 0$ be an element of the admissible set

$$\{\gamma \in \mathbb{R}^+ : \gamma > 2+\alpha,\ \gamma(\beta-2)+2>0,\ \gamma(\beta(1-\delta(r))-1)+1>0\}.$$

Then the problem (4.37)–(4.38) is local well-posed on $C([0,T]; L^2(\mathbb{R}^N)) \cap L^\gamma(0,T; L^r(\mathbb{R}^N))$.

By using the embedding

$$H^s(\mathbb{R}^N) \hookrightarrow L^2(\mathbb{R}^N) \cap L^\infty(\mathbb{R}^N), \quad \text{for } s > N/2,$$

one can prove the following result by employing the arguments used in the proof of Theorem 4.12.

Corollary 4.2 *Let* $s > N/2$. *Assume that for every* $R > 0$, *there exists* $C(R) < \infty$ *such that*

$$\|f(u)\|_{H^s(\mathbb{R}^N)} \le C(R)\|u\|_{H^s(\mathbb{R}^N)},$$

$$\|f(u) - f(v)\|_{L^2(\mathbb{R}^N)} \le C(R)\|u - v\|_{L^2(\mathbb{R}^N)},$$

for all $u, v \in H^s(\mathbb{R}^N)$ *and that of* $\|u\|_{L^\infty(\mathbb{R}^N)} \le R$, $\|v\|_{L^\infty(\mathbb{R}^N)} \le R$. *Then for* $(\phi, \psi) \in H^s(\mathbb{R}^N) \times H^{s-1}(\mathbb{R}^N)$, *the problem (4.37)–(4.38) is local well-posed on* $u \in (W_{T,R}, d)$ *for some* $T, R > 0$, *where* $(W_{T,R}, d)$ *is the metric space given by*

$$W_{T,R} = \left\{u \in L^\infty(0,T; H^s(\mathbb{R}^N)) : \|u\|_{L^\infty(0,T;H^s(\mathbb{R}^N))} \le R\right\},$$

equipped with the distance

$$d(u,v) = \|u - v\|_{L^\infty(0,T;L^2(\mathbb{R}^N))}, \quad \text{for } u, v \in W_{T,R}.$$

In the sequel, we consider a semilinear function of the form $f(u) = \lambda|u|^\alpha u$ for $\lambda \in \mathbb{R}$, and for $\alpha \ge 0$, a well-posedness result on Besov setting is also established. In order to complete the proof, we need the following result [33].

Lemma 4.14 *Let* $f(u) = \lambda|u|^\alpha u$ *for* $\lambda \in \mathbb{R}$ *and for* $\alpha \ge 0$. *If* $0 < s < N/2$, $2 \le \rho \le \rho^* = N(\alpha+2)/(N-2s)$, *then*

$$\|f(u)\|_{B^s_{\rho',2}} \lesssim \|u\|^{\alpha+1}_{B^s_{\rho,2}}$$

and

$$\|f(u) - f(v)\|_{L^{\rho'}(\mathbb{R}^N)} \lesssim \left(\|u\|^\alpha_{B^s_{\rho,2}} + \|v\|^\alpha_{B^s_{\rho,2}}\right)\|u - v\|_{L^\rho(\mathbb{R}^N)},$$

for any $u, v \in B^s_{\rho,2}$.

Proof Noting that

$$\big||u|^\alpha u - |v|^\alpha v\big| \le (\alpha+1)(|u|^\alpha + |v|^\alpha)|u-v|,$$

we obtain by the Hölder inequality and the embedding $\dot{B}^s_{\rho,2}(\mathbb{R}^N) \hookrightarrow L^{\rho^*}(\mathbb{R}^N)$ that

$$\|f(u)\|_{L^{\rho'}(\mathbb{R}^N)} \lesssim \|u\|^\alpha_{\dot{B}^s_{\rho,2}} \|u\|_{L^\rho(\mathbb{R}^N)}$$

and

$$\|f(u)-f(v)\|_{L^{\rho'}(\mathbb{R}^N)} \lesssim \left(\|u\|^\alpha_{\dot{B}^s_{\rho,2}} + \|v\|^\alpha_{\dot{B}^s_{\rho,2}}\right)\|u-v\|_{L^\rho(\mathbb{R}^N)}.$$

In view of the inequality

$$\|f(u)\|_{\dot{B}^s_{\rho',2}} \lesssim \|u\|^{\alpha+1}_{\dot{B}^s_{\rho,2}}$$

and the interpolation property $B^s_{\rho',2}(\mathbb{R}^N) = L^{\rho'}(\mathbb{R}^N) \cap \dot{B}^s_{\rho',2}(\mathbb{R}^N)$, the desired inequalities follow.

Theorem 4.13 *Let $N \ge 2$, $s \in (0, N/2)$, and $f(u) = \lambda|u|^\alpha u$ for $\alpha \ge 0$, $\lambda \in \mathbb{R}$. Let*

$$\frac{2N}{N-1} \le r \le \frac{2(N+1)}{N-1} \wedge \frac{N(\alpha+2)}{N-2s},$$

given $(\phi, \psi) \in H^{s+1,r'}(\mathbb{R}^N) \times H^{s,r'}(\mathbb{R}^N)$. Moreover, let $\gamma > 0$ being an element of the set

$$\{\gamma \in \mathbb{R}^+ : \gamma > \alpha+2,\ \gamma(\beta(2-2\delta(r))-2)+2>0\}. \tag{4.48}$$

Then the problem (4.37)–(4.38) is local well-posed on $L^\gamma(0,T; B^s_{r,2})$.

Proof Following the method of proof for Theorem 4.12, we just need to construct a local solution to the operator equation $(\Phi u)(t) = u(t)$ for $t \in [0,T]$ in a suitable ball in $L^\gamma(0,T; B^s_{r,2})$, where the ball is defined by

$$\mathscr{B}_M = \left\{u \in L^\gamma(0,T; B^s_{r,2}) : \|u\|_{L^\gamma(0,T;B^s_{r,2})} \le M\right\},$$

with the radius $M > 0$. Note that this space is not trivial. Indeed, for $\phi \in H^{s+1,r'}(\mathbb{R}^N)$, by virtue of Lemma 4.10, the embedding (4.44), and $H^{s_1,r'}(\mathbb{R}^N) \hookrightarrow H^{s_0,r'}(\mathbb{R}^N)$ for $s_1 \ge s_0$, $s_1, s_0 \in \mathbb{R}$, we have

$$\|\mathscr{C}_\sigma(t)\phi\|_{B^s_{r,2}} \lesssim \|\phi\|_{B^{s+\delta(r)}_{r',2}} \lesssim \|\phi\|_{H^{s+\delta(r),r'}(\mathbb{R}^N)} \lesssim \|\phi\|_{H^{s+1,r'}(\mathbb{R}^N)},$$

which shows that $\|\mathscr{C}_\sigma(\cdot)\phi\|_{L^\gamma(0,T;B^s_{r,2})} \lesssim \|\phi\|_{H^{s+1,r'}(\mathbb{R}^N)}$. Similarly, by $(g_{2-2\sigma} * g_\sigma)(t) = g_{2-\sigma}(t) \in L^\gamma(0,T;\mathbb{R})$, we also get $\|\mathscr{P}_\sigma(\cdot)\psi\|_{L^\gamma(0,T;B^s_{r,2})} \lesssim \|\psi\|_{H^{s,r'}(\mathbb{R}^N)}$. Thus, $u(t) = \mathscr{C}_\sigma(t)\phi$ is in $\mathscr{B}_M$ if $\phi \in H^{s+1,r'}(\mathbb{R}^N)$ and $\|\phi\|_{H^{s+1,r'}(\mathbb{R}^N)}$ is small enough. Endowed with the metric

$$d(u,v) = \|u - v\|_{L^\gamma(0,T;L^r(\mathbb{R}^N))},$$

then $(\mathscr{B}_M, d)$ is a complete metric space. Indeed, since $L^\gamma(0,T;B^s_{r,2})$ is reflexive, the closed ball of radius M is weakly compact; for details, see [33].

From Lemma 4.14, we get $\|f(u)\|_{B^s_{r',2}} \lesssim \|u\|^{\alpha+1}_{B^s_{r,2}}$, for all $u \in B^s_{r,2}$. Thus, it remains to consider the case $\Phi u \in L^\gamma(0,T;B^s_{r,2})$ for some $T > 0$. Noting that from the requirements of index r, it yields $1/2 \le \delta(r) \le N/(N+1)$, consider any $u \in B^s_{r,2}$, and then it follows from (4.48) that

$$\begin{aligned}\|(\Phi u)(t)\|_{B^s_{r,2}} \le& \|\mathscr{C}_\sigma(t)\phi\|_{B^s_{r,2}} + \|\mathscr{P}_\sigma(t)\psi\|_{B^s_{r,2}} + \|(\mathscr{Q}_\sigma f)(u)(t)\|_{B^s_{r,2}} \\ \lesssim& \|\mathscr{C}_\sigma(t)\phi\|_{H^{s+1,r'}(\mathbb{R}^N)} + \|\mathscr{P}_\sigma(t)\psi\|_{H^{s,r'}(\mathbb{R}^N)} \\ &+ \int_0^t \|\mathscr{S}_\sigma(t-s)f(u)(s)\|_{B^s_{r,2}} ds \\ \lesssim& \|\phi\|_{H^{s+1,r'}(\mathbb{R}^N)} \\ &+ \|\psi\|_{H^{s,r'}(\mathbb{R}^N)} + \int_0^t (t-s)^{\sigma(2-2\delta(r))-1} \|u(s)\|^{\alpha+1}_{B^s_{r,2}} ds.\end{aligned}$$

Noting that

$$\||u|^\alpha u\|_{L^{\gamma'}(0,t;B^s_{r,2})} \le t^{\frac{\gamma-(\alpha+2)}{\gamma}} \|u\|^{\alpha+1}_{L^\gamma(0,t;B^s_{r,2})},$$

we obtain

$$\begin{aligned}\|\Phi(u)\|_{L^\gamma(0,T;B^s_{r,2})} \lesssim& T^{\frac{1}{\gamma}}\|\phi\|_{H^{s+1,r'}(\mathbb{R}^N)} + T^{\frac{1}{\gamma}}\|\psi\|_{H^{s,r'}(\mathbb{R}^N)} \\ &+ T^{\sigma(2-2\delta(r))-\frac{\alpha+1}{\gamma}} \|u\|^{\alpha+1}_{L^\gamma(0,T;B^s_{r,2})}.\end{aligned}$$

On the other hand, from Lemma 4.14, for any $u, v \in \mathscr{B}_M$, we get

$$\|f(u) - f(v)\|_{L^{\gamma'}(0,T;L^{r'}(\mathbb{R}^N))} \lesssim T^{\frac{\gamma-(\alpha+2)}{\gamma}} M^\alpha \|u - v\|_{L^\gamma(0,T;L^r(\mathbb{R}^N))}.$$

Therefore, we have

$$\begin{aligned}\|\Phi(u)-\Phi(v)\|_{L^\gamma(0,T;L^r(\mathbb{R}^N))} =&\|(\mathcal{Q}_\sigma f)(u)-(\mathcal{Q}_\sigma f)(v)\|_{L^\gamma(0,T;L^r(\mathbb{R}^N))}\\ &\lesssim T^{\sigma_N-1+\frac{1}{\gamma}}\|f(u)-f(v)\|_{L^{\gamma'}(0,T;L^{r'}(\mathbb{R}^N))}\\ &\lesssim T^{\sigma_N-\frac{\alpha+1}{\gamma}}M^\alpha\|u-v\|_{L^\gamma(0,T;L^r(\mathbb{R}^N))}.\end{aligned}$$

This means that there exists a constant $C>0$ such that

$$\begin{aligned}\|\Phi(u)\|_{L^\gamma(0,T;B^s_{r,2})} \leq& CT^{\frac{1}{\gamma}}(\|\phi\|_{H^{s+1,r'}(\mathbb{R}^N)}+\|\psi\|_{H^{s,r'}(\mathbb{R}^N)})\\ &+CT^{\sigma(2-2\delta(r))-\frac{\alpha+1}{\gamma}}M^{\alpha+1}\end{aligned}$$

and

$$\|\Phi(u)-\Phi(v)\|_{L^\gamma(0,T;L^r(\mathbb{R}^N))}\leq CT^{\sigma_N-\frac{\alpha+1}{\gamma}}M^\alpha\|u-v\|_{L^\gamma(0,T;L^r(\mathbb{R}^N))}.$$

Hence, fixed $M>0$, let T be small enough such that

$$\begin{aligned}&CT^{\frac{1}{\gamma}}(\|\phi\|_{H^{s+1,r'}(\mathbb{R}^N)}+\|\psi\|_{H^{s,r'}(\mathbb{R}^N)})<M/2,\\ &CM^\alpha(T^{\sigma(2-2\delta(r))-\frac{\alpha+1}{\gamma}}+T^{\sigma_N-\frac{\alpha+1}{\gamma}})<1/2.\end{aligned}$$

Then Φ maps $\mathscr{B}_M$ into itself, and we obtain

$$d(\Phi(u),\Phi(v))<\frac{1}{2}d(u,v).$$

Thus, there exists a unique solution of the problem (4.37)–(4.38). The remaining proof is similar to that of Theorem 4.12. So we omit its details. The proof is completed.

Remark 4.7 Observe that the embedding $H^{s,r'}(\mathbb{R}^N)\hookrightarrow B^s_{r',2}(\mathbb{R}^N)$, and Lemma 4.10 yields

$$\|\mathscr{C}_\sigma(t)\phi\|_{B^s_{r,2}}\lesssim t^{-\beta\delta(r)}\|\phi\|_{H^{s,r'}(\mathbb{R}^N)},$$

which belongs to $L^\gamma(0,T;\mathbb{R})$. Hence, it shall possess a decay rate $\beta\delta(r)$, which implies that the solution does not belong to $C([0,T];B^s_{r,2})\cup L^\infty(0,T;B^s_{r,2})$.

Concerning with Remark 4.7, in the sequel, we establish the global well-posedness results for a special initial data.

Theorem 4.14 *Let $N\geq 2$, $\beta\in(1,4/3)$, $s\in(0,N/2)$, $\delta(r)=1/2$, and $f(u)=\lambda|u|^2u$ for $\lambda\in\mathbb{R}$. Given $\phi\in H^{s+\frac{1}{2},r'}(\mathbb{R}^N)$ with $\|\phi\|_{H^{s+\frac{1}{2},r'}(\mathbb{R}^N)}\leq\epsilon$ for some $\epsilon>0$ and $\psi\equiv 0$, then the problem (4.37)–(4.38) is global well-posed on Y_R, where (Y_R,d) $(R>0)$ is the metric space given by*

$$Y_R = \Big\{u \in C((0,\infty); B^s_{r,2}(\mathbb{R}^N)) : \sup_{t\in[0,\infty)} t^{\beta/4}\|u(t)\|_{B^s_{r,2}} \le R\Big\},$$

equipped with the distance

$$d(u,v) = \sup_{t\in[0,\infty)} t^{\beta/4}\|u(t)-v(t)\|_{L^r(\mathbb{R}^N)}, \quad \textit{for } u,v \in Y_R.$$

Proof Let $\sigma = \beta/2$ for $\beta \in (1,2)$. We next verify that the operator Φ maps Y_R into itself. Indeed, for $u \in Y_R$, by virtue of Lemma 4.10, we have

$$\begin{aligned}
\|(\Phi u)(t)\|_{B^s_{r,2}} &\lesssim t^{-\sigma/2}\|\phi\|_{H^{s+\frac{1}{2},r'}(\mathbb{R}^N)} + \int_0^t \|\mathscr{S}_\sigma(t-s)f(u)(s)\|_{B^s_{r,2}}ds\\
&\lesssim t^{-\sigma/2}\epsilon + \int_0^t (t-s)^{\sigma-1}\|f(u)(s)\|^3_{B^{s-\frac{1}{2}}_{r',2}}ds\\
&\lesssim t^{-\sigma/2}\epsilon + \int_0^t (t-s)^{\sigma-1}\|u(s)\|^3_{B^s_{r,2}}ds\\
&\lesssim t^{-\sigma/2}\epsilon + \int_0^t (t-s)^{\sigma-1}s^{-3\sigma/2}ds R^3,
\end{aligned}$$

where we have used the embedding $B^{s_0}_{r',2} \hookrightarrow B^{s_1}_{r',2}$ for $s_0 \ge s_1$, $s_0, s_1 \in \mathbb{R}$. Therefore, there exists a constant $C>0$ such that

$$t^{\sigma/2}\|(\Phi u)(t)\|_{B^s_{r,2}} \le C\epsilon + CR^3.$$

Taking $\epsilon \le R/(2C)$ for $R \le (1/(2C))^{1/2}$, due to the Fourier representation of operators and Lebesgue's dominated convergence theorem, the proof of the continuity of Φ is similar to Lemma 4.12. Hence, we deduce that $\Phi(u) \in Y_R$ for any $u \in Y_R$.

Next, we show that Φ is a contraction on Y_R. Indeed, for any $u, v \in Y_R$, combining the requirement of $\delta(r) = 1/2$ and Lemma 4.9 imply

$$\begin{aligned}
\|(\Phi u)(t)-(\Phi v)(t)\|_{L^r(\mathbb{R}^N)} &\le \int_0^t \|\mathscr{S}_\sigma(t-s)(f(u)(s)-f(v)(s))\|_{L^r(\mathbb{R}^N)}ds\\
&\lesssim \int_0^t (t-s)^{\sigma-1}\|f(u)(s)-f(v)(s)\|_{L^{r'}(\mathbb{R}^N)}ds.
\end{aligned}$$

Lemma 4.14 shows that

$$\begin{aligned}
\|(\Phi u)(t)-(\Phi v)(t)\|_{L^r(\mathbb{R}^N)} &\lesssim \int_0^t (t-s)^{\sigma-1}\Big(\|u\|^2_{B^s_{r,2}} + \|v\|^2_{B^s_{r,2}}\Big)\|u-v\|_{L^r(\mathbb{R}^N)}ds\\
&\lesssim \int_0^t (t-s)^{\sigma-1}s^{-3\sigma/2}ds R^2 d(u,v).
\end{aligned}$$

Therefore, there exists a constant $C > 0$ for $R \leq (2/C)^{1/2}$ (may be the same constant as above) such that

$$d(\Phi(u), \Phi(v)) \leq CR^2 d(u, v) \leq \frac{1}{2} d(u, v).$$

Consequently, Φ is a strict contraction on Y_R. This means that Φ has a unique fixed point $u \in Y_R$. The proof is completed.

4.3 Wave Equations with Exponential Nonlinearity

4.3.1 Introduction

In this section, we study a Cauchy problem for fractional wave equation

$$\partial_t^\alpha u - \Delta u = f(u), \quad (t, x) \in (0, \infty) \times \mathbb{R}^d, \tag{4.49}$$

with the initial conditions

$$u(0, x) = u_0(x), \quad \partial_t u(0, x) = 0, \quad x \in \mathbb{R}^d, \tag{4.50}$$

where the symbol Δ stands for the Laplace operator, f is an exponential nonlinear term which will be special later, and the notation ∂_t^α is the standard Caputo fractional derivative order $\alpha \in (1, 2)$.

Due to the applications in diffusion phenomena and the heat conduction with memory, for the linear case of Eq. (4.49), Fujita [64] obtained a representation formula of the fundamental solution by terms of probability density, and the author found that the fundamental solution has the similar property to that of the wave equation. Mainardi [151] has also proved that Eq. (4.49) controls the propagation of mechanical diffusion waves in viscoelastic media. More recently, Kim et al. [116] showed the BMO and L^p-L^q estimates of solutions, and they worked on the existence result for time fractional evolution equations with variable coefficients. Han et al. [79] concerned a Muckenhoupt A_p weights to the unique solvability of solutions. Dong and Liu [56] studied weighted mixed-norm estimates and the solvability of solutions. For the polynomial cases to Eq. (4.49), for example,

$$\partial_t^\alpha u - \Delta u = |u|^{p-1} u, \quad (t, x) \in (0, \infty) \times \mathbb{R}^d, \tag{4.51}$$

considered by [87], where they handled this equation on Hörmender spaces. Zhang and Li [238] discussed the well-posedness and blow-up results in the $L^p(\mathbb{R}^d)$ framework for nonlinearity $f(u) \sim |u|^p$. For a different polynomial nonlinearity and fractional Laplace operator, Djida et al. [54] proved the local existence of solutions by Fox H-functions and Fourier transform techniques.

In the semilinear cases of exponential nonlinearity, Ioku [94, 95] considered a heat equation, in which the global existence of solutions for the higher dimensions $d \geq 5$ and the lower dimensions $1 \leq d \leq 4$ on the small initial data in $\exp L^2(\mathbb{R}^d)$ were obtained, respectively. Furioli et al. [65] studied the asymptotic behavior and decay estimates for the critical nonlinearity. Suzuki [202] studied the existence/nonexistence of a local classical solution both in $\mathbb{R}^d$ and in a bounded domain. Ghoul et al. [71] investigated the existence of a stable blow-up solutions parabolic system. Fino and Kirane [62] concerned with the existence of local/global solutions with fractional Laplace operator. Furthermore, for the wave operators concerned with exponential nonlinearity, Ibrahim et al. [92] investigated the existence and asymptotic completeness of the wave operators with a defocusing exponential nonlinearity in two space dimensions. Saanouni [185] obtained a blowing-up result for arbitrary positive initial energy in two dimension, which is based on the previous global well-posedness results given by Mahouachi and Saanouni [148]. Struwe [199] established global well-posedness in the supercritical regime of large energies for smooth, radially symmetric data for similar nonlinearity $f(u) \sim ue^{u^2}$ in two space dimensions. However, the result of factional version is very rare. Bekkai et al. [17] concerned with local existence and blow-up of a unique solution to a time fractional diffusion equation with nonlocal exponent nonlinearity $f(u) \sim J_t^{1-\alpha}(e^u)$. It is thus a natural question whether there exist the local/global solvability results for the problem (4.49)–(4.50) of exponential nonlinearity.

For proving our main results, we will focus on the framework of L^p-L^q estimates and the analyticity of Mittag-Leffler functions, it is noticed that there are some differences in cases of the equation with polynomial nonlinear term, and we know that for the polynomial case of power nonlinearity $f(u) \sim |u|^{p-1}u$ with $p > 1$, the problem (4.49)–(4.50) satisfies a scale invariance property. In fact, for $\lambda > 0$, if u is a solution, then $u_\lambda(t,x) := \lambda^{\frac{2}{p-1}} u(\lambda^{\frac{2}{\alpha}} t, \lambda x)$ is also a solution with initial data $u_{0\lambda}(x) := \lambda^{\frac{2}{p-1}} u_0(\lambda x)$ and $\partial_t u(0, \lambda x) = 0$. However, one finds that there is no scale invariance property of exponential nonlinearity, likely growth $f(u) \sim e^{u^2}$. Additionally, for studying the existence of solution to the problem (4.49)–(4.50), the Orlicz space $\exp L^2(\mathbb{R}^d)$ shall be introduced, which is a subspace of $L^1_{loc}(\mathbb{R}^d)$, and it is more appropriate to handle the exponential nonlinear problem compared to $L^p(\mathbb{R}^d)$ framework. Note that $C_0^\infty(\mathbb{R}^d)$ is not dense in $\exp L^2(\mathbb{R}^d)$, in order to achieve the global solutions, it should be considered the definition of solutions under a weak mild sense, and we find that the order of fractional derivative is restricted in $\alpha \in (1, \frac{4}{3})$ for the lower dimensional case $1 \leq d \leq 3$. The minimum conditions are required to obtain a local solution for $d \geq 1$, and a dense subspace $\exp L_0^2(\mathbb{R}^d)$ is also introduced to overcome the missing problem for aforementioned denseness.

This section is organized as follows. In Sect. 4.3.2, we introduce some preliminaries for the problem (4.49)–(4.50), including fractional calculus, the Mittag-Leffler function, and the Orlicz spaces. Moreover, we discuss some estimates in the framework on the Orlicz spaces about solution operators derived by the problem (4.49)–(4.50). We establish the local/global existence criteria, and we show the relevant proofs in Sect. 4.3.3.

4.3.2 Preliminaries

In this section we present some well-known concepts and properties about fractional calculus and the Orlicz spaces. Moreover, we recall and prove the properties of the Mittag-Leffler functions. Firstly, let $\mathscr{S}(\mathbb{R}^d)$ stand for the Schwartz classification, and we also set the notations $\vee$ and $\wedge$ by $a \vee b = \max\{a, b\}$, $a \wedge b = \min\{a, b\}$ for any $a, b \in \mathbb{R}$, respectively. Let $T \in (0, +\infty]$ (if $T = +\infty$, we write $[0, T] := [0, \infty)$). We next recall the definitions of fractional calculus briefly; for more details, we refer to see [165, 180].

The Riemann-Liouville fractional integral of order $\alpha > 0$ for a function $u \in L^1(0, T; \mathscr{S}(\mathbb{R}^d))$ is defined by

$$J_t^\alpha u(t,x) = \frac{1}{\Gamma(\alpha)} \int_0^t (t-s)^{\alpha-1} u(s,x)ds, \quad t > 0,\ x \in \mathbb{R}^d,$$

where $\Gamma(\cdot)$ is the Gamma function. The Caputo fractional derivative of order $\alpha \in (1, 2)$ for a function $u \in L^1(0, T; \mathscr{S}(\mathbb{R}^d))$ such that fractional integral $J_t^{2-\alpha} u \in W^{2,1}(0, T; \mathscr{S}(\mathbb{R}^d))$ is defined by

$$\partial_t^\alpha u(t,x) = \partial_t^2 \big(J_t^{2-\alpha}(u(t,x) - u(0,x) - t\, \partial_t u(0,x)) \big), \quad t > 0,\ x \in \mathbb{R}^d.$$

Lemma 4.15 *Let* $\frac{d}{2} < p < +\infty$, $d \geq 1$, $\alpha \in (0, 2)$, $\beta \in \mathbb{R}$. *Then* $E_{\alpha,\beta}(-|\cdot|^2) \in L^p(\mathbb{R}^d)$.

Proof By virtue of Lemma 1.12, it is easy to check this argument. Indeed, from the fact

$$|E_{\alpha,\beta}(-|x|^2)| \leq C(1 + |x|^2)^{-1}, \quad \text{for all } x \in \mathbb{R}^d,$$

using the polar coordinates, we have

$$\begin{aligned}
\int_{\mathbb{R}^d} |E_{\alpha,\beta}(-|x|^2)|^p dx \leq & C^p \int_{\mathbb{R}^d} (1 + |x|^2)^{-p} dx \\
= & C^p w_{d-1} \int_0^\infty \frac{r^{d-1}}{(1+r^2)^p} dr \\
= & C^p w_{d-1} \int_0^{\pi/2} (\sin\theta)^{d-1} (\cos\theta)^{2p-d-1} d\theta \\
= & \frac{C^p \pi^{d/2} \Gamma\big(\frac{2p-d}{2}\big)}{\Gamma(p)},
\end{aligned}$$

where w_{d-1} is the surface of the unit sphere $\mathbf{S}^{d-1}$. Hence, we deduce the desired result.

Recall that an Orlicz space is defined as follows:

$$\exp L^p(\mathbb{R}^d) = \left\{ u \in L^1_{loc}(\mathbb{R}^d) : \int_{R^n} \left(\exp\left(\frac{|u(x)|^p}{\lambda^p} \right) - 1 \right) dx < +\infty \right\},$$

for some $\lambda > 0$ endowed with the Luxemburg norm

$$\|u\|_{\exp L^p(\mathbb{R}^d)} := \inf \left\{ \lambda > 0 : \int_{\mathbb{R}^d} \left(\exp\left(\frac{|u(x)|^p}{\lambda^p} \right) - 1 \right) dx < 1 \right\}.$$

Clearly, it is a Banach space. Since $C_0^\infty(\mathbb{R}^d)$ is not dense in $\exp L^p(\mathbb{R}^d)$, in order to analyze the local existence, we also use the dense subspace of $\exp L^p(\mathbb{R}^d)$ given by

$$\exp L_0^p(\mathbb{R}^d) = \Big\{ u \in \exp L^p(\mathbb{R}^d) : \text{there exists } \{u_n\}_{n=1}^\infty \subset C_0^\infty(\mathbb{R}^d),$$
$$\text{s.t. } \lim_{n\to\infty} \|u_n - u\|_{\exp L^p(\mathbb{R}^d)} = 0 \Big\},$$

which can be expressed by (see, e.g., [62, 95, 156, 157])

$$\exp L_0^p(\mathbb{R}^d) = \left\{ u \in L^1_{loc}(\mathbb{R}^d) : \int_{\mathbb{R}^d} \exp\left(\alpha |u(x)|^p - 1 \right) dx < +\infty \right\},$$

for every $\alpha > 0$.

Lemma 4.16 ([94]) *For every $2 \le p < \infty$, the following inequality holds:*

$$\|u\|_{L^p(\mathbb{R}^d)} \le (\Gamma(p/2+1))^{1/p} \|u\|_{\exp L^2(\mathbb{R}^d)}.$$

Lemma 4.17 ([62]) *For every $1 \le q \le p < \infty$, the embeddings $L^q(\mathbb{R}^d) \cap L^\infty(\mathbb{R}^d) \hookrightarrow \exp L_0^p(\mathbb{R}^d) \hookrightarrow \exp L^p(\mathbb{R}^d)$ hold, more precisely*

$$\|u\|_{\exp L^p(\mathbb{R}^d)} \le (\ln 2)^{-\frac{1}{p}} (\|u\|_{L^q(\mathbb{R}^d)} + \|u\|_{L^\infty(\mathbb{R}^d)}).$$

Lemma 4.18 ([65]) *For any $p \ge 1$ and $r \ge 1$, there exists a constant $C > 0$ (independent of p and r), such that*

$$\Gamma(rp+1)^{\frac{1}{p}} \le C\Gamma(r+1)p^r.$$

We next prove several useful estimates in the frameworks of L^∞-L^p and $\exp L^2(\mathbb{R}^d)$ estimates. For any $f \in \mathscr{S}(\mathbb{R}^d)$, we set the following two families of operators $\{\mathscr{Q}_\alpha(t)\}_{t\ge0}$ and $\{\mathscr{P}_\alpha(t)\}_{t\ge0}$:

$$\mathscr{Q}_\alpha(t)f(x) = \mathscr{F}^{-1}[E_{\alpha,1}(-t^\alpha|\xi|^2)\hat{f}(\xi)](x),$$
$$\mathscr{P}_\alpha(t)f(x) = t^{\alpha-1}\mathscr{F}^{-1}[E_{\alpha,\alpha}(-t^\alpha|\xi|^2)\hat{f}(\xi)](x).$$

In fact, because of Lemma 1.12, these operator families are well defined. In particular, we have the next framework of L^p-L^q estimate of these operator families.

Lemma 4.19 ([45]) *Let $1 < p \leq q < +\infty$. Then, there is a constant $C > 0$ such that for all $f \in L^p(\mathbb{R}^d)$, $t > 0$,*

$$\|\mathscr{Q}_\alpha(t)f\|_{L^q(\mathbb{R}^d)} \leq Ct^{-\frac{\alpha d}{2}\left(\frac{1}{p}-\frac{1}{q}\right)}\|f\|_{L^p(\mathbb{R}^d)}, \quad \frac{d}{2}\left(\frac{1}{p}-\frac{1}{q}\right) < 1$$

and

$$\|\mathscr{P}_\alpha(t)f\|_{L^q(\mathbb{R}^d)} \leq Ct^{\alpha-1-\frac{\alpha d}{2}\left(\frac{1}{p}-\frac{1}{q}\right)}\|f\|_{L^p(\mathbb{R}^d)}, \quad 1-\frac{1}{\alpha} < \frac{d}{2}\left(\frac{1}{p}-\frac{1}{q}\right) < 1.$$

Note that the above estimates of constant $C > 0$ can be removed at the first inequality and be modified by $1/\Gamma(\alpha)$ at the second inequality in the framework of L^p-L^p estimates. Indeed, from the definitions of $\mathscr{Q}_\alpha(t)$ and $\mathscr{P}_\alpha(t)$, by using (vii) in Lemma 1.14 and $0 < \Gamma(\alpha_0) \leq \Gamma(\alpha) < 1$ for $\alpha \in (1,2)$ with $\Gamma(\alpha_0) = \min_{\alpha\in(1,2)} \Gamma(\alpha)$, it is easy to check this argument. We next fill the gap of $L^\infty(\mathbb{R}^d)$ for the families of operators $\{\mathscr{Q}_\alpha(t)\}_{t\geq 0}$ and $\{\mathscr{P}_\alpha(t)\}_{t\geq 0}$.

Lemma 4.20 *Let $1 \vee \frac{d}{2} \leq p \leq 2$ for $d = 1, 2, 3$. Then, there is a constant $C > 0$ such that for all $f \in L^p(\mathbb{R}^d)$, $t > 0$,*

$$\|\mathscr{Q}_\alpha(t)f\|_{L^\infty(\mathbb{R}^d)} \leq Ct^{-\frac{\alpha d}{2p}}\|f\|_{L^p(\mathbb{R}^d)}, \quad \text{and} \quad \|\mathscr{P}_\alpha(t)f\|_{L^\infty(\mathbb{R}^d)} \leq Ct^{\alpha-1-\frac{\alpha d}{2p}}\|f\|_{L^p(\mathbb{R}^d)}.$$

Proof Using the definitions of $\mathscr{Q}_\alpha(t)$ and $\mathscr{P}_\alpha(t)$, it suffices to prove the first estimate since the proof of second estimate is similar. In fact, by Hölder's inequality, we have

$$\begin{aligned}\|\mathscr{Q}_\alpha(t)f\|_{L^\infty(\mathbb{R}^d)} =& \|\mathscr{F}^{-1}[E_{\alpha,1}(-|\xi|^2t^\alpha)\mathscr{F}f]\|_{L^\infty(\mathbb{R}^d)}\\ \leq& \int_{\mathbb{R}^d}|E_{\alpha,1}(-|\xi|^2t^\alpha)\hat{f}(\xi)|d\xi\\ \leq& \|\hat{f}\|_{L^{p'}(\mathbb{R}^d)}\left(\int_{\mathbb{R}^d}|E_{\alpha,1}(-|\xi|^2t^\alpha)|^p d\xi\right)^{1/p}.\end{aligned}$$

The Hausdorff-Young inequality implies that

$$\|\hat{f}\|_{L^{p'}(\mathbb{R}^d)} \leq \|f\|_{L^p(\mathbb{R}^d)}, \quad \text{for } 1 \leq p \leq 2.$$

Therefore, there exists a positive constant $C = C(\alpha, d, p)$ such that

$$\begin{aligned}\|\mathscr{Q}_\alpha(t)f\|_{L^\infty(\mathbb{R}^d)} \le & t^{-\frac{\alpha d}{2p}}\|f\|_{L^p(\mathbb{R}^d)}\left(\int_{\mathbb{R}^d}|E_{\alpha,1}(-|\xi|^2)|^p d\xi\right)^{1/p}\\ \le & Ct^{-\frac{\alpha d}{2p}}\|f\|_{L^p(\mathbb{R}^d)},\end{aligned}$$

where we have used the condition $d/2 \le p \le 2$ and Lemma 4.15. The proof is completed.

Lemma 4.21 *Let* $1 \vee \frac{d}{2} \le p, q \le 2$ *for* $d = 1, 2, 3$. *Then the following* L^p-$\exp L^2$ *estimates hold for* $t > 0$*:*

$$\|\mathscr{Q}_\alpha(t)f\|_{\exp L^2(\mathbb{R}^d)} \le \|f\|_{\exp L^2(\mathbb{R}^d)}, \quad \text{for } f \in \exp L^2(\mathbb{R}^d),$$

$$\|\mathscr{Q}_\alpha(t)f\|_{\exp L^2(\mathbb{R}^d)} \le Ct^{-\frac{\alpha d}{2p}}(\ln(t^{-\frac{\alpha d}{2}}+1))^{-\frac{1}{2}}\|f\|_{L^p(\mathbb{R}^d)}, \quad \text{for } f \in L^p(\mathbb{R}^d),$$

$$\|\mathscr{Q}_\alpha(t)f\|_{\exp L^2(\mathbb{R}^d)} \le Ct^{-\frac{\alpha d}{2q}}\|f\|_{L^q(\mathbb{R}^d)}+2\|f\|_{L^2(\mathbb{R}^d)}, \quad \text{for } f \in L^q(\mathbb{R}^d)\cap L^2(\mathbb{R}^d),$$

where the constant $C > 0$ *may depend on* α, d, p, q.

Proof The first inequality is showed by the standard L^p-L^p estimates of Lemma 4.19 and the Taylor expansion. In fact, for any $\lambda > 0$, we have

$$\begin{aligned}\int_{\mathbb{R}^d}\left(\exp(|\mathscr{Q}_\alpha(t)f|/\lambda)^2-1\right)dx =& \sum_{k=1}^{\infty}\frac{\|\mathscr{Q}_\alpha(t)f\|_{L^{2k}(\mathbb{R}^d)}^{2k}}{k!\lambda^{2k}}\\ \le & \sum_{k=1}^{\infty}\frac{\|f\|_{L^{2k}(\mathbb{R}^d)}^{2k}}{k!\lambda^{2k}} = \int_{\mathbb{R}^d}\left(\exp(|f|/\lambda)^2-1\right)dx.\end{aligned}$$

Therefore we have

$$\begin{aligned}\|\mathscr{Q}_\alpha(t)f\|_{\exp L^2(\mathbb{R}^d)} =& \inf\left\{\lambda>0:\int_{\mathbb{R}^d}\left(\exp(|\mathscr{Q}_\alpha(t)f|/\lambda)^2-1\right)dx \le 1\right\}\\ \le & \inf\left\{\lambda>0:\int_{\mathbb{R}^d}\left(\exp(|f|/\lambda)^2-1\right)dx \le 1\right\}\\ =& \|f\|_{\exp L^2(\mathbb{R}^d)},\end{aligned}$$

which proves the first inequality. Next, by Lemma 4.19, we have

$$\int_{\mathbb{R}^d}\left(\exp(|\mathscr{Q}_\alpha(t)f|/\lambda)^2-1\right)dx = \sum_{k=1}^{\infty}\frac{\|\mathscr{Q}_\alpha(t)f\|_{L^{2k}(\mathbb{R}^d)}^{2k}}{k!\lambda^{2k}}$$

$$\leq \sum_{k=1}^{\infty} \frac{C^{2k} t^{-\frac{\alpha d}{2}\left(\frac{1}{p}-\frac{1}{2k}\right)2k} \|f\|_{L^p(\mathbb{R}^d)}^{2k}}{k!\lambda^{2k}}$$

$$= t^{\frac{\alpha d}{2}} \left(\exp\left(Ct^{-\frac{\alpha d}{2p}} \|f\|_{L^p(\mathbb{R}^d)}/\lambda \right)^2 - 1 \right),$$

and then

$$\|\mathcal{Q}_\alpha(t) f\|_{\exp L^2(\mathbb{R}^d)} = \inf\left\{ \lambda > 0 : \int_{\mathbb{R}^d} \left(\exp(|\mathcal{Q}_\alpha(t) f|/\lambda)^2 - 1 \right) dx \leq 1 \right\}$$

$$\leq \inf\left\{ \lambda > 0 : t^{\frac{\alpha d}{2}} \left(\exp\left(Ct^{-\frac{\alpha d}{2p}} \|f\|_{L^p(\mathbb{R}^d)}/\lambda \right)^2 - 1 \right) \leq 1 \right\}$$

$$= Ct^{-\frac{\alpha d}{2p}} (\ln(t^{-\frac{\alpha d}{2}} + 1)^{-\frac{1}{2}} \|f\|_{L^p(\mathbb{R}^d)}.$$

We have shown the second inequality. The last inequality is proved easily by Lemma 4.20, the standard L^p-L^q estimate, and the embedding $L^2(\mathbb{R}^d) \cap L^\infty(\mathbb{R}^d) \hookrightarrow \exp L^2(\mathbb{R}^d)$ in Lemma 4.17 for $(\ln 2)^{-\frac{1}{2}} < 2$ as

$$\|\mathcal{Q}_\alpha(t) f\|_{\exp L^2(\mathbb{R}^d)} \leq 2(\|\mathcal{Q}_\alpha(t) f\|_{L^2(\mathbb{R}^d)} + \|\mathcal{Q}_\alpha(t) f\|_{L^\infty(\mathbb{R}^d)})$$

$$\leq Ct^{-\frac{\beta d}{2q}} \|f\|_{L^q(\mathbb{R}^d)} + 2\|f\|_{L^2(\mathbb{R}^d)}.$$

The proof is completed.

Lemma 4.22 *Let* $\alpha \in (1, \frac{4}{3})$, $1 \vee \frac{d}{2} \leq p < \frac{2\alpha d}{(4+d)\alpha - 4}$, *and* $1 \vee \frac{d}{2} < q \leq 2$ *for* $d = 1, 2, 3$. *Then the following estimates hold for* $t > 0$*:*

$$\|\mathcal{P}_\alpha(t) f\|_{\exp L^2(\mathbb{R}^d)} \leq \frac{t^{\alpha-1}}{\Gamma(\alpha)} \|f\|_{\exp L^2(\mathbb{R}^d)}, \quad for\ f \in \exp L^2(\mathbb{R}^d),$$

$$\|\mathcal{P}_\alpha(t) f\|_{\exp L^2(\mathbb{R}^d)} \leq Ct^{\alpha-1-\frac{\alpha d}{2p}} (\ln(t^{-\frac{\alpha d}{2}} + 1))^{-\frac{1}{2}} \|f\|_{L^p(\mathbb{R}^d)}, \quad for\ f \in L^p(\mathbb{R}^d),$$

$$\|\mathcal{P}_\alpha(t) f\|_{\exp L^2(\mathbb{R}^d)} \leq Ct^{\alpha-1-\frac{\alpha d}{2q}} \|f\|_{L^q(\mathbb{R}^d)} + 2\|f\|_{L^2(\mathbb{R}^d)}, \quad for\ f \in L^q(\mathbb{R}^d) \cap L^2(\mathbb{R}^d),$$

where the constant $C > 0$ *may depend on* α, d, p, q.

Proof By applying the similar arguments as in Lemma 4.21, it is not difficult to show the first and third L^p-expL^2 estimates for the family of operator $\{\mathcal{P}_\alpha(t)\}_{t \geq 0}$, while there is few slight revision on previous results for the second estimate. In fact, we have

$$\begin{aligned}\int_{\mathbb{R}^d}\left(\exp\left(|\mathscr{P}_\alpha(t)f|/\lambda\right)^2-1\right)dx &= \sum_{k=1}^{\infty}\frac{\|\mathscr{P}_\alpha(t)f\|_{L^{2k}(\mathbb{R}^d)}^{2k}}{k!\lambda^{2k}}\\ &\leq \sum_{k=1}^{\infty}\frac{C^{2k}t^{(\alpha-1)2k-\frac{\alpha d}{2}\left(\frac{1}{p}-\frac{1}{2k}\right)2k}\|f\|_{L^p(\mathbb{R}^d)}^{2k}}{k!\lambda^{2k}}\\ &= t^{\frac{\alpha d}{2}}\left(\exp\left(Ct^{\alpha-1-\frac{\alpha d}{2p}}\|f\|_{L^p(\mathbb{R}^d)}/\lambda\right)^2-1\right),\end{aligned}$$

for any $1-\frac{1}{\alpha}<\frac{d}{2}\left(\frac{1}{p}-\frac{1}{2k}\right)<1$ which holds for the condition $\frac{d}{2}\leq p<\frac{2\alpha d}{(4+d)\alpha-4}$ and all $k\in\mathbb{N}^+$, and then

$$\begin{aligned}\|\mathscr{P}_\alpha(t)f\|_{\exp L^2(\mathbb{R}^d)} &= \inf\left\{\lambda>0:\int_{\mathbb{R}^n}\left(\exp\left(|\mathscr{P}_\alpha(t)f|/\lambda\right)^2-1\right)dx\right\}\\ &\leq \inf\left\{\lambda>0: t^{\frac{\alpha d}{2}}\left(\exp\left(Ct^{\alpha-1-\frac{\alpha d}{2p}}\|f\|_{L^p(\mathbb{R}^d)}/\lambda\right)^2-1\right)\leq 1\right\}\\ &= Ct^{\alpha-1-\frac{\alpha d}{2p}}\left(\ln(t^{-\frac{\alpha d}{2}}+1)\right)^{-\frac{1}{2}}\|f\|_{L^p(\mathbb{R}^d)}.\end{aligned}$$

We thus prove the second inequality.

Lemma 4.23 *Let $g\in L^1(0,T;\exp L^2(\mathbb{R}^d))$; then*

$$\int_0^t \mathscr{P}_\alpha(t-s)g(s)ds\in C([0,T];\exp L^2(\mathbb{R}^d)).$$

Moreover,

$$\left\|\int_0^t \mathscr{P}_\alpha(t-s)g(s)ds\right\|_{L^\infty(0,T;\exp L^2(\mathbb{R}^d))}\leq C\|g\|_{L^1(0,T;\exp L^2(\mathbb{R}^d))}, \tag{4.52}$$

where the constant $C>0$ may depend on α, d, p, q, T.

Proof By Lemma 4.22, it is not difficult to check the estimate (4.52). Hence, we next show the continuity. Let $h>0$ such that $0\leq t<t+h\leq T$, and then

$$\begin{aligned}&\int_0^{t+h}\mathscr{P}_\alpha(t+h-s)g(s)ds-\int_0^t\mathscr{P}_\alpha(t-s)g(s)ds\\ &=\int_t^{t+h}\mathscr{P}_\alpha(t+h-s)g(s)ds+\int_0^t(\mathscr{P}_\alpha(t+h-s)-\mathscr{P}_\alpha(t-s))g(s)ds\\ &=:g_1(t)+g_2(t).\end{aligned}$$

Therefore, by Lemma 4.22 for $\alpha \in (1, 2)$, we have

$$
\begin{aligned}
\|g_1(t)\|_{\exp L^2(\mathbb{R}^d)} \le & \int_t^{t+h} \|\mathscr{P}_\alpha(t+h-s)g(s)\|_{\exp L^2(\mathbb{R}^d)} ds \\
\le & \frac{1}{\Gamma(\alpha)} \int_t^{t+h} (t+h-s)^{\alpha-1} \|g(s)\|_{\exp L^2(\mathbb{R}^d)} ds \\
\le & \frac{h^{\alpha-1}}{\Gamma(\alpha)} \|g\|_{L^1(0,T;\exp L^2(\mathbb{R}^d))} \to 0, \quad as\ h \to 0.
\end{aligned}
$$

On the other hand, for any $f \in \mathscr{S}(\mathbb{R}^d)$, from the Fubini theorem and the definition of $\mathscr{P}_\alpha(t)$, Proposition 1.14 implies that

$$
\begin{aligned}
(\mathscr{P}_\alpha(t) - \mathscr{P}_\alpha(s))f = & \mathscr{F}^{-1}[t^{\alpha-1}E_{\alpha,\alpha}(-t^\alpha|\xi|^2) - s^{\alpha-1}E_{\alpha,\alpha}(-s^\alpha|\xi|^2)\mathscr{F}(f)] \\
= & \mathscr{F}^{-1}\left[\int_s^t \tau^{\alpha-2}E_{\alpha,\alpha-1}(-\tau^\alpha|\xi|^2)d\tau\mathscr{F}(f)\right] \\
= & \int_s^t \tau^{\alpha-2}\mathscr{F}^{-1}\left[E_{\alpha,\alpha-1}(-\tau^\alpha|\xi|^2)\mathscr{F}(f)\right]d\tau.
\end{aligned}
$$

Hence, in view of Proposition 1.12, one has

$$
\begin{aligned}
\|g_2(t)\|_{\exp L^2(\mathbb{R}^d)} \le & \int_0^t \|(\mathscr{P}_\alpha(t+h-s) - \mathscr{P}_\alpha(t-s))g(s)\|_{\exp L^2(\mathbb{R}^d)} ds \\
\le & \frac{1}{\Gamma(\alpha)} \int_0^t \int_t^{t+h} (\tau - s)^{\alpha-2} \|g(s)\|_{\exp L^2(\mathbb{R}^d)} d\tau ds \\
\le & \frac{1}{(\alpha-1)\Gamma(\alpha)} \int_0^t ((t+h-s)^{\alpha-1} - (t-s)^{\alpha-1}) \|g(s)\|_{\exp L^2(\mathbb{R}^d)} ds \\
\le & \frac{h^{\alpha-1}}{(\alpha-1)\Gamma(\alpha)} \|g\|_{L^1(0,T;\exp L^2(\mathbb{R}^d))} \to 0, \quad as\ h \to 0,
\end{aligned}
$$

where we have used $0 \le a^c - b^c \le (a-b)^c$ for any $0 \le b \le a$ and $c \in [0, 1]$ in the above last inequality. We thus get the desired results. The proof is completed.

4.3.3 Existence Analysis

In this subsection, we prove the local and global existence results. In order to achieve this goal, we first consider the linear case of the problem (4.49)–(4.50), that is,

$$
\begin{cases} \partial_t^\alpha u(t,x) - \Delta u(t,x) = f(t,x), & (t,x) \in (0,T) \times \mathbb{R}^d, \\ u(0,x) = u_0(x), \ u_t(0,x) = 0. \end{cases} \tag{4.53}
$$

Next, we establish the solution representation of linear problem (4.53). Now, let $u_0 \in \mathscr{S}(\mathbb{R}^d)$ and $f \in L^1(0,T;\mathscr{S}(\mathbb{R}^d))$, and firstly taking the Laplace transform $\mathscr{L}$ of both sides of (4.53) with respect to $t \in [0,T)$, it follows that

$$
\lambda^\alpha \mathscr{L}[u(\cdot,x)](\lambda) - \lambda^{\alpha-1} u_0(x) - \mathscr{L}[\Delta u(\cdot,x)](\lambda) = \mathscr{L}[f(\cdot,x)](\lambda).
$$

Next, taking the Fourier transform $\mathscr{F} = \widehat{\cdot}$ of both sides of (4.53) with respect to $x \in \mathbb{R}^d$ and using the properties of Mittag-Leffler functions, we obtain

$$
\hat{u}(t,\xi) = E_{\alpha,1}(-t^\alpha|\xi|^2)\hat{u_0}(\xi) + \int_0^t (t-s)^{\alpha-1} E_{\alpha,\alpha}(-(t-s)^\alpha|\xi|^2)\hat{f}(s,\xi)ds.
$$

Therefore, similarly to [87, Proposition 2.1], there exists a solution u belonging to space $C([0,T);\mathscr{S}(\mathbb{R}^d))$ as follows:

$$
u(t) = \mathscr{Q}_\alpha(t)u_0 + \int_0^t \mathscr{P}_\alpha(t-s)f(s)ds.
$$

In the sequel, we concern with the exponential nonlinearity, which is also considered in [94] for the heat equations as follows:

$$
f(u) = \begin{cases} u(e^{u^2} - 1), & d = 2,3, \\ u(e^{u^2} - 1 - u^2), & d = 1. \end{cases} \tag{4.54}
$$

4.3.3.1 Local Existence of Solutions

In this sequel, we prove the local existence result. To begin with, we introduce a definition of mild solutions to the problem (4.49)–(4.50).

Definition 4.5 Let $u_0 \in \exp L_0^2(\mathbb{R}^d)$ and $T > 0$. We say that u is a mild solution for the problem (4.49)–(4.50) if $u \in C([0,T];\exp L_0^2(\mathbb{R}^d))$ satisfying

$$
u(t) = \mathscr{Q}_\alpha(t)u_0 + \int_0^t \mathscr{P}_\alpha(t-s)f(u(s))ds. \tag{4.55}
$$

In the sequel, according to the technique as in [62], we shall split the initial data $u_0 \in \exp L_0^2(\mathbb{R}^d)$ into two parts: a small data in $\exp L^2(\mathbb{R}^d)$ and another smooth part from the density of $C_0^\infty(\mathbb{R}^d)$ in $\exp L_0^2(\mathbb{R}^d)$. First we solve the initial value problem with smooth initial data to obtain a local and bounded solution v. Then we consider the perturbed equation satisfied by $w := u - v$ with small initial data. In fact, for $\varepsilon > 0$ to be chosen later, we write $u_0 = v_0 + w_0$, where $v_0 \in C_0^\infty(\mathbb{R}^d)$ and $\|w_0\|_{\exp L^2(\mathbb{R}^d)} \le \varepsilon$. Hence, the problem (4.49)–(4.50) may split into the following two Cauchy problems:

$$\begin{cases} \partial_t^\alpha v - \Delta v = f(v), \\ v(0) = v_0, \quad \partial_t v(0) = 0 \end{cases} \tag{4.56}$$

and

$$\begin{cases} \partial_t^\alpha w - \Delta w = f(w+v) - f(v), \\ w(0) = w_0, \quad \partial_t w(0) = 0. \end{cases} \tag{4.57}$$

Lemma 4.24 *If $\phi \in \exp L_0^2(\mathbb{R}^d)$, then $\mathcal{Q}_\alpha(\cdot)\phi \in C([0,\infty); \exp L_0^2(\mathbb{R}^d))$.*

Proof Let $\phi \in \exp L_0^2(\mathbb{R}^d)$. By (i) of Lemma 4.21 and the definition of $\exp L_0^2(\mathbb{R}^d)$, we have $\mathcal{Q}_\alpha(t)\phi \in \exp L_0^2(\mathbb{R}^d)$ for every $t > 0$. Thus, by the linearity of $\mathcal{Q}_\alpha(t)$, it remains to prove the continuity at $t = 0$ in $\exp L_0^2(\mathbb{R}^d)$, i.e., it remains to get

$$\lim_{t\to 0} \|\mathcal{Q}_\alpha(t)\phi - \phi\|_{\exp L^2(\mathbb{R}^d)} = 0.$$

Since $\phi \in \exp L_0^2(\mathbb{R}^d)$, there exists a sequence $\{\phi_n\}_{n=1}^\infty \subset C_0^\infty(\mathbb{R}^d)$ such that

$$\lim_{n\to\infty} \|\phi_n - \phi\|_{\exp L^2(\mathbb{R}^d)} = 0.$$

By embeddings $L^2(\mathbb{R}^d) \cap L^\infty(\mathbb{R}^d) \hookrightarrow \exp L_0^2(\mathbb{R}^d) \hookrightarrow \exp L^2(\mathbb{R}^d)$ in Lemma 4.17 and estimation (i) of Lemma 4.21, it yields

$$\begin{aligned} &\|\mathcal{Q}_\alpha(t)\phi - \phi\|_{\exp L^2(\mathbb{R}^d)} \\ \le &\|\mathcal{Q}_\alpha(t)(\phi - \phi_n)\|_{\exp L^2(\mathbb{R}^d)} + \|\mathcal{Q}_\alpha(t)\phi_n - \phi_n\|_{\exp L^2(\mathbb{R}^d)} + \|\phi_n - \phi\|_{\exp L^2(\mathbb{R}^d)} \\ \le &2\|\mathcal{Q}_\alpha(t)\phi_n - \phi_n\|_{L^2(\mathbb{R}^d)} + 2\|\mathcal{Q}_\alpha(t)\phi_n - \phi_n\|_{L^\infty(\mathbb{R}^d)} + 2\|\phi_n - \phi\|_{\exp L^2(\mathbb{R}^d)}. \end{aligned}$$

Since $\phi_n \in C_0^\infty(\mathbb{R}^d)$, Plancherel's identity shows that

$$\begin{aligned} \|\mathcal{Q}_\alpha(t)\phi_n - \phi_n\|_{L^2(\mathbb{R}^d)} &= \left\| \mathcal{F}^{-1}[E_{\alpha,1}(-t^\alpha|\xi|^2)\mathcal{F}(\phi_n)(\xi)](x) - \phi_n(x) \right\|_{L^2(\mathbb{R}^d)} \\ &= \left\| \mathcal{F}^{-1}[(E_{\alpha,1}(-t^\alpha|\xi|^2) - 1)\mathcal{F}(\phi_n)(\xi)](x) \right\|_{L^2(\mathbb{R}^d)} \\ &= \left\| (E_{\alpha,1}(-t^\alpha|\xi|^2) - 1)\mathcal{F}(\phi_n)(\xi) \right\|_{L^2(\mathbb{R}^d)}. \end{aligned}$$

From the Hausdorff-Young inequality and Proposition 1.12, it follows that

$$\left\|(E_{\alpha,1}(-t^\alpha|\xi|^2)-1)\mathscr{F}(\phi_n)(\xi)\right\|_{L^2(\mathbb{R}^d)} \leq 2\|\phi_n\|_{L^2(\mathbb{R}^d)}.$$

By the Lebesgue dominated convergence theorem, we have

$$\|\mathscr{Q}_\alpha(t)\phi_n-\phi_n\|_{L^2(\mathbb{R}^d)} \leq \left\|(E_{\alpha,1}(-t^\alpha|\xi|^2)-1)\mathscr{F}(\phi_n)(\xi)\right\|_{L^2(\mathbb{R}^d)} \to 0,$$

as $t\to 0$. Since $|E_{\alpha,1}(-t^\alpha|\xi|^2)|\leq 1$ for $t>0$, $|\xi|>0$ from Proposition 1.14, and $E_{\alpha,1}(-t^\alpha|\xi|^2)$ is close to 1 for small enough $0<t\ll 1$ and fixed $|\xi|>0$, it yields that

$$1-E_{\alpha,1}(-t^\alpha|\xi|^2)\leq E_{\alpha,1}(-t^\alpha|\xi|^2).$$

By virtue of the fact that $C_0^\infty(\mathbb{R}^d)$ is dense in $L^2(\mathbb{R}^d)$ and then

$$\begin{aligned}\|\mathscr{Q}_\alpha(t)\phi_n-\phi_n\|_{L^\infty(\mathbb{R}^d)} &\leq \int_{\mathbb{R}^d}\left|(E_{\alpha,1}(-t^\alpha|\xi|^2)-1)\mathscr{F}(\phi_n)(\xi)\right|d\xi\\ &\leq \left(\int_{\mathbb{R}^d}|E_{\alpha,1}(-t^\alpha|\xi|^2)|^2d\xi\right)^{1/2}\|\hat{\phi}_n\|_{L^2(\mathbb{R}^d)},\end{aligned}$$

in which the Plancherel identity and the Hausdorff-Young inequality deduce the boundedness of the above last right-hand inequality for small enough $0<t\ll 1$, moreover Lebesgue's dominated convergence theorem implies $\|\mathscr{Q}_\alpha(t)\phi_n-\phi_n\|_{L^\infty(\mathbb{R}^d)}\to 0$ as $t\to 0$. Hence, $\mathscr{Q}_\alpha(t)$ is strongly continuous on $L^2(\mathbb{R}^d)\cap L^\infty(\mathbb{R}^d)$; it means that

$$\lim_{t\to 0}\left(\|\mathscr{Q}_\alpha(t)\phi_n-\phi_n\|_{L^2(\mathbb{R}^d)}+\|\mathscr{Q}_\alpha(t)\phi_n-\phi_n\|_{L^\infty(\mathbb{R}^d)}\right)=0.$$

Hence, we have

$$\limsup_{t\to 0}\|\mathscr{Q}_\alpha(t)\phi-\phi\|_{\exp L^2(\mathbb{R}^d)}\leq 2\|\phi_n-\phi\|_{\exp L^2(\mathbb{R}^d)},$$

for every $n\in\mathbb{N}^+$. This finishes the proof of the lemma.

Next, we obtain the local existence of solution for the problem (4.56) as follows.

Lemma 4.25 *Let $v_0\in L^2(\mathbb{R}^d)\cap L^\infty(\mathbb{R}^d)$. Then there exist a time $T>0$ and a mild solution $v\in C([0,T];\exp L_0^2(\mathbb{R}^d))\cap L^\infty(0,T;L^\infty(\mathbb{R}^d))$.*

Proof We first introduce a Banach space as follows:

$$Y_M:=\{v\in C([0,T];\exp L_0^2(\mathbb{R}^d))\cap L^\infty(0,T;L^\infty(\mathbb{R}^d)):\ \|v\|_{Y_M}\leq 4\|v_0\|_{L^2\cap L^\infty}\},$$

where

$$\|v\|_{Y_M} := \|v\|_{L^\infty(0,T;\exp L_0^2(\mathbb{R}^d))} + \|v\|_{L^\infty(0,T;L^\infty(\mathbb{R}^d))},$$

$$\|v_0\|_{L^2\cap L^\infty} := \|v_0\|_{L^2(\mathbb{R}^d)} + \|v_0\|_{L^\infty(\mathbb{R}^d)}.$$

For any $v \in Y_M$, define an operator $\mathscr{T}$ on Y_M by

$$(\mathscr{T}v)(t) = \mathscr{Q}_\alpha(t)v_0 + \int_0^t \mathscr{P}_\alpha(t-s)f(v(s))ds.$$

In order to use the Banach fixed point theorem, it suffices to show that $\mathscr{T}$ is a contraction from Y_M into itself for small $T > 0$.

Claim I. $\mathscr{T} : Y_M \to Y_M$. Since $v_0 \in L^2(\mathbb{R}^d) \cap L^\infty(\mathbb{R}^d)$, it follows from Lemma 4.17 that $v_0 \in \exp L_0^2(\mathbb{R}^d)$ and from Lemma 4.24 that $\mathscr{Q}_\alpha(t)v_0 \in C([0,T]; \exp L_0^2(\mathbb{R}^d))$.
In addition, from (4.54) we have

$$\|f(v)\|_{L^2(\mathbb{R}^d)} \le Ce^{\|v\|^2_{L^\infty(\mathbb{R}^d)}}\|v\|_{L^2(\mathbb{R}^d)} \le Ce^{\|v\|^2_{L^\infty(\mathbb{R}^d)}}\|v\|_{L^2\cap L^\infty},$$

which implies that $f(v) \in \exp L_0^2(\mathbb{R}^d)$ by Lemma 4.17 for any $v \in L^\infty(\mathbb{R}^d)$. Moreover, we also have $f(v) \in L^1(0,T; \exp L_0^2(\mathbb{R}^d))$. Lemma 4.23 implies that

$$\int_0^t \mathscr{P}_\alpha(t-s)f(v(s))ds \in C([0,T]; \exp L_0^2(\mathbb{R}^d)).$$

This means that $\mathscr{T}(v) \in Y_M$ enjoys the following estimates:

$$\begin{aligned}
\|\mathscr{T}(v)\|_{\exp L^2(\mathbb{R}^d)} \le& \|v_0\|_{\exp L^2(\mathbb{R}^d)} + \frac{1}{\Gamma(\alpha)}\int_0^t (t-s)^{\alpha-1}\|f(v)\|_{\exp L^2(\mathbb{R}^d)}ds\\
\le& 2\|v_0\|_{L^2\cap L^\infty} + C\int_0^t (t-s)^{\alpha-1}e^{\|v(s)\|^2_{L^\infty(\mathbb{R}^d)}}\|v(s)\|_{L^2\cap L^\infty}ds\\
\le& 2\|v_0\|_{L^2\cap L^\infty} + C\int_0^t (t-s)^{\alpha-1}e^{4\|v_0\|_{L^2\cap L^\infty}}\|v_0\|_{L^2\cap L^\infty}ds,
\end{aligned}$$

and then

$$\begin{aligned}
\|\mathscr{T}(v)\|_{\exp L^2(\mathbb{R}^d)} \le& 2\|v_0\|_{L^2\cap L^\infty} + Ct^\alpha e^{4\|v_0\|_{L^2\cap L^\infty}}\|v_0\|_{L^2\cap L^\infty}\\
\le& 4\|v_0\|_{L^2\cap L^\infty},
\end{aligned}$$

for some small $T > 0$ such that $CT^\alpha e^{4\|v_0\|_{L^2\cap L^\infty}} \le 2$. Therefore, using Lemma 4.20 for any $v \in Y_M$, similarly we have

$$
\begin{aligned}
\|\mathscr{T}(v)\|_{L^\infty(\mathbb{R}^d)} \leq & 2\|v_0\|_{L^2\cap L^\infty} + \frac{1}{\Gamma(\alpha)}\int_0^t (t-s)^{\alpha-1-\frac{\alpha d}{4}}\|f(v)\|_{L^2(\mathbb{R}^d)}ds \\
\leq & 2\|v_0\|_{L^2\cap L^\infty} + C\int_0^t (t-s)^{\alpha-1-\frac{\alpha d}{4}} e^{\|v(s)\|_{L^\infty(\mathbb{R}^d)}}\|v(s)\|_{L^2\cap L^\infty}ds \\
\leq & 2\|v_0\|_{L^2\cap L^\infty} + Ct^{\alpha(4-d)/4}e^{4\|v_0\|_{L^2\cap L^\infty}}\|v_0\|_{L^2\cap L^\infty} \\
\leq & 4\|v_0\|_{L^2\cap L^\infty},
\end{aligned}
$$

for some small $T > 0$. This proves that $\mathscr{T}(v) \in Y_M$.

Claim II. $\mathscr{T}$ is a contraction. From the assumption of f, we first obtain

$$
f(0) = 0, \quad |f(u) - f(v)| \leq C(|u^2 e^{\lambda u^2} + v^2 e^{\lambda v^2}||u - v|), \tag{4.58}
$$

for some $C > 0$ and $\lambda > 0$. For any $v_1, v_2 \in Y_M$, by (4.58) it follows that

$$
\begin{aligned}
&\|f(v_1) - f(v_2)\|_{L^2(\mathbb{R}^d)} \\
\leq & C\|v_1 - v_2\|_{L^2(\mathbb{R}^d)}(\|v_1\|^2_{L^\infty(\mathbb{R}^d)} e^{\lambda\|v_1\|^2_{L^\infty(\mathbb{R}^d)}} + \|v_2\|^2_{L^\infty(\mathbb{R}^d)} e^{\lambda\|v_2\|^2_{L^\infty(\mathbb{R}^d)}}) \\
\leq & C\|v_1 - v_2\|_{L^2(\mathbb{R}^d)}\|v_0\|^2_{L^2\cap L^\infty} e^{16\lambda\|v_0\|^2_{L^2\cap L^\infty}} \\
\leq & C\|v_1 - v_2\|_{Y_M}\|v_0\|^2_{L^2\cap L^\infty} e^{16\lambda\|v_0\|^2_{L^2\cap L^\infty}},
\end{aligned}
$$

which means that

$$
\begin{aligned}
\|\mathscr{T}(v_1) - \mathscr{T}(v_2)\|_{Y_M} \leq & \frac{1}{\Gamma(\alpha)}\int_0^t (t-s)^{\alpha-1}\|f(v_1) - f(v_2)\|_{Y_M}ds \\
\leq & C\int_0^t (t-s)^{\alpha-1}\|v_0\|^2_{L^2\cap L^\infty} e^{16\lambda\|v_0\|^2_{L^2\cap L^\infty}}\|v_1 - v_2\|_{Y_M}ds \\
\leq & CT^\alpha\|v_0\|^2_{L^2\cap L^\infty} e^{16\lambda\|v_0\|^2_{L^2\cap L^\infty}}\|v_1 - v_2\|_{Y_M},
\end{aligned}
$$

for $T > 0$ small enough such that $CT^\alpha\|v_0\|^2_{L^2\cap L^\infty} e^{16\lambda\|v_0\|^2_{L^2\cap L^\infty}} < \frac{1}{2}$. Then

$$
\|\mathscr{T}(v_1) - \mathscr{T}(v_2)\|_{Y_M} \leq \frac{1}{2}\|v_1 - v_2\|_{Y_M}.
$$

This shows the desired conclusion.

Lemma 4.26 *Let $v \in L^\infty(0, T; L^\infty(\mathbb{R}^d))$ for some $T > 0$. Let $2 \leq r < +\infty$, and $w_1, w_2 \in \exp L^2(\mathbb{R}^d)$ with sufficiently small $\|w_1\|_{\exp L^2(\mathbb{R}^d)} \leq M$, $\|w_2\|_{\exp L^2(\mathbb{R}^d)} \leq M$ for some $M > 0$ and $\lambda > 0$ such that $2\lambda M^2 r < 1$. Then, there exists a constant $C > 0$ such that*

$$
\begin{aligned}
&\|f(w_1+v)-f(w_2+v)\|_{L^r(\mathbb{R}^d)}\\
\leq &Cr^{\frac{1}{2}}\left(\sum_{k=1}^{\infty}(2\lambda)^k M^{2k} r^k + e^{2\lambda\|v\|^2_{L^\infty(\mathbb{R}^d)}}\right)\|w_1-w_2\|_{\exp L^2(\mathbb{R}^d)}.
\end{aligned}
$$

Proof By the assumption on f in (4.58), there is a $\lambda > 0$ such that

$$
|f(u)-f(v)| \leq C|u-v|\sum_{k=1}^{\infty}\frac{\lambda^k}{k!}(|u|^{2k}+|v|^{2k}).
$$

Therefore, we have

$$
\begin{aligned}
|f(w_1+v)-f(w_2+v)| \leq &C|w_1-w_2|\sum_{k=1}^{\infty}\frac{\lambda^k}{k!}\left(|w_1+v|^{2k}+|w_2+v|^{2k}\right)\\
\leq &C|w_1-w_2|\sum_{k=1}^{\infty}\frac{(2\lambda)^k}{2k!}\left(|w_1|^{2k}+|w_2|^{2k}+2|v|^{2k}\right)\\
\leq &C|w_1-w_2|\left(\sum_{k=1}^{\infty}\frac{(2\lambda)^k}{2k!}(|w_1|^{2k}+|w_2|^{2k})+e^{2\lambda|v|^2}\right),
\end{aligned}
$$

where we have used the inequality

$$
(a+b)^r \leq 2^{r-1}(a^r+b^r), \quad a,b\geq 0, \quad r\geq 1.
$$

The Hölder inequality shows that

$$
\left\||w_1-w_2|(|w_1|^{2k}+|w_2|^{2k})\right\|_{L^r(\mathbb{R}^d)} \leq \|w_1-w_2\|_{L^{2r}(\mathbb{R}^d)}\||w_1|^{2k}+|w_2|^{2k}\|_{L^{2r}(\mathbb{R}^d)},
$$

and then for $v \in L^\infty(\mathbb{R}^d)$, we have

$$
\begin{aligned}
&\|f(w_1+v)-f(w_2+v)\|_{L^r(\mathbb{R}^d)}\\
\leq &C\sum_{k=1}^{\infty}\frac{(2\lambda)^k}{2k!}\|w_1-w_2\|_{L^{2r}(\mathbb{R}^d)}\||w_1|^{2k}\\
&+|w_2|^{2k}\|_{L^{2r}(\mathbb{R}^d)}+Ce^{2\lambda\|v\|^2_{L^\infty(\mathbb{R}^d)}}\|w_1-w_2\|_{L^r(\mathbb{R}^d)}.
\end{aligned}
$$

By virtue of Lemma 4.16, we have

$$
\|w_1-w_2\|_{L^{2r}(\mathbb{R}^d)} \leq (\Gamma(r+1))^{\frac{1}{2r}}\|w_1-w_2\|_{\exp L^2(\mathbb{R}^d)}
$$

and

$$\|w_1 - w_2\|_{L^r(\mathbb{R}^d)} \le (\Gamma(r/2+1))^{\frac{1}{r}} \|w_1 - w_2\|_{\exp L^2(\mathbb{R}^d)}.$$

Together with Lemma 4.16 again and Lemma 4.18, we have the estimates

$$\begin{aligned}
&\|f(w_1+v) - f(w_2+v)\|_{L^r(\mathbb{R}^d)} \\
\le& Cr^{\frac{1}{2}} \left(\sum_{k=1}^{\infty} \frac{(2\lambda)^k}{2k!} (\|w_1\|^{2k}_{L^{2kr}(\mathbb{R}^d)} + \|w_2\|^{2k}_{L^{2rk}(\mathbb{R}^d)}) + e^{2\lambda \|v\|^2_{L^\infty(\mathbb{R}^d)}} \right) \|w_1 - w_2\|_{\exp L^2(\mathbb{R}^d)} \\
\le& Cr^{\frac{1}{2}} \left(\sum_{k=1}^{\infty} \frac{(2\lambda)^k}{2k!} \Gamma(kr+1)^{\frac{1}{r}} (\|w_1\|^{2k}_{\exp L^2(\mathbb{R}^d)} + \|w_2\|^{2k}_{\exp L^2(\mathbb{R}^d)}) + e^{2\lambda \|v\|^2_{L^\infty(\mathbb{R}^d)}} \right) \\
&\times \|w_1 - w_2\|_{\exp L^2(\mathbb{R}^d)} \\
\le& Cr^{\frac{1}{2}} \left(\sum_{k=1}^{\infty} (2\lambda)^k M^{2k} r^k + e^{2\lambda \|v\|^2_{L^\infty(\mathbb{R}^d)}} \right) \|w_1 - w_2\|_{\exp L^2(\mathbb{R}^d)}.
\end{aligned}$$

Consequently, the proof is completed.

Next, we establish the existence of solution for the problem (4.57) as follows.

Lemma 4.27 *Let $w_0 \in \exp L^2_0(\mathbb{R}^d)$. Let $T > 0$, and let $v \in L^\infty(0, T; L^\infty(\mathbb{R}^d))$ be given by Lemma 4.25. Then, for $\|w_0\|_{\exp L^2(\mathbb{R}^d)} \le \varepsilon$, with $\varepsilon > 0$ small enough, there exists a mild solution $w \in C([0, T_*]; \exp L^2_0(\mathbb{R}^d))$ for some $T_* = T(w_0, \varepsilon, v) > 0$.*

Proof For any $\varepsilon > 0$, let $T > 0$, we define the following complete metric space:

$$Z_M := \{w \in C([0,T]; \exp L^2_0(\mathbb{R}^d)) : \|w\|_{L^\infty(0,T;\exp L^2_0(\mathbb{R}^d))} \le 2\varepsilon\},$$

and we consider a mapping $\mathscr{F}$ on Z_M by

$$(\mathscr{F} w)(t) = \mathscr{Q}_\alpha(t) w_0 + \int_0^t \mathscr{P}_\alpha(t-s)(f(w(s)+v(s)) - f(v(s)))ds.$$

Claim I. $\mathscr{F}$ is a contraction. Let $w_1, w_2 \in Z_M$. By Lemma 4.17, we first have

$$\|\mathscr{F}(w_1) - \mathscr{F}(w_2)\|_{\exp L^2(\mathbb{R}^d)} \le 2\|\mathscr{F} w_1 - \mathscr{F} w_2\|_{L^2(\mathbb{R}^d)} + 2\|\mathscr{F} w_1 - \mathscr{F} w_2\|_{L^\infty(\mathbb{R}^d)}.$$

Hence, one yields from Lemma 4.20

$$\begin{aligned}
&\|\mathscr{F}(w_1) - \mathscr{F}(w_2)\|_{L^\infty(\mathbb{R}^d)} \\
&= \left\| \int_0^t \mathscr{P}_\alpha(t-s)(f(w_1(s)+v(s)) - f(w_2(s)+v(s)))ds \right\|_{L^\infty(\mathbb{R}^d)}
\end{aligned}$$

$$\leq C \int_0^t (t-s)^{\alpha-1-\frac{\alpha d}{4}} \|f(w_1(s)+v(s)) - f(w_2(s)+v(s))\|_{L^2(\mathbb{R}^d)} ds.$$

Applying Lemma 4.26 with $r = 2$ and under the condition $16\lambda\varepsilon^2 < 1$ for $\lambda > 0$ given in (4.58), we obtain

$$\begin{aligned}
&\|\mathcal{F}(w_1) - \mathcal{F}(w_2)\|_{L^\infty(\mathbb{R}^d)} \\
\leq& C \left(\sum_{k=1}^{\infty} (4\lambda)^k (2\varepsilon)^{2k} + e^{2\lambda \|v\|^2_{L^\infty(0,T;L^\infty(\mathbb{R}^d))}} \right) \\
&\times \int_0^t (t-s)^{\alpha-1-\frac{\alpha d}{4}} \|w_1(s) - w_2(s)\|_{\exp L^2(\mathbb{R}^d)} ds \\
\leq& C_\alpha T^{\alpha\left(1-\frac{d}{4}\right)} \|w_1 - w_2\|_{L^\infty(0,T;\exp L^2(\mathbb{R}^d))},
\end{aligned}$$

where positive constant C_α may depend on $\alpha, \lambda, \varepsilon, d$, and $\|v\|_{L^\infty(0,T;L^\infty(\mathbb{R}^d))}$.

On the other hand, by Lemma 4.21 and Lemma 4.26 with $r = 2$, we also have

$$\begin{aligned}
&\|\mathcal{F}(w_1) - \mathcal{F}(w_2)\|_{L^2(\mathbb{R}^d)} \\
=& \left\| \int_0^t \mathcal{P}_\alpha(t-s)(f(w_1(s)+v(s)) - f(w_2(s)+v(s)))ds \right\|_{L^2(\mathbb{R}^d)} \\
\leq& \frac{1}{\Gamma(\alpha)} \int_0^t (t-s)^{\alpha-1} \|f(w_1(s)+v(s)) - f(w_2(s)+v(s))\|_{L^2(\mathbb{R}^d)} ds \\
\leq& C \left(\sum_{k=1}^{\infty} (4\lambda)^k (2\varepsilon)^{2k} + e^{2\lambda \|v\|^2_{L^\infty(0,T;L^\infty(\mathbb{R}^d))}} \right) \int_0^t (t-s)^{\alpha-1} \|w_1(s) \\
&\quad - w_2(s)\|_{\exp L^2(\mathbb{R}^d)} ds \\
\leq& C'_\alpha T^\alpha \|w_1 - w_2\|_{L^\infty(0,T;\exp L^2(\mathbb{R}^d))},
\end{aligned}$$

where positive constant C'_α may depend on $\alpha, \lambda, \varepsilon, d$, and $\|v\|_{L^\infty(0,T;L^\infty(\mathbb{R}^d))}$. Therefore, it follows that

$$\|\mathcal{F}(w_1) - \mathcal{F}(w_2)\|_{\exp L^2(\mathbb{R}^d)} \leq (C_\alpha T^{\alpha\left(1-\frac{d}{4}\right)} + C'_\alpha T^\alpha) \|w_1 - w_2\|_{L^\infty(0,T;\exp L^2(\mathbb{R}^d))}.$$

By choosing T small enough such that

$$C_\alpha T^{\alpha\left(1-\frac{d}{4}\right)} + C'_\alpha T^\alpha \leq \frac{1}{2}, \tag{4.59}$$

then, one finds $\|\mathcal{F}(w_1) - \mathcal{F}(w_2)\|_{L^\infty(0,T;\exp L^2(\mathbb{R}^d))} \leq \frac{1}{2} \|w_1 - w_2\|_{L^\infty(0,T;\exp L^2(\mathbb{R}^d))}$.

Claim II. $\mathscr{F} : Z_M \to Z_M$. Let $w \in Z_M$. As $w_0 \in \exp L^2_0(\mathbb{R}^d)$ and Lemma 4.24, it follows that

$$\mathscr{Q}_\alpha(t)w_0 \in C([0, T]; \exp L^2_0(\mathbb{R}^d)).$$

Next, let $w_1 = w$ and $w_2 = 0$ in the above estimates in Claim I. Under the condition $16\lambda\varepsilon^2 < 1$, the nonlinear term satisfies

$$\mathscr{F}(w) - \mathscr{Q}_\alpha(t)w_0 \in L^\infty(0, T; \exp L^2(\mathbb{R}^d)).$$

By the standard smoothing effects of $\mathscr{Q}_\alpha(t)$ and $\mathscr{P}_\alpha(t)$, we obtain

$$\mathscr{F}(w) - \mathscr{Q}_\alpha(t)w_0 \in C([0, T]; \exp L^2(\mathbb{R}^d)).$$

This means that for T small enough

$$\|\mathscr{F}(w_1)\|_{\exp L^2(\mathbb{R}^d)} \le \|w_0\|_{\exp L^2(\mathbb{R}^d)} + \frac{1}{2}\|w_1\|_{L^\infty(0,T;\exp L^2(\mathbb{R}^d))} \le \varepsilon + \frac{1}{2}(2\varepsilon) \le 2\varepsilon.$$

This proves that $\mathscr{F}$ maps Z_M into itself.

Under the assumption of f satisfying (4.54), we get the local existence result of this section as follows.

Theorem 4.15 (Local Existence) *Let $u_0 \in \exp L^2_0(\mathbb{R}^d)$; then for a time $T = T(u_0) > 0$, there exists a mild solution $u \in C([0, T]; \exp L^2_0(\mathbb{R}^d))$ of the problem (4.49)–(4.50).*

Proof We choose T and ε small enough in the following sense and fix $\varepsilon > 0$ such that $16\lambda\varepsilon^2 < 1$ with $\lambda > 0$ given in (4.58).

By analyzing arguments above, one can decompose $u_0 = w_0 + v_0$ with $w_0 \in C_0^\infty(\mathbb{R}^d)$ and $\|v_0\|_{\exp L^2(\mathbb{R}^d)} \le \varepsilon$. By Lemma 4.25, we know that there exist a time $0 < T_1 = T_1(\|w_0\|_{L^2\cap L^\infty})$ and a mild solution $v \in C([0, T_1]; \exp L^2_0(\mathbb{R}^d)) \cap L^\infty(0, T_1; L^\infty(\mathbb{R}^d))$ such that $\|v\|_{L^\infty(0,T;L^2\cap L^\infty)} \le 4\|v_0\|_{L^2\cap L^\infty}$. By choosing $T \in (0, T_1)$ small enough such that (4.59) satisfies, by using Lemma 4.27, it follows that there exists a mild solution $w \in C([0, T]; \exp L^2_0(\mathbb{R}^d))$. Therefore, we derive that $u := w + v$ is a mild solution in $C([0, T]; \exp L^2_0(\mathbb{R}^d))$. The proof is completed.

4.3.3.2 Global Existence of Solutions

In the sequel, we prove the global existence result. In order to do this, for $M > 0$, set operator

$$F(u)(t) := \mathscr{Q}_\alpha(t)u_0 + \int_0^t \mathscr{P}_\alpha(t-s)f(u(s))ds,$$

and we introduce a space

$$X_M := \{u \in L^\infty(0,\infty;\exp L^2(\mathbb{R}^d)) : \sup_{t>0} t^\sigma \|u(t)\|_{L^r(\mathbb{R}^d)} + \|u\|_{L^\infty(0,\infty;\exp L^2(\mathbb{R}^d))} \le M\},$$

where $\sigma = \frac{\alpha d}{2}(\frac{1}{2} - \frac{1}{r})$ for some $r > 3$. The space X_M is a complete metric space with the distance

$$d(u, v) = \sup_{t>0} t^\sigma \|u(t) - v(t)\|_{L^r(\mathbb{R}^d)}.$$

It needs to point out that $C_0^\infty(\mathbb{R}^d)$ is not dense in $\exp L^2(\mathbb{R}^d)$, which implies that $\mathcal{Q}_\alpha(t)u_0$ is not continuous at $t = 0$ in $\exp L^2(\mathbb{R}^d)$, i.e., for any $u_0 \in \exp L^2(\mathbb{R}^d)$, the following inequality may hold:

$$\lim_{t\to 0} \|\mathcal{Q}_\alpha(t)u_0 - u_0\|_{\exp L^2(\mathbb{R}^d)} \ge 1. \tag{4.60}$$

In fact, for any $\lambda > 0$, let $\mu_v(\lambda) := |\{x \in \mathbb{R}^d : |v(x)| > \lambda\}|$ be a distribution function of v. We next use the rearrangement technique developing by [94], i.e., it yields

$$\|\mathcal{Q}_\alpha(t)u_0 - u_0\|_{\exp L^2(\mathbb{R}^d)} \ge C \sup_{0<r<1} \frac{(\mathcal{Q}_\alpha(t)u_0 - u_0)^{**}(r)}{\sqrt{\ln \frac{e}{r}}},$$

where $C > 0$ and we have used the properties (see, e.g., [94, Lemma 5.2], [120])

$$C_1\|v\|_{\exp L^2(\mathbb{R}^d)} \le \sup_{0<r<1} \frac{v^{**}(r)}{\sqrt{\ln \frac{e}{r}}} + \|v\|_{L^2(\mathbb{R}^d)} \le C_2\|v\|_{\exp L^2(\mathbb{R}^d)},$$

for some constants $C_1, C_2 > 0$, and notation v^{**} stands for the maximal function of v^* given by $v^{**}(r) = \frac{1}{r}\int_0^r v^*(s)ds$, with v^* being a nonincreasing rearrangement of v by $v^*(r) := \inf\{\lambda > 0 : \mu_v(\lambda) \le r\}$. Therefore, due to the triangle inequality of v^{**}, we have

$$\frac{(u_0 - \mathcal{Q}_\alpha(t)u_0)^{**}(r)}{\sqrt{\ln \frac{e}{r}}} \ge \frac{(u_0)^{**}(r) - (\mathcal{Q}_\alpha(t)u_0)^{**}(r)}{\sqrt{\ln \frac{e}{r}}},$$

which implies from the nonnegative property of v^* that

$$\|\mathcal{Q}_\alpha(t)u_0 - u_0\|_{\exp L^2(\mathbb{R}^d)} \ge C \sup_{0<r<1} \frac{(\mathcal{Q}_\alpha(t)u_0 - u_0)^{**}(r)}{\sqrt{\ln \frac{e}{r}}}$$

$$\geq C \lim_{t\to 0} \frac{(\mathcal{Q}_\alpha(t)u_0 - u_0)^{**}(r)}{\sqrt{\ln \frac{e}{r}}}$$

$$\geq C \lim_{t\to 0} \frac{(u_0)^{**}(r) - (\mathcal{Q}_\alpha(t)u_0)^{**}(r)}{\sqrt{\ln \frac{e}{r}}}.$$

According to Lemma 4.20, since $\mathcal{Q}_\alpha(t)u_0 \in L^\infty(\mathbb{R}^d)$ for all $t \geq \varepsilon$ with any $\varepsilon > 0$, it yields that $(\mathcal{Q}_\alpha(t)u_0)^{**}(r) \in L^\infty(0, \infty)$ for all $t \geq \varepsilon$. Therefore one has

$$\lim_{r\to 0} \frac{(\mathcal{Q}_\alpha(t)u_0)^{**}(r)}{\sqrt{\ln \frac{e}{r}}} = 0.$$

Let $u_0(x) = \frac{1}{\sqrt{dB_d|x|^d}}$ for $0 < |x| < \frac{1}{dB_d}$, and it is zero for otherwise, where B_d is the measure of the unit ball in $\mathbb{R}^d$; it yields $(u_0)^{**}(r) = \frac{1}{\sqrt{r}}$ for $0 < r < 1$. Hence we get

$$\|\mathcal{Q}_\alpha(t)u_0 - u_0\|_{\exp L^2(\mathbb{R}^d)} \geq C \lim_{r\to 0} \frac{(u_0)^{**}(r)}{\sqrt{\ln \frac{e}{r}}} = +\infty \gg 1.$$

The inequality (4.60) is showed. Therefore we shall consider the global existence of solution to the Cauchy problem (4.49)–(4.50) by a weak form at continuity on initial value term as follows.

Definition 4.6 We say that $u \in L^\infty(0, \infty; \exp L^2(\mathbb{R}^d))$ is a weak mild solution for the problem (4.49)–(4.50) if u satisfies the integral equation (4.55) in $\exp L^2(\mathbb{R}^d)$ for almost all $t > 0$, and it also satisfies the continuity to initial data in the sense that

$$w^* - \lim_{t\to 0} u(t) = u_0, \ \text{ in } \exp L^2(\mathbb{R}^d).$$

Here we say that $u(t) \to u_0$ in weak* sense in $\exp L^2(\mathbb{R}^d)$ if and only if

$$\lim_{t\to 0} \int_{\mathbb{R}^d} [u(t,x)\phi(x) - u_0(x)\phi(x)]dx = 0, \quad \text{for } \phi \in L^1(\ln L)^{1/2}(\mathbb{R}^d),$$

where $L^1(\ln L)^{1/2}(\mathbb{R}^d)$ is a predual space of $\exp L^2(\mathbb{R}^d)$ given by

$$L^1(\ln L)^{1/2}(\mathbb{R}^d) = \left\{ f \in L^1_{loc}(\mathbb{R}^d) : \int_{\mathbb{R}^d} |f(x)|(\ln(2 + |f(x)|))^{1/2} dx < +\infty \right\}.$$

Lemma 4.28 *Let* $\alpha \in (1, \frac{4}{3})$. *Then*

$$\left\| \int_0^t \mathscr{P}_\alpha(t-s) f(u(s)) ds \right\|_{L^\infty(0,\infty;\exp L^2(\mathbb{R}^d))} \le H(M)$$

and

$$\sup_{t>0} t^\sigma \left\| \int_0^t \mathscr{P}_\alpha(t-s)(f(u(s)) - f(v(s))) ds \right\|_{L^r(\mathbb{R}^d)} \le C(M) \sup_{t>0} t^\sigma \|u(t)-v(t)\|_{L^r(\mathbb{R}^d)},$$

where

$$H(M) = o(M), \quad C(M) = \begin{cases} O(M^2), & d = 2, 3, \\ O(M^4), & d = 1. \end{cases}$$

Proof We first concern the dimensions $d = 2, 3$. Let $1 \vee \frac{d}{2} \le p < \frac{2\alpha d}{(4+d)\alpha - 4}$, and for each $k \in \mathbb{N}^+$,

$$\varpi = \frac{2d((2k+1)(r-2) - br)}{(d+4)(r-2) - dbr}, \quad \theta = \frac{br}{(r-2)(2k+1)}, \quad \frac{1}{p} = \frac{1}{2} + \frac{2}{d} - \frac{b}{2},$$

for some $\frac{4-d}{d} < b < \frac{4}{\alpha d} \wedge \frac{(d+4)(r-2)}{dr}$ with $r > 3$. Therefore, $\varpi \ge 2$ and $\theta \in [0, 1]$. In addition, let a be the smallest positive number satisfying $a = 2\ln(a+1)$. We first have

$$(\ln((t-s)^{-\frac{\alpha d}{2}} + 1))^{-\frac{1}{2}} \le \sqrt{2}(t-s)^{\frac{\alpha d}{4}}, \quad \text{for } 0 \le s \le t - a^{-\frac{2}{\alpha d}}$$

and

$$(\ln((t-s)^{-\frac{\alpha d}{2}} + 1))^{-\frac{1}{2}} \le \frac{\sqrt{2a}}{2}, \quad \text{for } t - a^{-\frac{2}{\alpha d}} \le s \le t.$$

Therefore, by the second estimate in Lemma 4.22, we have

$$\begin{aligned}
&\left\| \int_0^t \mathscr{P}_\alpha(t-s) f(u(s)) ds \right\|_{\exp L^2(\mathbb{R}^d)} \\
\le& C \int_0^t (t-s)^{\alpha - 1 - \frac{\alpha d}{2p}} (\ln((t-s)^{-\frac{\alpha d}{2}} + 1))^{-\frac{1}{2}} \|f(u(s))\|_{L^p(\mathbb{R}^d)} ds \\
=& C \int_0^{t - a^{-\frac{2}{\alpha d}}} (t-s)^{\alpha - 1 - \frac{\alpha d}{2p}} (\ln((t-s)^{-\frac{\alpha d}{2}} + 1))^{-\frac{1}{2}} \|f(u(s))\|_{L^p(\mathbb{R}^d)} ds \\
&+ C \int_{t - a^{-\frac{2}{\alpha d}}}^t (t-s)^{\alpha - 1 - \frac{\alpha d}{2p}} (\ln((t-s)^{-\frac{\alpha d}{2}} + 1))^{-\frac{1}{2}} \|f(u(s))\|_{L^p(\mathbb{R}^d)} ds
\end{aligned}$$

$$\leq \sqrt{2}C \int_0^{t-a^{-\frac{2}{\alpha d}}} (t-s)^{\alpha-1-\frac{\alpha d}{2p}+\frac{\alpha d}{4}} \|f(u(s))\|_{L^p(\mathbb{R}^d)} ds$$

$$+ \frac{\sqrt{2a}C}{2} \int_{t-a^{-\frac{2}{\alpha d}}}^t (t-s)^{\alpha-1-\frac{\alpha d}{2p}} \|f(u(s))\|_{L^p(\mathbb{R}^d)} ds$$

$$\leq C \int_0^t (t-s)^{\alpha-1-\frac{\alpha d}{2p}+\frac{\alpha d}{4}} \|f(u(s))\|_{L^p(\mathbb{R}^d)} ds + C \sup_{t>0} \|f(u(t))\|_{L^p(\mathbb{R}^d)} =: I + II.$$

By estimating f it yields

$$|f(u)| \leq \sum_{k=1}^{\infty} \frac{1}{k!} |u|^{2k+1}.$$

The Hölder interpolation inequality implies for each $k \in \mathbb{N}^+$ that

$$\|u\|_{L^{(2k+1)p}(\mathbb{R}^d)} \leq C \|u\|_{L^r(\mathbb{R}^d)}^{\theta} \|u\|_{L^{\varpi}(\mathbb{R}^d)}^{1-\theta}, \qquad \frac{1}{(2k+1)p} = \frac{\theta}{r} + \frac{1-\theta}{\varpi}.$$

Using Lemma 4.16, we thus obtain

$$\int_0^t (t-s)^{\alpha-1-\frac{\alpha d}{2p}+\frac{\alpha d}{4}} \|u(s)\|_{L^{(2k+1)p}(\mathbb{R}^d)}^{2k+1} ds$$

$$\leq \Gamma\left(\frac{\varpi}{2}+1\right)^{\frac{(2k+1)(1-\theta)}{\varpi}} \int_0^t (t-s)^{\alpha-1-\frac{\alpha d}{2p}+\frac{\alpha d}{4}} \|u(s)\|_{L^r(\mathbb{R}^d)}^{(2k+1)\theta} \|u(s)\|_{\exp L^2(\mathbb{R}^d)}^{(2k+1)(1-\theta)} ds$$

$$\leq \Gamma\left(\frac{\varpi}{2}+1\right)^{\frac{(2k+1)(1-\theta)}{\varpi}} M^{2k+1} \int_0^t (t-s)^{\alpha-1-\frac{\alpha d}{2p}+\frac{\alpha d}{4}} s^{-(2k+1)\theta\sigma} ds$$

$$\leq C^k k! M^{2k+1} t^{\alpha-\frac{\alpha d}{2p}+\frac{\alpha d}{4}-(2k+1)\theta\sigma} B\left(\alpha - \frac{\alpha d}{2p} + \frac{\alpha d}{4}, 1-(2k+1)\theta\sigma\right),$$

for $u \in X_M$ and $\Gamma(\varpi/2+1)^{\frac{2k(1-\theta)}{\varpi}} \leq C^k k!$ by Lemma 4.18 (see, e.g., [94]). Therefore, we obtain

$$I \leq \sum_{k=1}^{\infty} C^{k+1} M^{2k+1} t^{\alpha-\frac{\alpha d}{2p}+\frac{\alpha d}{4}-(2k+1)\theta\sigma} B\left(\alpha - \frac{\alpha d}{2p} + \frac{\alpha d}{4}, 1-(2k+1)\theta\sigma\right).$$

Together with exponents $\varpi, p, \alpha, r, \theta$, and σ, one finds that these exponents are ready to satisfy

$$\alpha - \frac{\alpha d}{2p} + \frac{\alpha d}{4} - (2k+1)\theta\sigma = 0, \quad 0 < \alpha - \frac{\alpha d}{2p} + \frac{\alpha d}{4}, \quad \varpi \geq 2, \quad 0 \leq \theta \leq 1,$$

$$0 < 1-(2k+1)\theta\sigma, \quad 1 < p < \frac{2\alpha d}{(4+d)\alpha-4}, \quad \frac{1}{(2k+1)p} = \frac{\theta}{r} + \frac{1-\theta}{\varpi}.$$

Therefore, we get

$$B\left(\alpha - \frac{\alpha d}{2p} + \frac{\alpha d}{4}, 1-(2k+1)\theta\sigma\right) = \Gamma\left(\alpha - \frac{\alpha d}{2p} + \frac{\alpha d}{4}\right)\Gamma\left(1 - \frac{a\alpha d}{4}\right) \leq C,$$

and then we have

$$I \leq C\sum_{k=1}^{\infty}\frac{1}{k!}M^{2k+1}C^k k! \leq CM^3,$$

for some small $M > 0$.

Let us prove the second Claim II. In fact, by the assumption of f, the Hölder inequality also implies

$$\|f(u)\|_{L^p(\mathbb{R}^d)} \leq \|u(e^{u^2}-1)\|_{L^p(\mathbb{R}^d)} \leq C\|u\|_{L^{2p}(\mathbb{R}^d)}\|e^{u^2}-1\|_{L^{2p}(\mathbb{R}^d)}. \tag{4.61}$$

Therefore, we have

$$\|e^{u^2}-1\|_{L^{2p}(\mathbb{R}^d)} \leq \sum_{k=1}^{\infty}\frac{1}{k!}\|u\|^{2k}_{L^{4kp}(\mathbb{R}^d)} \leq \sum_{k=1}^{\infty}\frac{1}{k!}\Gamma(2kp+1)^{\frac{1}{2p}}\|u\|^{2k}_{\exp L^2(\mathbb{R}^d)},$$

which from Lemma 4.18 yields that

$$\|e^{u^2}-1\|_{L^{2p}(\mathbb{R}^d)} \leq C\sum_{k=1}^{\infty}(2p)^k\|u\|^{2k}_{\exp L^2(\mathbb{R}^d)} \leq C\sum_{k=1}^{\infty}(2p)^kM^{2k}.$$

Let $M > 0$ small enough, and it yields that

$$\sup_{t>0}\|f(u)\|_{L^p(\mathbb{R}^d)} \leq \Gamma(p+1)^{\frac{1}{2p}}\|u\|_{\exp L^2(\mathbb{R}^d)} \cdot C\sum_{k=1}^{\infty}(2p)^kM^{2k} \leq CM^3.$$

For the case $d = 1$, the proof is similar to that under assumption of the function f given by

$$f(u) = u(e^{u^2}-1-u^2) = \sum_{k=2}^{\infty}\frac{1}{k!}u^{2k+1},$$

so that the exponent $\varpi \geq 2$, i.e., $\exp L^2(\mathbb{R}^d) \hookrightarrow L^{\varpi}(\mathbb{R}^d)$.

From (4.58), clearly for some $\lambda > 0$ and $C > 0$ we also have

$$|f(u) - f(v)| \leq C|u - v| \sum_{k=1}^{\infty} \frac{\lambda^k}{k!}(|u|^{2k} + |v|^{2k}).$$

By virtue of the Hölder inequality and the standard estimates in Lemma 4.19, we have

$$\begin{aligned}
&\left\| \int_0^t \mathscr{P}_\alpha(t-s)(f(u) - f(v))ds \right\|_{L^r(\mathbb{R}^d)} \\
\leq & C \int_0^t (t-s)^{\alpha - 1 - \frac{\alpha d}{2}\left(\frac{1}{l} - \frac{1}{r}\right)} \|f(u) - f(v)\|_{L^l(\mathbb{R}^d)} ds \\
\leq & C \sum_{k=1}^{\infty} \frac{\lambda^k}{k!} \int_0^t (t-s)^{\alpha - 1 - \frac{\alpha d}{2}\left(\frac{1}{l} - \frac{1}{r}\right)} \||u - v|(|u|^{2k} + |v|^{2k})\|_{L^l(\mathbb{R}^d)} ds \\
\leq & C \sum_{k=1}^{\infty} \frac{\lambda^k}{k!} \int_0^t (t-s)^{\alpha - 1 - \frac{\alpha d}{2}\left(\frac{1}{l} - \frac{1}{r}\right)} \|u - v\|_{L^r(\mathbb{R}^d)} \left(\|u\|_{L^{2k\gamma}(\mathbb{R}^d)}^{2k} + \|v\|_{L^{2k\gamma}(\mathbb{R}^d)}^{2k} \right) ds,
\end{aligned}$$

where

$$1 - \frac{1}{\alpha} < \frac{d}{2}\left(\frac{1}{l} - \frac{1}{r}\right) < 1, \quad \frac{1}{l} = \frac{1}{r} + \frac{1}{\gamma}, \quad \text{for} \quad \frac{d}{2} < \gamma < \frac{\alpha d}{2(\alpha - 1)}.$$

Applying the Hölder interpolation inequality, we get

$$\|u\|_{L^{2k\gamma}(\mathbb{R}^d)} \leq \|u\|_{L^r(\mathbb{R}^d)}^{\vartheta} \|u\|_{L^\zeta(\mathbb{R}^d)}^{1-\vartheta}, \quad \frac{1}{2k\gamma} = \frac{\vartheta}{r} + \frac{1 - \vartheta}{\zeta},$$

where we set

$$\zeta = \frac{2d(2k(r-2) - cr)}{(4 - cd)r - 8}, \quad \vartheta = \frac{cr}{2k(r-2)}, \quad \frac{1}{\gamma} = \frac{2}{d} - \frac{c}{2},$$

for some $0 < c < \frac{(4 - \alpha d)r + 2\alpha d}{\alpha d r}$ with $d = 2, 3$. Therefore, Lemma 4.17 shows that

$$\begin{aligned}
&\left\| \int_0^t \mathscr{P}_\alpha(t-s)(f(u) - f(v))ds \right\|_{L^r(\mathbb{R}^d)} \\
\leq & C \sum_{k=1}^{\infty} \frac{\lambda^k}{k!} \int_0^t (t-s)^{\alpha - 1 - \frac{\alpha d}{2}\left(\frac{1}{l} - \frac{1}{r}\right)} \|u - v\|_{L^r(\mathbb{R}^d)} \\
& \times \left(\|u\|_{L^r(\mathbb{R}^d)}^{2k\vartheta} \|u\|_{L^\zeta(\mathbb{R}^d)}^{2k(1-\vartheta)} + \|v\|_{L^r(\mathbb{R}^d)}^{2k\vartheta} \|v\|_{L^\zeta(\mathbb{R}^d)}^{2k(1-\vartheta)} \right) ds
\end{aligned}$$

$$
\leq C\sum_{k=1}^{\infty}\frac{\lambda^k}{k!}\int_0^t (t-s)^{\alpha-1-\frac{\alpha d}{2}\left(\frac{1}{l}-\frac{1}{r}\right)}\|u-v\|_{L^r(\mathbb{R}^d)}
$$

$$
\times\left(\Gamma\left(\frac{\zeta}{2}+1\right)\right)^{\frac{2k(1-\vartheta)}{\zeta}}\left(\|u\|_{L^r(\mathbb{R}^d)}^{2k\vartheta}\|u\|_{\exp L^2(\mathbb{R}^d)}^{2k(1-\vartheta)}+\|v\|_{L^r(\mathbb{R}^d)}^{2k\vartheta}\|v\|_{\exp L^2(\mathbb{R}^d)}^{2k(1-\vartheta)}\right)ds,
$$

for $\zeta \geq 2$. Now, for any $u \in X_M$, it yields

$$
t^\sigma\left\|\int_0^t \mathscr{P}_\alpha(t-s)(f(u)-f(v))ds\right\|_{L^r(\mathbb{R}^d)}
$$

$$
\leq Ct^\sigma\sum_{k=1}^{\infty}\frac{\lambda^k}{k!}\Gamma\left(\frac{\zeta}{2}+1\right)^{\frac{2k(1-\vartheta)}{\zeta}}\int_0^t (t-s)^{\alpha-1-\frac{\alpha d}{2}\left(\frac{1}{l}-\frac{1}{r}\right)}s^{-\sigma(1+2k\vartheta)}ds
$$

$$
\times M^{2k}\sup_{s>0}s^\sigma\|u(s)-v(s)\|_{L^r(\mathbb{R}^d)}
$$

$$
\leq\sum_{k=1}^{\infty}C^{k+1}\lambda^k B\left(\alpha-\frac{\alpha d}{2}\left(\frac{1}{l}-\frac{1}{r}\right),1-\sigma(1+2k\vartheta)\right)
$$

$$
\times M^{2k}\sup_{s>0}s^\sigma\|u(s)-v(s)\|_{L^r(\mathbb{R}^d)},
$$

where $\Gamma(\zeta/2+1)^{\frac{2k(1-\vartheta)}{\zeta}} \leq C^k k!$ and the exponents $r, \gamma, l, \vartheta, \zeta$ satisfy

$$
1<l\leq r<+\infty,\quad \alpha+\sigma-\frac{\alpha d}{2}\left(\frac{1}{l}-\frac{1}{r}\right)-\sigma(1+2k\vartheta)=0,
$$

$$
1-\sigma(1+2k\vartheta)>0,\quad \alpha-\frac{\alpha d}{2}\left(\frac{1}{l}-\frac{1}{r}\right)>0,\quad 0\leq\vartheta\leq 1,
$$

$$
\zeta\geq 2,\quad 1-\frac{1}{\alpha}<\frac{d}{2}\left(\frac{1}{l}-\frac{1}{r}\right)<1,\quad \frac{1}{l}=\frac{1}{r}+\frac{1}{\gamma},\quad \frac{1}{2k\gamma}=\frac{\vartheta}{r}+\frac{1-\vartheta}{\zeta}.
$$

For those exponents, we obtain that

$$
B\left(\alpha-\frac{\alpha d}{2}\left(\frac{1}{l}-\frac{1}{r}\right),1-\sigma(1+2k\vartheta)\right)=\frac{\Gamma(\alpha-\frac{\alpha d}{2\gamma})\Gamma\left(\frac{(1-\alpha d-c\alpha d)r+2\alpha d}{4r}\right)}{\Gamma(1-\sigma)}\leq C,
$$

for some constant $C > 0$. Hence, it follows that

$$
t^\sigma\left\|\int_0^t \mathscr{P}_\alpha(t-s)(f(u)-f(v))ds\right\|_{L^r(\mathbb{R}^d)}
$$

$$
\leq\sum_{k=1}^{\infty}\lambda^k C^{k+1}B\left(\alpha-\frac{\alpha d}{2\gamma},1-\sigma(1+2k\vartheta)\right)M^{2k}\sup_{s>0}s^\sigma\|u(s)-v(s)\|_{L^r(\mathbb{R}^d)}
$$

$$\leq C \sum_{k=1}^{\infty} \lambda^k C^k M^{2k} \sup_{s>0} s^{\sigma} \|u(s) - v(s)\|_{L^r(\mathbb{R}^d)},$$

for some small $M > 0$. This proves the case of $d = 2, 3$.

For the case $d = 1$, the method in the proof is similar to the assumption of function f so that the exponent $\zeta \geq 2$ and $\exp L^2(\mathbb{R}^d) \hookrightarrow L^{\zeta}(\mathbb{R}^d)$. The proof is completed.

Theorem 4.16 (Global Existence) *Let $\alpha \in (1, 4/3)$ and $d = 1, 2, 3$. If there exists $\varepsilon > 0$ such that for all $u_0 \in \exp L^2(\mathbb{R}^d)$ with $\|u_0\|_{\exp L^2(\mathbb{R}^d)} \leq \varepsilon$, then there exists a solution $u \in L^{\infty}(0, \infty; \exp L^2(\mathbb{R}^d))$ satisfying*

$$\lim_{t \to 0} \|u(t) - \mathcal{Q}_{\alpha}(t) u_0\|_{\exp L^2(\mathbb{R}^d)} = 0.$$

Proof By Lemma 4.28, the map $F(u)$ is onto X_M for any $u \in X_M$. In fact, we first obtain

$$\begin{aligned}\|F(u)(t)\|_{\exp L^2(\mathbb{R}^d)} \leq & \|\mathcal{Q}_{\alpha}(t) u_0\|_{\exp L^2(\mathbb{R}^d)} + \left\| \int_0^t \mathcal{P}_{\alpha}(t-s) f(u(s)) ds \right\|_{\exp L^2(\mathbb{R}^d)} \\ \leq & \|u_0\|_{\exp L^2(\mathbb{R}^d)} + H(M),\end{aligned}$$

which follows from Lemma 4.21. Moreover, Lemma 4.16 and Lemma 4.19 show that

$$\begin{aligned}\|F(u)(t)\|_{L^r(\mathbb{R}^d)} \leq & \|\mathcal{Q}_{\alpha}(t) u_0\|_{L^r(\mathbb{R}^d)} + \left\| \int_0^t \mathcal{P}_{\alpha}(t-s) f(u(s)) ds \right\|_{L^r(\mathbb{R}^d)} \\ \leq & t^{-\sigma} \|u_0\|_{\exp L^2(\mathbb{R}^d)} + C(M) t^{-\sigma} \sup_{t>0} t^{\sigma} \|u(t)\|_{L^r(\mathbb{R}^d)}.\end{aligned}$$

Thus, together with the arguments mentioned above, we obtain that F is a map on X_M to itself if M and $\|u_0\|_{\exp L^2(\mathbb{R}^d)}$ are small enough. Similarly, it is not difficult to show that F is also a contraction mapping by the second inequality in Lemma 4.28. Consequently, the contraction mapping principle derives that F has a fixed point to the current problem.

In the sequel, let u be the fixed point of operator F defined on X_M; then by using Lemma 4.17 and according to the third inequality in Lemma 4.22, we have

$$\begin{aligned}& \|u(t) - \mathcal{Q}_{\alpha}(t) u_0\|_{\exp L^2(\mathbb{R}^d)} \\ \leq & \int_0^t \|\mathcal{P}_{\alpha}(t-s) f(u)\|_{\exp L^2(\mathbb{R}^d)} ds \\ \leq & C \int_0^t (t-s)^{\alpha - 1 - \frac{\alpha d}{4}} \|f(u)\|_{L^2(\mathbb{R}^d)} ds + 2 \int_0^t \|f(u)\|_{L^2(\mathbb{R}^d)} ds.\end{aligned}$$

To estimate $\|f(u)\|_{L^2(\mathbb{R}^d)}$, let $p = 2$ in (4.61), and we have for $d = 2, 3$,

$$\|f(u)\|_{L^2(\mathbb{R}^d)} \leq C\|u\|_{L^4(\mathbb{R}^d)}\|e^{u^2} - 1\|_{L^4(\mathbb{R}^d)} \leq C\|u\|_{\exp L^2(\mathbb{R}^d)},$$

and also for $d = 1$, we have

$$\|f(u)\|_{L^2(\mathbb{R}^d)} \leq C\|u\|_{L^4(\mathbb{R}^d)}\|e^{u^2} - 1 - u^2\|_{L^4(\mathbb{R}^d)} \leq C\|u\|_{\exp L^2(\mathbb{R}^d)}.$$

Therefore, it follows that

$$\begin{aligned}
&\|u(t) - \mathscr{Q}_\alpha(t)u_0\|_{\exp L^2(\mathbb{R}^d)}\\
\leq &C\int_0^t (t-s)^{\alpha-1-\frac{\alpha d}{4}}\|u(s)\|_{\exp L^2(\mathbb{R}^d)}ds + C\int_0^t \|u(s)\|_{\exp L^2(\mathbb{R}^d)}ds\\
\leq &Ct^{\alpha\left(1-\frac{d}{4}\right)}\|u\|_{L^\infty(0,\infty;\exp L^2(\mathbb{R}^d))} + Ct\|u\|_{L^\infty(0,\infty;\exp L^2(\mathbb{R}^d))}\\
\to &0, \quad as\ t \to 0.
\end{aligned}$$

The proof of continuity of solution on initial data is completed.

To finish the proof, let $X := L^1(\ln L)^{1/2}(\mathbb{R}^d)$, and it is known that X is a Banach space and $C_0^\infty(\mathbb{R}^d)$ is dense in X. The problem is continuous at $t = 0$ in the weak* sense. In fact, let $\phi \in X$. By the Hölder inequality of the Orlicz space, we have

$$\begin{aligned}
\left|\int_{\mathbb{R}^d} (\mathscr{Q}_\alpha(t)u_0(x) - u_0(x))\phi(x)dx\right| &= \left|\int_{\mathbb{R}^d} u_0(x)(\mathscr{Q}_\alpha(t)\phi(x) - \phi(x))dx\right|\\
&\leq \|u_0\|_{\exp L^2(\mathbb{R}^d)}\|\mathscr{Q}_\alpha(t)\phi - \phi\|_X.
\end{aligned}$$

As for the similar manner as in Lemma 4.24, by the density of $C_0^\infty(\mathbb{R}^d)$ in X, it is easy to check that

$$\lim_{t\to 0}\|\mathscr{Q}_\alpha(t)\phi - \phi\|_X = 0.$$

This completes the weak* convergence. Thus, the proof is completed.

Chapter 5
Inverse Problems of Fractional Wave Equations

5.1 Backward Problem

5.1.1 Introduction

Let $\Omega \subset \mathbb{R}^n$ be a bounded domain with sufficient smooth boundary $\partial\Omega$. We shall consider the following backward problem for time fractional wave equation

$$\begin{cases} \partial_t^\alpha u - \Delta u = f(t, x, u), & x \in \Omega,\ t \in (0, T], \\ u(t, x) = 0, & x \in \partial\Omega,\ t \in (0, T], \\ \partial_t u(0, x) = 0, & x \in \Omega, \\ u(T, x) = g(x), & x \in \Omega, \end{cases} \tag{5.1}$$

where $\partial_t = \partial/\partial t$ and ∂_t^α is the Caputo fractional derivative of order $\alpha \in (1, 2)$ defined by

$$\partial_t^\alpha u(t, x) = \frac{1}{\Gamma(2-\alpha)} \int_0^t (t-s)^{1-\alpha} \frac{\partial^2}{\partial s^2} u(s, x) ds, \quad t > 0,$$

provided that the right-hand side is pointwise defined on $[0, \infty)$, where $\Gamma(\cdot)$ stands for the Gamma function. f is a nonlinear function that will satisfy some suitable assumptions.

As for the fractional wave equation while sometimes it is called the superdiffusion equation, it is also substituted for modeling the propagation of diffusive waves in viscoelastics solids frequently, see e.g., [151, 154]. This is one of the reasons that many researchers pay attention to study these problems, see, e.g., [10, 81, 116, 186] and the related references therein.

Y. Zhou, *Fractional Diffusion and Wave Equations*,
https://doi.org/10.1007/978-3-031-74031-2_5

If the final value condition $u(T, x) = g(x)$ in the problem (5.1) is replaced by an initial value condition $u(0, x) = y(x)$, then the problem (5.1) is called the forward problem of time fractional wave equation. As we know, there are many papers coping with forward problems of time fractional wave equation, for examples, Li and Wang [133] studied some regularity properties of time fractional stochastic wave equation which is forced by an additive space-time white noise. The regularity of weak solutions for time fractional wave equation has been studied by Otárola and Salgado [174]. As for backward problem, which is one of the main topics of inverse problem, we find that there are still few papers about backward problem for time fractional wave equation, Wei and Zhang [226] studied the existence, uniqueness, and conditional stability for the backward problem, and the Tikhonov regularization method has been used to solve regularized solution. Following this paper, Tuan et al. [208] considered some existence and regularity results for final value problems (also called backward problems) with respect to linear function. Huynh et al. [91] studied the regularized solution for an inhomogeneous problem in a general bounded domain by applying the fractional Landweber regularization method. Inspired by the above research studies, we will consider some existence results under some different conditions of nonlinear function.

On the contrary, another issue which is worthy of consideration for the backward problem about fractional wave equation is seriously ill-posed in the sense of Hadamard, that is, even if a solution will exist and it is unique, but it is not stable, in a word, it does not depend continuously on the given data. In order to achieve it at practical applications, some numerical methods are proposed to study the ill-posed behavior, the regularization solution, and some error analyses are given. Additionally, one can find that the backward problem has emerged in optimal control, mathematical finance, and then some theoretical analyses are established to study these problems contained with the properties of solution of existence, uniqueness, regularity, and convergence that reflect the nature of the problem more clearly. In fact, our problem is seriously ill-posed, and it urges us to prove the convergence rate for the regularized solutions. For more details about backward problems, we refer to [38, 90, 101, 177, 207, 222, 245] and the related references therein.

The rest of this section is as follows. In Sect. 5.1.2, we introduce some concepts, preliminaries, and the properties of the Mittag–Leffler functions. In Sect. 5.1.3, we derive the solution representation of the problem (5.1), and some useful properties of solution operators will also be discussed. Furthermore, several existence results without Lipschitz condition are obtained in Sect. 5.1.4. In Sect. 5.1.5, a regularization method is proposed to approximate the solutions.

5.1.2 Preliminaries

In this subsection, some preliminaries will be presented in order to derive the solution representation as well as our main results.

We adopt the eigenvalues of the Laplacian operator $L = -\Delta$. Since the operator L is nonnegative and self-adjoint in Sobolev space $H_0^1(\Omega)$, there exists an orthonormal basis of $L^2(\Omega)$ consisting of eigenfunctions $\phi_k \in H_0^1(\Omega)$, $k = 1, 2, \ldots$, that correspond to eigenvalues

$$0 < \lambda_1 \le \lambda_2 \le \ldots \le \lambda_k \le \ldots \to \infty,$$

which satisfies

$$L\phi_k = \lambda_k \phi_k, \quad \text{in } \Omega, \qquad \phi_k = 0, \quad \text{on } \partial\Omega.$$

We first take the domain $\mathscr{H}^s(\Omega) = D(L^s)$ of the fractional power operator L^s, for $s \ge 0$, and the space is introduced by

$$\mathscr{H}^s(\Omega) = \left\{ u \in L^2(\Omega) : \sum_{k=1}^{\infty} \lambda_k^{2s} |u_k|^2 < \infty \right\},$$

as the Hilbert space of functions

$$u := \sum_{k=1}^{\infty} u_k \phi_k = \sum_{k=1}^{\infty} (u, \phi_k)\phi_k \in L^2(\Omega), \tag{5.2}$$

equipped with norm

$$\|u\|^2_{\mathscr{H}^s(\Omega)} = \sum_{k=1}^{\infty} \lambda_k^{2s} |u_k|^2.$$

Let E, Y be two Banach spaces, and $\mathscr{B}(E, Y)$ stands for the space of all linear bounded operators from E into Y. Now, we consider a space E with the norm $|\cdot|$, specially, let the norm of space $L^2(\Omega)$ be given by $\|\cdot\|$, and inner product is defined as $(\cdot, \cdot)$. We denote by $C([0, T]; E)$ a Banach space of continuous maps from $[0, T]$ into E with $\sup_{t\in[0,T]} |u(t)| < \infty$, $C^\eta((0, T]; E)$ stands for a Banach space of weight continuous functions on $(0, T]$ into E with exponent $\eta \in (0, \alpha]$ as follows:

$$C^\eta((0, T]; E) = \left\{ u \in C((0, T]; E) : \lim_{t\to 0+} t^\eta |u(t)| \text{ exists and finite} \right\},$$

equipped with the norm

$$\|u\|_{C^\eta((0,T];E)} = \sup_{0\le t\le T} t^\eta |u(t)|.$$

Let $1 \le p < \infty$ and let $L^p(0, T; E)$ denote all the p-integrable Lebesgue measure functions equipped with the norm

$$\|u\|_{L^p(0,T;E)} = \left(\int_0^T |u(t)|^p dt\right)^{\frac{1}{p}} < \infty.$$

Let $E_{\alpha,\beta}(z)$ be the Mittag–Leffler function as in Definition 1.7.

Lemma 5.1 ([10]) *Let $1 < \beta < 2$, $\beta' \in \mathbb{R}$, and $\lambda > 0$. Assume that $0 \le \mu \le 1$, $0 < \nu < \beta$. Then there exists a positive constant C_1 such that*

$$\left|\lambda^{\mu} t^{\nu} E_{\beta,\beta'}(-\lambda t^{\beta})\right| \le C_1 t^{\nu-\beta\mu}, \quad t > 0.$$

Lemma 5.2 ([226, Lemma 3.2]) *For $1 < \alpha < 2$ and any fixed $T > 0$, there is at most a finite index set $\Theta = \{k_1, k_2, ..., k_n\}$ such that $E_{\alpha,1}(-\lambda_k T^{\alpha}) = 0$ for $k \in \Theta$ and $E_{\alpha,1}(-\lambda_k T^{\alpha}) \ne 0$ for $k \in \mathbb{N}^+ \setminus \Theta$.*

Lemma 5.3 ([226, Lemma 3.6]) *Let $1 < \alpha < 2$. Then there exist positive constants M_-, M_+ depending on α, T, and finite eigenvalues λ_k with $k \in \{k_1, k_2, \ldots, k_{n+m}\} \setminus \Theta$, $m \in \mathbb{N}_0$ such that*

$$\frac{M_-}{\lambda_k} \le |E_{\alpha,1}(-\lambda_k T^{\alpha})| \le \frac{M_+}{\lambda_k}, \quad k \in \mathbb{N}^+ \setminus \Theta.$$

Noting that, in view of the approximation form of the Mittag–Leffler function (1.12) and Lemma 5.2, there exists $L_0 > 0$ such that

$$E_{\alpha,1}(-\lambda_k T^{\alpha}) \le \frac{1}{2\Gamma(1-\alpha)\lambda_k T^{\alpha}} < 0, \quad \lambda_k T^{\alpha} > L_0, \tag{5.3}$$

for $1 < \alpha < 2$, and thus $E_{\alpha,1}(-\lambda_k T^{\alpha}) = 0$ only if $\lambda_k T^{\alpha} \le L_0$. Since $\lim_{k\to\infty} \lambda_k = +\infty$, there are only finite λ_k satisfying $\lambda_k T^{\alpha} \le L_0$ with $k \in \Theta$. According to the abovementioned discussions and related lemmas, we know that there exist some finite λ_k and T such that $E_{\alpha,1}(-\lambda_k T^{\alpha}) = 0$, for every $k \in \Theta$. Thus, throughout this section, we shall get rid of the part of $k \in \Theta$ in λ_k and set $\mathscr{E}_{\alpha}(-\lambda_k T^{\alpha}) := E_{\alpha,1}(-\lambda_k T^{\alpha}) \ne 0$ for $k \in \mathbb{N}^+ \setminus \Theta$, with using these notations $\mathscr{E}_{\alpha}(-\lambda_k t^{\alpha}) := E_{\alpha,1}(-\lambda_k t^{\alpha})$ and $\mathscr{E}_{\alpha,\beta}(-\lambda_k t^{\alpha}) := E_{\alpha,\beta}(-\lambda_k t^{\alpha})$, for $k \in \mathbb{N}^+ \setminus \Theta$. Consequently, similarly to Lemma 5.2, Lemma 5.3, and (5.3), together with above arguments as well as (i) and (ii) of Proposition 1.13, one can check the following inequalities obviously, for $k \in \mathbb{N}^+ \setminus \Theta$, $t \ge 0$,

$$\frac{c_-}{1+\lambda_k T^{\alpha}} \le |\mathscr{E}_{\alpha}(-\lambda_k T^{\alpha})| \le \frac{c_+}{1+\lambda_k T^{\alpha}}, \quad |\mathscr{E}_{\alpha,\zeta}(-\lambda_k t^{\alpha})| \le \frac{c_+}{1+\lambda_k t^{\alpha}}, \tag{5.4}$$

where $\zeta \in \mathbb{R}$,

$$c_- := \min\left\{(2|\Gamma(1-\alpha)|)^{-1}, M_-, M_- T^{\alpha}\right\}$$

and

$$c_+ := \max\{M, C_1, M_+, M_+(T^\alpha + \lambda_1^{-1})\}.$$

Clearly, we have $c_- < c_+$.

In the sequel, set $\Pi = \mathbb{N}^+ \setminus \Theta$, and we will assume $k \in \Pi$ all the time.

5.1.3 Solution Representation

In the subsection, we first give a suitable definition of mild solution of the problem (5.1), and we will further study the properties of solution operators that are derived from solution representation.

5.1.3.1 Definition of Mild Solution

Let u be the solution of initial-boundary value problems with respect to forward time fractional wave equation, and multiplying ϕ_k of both sides of the problem (5.1) yields

$${}_0^C D_t^\alpha u_k(t) + \lambda_k u_k(t) = f_k(t, u)$$

associated with the initial conditions $u_k(0) = u_{0k} = (u_0, \phi_k)$, $u_k'(0) = u_{1k} = (u_1, \phi_k) = 0$ and $f_k(t, u) = (f(t, u), \phi_k)$. Then, from Theorem 5.15 in [115], one obtains

$$u_k(t) = E_\alpha(-\lambda_k t^\alpha) u_{0k} + \int_0^t (t - \tau)^{\alpha - 1} E_{\alpha,\alpha}(-\lambda_k (t - \tau)^\alpha) f_k(\tau, u) d\tau. \tag{5.5}$$

By substituting $t = T$ into (5.5), it yields

$$u_k(T) = E_\alpha(-\lambda_k T^\alpha) u_{0k} + \int_0^T (T - \tau)^{\alpha - 1} E_{\alpha,\alpha}(-\lambda_k (T - \tau)^\alpha) f_k(\tau, u) d\tau.$$

Let $g_k = (g, \phi_k)$, since $E_\alpha(-\lambda_k T^\alpha)$ may be equal to zero for some $k \in \Theta$, and then, for $k \in \mathbb{N}^+ \setminus \Theta$, we have

$$\begin{aligned} u_k(t) = &\frac{\mathscr{E}_\alpha(-\lambda_k t^\alpha)}{\mathscr{E}_\alpha(-\lambda_k T^\alpha)} \left(g_k - \int_0^T (T - \tau)^{\alpha - 1} \mathscr{E}_{\alpha,\alpha}(-\lambda_k (T - \tau)^\alpha) f_k(\tau, u) d\tau \right) \\ &+ \int_0^t (t - \tau)^{\alpha - 1} \mathscr{E}_{\alpha,\alpha}(-\lambda_k (t - \tau)^\alpha) f_k(\tau, u) d\tau. \end{aligned}$$

Let us simplify $u(t,x)$ with $u(t)$. For any $v \in L^2(\Omega)$, denote two operators by

$$\mathscr{S}_\alpha(t)v = \sum_{k=1,k\in\Pi}^{\infty} \frac{\mathscr{E}_\alpha(-\lambda_k t^\alpha)}{\mathscr{E}_\alpha(-\lambda_k T^\alpha)}(v,\phi_k)\phi_k$$

and

$$\mathscr{P}_\alpha(t)v = t^{\alpha-1} \sum_{k=1,k\in\Pi}^{\infty} \mathscr{E}_{\alpha,\alpha}(-\lambda_k t^\alpha)(v,\phi_k)\phi_k.$$

From the above arguments, let symbol $\circ$ be a composition operator as follows:

$$\mathscr{S}_\alpha(t)\circ\mathscr{P}_\alpha(\varsigma)v = \varsigma^{\alpha-1} \sum_{k=1,k\in\Pi}^{\infty} \frac{\mathscr{E}_\alpha(-\lambda_k t^\alpha)}{\mathscr{E}_\alpha(-\lambda_k T^\alpha)}\mathscr{E}_{\alpha,\alpha}(-\lambda_k \varsigma^\alpha)(v,\phi_k)\phi_k, \quad v \in L^2(\Omega),$$

for $t, \varsigma \in [0,T]$. Hence, we can find a mild solution of the problem (5.1) in which its definition is given below.

Definition 5.1 For every $\eta \in (0,\alpha]$, a function u is called a mild solution of the problem (5.1) if $u \in C^\eta((0,T];L^2(\Omega))$ and it satisfies the integral equation

$$u(t) = \mathscr{S}_\alpha(t)g - \int_0^T \mathscr{S}_\alpha(t)\circ\mathscr{P}_\alpha(T-\tau)f(\tau,u)d\tau + \int_0^t \mathscr{P}_\alpha(t-\tau)f(\tau,u)d\tau. \tag{5.6}$$

5.1.3.2 Some Properties

Property 5.1 $\mathscr{S}_\alpha(t)$ is a unbounded operator at the time $t=0$ in $L^2(\Omega)$, while it belongs to $\mathscr{B}(\mathscr{H}^1(\Omega), L^2(\Omega))$.

Proof Obviously, $\mathscr{S}_\alpha(t)$ is linear operator. If taking $v_n = \phi_n(x)$, $n \in \Pi$, in view of (5.4) and $\lim_{t\to 0}\mathscr{E}_\alpha(-\lambda_k t^\alpha) = 1$, we deduce that $\|v_n\| = 1$ and

$$\begin{aligned}
\|\mathscr{S}_\alpha(0)v_n\|^2 &= \sum_{k=1,k\in\Pi}^{\infty} \frac{1}{|\mathscr{E}_\alpha(-\lambda_k T^\alpha)|^2}|(v_n,\phi_k)|^2 \\
&\geq \frac{1}{c_+^2}\sum_{k=1,k\in\Pi}^{\infty}(1+\lambda_k T^\alpha)^2|(v_n,\phi_k)|^2 \\
&\geq \frac{1}{c_+^2} + \frac{T^{2\alpha}}{c_+^2}\lambda_n^2.
\end{aligned}$$

Therefore, it follows that $\|\mathscr{S}_\alpha(0)v_n\| > T^\alpha\lambda_n/c_+$. From $\lambda_n \to \infty$ as $n \to \infty$, it shows that $\mathscr{S}_\alpha(t)$ is unbounded in $L^2(\Omega)$ at time $t = 0$. On the contrary, for any $v \in \mathscr{H}^1(\Omega)$, the Sobolev embedding $\mathscr{H}^1(\Omega) \hookrightarrow L^2(\Omega)$ implies

$$\begin{aligned}
\|\mathscr{S}_\alpha(0)v\|^2 \leq & \frac{1}{c_-^2} \sum_{k=1,k\in\Pi}^{\infty} \left(1+\lambda_k T^\alpha\right)^2 |(v,\phi_k)|^2 \\
\leq & \frac{2}{c_-^2}\|v\|^2 + \frac{2T^{2\alpha}}{c_-^2} \sum_{k=1,k\in\Pi}^{\infty} \lambda_k^2 |(v,\phi_k)|^2 \\
\leq & \frac{2C_2^2}{c_-^2}\|v\|^2_{\mathscr{H}^1(\Omega)} + \frac{2T^{2\alpha}}{c_-^2}\|v\|^2_{\mathscr{H}^1(\Omega)},
\end{aligned}$$

where C_2 is a positive constant, and in addition, we use the inequality $(1+a)^2 \leq 2(1+a^2)$, $a \in \mathbb{R}$. Thus, we show the desired results.

Property 5.2 Let $v \in L^2(\Omega)$. Then $\mathscr{S}_\alpha(t)v$ is continuous on $L^2(\Omega)$ for all $t \in (0, T]$, that is $\mathscr{S}_\alpha(t)v \in C((0, T]; L^2(\Omega))$.

Proof In fact, we just need to show that the series

$$\sum_{k=1,k\in\Pi}^{\infty} \frac{\mathscr{E}_\alpha(-\lambda_k t^\alpha)}{\mathscr{E}_\alpha(-\lambda_k T^\alpha)}(v,\phi_k)\phi_k(x)$$

is uniformly convergent on $L^2(\Omega)$ for any $v \in L^2(\Omega)$ and any $t \in [\delta, T]$ with $\delta > 0$.

By virtue of (5.4), we get

$$\left|\frac{\mathscr{E}_\alpha(-\lambda_k t^\alpha)}{\mathscr{E}_\alpha(-\lambda_k T^\alpha)}\right| \leq \frac{c_+}{c_-}\frac{1+\lambda_k T^\alpha}{1+\lambda_k t^\alpha} \leq \frac{c_+ T^\alpha}{c_- t^\alpha}, \quad \text{for all } t \in (0, T]. \tag{5.7}$$

In the following, we know that $\mathscr{E}_\alpha(-\lambda_k t^\alpha)$ is uniform continuous since the identity

$$\mathscr{E}_{\alpha,1}(-z^2) = \int_0^\infty \mathscr{M}_{\alpha/2}(\theta)\cos(z\theta)d\theta, \quad z \in \mathbb{C},\ \alpha \in (1,2). \tag{5.8}$$

This above identity can be founded in [151], where $\mathscr{M}_\varrho(\cdot)(\varrho \in (0,1))$ is the Wright-type function defined as in Definition 1.8.

In fact, from the uniform continuity of $\cos(\sqrt{z})$ for $z \in \mathbb{R}_+$, we know that for any $\varepsilon > 0$ and each $k \in \mathbb{N}^+$, there exists a $\delta > 0$ such that, for $t_1, t_2 \in \mathbb{R}_+$ with $|t_2 - t_1| < \delta$,

$$\left|\cos\left(\sqrt{\lambda_k t_2^\alpha}\theta\right) - \cos\left(\sqrt{\lambda_k t_1^\alpha}\theta\right)\right| < \varepsilon.$$

Therefore, by virtue of (5.8), we have

$$\begin{aligned}
&|\mathscr{E}_\alpha(-\lambda_k t_2^\alpha) - \mathscr{E}_\alpha(-\lambda_k t_1^\alpha)| \\
&= \left| \int_0^\infty \mathscr{M}_{\alpha/2}(\theta) \left(\cos\left(\sqrt{\lambda_k t_2^\alpha}\theta\right) - \cos\left(\sqrt{\lambda_k t_1^\alpha}\theta\right)\right) d\theta \right| \\
&< \varepsilon,
\end{aligned}$$

where we use the property

$$\mathscr{M}_\varrho(\theta) \geq 0, \quad \int_0^\infty \mathscr{M}_\varrho(\theta) d\theta = 1.$$

It allows us to obtain that the desired series is uniformly convergent on $[\delta, T]$ by using the Cauchy convergence criterion. Now, for any $\varepsilon > 0$, there exists $M > 0$ such that for all positive integers p whereas $m \in \Pi$ and $m > M$

$$\sum_{k=m+1}^{m+p} |(v, \phi_k)|^2 < \left(\frac{c_- t^\alpha}{c_+ T^\alpha} \right)^2 \varepsilon, \quad \text{for all } t \in [\delta, T].$$

Let

$$S_m(t)v = \sum_{k=1}^{m} \frac{\mathscr{E}_\alpha(-\lambda_k t^\alpha)}{\mathscr{E}_\alpha(-\lambda_k T^\alpha)} (v, \phi_k)\phi_k(x).$$

Therefore, it yields that

$$\begin{aligned}
\|S_{m+p}(t)v - S_m(t)v\|^2 &= \sum_{k=m+1}^{m+p} \left| \frac{\mathscr{E}_\alpha(-\lambda_k t^\alpha)}{\mathscr{E}_\alpha(-\lambda_k T^\alpha)} (v, \phi_k) \right|^2 \\
&\leq \left(\frac{c_+ T^\alpha}{c_- t^\alpha} \right)^2 \sum_{k=m+1}^{m+p} |(v, \phi_k)|^2 < \varepsilon.
\end{aligned}$$

By the arbitrariness of ε, we deduce that the conclusion holds.

In the sequel, for convenience, we set $X := C^\eta((0, T]; L^2(\Omega))$, $\eta = \alpha\gamma$ for $\gamma \in (0, 1]$, and let operator $\mathscr{T}_\alpha$ be defined by

$$(\mathscr{T}_\alpha v)(t) = \int_0^T \mathscr{S}_\alpha(t) \circ \mathscr{P}_\alpha(T - \tau) v(\tau) d\tau, \quad v \in X.$$

Lemma 5.4 *Let $\gamma \in \left(0, \frac{1}{\alpha}\right)$ such that $\eta \in (0, 1)$. Then operator $\mathscr{T}_\alpha$ is a completely continuous operator from X into X.*

Proof For every $n \in \Pi$, let $\Phi_n = \text{span}\{\phi_1(x), ..., \phi_n(x)\}$, and since $\{\phi_k\}_{k=1}^{\infty}$ is an orthonormal basis in $L^2(\Omega)$, one finds that $L^2(\Omega)$ can be expressed by $\text{span}\{\phi_1(x), ..., \phi_n(x), ...\}$. Obviously, Φ_n is a finite dimensional subspace of $L^2(\Omega)$. For every $n \in \Pi$, denote operators $\mathscr{S}_{\alpha,n}(t, \varsigma) : L^2(\Omega) \to \Phi_n$ by

$$\mathscr{S}_{\alpha,n}(t, \varsigma)v = \varsigma^{\alpha-1} \sum_{k=1, k\in\Pi}^{n} \frac{\mathscr{E}_\alpha(-\lambda_k t^\alpha)}{\mathscr{E}_\alpha(-\lambda_k T^\alpha)} \mathscr{E}_{\alpha,\alpha}(-\lambda_k \varsigma^\alpha)(v, \phi_k)\phi_k(x),$$

for $t \in (0, T]$, $\varsigma := T - \tau$ with $\tau \in [0, T]$. Observe that $\mathscr{S}_{\alpha,n}(t, \varsigma)$ are linear finite dimensional operators. Next, for every $n \in \Pi$, we define linear operators $\mathscr{T}_{\alpha,n}$ in the same way by

$$(\mathscr{T}_{\alpha,n}v)(t) = \int_0^T \mathscr{S}_{\alpha,n}(t, T - \tau)v(\tau)d\tau, \quad v \in X.$$

Obviously, $\mathscr{T}_{\alpha,n}v$ are well defined on X. Denote a bounded set on X by $U_r = \{v \in X : \|v\|_X \le r\}$ for each positive constant r. We shall prove that for any positive constant r, the set $\{t^\eta(\mathscr{T}_{\alpha,n}v)(t),\ v \in U_r\}$ is relatively compact in X.

For any $v \in X$, it follows from the fact $|\mathscr{E}_{\alpha,\zeta}(-\lambda_k z^\alpha)| \le c_+$, $\zeta \in \mathbb{R}$, $z > 0$ and (5.4) that

$$\begin{aligned}
&\|\mathscr{S}_{\alpha,n}(t, \varsigma)v(\tau)\|^2 \\
=&\varsigma^{2(\alpha-1)} \sum_{k=1}^{n} \left| \frac{\mathscr{E}_\alpha^\gamma(-\lambda_k t^\alpha)}{\mathscr{E}_\alpha^\gamma(-\lambda_k T^\alpha)} \frac{\mathscr{E}_\alpha^{1-\gamma}(-\lambda_k t^\alpha)}{\mathscr{E}_\alpha^{1-\gamma}(-\lambda_k T^\alpha)} \mathscr{E}_{\alpha,\alpha}^\gamma(-\lambda_k \varsigma^\alpha)\mathscr{E}_{\alpha,\alpha}^{1-\gamma}(-\lambda_k \varsigma^\alpha) \right|^2 \\
&|(v(\tau), \phi_k)|^2 \\
\le&\frac{c_+^4}{c_-^2} \varsigma^{2(\alpha-1)} \sum_{k=1}^{n} \left(\frac{1 + \lambda_k T^\alpha}{1 + \lambda_k t^\alpha} \right)^{2\gamma} \left(\frac{1 + \lambda_k T^\alpha}{1 + \lambda_k \varsigma^\alpha} \right)^{2-2\gamma} |(v(\tau), \phi_k)|^2 \\
\le&\frac{c_+^4}{c_-^2} T^{2\alpha} t^{-2\alpha\gamma} \varsigma^{2(\alpha\gamma-1)} \|v(\tau)\|^2.
\end{aligned} \tag{5.9}$$

Therefore, by the assumption of $\eta \in (0, 1)$, we have

$$\begin{aligned}
\|(\mathscr{T}_{\alpha,n}v)(t)\| \le& \int_0^T \|\mathscr{S}_{\alpha,n}(t, T - \tau)v(\tau)\|d\tau \\
\le&\frac{c_+^2}{c_-} T^\alpha t^{-\eta} \int_0^T (T - \tau)^{\eta-1} \|v(\tau)\|d\tau \\
\le&\frac{c_+^2 \pi}{c_- \sin(\pi\eta)} T^\alpha t^{-\eta} \|v\|_X,
\end{aligned} \tag{5.10}$$

where we use $B(\eta, 1-\eta) = \pi/\sin(\pi\eta)$, and $B(\cdot,\cdot)$ is the Beta function. To begin with, we deduce that $\lim_{t\to 0+} t^\eta(\mathscr{T}_{\alpha,n}v)(t)$ exists and is finite. In fact, from (5.9), one can see that the representation

$$t^\eta \varsigma^{\alpha-1} \sum_{k=1,k\in\Pi}^{n} \frac{\mathscr{E}_\alpha(-\lambda_k t^\alpha)}{\mathscr{E}_\alpha(-\lambda_k T^\alpha)} \mathscr{E}_{\alpha,\alpha}(-\lambda_k \varsigma^\alpha)(v,\phi_k)\phi_k(x)$$

is bounded and integrable for $\varsigma = T-\tau$ with respect to a.e. $\tau \in [0,T]$ on $L^2(\Omega)$, which implies that $t^\eta(\mathscr{T}_{\alpha,n}v)(t)$ is uniformly bounded on $L^2(\Omega)$. It follows from the uniform continuity of $\mathscr{E}_\alpha(-\lambda_k t^\alpha)$ that $t^\eta \mathscr{S}_{\alpha,n}(t,\cdot)v$ is uniformly continuous for $t \in (0,T]$. Thus $t^\eta(\mathscr{T}_{\alpha,n})v(t)$ is uniformly continuous for $t \in (0,T]$, which implies from (5.10) that $\lim_{t\to 0+} t^\eta(\mathscr{T}_{\alpha,n}v)(t)$ exists and is finite. Let $z(0) := \lim_{t\to 0+} t^\eta(\mathscr{T}_{\alpha,n}v)(t)$; hence we deduce $z(0)$ is well defined.

For any $w \in \mathfrak{U}_r := \{y \in C([0,T]; L^2(\Omega)) : \|y\|_{C([0,T];L^2(\Omega))} \le r\}, r > 0$, set

$$v(t) = t^{-\eta} w(t), \quad \text{for } t \in (0,T].$$

Thus, $v \in U_r$. Define

$$(\mathfrak{T}_{\alpha,n}w)(t) = \begin{cases} t^\eta(\mathscr{T}_{\alpha,n}v)(t), & \text{for } t \in (0,T], \\ z(0), & \text{for } t = 0. \end{cases}$$

Thenceforth, it remains to show that $\{\mathfrak{T}_{\alpha,n}w : w \in \mathfrak{U}_r\}$ is relatively compact. Observe that, from (5.10), $\|\mathfrak{T}_{\alpha,n}w\|_{C([0,T];L^2(\Omega))} \le c_+^2 T^\alpha r\pi/(c_- \sin(\pi\eta))$. Thus, we conclude that the set $\mathfrak{U}_r$ is uniformly bounded. In order to prove that the set $\{\mathfrak{T}_{\alpha,n}w,\ w \in \mathfrak{U}_r\}$ is equicontinuous, we need to prove that $\mathscr{S}_{\alpha,n}(t,\cdot)$ is continuous in the uniform operator topology on $L^2(\Omega)$ for all $t > 0$. For this purpose, we need to show the compactness and strong continuity of this operator. By applying (5.4), for any $v \in L^2(\Omega)$ and any $\delta > 0$ such that $t \in [\delta, T]$, it follows that

$$\begin{aligned} \|\mathscr{S}_{\alpha,n}(t,\cdot)v\|^2 \le & T^{2(\alpha-1)} \sum_{k=1,k\in\Pi}^{n} \frac{c_+^4}{c_-^2} \left(\frac{1+\lambda_k T^\alpha}{1+\lambda_k t^\alpha}\right)^2 |(v,\phi_k)|^2 \\ \le & \left(\frac{c_+^2 T^{2\alpha-1}}{c_-\delta^\alpha}\right)^2 \|v\|^2. \end{aligned} \tag{5.11}$$

By virtue of the range $R(\mathscr{S}_{\alpha,n}(t,\cdot)v)$ finite, we conclude that the operator $\mathscr{S}_{\alpha,n}(t,\cdot)$ is compact operator on $L^2(\Omega)$ for every $n \in \Pi$. In addition, in view of (5.8) and (5.11), for any $v \in L^2(\Omega)$ and any $\delta > 0$, for $t_1, t_2 \in [\delta, T]$ with $t_1 < t_2$, we get

$$
\begin{aligned}
&\|\mathscr{S}_{\alpha,n}(t_2,\cdot)v - \mathscr{S}_{\alpha,n}(t_1,\cdot)v\|^2 \\
\leq & c_+^2 T^{2(\alpha-1)} \sum_{k=1,k\in\Pi}^{n} \left| \frac{\mathscr{E}_\alpha(-\lambda_k t_2^\alpha) - \mathscr{E}_\alpha(-\lambda_k t_1^\alpha)}{\mathscr{E}_\alpha(-\lambda_k T^\alpha)} \right|^2 |(v,\phi_k)|^2 \\
\leq & 2\left(\frac{c_+^2 T^{2\alpha-1}}{c_-\delta^\alpha}\right)^2 \|v\|^2.
\end{aligned}
$$

Therefore, in view of uniform continuity of $\mathscr{E}_\alpha(-\lambda_k t^\alpha)$, by applying the property of series, we thus obtain

$$
\|\mathscr{S}_{\alpha,n}(t_2,\cdot)v - \mathscr{S}_{\alpha,n}(t_1,\cdot)v\| \to 0, \quad \text{as } t_2 \to t_1,
$$

which shows that $\mathscr{S}_{\alpha,n}(t,\cdot)v$ are strongly continuous for all $t \in [\delta, T]$. Consequently, we conclude the desired result.

Now, for $t_1 = 0$, $0 < t_2 \leq T$, it is easy to see that

$$
\|(\mathfrak{T}_{\alpha,n}w)(t_2) - (\mathfrak{T}_{\alpha,n}w)(0)\| \to 0, \quad \text{as } t_2 \to 0.
$$

For any $0 < t_1 < t_2 \leq T$, we have

$$
\begin{aligned}
&\|(\mathfrak{T}_{\alpha,n}w)(t_2) - (\mathfrak{T}_{\alpha,n}w)(t_1)\| \\
=&\|t_2^\eta(\mathscr{T}_{\alpha,n}v)(t_2) - t_1^\eta(\mathscr{T}_{\alpha,n}v)(t_1)\| \\
\leq&|t_2^\eta - t_1^\eta|\|(\mathscr{T}_{\alpha,n}v)(t_2)\| + t_1^\eta\|(\mathscr{T}_{\alpha,n}v)(t_2) - (\mathscr{T}_{\alpha,n}v)(t_1)\| \\
=:&I_1 + I_2.
\end{aligned}
$$

By the inequality $a^\rho - b^\rho \leq (a-b)^\rho$ for $0 < b < a$ and $\rho \in [0,1]$, obviously, from (5.10) and $|t_2^\eta - t_1^\eta| \leq (t_2 - t_1)^\eta$, we get $I_1 \to 0$ as $t_2 \to t_1$. As for I_2, by virtue of the continuity in the uniform operator topology of $\mathscr{S}_{\alpha,n}(t,\cdot)$ for all $t > 0$, we obtain

$$
\begin{aligned}
\|(\mathscr{T}_{\alpha,n}v)(t_2) - (\mathscr{T}_{\alpha,n}v)(t_1)\| \leq& \int_0^T \|(\mathscr{S}_{\alpha,n}(t_2, T-\tau) - \mathscr{S}_{\alpha,n}(t_1, T-\tau))v(\tau)\|d\tau \\
\leq& \frac{T^{1-\eta_r}}{1-\eta} \sup_{\varsigma\in[0,T]} \|\mathscr{S}_{\alpha,n}(t_2,\varsigma) - \mathscr{S}_{\alpha,n}(t_1,\varsigma)\|_{\mathscr{B}(L^2(\Omega))} \\
\to& 0, \quad \text{as } t_2 \to t_1.
\end{aligned}
$$

That means $I_2 \to 0$ as $t_2 \to t_1$. Therefore, the right-hand side of the aforementioned inequality tends to zero. From above arguments, one can easily deduce that the set $\{\mathfrak{T}_{\alpha,n}w,\ w \in \mathfrak{U}_r\}$ is equicontinuous. Thus, according to the Arzelà–Ascoli theorem, we conclude that operators $\mathfrak{T}_{\alpha,n}$ are compact on $C([0,T]; L^2(\Omega))$ as well as operators $\mathscr{T}_{\alpha,n}$ are compact on X.

Now, we prove that $\mathscr{T}_{\alpha,n}$ converge uniformly to $\mathscr{T}_\alpha$ whenever n tends to infinite. Indeed, for any $v \in L^2(\Omega)$ and any $\gamma \in (0,1)$, by applying Lemma 5.1 with respect to $\mu \in (0,1)$ and $\varsigma = T - \tau \in (0,T]$, we first have

$$
\begin{aligned}
&\|\mathscr{S}_{\alpha,n}(t,\varsigma)v - \mathscr{S}_\alpha(t)\circ\mathscr{P}_\alpha(\varsigma)v\|^2 \\
\le&\varsigma^{2(\alpha-1)}\sum_{k=n+1}^{\infty}\left|\frac{\mathscr{E}_\alpha(-\lambda_k t^\alpha)}{\mathscr{E}_\alpha(-\lambda_k T^\alpha)}\mathscr{E}_{\alpha,\alpha}(-\lambda_k\varsigma^\alpha)\right|^2|(v,\phi_k)|^2 \\
\le&\frac{c_+^{4-2\gamma}}{c_-^2}T^{2\alpha}t^{-2\eta}\varsigma^{2(\eta-1)}\sum_{k=n+1}^{\infty}\left|\lambda_k^{-\mu\gamma}(\lambda_k^\mu\mathscr{E}_{\alpha,\alpha}(-\lambda_k\varsigma^\alpha))^\gamma\right|^2|(v,\phi_k)|^2 \\
\le&\frac{c_+^{6-2\gamma}}{c_-^2}T^{2\alpha}t^{-2\eta}\varsigma^{2(\eta-1)-2\eta\mu}\lambda_{n+1}^{-2\mu\gamma}\|v\|^2.
\end{aligned}
$$

Therefore, for any $v \in X$, by the Hölder's inequality, we get

$$
\begin{aligned}
&\|(\mathscr{T}_\alpha v)(t) - (\mathscr{T}_{\alpha,n}v)(t)\| \\
\le&\int_0^T\|(\mathscr{S}_{\alpha,n}(t,T-\tau) - \mathscr{S}_\alpha(t)\circ\mathscr{P}_\alpha(T-\tau))v(\tau)\|d\tau \\
\le&\frac{c_+^{3-\gamma}}{c_-}T^\alpha t^{-\eta}\lambda_{n+1}^{-\mu\gamma}\int_0^T(T-\tau)^{\eta(1-\mu)-1}\|v(\tau)\|d\tau \\
\le&\frac{c_+^{3-\gamma}}{c_-}T^{\alpha-\eta\mu}t^{-\eta}\lambda_{n+1}^{-\mu\gamma}B\big(\eta(1-\mu),1-\eta\big)\|v\|_X,
\end{aligned}
\tag{5.12}
$$

which implies that

$$
\|\mathscr{T}_\alpha v - \mathscr{T}_{\alpha,n}v\|_X \le \frac{c_+^{3-\gamma}}{c_-}T^{\alpha-\eta\mu}\lambda_{n+1}^{-\mu\gamma}B\big(\eta(1-\mu),1-\eta\big)\|v\|_X.
$$

Noting the asymptotic property of the eigenvalue with $\lambda_{n+1} \to \infty$ as $n \to \infty$, we thus get $\mathscr{T}_{\alpha,n}v \to \mathscr{T}_\alpha v$ in X. It means that the operator $\mathscr{T}_\alpha$ is a compact operator from X into X.

Next, we show that $\mathscr{T}_\alpha$ is a continuous operator. In fact, let $\{v_i\}_{i=1}^\infty \subset U_r$ and $v \in U_r$ with $\lim_{i\to\infty} v_i = v$ in U_r. Similarly to (5.9), it yields

$$
\|\mathscr{S}_\alpha(t)\circ\mathscr{P}_\alpha(\varsigma)v(\tau)\| \le \frac{c_+^2}{c_-}T^\alpha t^{-\eta}\varsigma^{\eta-1}\|v(\tau)\|. \tag{5.13}
$$

Hence, we get

$$\|\mathscr{S}_\alpha(t) \circ \mathscr{P}_\alpha(T-\tau)(v_i(\tau)-v(\tau))\|^2$$
$$\leq \frac{2c_+^2}{c_-^2} T^\alpha t^{-\eta}(T-\tau)^{\eta-1}(\|v_i(\tau)\|^2 + \|v(\tau)\|^2),$$

together with the Lebesgue's dominated convergence theorem and the same way as in above arguments, we get

$$\|(\mathscr{T}_\alpha v_i)(t) - (\mathscr{T}_\alpha v)(t)\| \leq \int_0^T \|\mathscr{S}_\alpha(t) \circ \mathscr{P}_\alpha(T-\tau)(v_i(\tau)-v(\tau))\| d\tau$$
$$\to 0, \quad \text{as } i \to \infty.$$

Hence, $\mathscr{T}_\alpha v_i \to \mathscr{T}_\alpha v$ as $i \to \infty$. Then, $\mathscr{T}_\alpha$ is continuous, and we conclude that $\mathscr{T}_\alpha$ is completely continuous. The proof is completed.

Similarly to Property 5.2 and Lemma 5.4, it is not difficult to check the following lemma.

Lemma 5.5 *The operator $\mathscr{S}_\alpha(t)$ is compact for every $t \in (0, T]$ and continuous in the uniform operator topology on $\mathscr{B}(L^2(\Omega))$ for all $t \in (0, T]$, and $\mathscr{P}_\alpha(t)$ is compact for every $t \geq 0$ and continuous in the uniform operator topology on $\mathscr{B}(L^2(\Omega))$ for all $t \geq 0$, respectively.*

5.1.4 Existence and Uniqueness

In this subsection, the existence and uniqueness results of mild solutions for the present problem are considered. To achieve this goal, we need the following assumption.

(Hf1) $f : (0, T] \times L^2(\Omega) \to L^2(\Omega)$ is continuous with respect to u and is measurable with respect to t, and there exists a positive constant L_f such that

$$\|f(t, u)\| \leq L_f \|u\|, \quad \forall u \in L^2(\Omega).$$

Theorem 5.1 *Let $\gamma \in \left(0, \frac{1}{\alpha}\right)$ and $g \in \mathscr{H}^{1-\gamma}(\Omega)$. Assume that (Hf1) holds. Then the problem (5.1) has at least one mild solution provided with*

$$\kappa := c_+ L_f T^{\alpha-\eta} B(\alpha, 1-\eta) + \frac{c_+^2 \pi}{c_- \sin(\pi\eta)} L_f T^\alpha \leq \frac{1}{2}.$$

Proof For each $r > 0$, denote a set

$$B_r = \{u \in X : \|u\|_X \leq r\}.$$

Clearly, B_r is a bounded closed and convex subset of X. To achieve the aim of this theorem, we need to show that the operator equation $u = \mathscr{F}u$ has a solution in B_r, where $\mathscr{F}$ is defined as

$$(\mathscr{F}u)(t) = \mathscr{S}_\alpha(t)g - (\mathscr{T}_\alpha f)(t) + (\mathscr{Q}_\alpha f)(t)$$

and $\mathscr{Q}_\alpha$ is defined by

$$(\mathscr{Q}_\alpha f)(t) = \int_0^t \mathscr{P}_\alpha(t-\tau)f(\tau, u(\tau))d\tau.$$

Claim I. The operator $\mathscr{F} : C((0,T]; L^2(\Omega)) \to C((0,T]; L^2(\Omega))$ is well defined.

Indeed, from Property 5.2, we know that $\mathscr{S}_\alpha(t)g \in C((0,T]; L^2(\Omega))$ for $g \in L^2(\Omega)$, so it still holds for $g \in \mathscr{H}^{1-\gamma}(\Omega)$. In view of Lemma 5.4, we see that $(\mathscr{T}_\alpha f)(t) \in C((0,T]; L^2(\Omega))$. From the assumption of f, for any $t_1, t_2 \in [0,T]$ with $t_1 < t_2$, it yields

$$\begin{aligned}\|(\mathscr{Q}_\alpha f)(t_2) - (\mathscr{Q}_\alpha f)(t_1)\| \le & \left\| \int_0^{t_1} (\mathscr{P}_\alpha(t_2-\tau) - \mathscr{P}_\alpha(t_1-\tau))f(\tau, u(\tau))d\tau \right\| \\ & + \left\| \int_{t_1}^{t_2} \mathscr{P}_\alpha(t_2-\tau)f(\tau, u(\tau))d\tau \right\| \\ =: & J_1 + J_2.\end{aligned}$$

For J_1, from Lemma 5.5, we have

$$\begin{aligned}J_1 \le & \int_0^{t_1} \|(\mathscr{P}_\alpha(t_2-\tau) - \mathscr{P}_\alpha(t_1-\tau))f(\tau, u(\tau))\|d\tau \\ \le & \frac{L_f t_1^{1-\eta}}{1-\eta}\|u\|_X \sup_{\tau\in[0,t_1]} \|\mathscr{P}_\alpha(t_2-\tau) - \mathscr{P}_\alpha(t_1-\tau)\|_{\mathscr{B}(L^2(\Omega))},\end{aligned}$$

which implies that $J_1 \to 0$ as $t_2 \to t_1$. As for J_2, we have

$$\begin{aligned}J_2 \le & \int_{t_1}^{t_2} (t_2-\tau)^{\alpha-1} \sqrt{\sum_{k=1,k\in\Pi}^{\infty} \left|\mathscr{E}_{\alpha,\alpha}(-\lambda_k(t_2-\tau)^\alpha)\right|^2 |f_k(\tau, u(\tau))|^2} d\tau \\ \le & L_f c_+ \int_{t_1}^{t_2} (t_2-\tau)^{\alpha-1}\|u(\tau)\|d\tau \\ \le & \frac{L_f c_+}{1-\eta}\|u\|_X (t_2-t_1)^{\alpha-\eta} \to 0, \quad \text{as } t_2 \to t_1.\end{aligned}$$

Thus, $\mathscr{Q}_\alpha f \in C([0,T]; L^2(\Omega))$. Combining with above arguments, for any $u \in C((0,T]; L^2(\Omega))$, we obtain $\mathscr{F}u \in C((0,T]; L^2(\Omega))$.

Claim II. The operator $\mathscr{F}u \in B_r$ for any $u \in B_r$.

From the inequality $(1+a)^b \leq 1+a^b$ and $(1+c)^2 \leq 2(1+c^2)$ for $a, c \geq 0$ and $b \in [0,1]$, it is clear from Lemma 5.1 and (5.4) that

$$
\begin{aligned}
\|\mathscr{S}_\alpha(t)g\|^2 \leq & \frac{c_+^2}{c_-^2} \sum_{k=1,k\in\Pi}^{\infty} \left(\frac{1+\lambda_k T^\alpha}{1+\lambda_k t^\alpha}\right)^2 |(g,\phi_k)|^2 \\
\leq & \frac{c_+^2}{c_-^2} T^{2\eta} t^{-2\eta} \sum_{k=1,k\in\Pi}^{\infty} \left(\frac{1+\lambda_k T^\alpha}{1+\lambda_k t^\alpha}\right)^{2(1-\gamma)} \lambda_k^{-2(1-\gamma)} \lambda_k^{2(1-\gamma)} |(g,\phi_k)|^2 \\
\leq & \frac{2c_+^2}{c_-^2} T^{2\eta} t^{-2\eta} \sum_{k=1,k\in\Pi}^{\infty} \left(\lambda_k^{-2(1-\gamma)} + T^{2\alpha(1-\gamma)}\right) \lambda_k^{2(1-\gamma)} |(g,\phi_k)|^2 \\
\leq & C_T^2 t^{-2\eta} \|g\|^2_{\mathscr{H}^{1-\gamma}(\Omega)},
\end{aligned}
\tag{5.14}
$$

where $C_T = \frac{c_+ T^\eta}{c_-}\sqrt{2(\lambda_1^{-2(1-\gamma)} + T^{2\alpha(1-\gamma)})}$. Therefore, we deduce that $\|\mathscr{S}_\alpha(\cdot)g\|_X \leq C_T \|g\|_{\mathscr{H}^{1-\gamma}(\Omega)}$.

It is similar to (5.10), for any $u \in B_r$, that

$$
\begin{aligned}
\|(\mathscr{T}_\alpha f)(t)\| \leq & \frac{c_+^2}{c_-} T^\alpha L_f t^{-\eta} \int_0^T (T-\tau)^{\eta-1} \|u(\tau)\| d\tau \\
\leq & \frac{c_+^2 \pi}{c_- \sin(\pi\eta)} T^\alpha L_f t^{-\eta} \|u\|_X,
\end{aligned}
$$

which deduces $\|\mathscr{T}_\alpha f\|_X \leq c_+^2 L_f T^\alpha \pi r/(c_- \sin(\pi\eta))$. Moreover, from (5.4), we have

$$
\|(\mathscr{Q}_\alpha f)(t)\| \leq c_+ L_f \int_0^t (t-\tau)^{\alpha-1} \|u(\tau)\| d\tau \leq c_+ L_f T^{\alpha-\eta} B(\alpha, 1-\eta) r.
$$

Therefore, one can choose r large enough such that

$$
C_T \|g\|_{\mathscr{H}^{1-\gamma}(\Omega)} + \frac{c_+^2 \pi}{c_- \sin(\pi\eta)} L_f T^\alpha r + c_+ L_f T^\alpha B(\alpha, 1-\eta) r \leq r,
$$

and then we get

$$
\|\mathscr{F}u\|_X \leq \|\mathscr{S}_\alpha(\cdot)g\|_X + \|\mathscr{T}_\alpha f\|_X + \|\mathscr{Q}_\alpha f\|_X \leq r.
$$

This implies that $\mathscr{F}(B_r) \subseteq B_r$.

Claim III. Operator $\mathscr{F}$ is completely continuous.

Obviously, from Lemmas 5.4 and 5.5, we just need to show that $\mathscr{F}$ is a completely continuous operator. Indeed, by Lemma 5.4, for every $t \in [0, T]$, it is sufficient to prove that $\mathscr{Q}_\alpha$ is completely continuous in X. Since $\mathscr{P}_\alpha(t)$ is compact for every $t \in [0, T]$ in view of Lemma 5.5, we can structure a family of finite dimensional compact operators as the same way in Lemma 5.4 by

$$(\mathscr{Q}_{\alpha,n} f)(t) = \int_0^t \mathscr{P}_{\alpha,n}(t-\tau) f(\tau, u(\tau)) d\tau$$

for every $n \in \Pi$, in which

$$\mathscr{P}_{\alpha,n}(t)v = t^{\alpha-1} \sum_{k=1, k\in\Pi}^{n} \mathscr{E}_{\alpha,\alpha}(-\lambda_k t^\alpha)(v, \phi_k)\phi_k, \quad v \in L^2(\Omega).$$

It is clear that the $H^n(t) = \{t^\eta (\mathscr{Q}_{\alpha,n} f)(t) : u \in B_r\}$ are relatively compact for every $t \in [0, T]$.

On the contrary, applying Lemma 5.1 with respect to $\mu \in (0, 1)$, and (5.4), we get

$$\begin{aligned}
\|(\mathscr{Q}_\alpha f)(t) - (\mathscr{Q}_{\alpha,n} f)(t)\| &= \left\| \int_0^t (\mathscr{P}_\alpha(t-\tau) - \mathscr{P}_{\alpha,n}(t-\tau)) f(\tau, u(\tau)) d\tau \right\| \\
&\le c_+ L_f \lambda_{n+1}^{-\mu} \int_0^t (t-\tau)^{\alpha-1-\alpha\mu} \|u(\tau)\| \, d\tau \\
&\le c_+ L_f B(\alpha(1-\mu), 1-\eta) t^{\alpha(1-\mu)-\eta} \lambda_{n+1}^{-\mu} r,
\end{aligned}$$

which implies that $\|\mathscr{Q}_\alpha f - \mathscr{Q}_{\alpha,n} f\|_X \to 0$, as $n \to \infty$. Consequently, we derive that $\{H^n(t)\}$ is arbitrarily close to the set $H(t) = \{t^\eta (\mathscr{Q}_\alpha f)(t) : u \in B_r\}$. Thus, $H(t)$ is relatively compact in X for every $t \in [0, T]$. Therefore, $\mathscr{Q}_\alpha$ is a compact operator.

Furthermore, by using the same ways as in Claim I and Lemma 5.4, one can check that the set $H(t)$ is equicontinuous. Next, we will show that $\mathscr{Q}_\alpha$ is continuous.

Let $\{u_m\}_{m=1}^\infty \subset B_r$ and $u \in B_r$ such that $\lim_{m\to\infty} u_m = u$; hence from the continuity assumption of f, it yields

$$\lim_{m\to\infty} f(t, u_m(t)) = f(t, u(t)), \quad t \in (0, T]$$

and

$$\|f(\tau, u_m(\tau)) - f(\tau, u(\tau))\| \le L_f \|u_m(\tau)\| + L_f \|u(\tau)\| \le 2L_f \tau^{-\eta} r.$$

Therefore, from the fact that $(t-\tau)^{\alpha-1}\tau^{-\eta} \in L^1(0, t)$ for a.e., $\tau \in (0, t)$ and the Lebesgue's dominated convergence theorem implies

$$
\begin{aligned}
\|(\mathscr{Q}_\alpha f_m)(t) - (\mathscr{Q}_\alpha f)(t)\| &= \left\| \int_0^t \mathscr{P}_\alpha(t-\tau)(f(\tau, u_m(\tau)) - f(\tau, u(\tau)))d\tau \right\| \\
&\le c_+ \int_0^t (t-\tau)^{\alpha-1} \|f(\tau, u_m(\tau)) - f(\tau, u(\tau))\| \, d\tau \\
&\to 0, \quad \text{as } m \to \infty.
\end{aligned}
$$

Hence, $\|\mathscr{Q}_\alpha f_m - \mathscr{Q}_\alpha f\|_X \to 0$ as $m \to \infty$, which shows that $\mathscr{Q}_\alpha$ is continuous, and then operator $\mathscr{F}$ is also continuous. By the Arzelà–Ascoli theorem, we know that $\mathscr{F}$ is completely continuous. Consequently, the Schauder fixed point theorem shows that $\mathscr{F}$ has at least one fixed point on B_r, and then problem (5.1) has a mild solution. The proof is completed.

Remark 5.1 Noting that the above existence result does not need to assume the Lipschitz type condition or smoothness of nonlinear functions, that is, the assumed condition of existence result is weaker than which in paper [91]. On the other hand, PDE can be abstracted to an abstract differential equation. We pay attention to study the existence of mild solutions of such abstract problem. From this point of view, it is not necessary to assume that the function f is continuously differentiable shown in [10].

Remark 5.2 If the following condition

$$
\|f(t, u)\| \le L'_f \|u\|_X, \quad \forall u \in X, \tag{5.15}
$$

substitutes for (Hf1) for some constant $L'_f > 0$, then the operator $\mathscr{T}_\alpha$ is also completely continuous. Obviously, the above condition is stronger than the condition (Hf1). However, we can pick a different range of η by $\eta \in [1, \alpha]$, that is, $\gamma \in [\frac{1}{\alpha}, 1]$. By repeating the above proof process in Theorem 5.1, we also establish an existence result of mild solutions (see below). In addition, we also remark that there exists a solution on $C^\alpha((0, T]; L^2(\Omega))$ for $\eta = \alpha$ ($\gamma = 1$).

Theorem 5.2 *Let $\gamma \in \left[\frac{1}{\alpha}, 1\right]$ and $g \in \mathscr{H}^{1-\gamma}(\Omega)$. Assume that (Hf1) holds with respect to f satisfying (5.15). Then the problem (5.1) has at least one mild solution provided with*

$$
\frac{c_+^2}{c_-\eta} L'_f T^{\eta+\alpha} + c_+ L'_f T^{\alpha+\eta} \le \frac{1}{2}.
$$

(Hf2) There exists a positive constant L''_f such that $f : (0, T] \times L^2(\Omega) \to L^2(\Omega)$ satisfies the following condition

$$
\|f(t, u) - f(t, v)\| \le L''_f \|u - v\|, \quad \forall u, v \in L^2(\Omega).
$$

Theorem 5.3 *Assume that the hypotheses of Theorem 5.1 and (Hf2) hold, and*

$$C'_T = c_+ L''_f T^\alpha B(\alpha, 1-\eta) + c_+^2 L''_f \pi/(c_- \sin(\pi\eta)) < 1.$$

Then the problem (5.1) has a unique mild solution.

Proof Define operator $\mathscr{F}$ as in Theorem 5.1; obviously, $\mathscr{F}$ maps B_r into itself. For any $u, v \in X$, similarly to (5.10), we have

$$\begin{aligned}
\|(\mathscr{F}u)(t) - (\mathscr{F}v)(t)\| = & \left\| \int_0^t \mathscr{P}_\alpha(t-\tau)(f(\tau, u(\tau)) - f(\tau, v(\tau)))d\tau \right\| \\
& + \left\| \int_0^T \mathscr{S}_\alpha(t) \circ \mathscr{P}_\alpha(T-\tau)(f(\tau, u(\tau)) \right. \\
& - f(\tau, v(\tau)))d\tau \| \\
\le & c_+ L''_f \int_0^t (t-\tau)^{\alpha-1} \|u(\tau) - v(\tau)\| \, d\tau \\
& + \frac{c_+^2}{c_-} T^\alpha L''_f t^{-\eta} \int_0^T (T-\tau)^{\eta-1} \|u(\tau) - v(\tau)\| \, d\tau,
\end{aligned}$$

which implies that $\|\mathscr{F}u - \mathscr{F}v\|_X \le C'_T \|u - v\|_X$, where $C'_T < 1$. Thus, $\mathscr{F}$ is a contraction operator. Therefore, problem (5.1) has a unique mild solution.

(Hf3) $f : (0, T] \times L^2(\Omega) \to L^2(\Omega)$ is continuous with respect to u, and it is measurable with respect to t. There exist a positive constant p with $p > \max\{1/\eta, 1\}$ and a positive function $\vartheta(\cdot) \in L^p(0, T)$ such that

$$\|f(t, u)\| \le \vartheta(t), \quad \forall u \in X, \ t \in (0, T]. \tag{5.16}$$

Theorem 5.4 *Let $g \in \mathscr{H}^{1-\gamma}(\Omega)$ for $\gamma \in (0, 1]$. Assume that (Hf3) holds. Then problem (5.1) has at least one mild solution $u \in C^\eta((0, T]; L^2(\Omega)) \cap L^q(0, T; L^2(\Omega))$ for $1 < q < 1/\eta$.*

Proof Let us begin the proof of the compactness of $\mathscr{T}_\alpha$. By repeating the proving process of Lemma 5.4, there is a similar method to show that $\mathscr{T}_{\alpha,n}$ converge uniformly to $\mathscr{T}_\alpha$ as $n \to \infty$. Indeed, for any $u \in L^2(\Omega)$, by applying Lemma 5.1 with respect to $\mu \in \left(0, 1 - \frac{1}{p\eta}\right)$, one can use the same way as in (5.12) that

$$\begin{aligned}
& \|(\mathscr{T}_\alpha f)(t) - (\mathscr{T}_{\alpha,n} f)(t)\| \\
\le & \frac{c_+^2}{c_-} T^\alpha t^{-\eta} \lambda_{n+1}^{-\gamma\mu} \int_0^T (T-\tau)^{\eta(1-\mu)-1} \|f(\tau, u(\tau))\| d\tau \\
\le & \frac{c_+^2}{c_-} T^\alpha t^{-\eta} \lambda_{n+1}^{-\gamma\mu} \int_0^T (T-\tau)^{\eta(1-\mu)-1} \vartheta(\tau) d\tau \\
\le & \frac{c_+^2}{c_-} \left(\frac{p-1}{\eta p(1-\mu) - 1} \right)^{1-\frac{1}{p}} T^{\alpha+\eta(1-\mu)-\frac{1}{p}} t^{-\eta} \lambda_{n+1}^{-\gamma\mu} \|\vartheta\|_{L^p(0,T)},
\end{aligned} \tag{5.17}$$

which implies that $\|\mathscr{T}_\alpha f - \mathscr{T}_{\alpha,n} f\|_X \to 0$ as $n \to \infty$.

Next, we just show that the operator $\mathscr{F}$ maps B_r into itself. In fact, for any $u \in B_r$, in view of (5.13), we have

$$\begin{aligned}\|(\mathscr{T}_\alpha f)(t)\| \leq & \frac{c_+^2}{c_-} T^\alpha t^{-\eta} \int_0^T (T-\tau)^{\eta-1} \vartheta(\tau) d\tau \\ \leq & \frac{c_+^2}{c_-} \left(\frac{p-1}{\eta p-1}\right)^{1-\frac{1}{p}} T^{\alpha+\eta-\frac{1}{p}} t^{-\eta} \|\vartheta\|_{L^p(0,T)}.\end{aligned} \tag{5.18}$$

Moreover, from (5.4), we get

$$\|(\mathscr{Q}_\alpha f)(t)\| \leq c_+ \int_0^t (t-\tau)^{\alpha-1} \vartheta(\tau) d\tau \leq c_+ t^{\alpha-\frac{1}{p}} \|\vartheta\|_{L^p(0,T)}. \tag{5.19}$$

Therefore, one can select r large enough such that

$$\begin{aligned}& C_T \|g\|_{\mathscr{H}^{s(1-\gamma)}(\Omega)} + \frac{c_+^2}{c_-} \left(\frac{p-1}{\eta p-1}\right)^{1-\frac{1}{p}} T^{\alpha+\eta-\frac{1}{p}} \|\vartheta\|_{L^p(0,T)} \\ & + c_+ T^{\alpha+\eta-\frac{1}{p}} \|\vartheta\|_{L^p(0,T)} \leq r,\end{aligned}$$

and then we get that $\mathscr{F}(B_r) \subseteq B_r$. The remains of the proof of existence result are similar to Theorem 5.1.

Finally, we will check that $u \in L^q(0,T; L^2(\Omega))$. In fact, one sees from (5.4) the assumptions of f, (5.14), (5.18), and (5.19), that

$$\begin{aligned}& \|\mathscr{S}_\alpha(t) g\|_{L^q(0,T;L^2(\Omega))} + \|(\mathscr{T}_\alpha f)(t)\|_{L^q(0,T;L^2(\Omega))} + \|(\mathscr{Q}_\alpha f)(t)\|_{L^q(0,T;L^2(\Omega))} \\ \leq & \left(\frac{1}{1-\eta q}\right)^{1/q} \left(C_T T^{\frac{1}{q}-\eta} \|g\|_{\mathscr{H}^{1-\gamma}(\Omega)} + c_+ T^{\alpha+\frac{1}{q}-\frac{1}{p}} \|\vartheta\|_{L^p(0,T)} \right. \\ & \left. + \frac{c_+^2}{c_-} \left(\frac{p-1}{\eta p-1}\right)^{1-\frac{1}{p}} T^{\alpha+\frac{1}{q}-\frac{1}{p}} \|\vartheta\|_{L^p(0,T)} \right) < \infty,\end{aligned}$$

which implies $u \in L^q(0,T; L^2(\Omega))$. Hence, the proof is completed.

(Hρ) There exists a positive function $\rho(t) \in L^1(0,T)$ such that

$$\Lambda_\rho := \int_0^T (T-\tau)^{-1} \rho(\tau) d\tau < \infty.$$

Noting that this function of (Hρ) will exist, for example, $\rho(t) = T - t$ for $t \in (0,T]$.

Theorem 5.5 *Let $g \in \mathscr{H}^1(\Omega)$. Suppose that there exists a positive function $\vartheta(\cdot)$ satisfying (Hρ) such that $f : (0,T] \times L^2(\Omega) \to L^2(\Omega)$ is continuous with respect to*

u and it is measurable with respect to t and satisfies (5.16). Then the mild solutions belong to $C([0,T];L^2(\Omega))$ for some $\eta \in (0,\alpha)$.

Proof According to the assumptions of f, it is not difficult to check that there exists at least one mild solution $u \in C^{\eta}((0,T];L^2(\Omega))$. In the sequel, we shall show $u \in C([0,T];L^2(\Omega))$. Now, for any $0 \le t_1 < t_2 \le T$, it follows that

$$\begin{aligned}\|u(t_2)-u(t_1)\| \le &\|\mathscr{S}_\alpha(t_2)g - \mathscr{S}_\alpha(t_1)g\| + \|(\mathscr{T}_\alpha f)(t_1) - (\mathscr{T}_\alpha f)(t_2)\| \\ &+ \|(\mathscr{Q}_\alpha f)(t_2) - (\mathscr{Q}_\alpha f)(t_1)\|.\end{aligned} \tag{5.20}$$

Noting that if $g \in \mathscr{H}^1(\Omega)$, then by Property 5.1, $\mathscr{S}_\alpha(t)g$ is bounded in $L^2(\Omega)$ for all $t \in [0,T]$. Hence, we first obtain that $\mathscr{S}_\alpha(\cdot)g \in C([0,T];L^2(\Omega))$.

On the contrary, by using (i) in Proposition 1.14, we have

$$\begin{aligned}\|(\mathscr{T}_\alpha f)(t_1) - (\mathscr{T}_\alpha f)(t_2)\| \le &C\left|\int_{t_1}^{t_2} (t_2-z)^{1-\alpha} z^{\alpha-2} dz\right| \\ &\times \int_0^T (T-\tau)^{-1}\|f(\tau,u(\tau))\|d\tau \\ \le &C\int_{t_1}^{t_2} (t_2-z)^{1-\alpha} z^{\alpha-2} dz \Lambda_\vartheta,\end{aligned}$$

for constant $C = 2c_+/\Gamma(2-\alpha)$, where we have used the estimate

$$|\mathscr{E}_\alpha(-\lambda_k t_2^\alpha) - \mathscr{E}_\alpha(-\lambda_k t_1^\alpha)| \le \frac{2c_+}{\Gamma(2-\alpha)}\int_{t_1}^{t_2}(t_2-z)^{1-\alpha} z^{\alpha-2} dz.$$

Next, we shall estimate the last term in (5.20). To begin with, by using (ii) in Proposition 1.14, it follows that

$$\begin{aligned}\left\|(\mathscr{Q}_\alpha f)(t_2) - (\mathscr{Q}_\alpha f)(t_1)\right\| \le &c_+\int_{t_1}^{t_2}(t_2-s)^{\alpha-1}\vartheta(\tau)d\tau \\ &+ c_+\int_0^{t_1}\left|\int_{t_1-\tau}^{t_2-\tau} z^{\alpha-2}dz\right|\vartheta(\tau)d\tau \\ \le &\frac{c_+\alpha}{\alpha-1}(t_2-t_1)^{\alpha-1}\|\vartheta\|_{L^1(0,T)}.\end{aligned}$$

Thus, together with arguments above, let $t_2 \to t_1$, and it is clear that $u(t_2) \to u(t_1)$ in $L^2(\Omega)$.

Moreover, setting $\varsigma = T-\tau$, for any $v \in L^2(\Omega)$, it yields

$$\|\mathscr{S}_\alpha(t)\circ\mathscr{P}_\alpha(\varsigma)v\|^2 \le \frac{c_+^4}{c_-^2}\varsigma^{2(\alpha-1)}\sum_{k=1,k\in\Pi}^{\infty}\left(\frac{1+\lambda_k T^\alpha}{1+\lambda_k\varsigma^\alpha}\right)^2|(v,\phi_k)|^2$$

$$\leq \frac{c_+^4}{c_-^2} T^{2\alpha} \varsigma^{-2} \|v\|^2.$$

Therefore, we have

$$\|(\mathscr{T}_\alpha f)(t)\| \leq \frac{c_+^2}{c_-} T^\alpha \int_0^T (T-\tau)^{-1} \|f(\tau, u(\tau))\| \, d\tau \leq \frac{c_+^2}{c_-} T^\alpha \Lambda_\vartheta.$$

In addition, one has

$$\|(\mathscr{Q}_\alpha f)(t)\| \leq c_+ \int_0^t (t-\tau)^{\alpha-1} \vartheta(\tau) d\tau \leq c_+ T^\alpha \Lambda_\vartheta.$$

Consequently, we deduce that $u \in C([0, T]; L^2(\Omega))$. The proof is completed.

5.1.5 Regularization

Let $\mathbf{R}(\epsilon, \lambda_k)$ be identity to

$$\mathbf{R}(\epsilon, \lambda_k) = \frac{|\mathscr{E}_\alpha(-\lambda_k T^\alpha)|^2}{|\mathscr{E}_\alpha(-\lambda_k T^\alpha)|^2 + \epsilon \lambda_k^2}, \qquad \epsilon > 0, \ k \in \Pi = \mathbb{N}^+ \setminus \Theta$$

and let

$$C_\epsilon = \begin{cases} \mathbf{C}_1 \epsilon^{\frac{\sigma}{4}}, & 0 < \sigma < 4, \\ \mathbf{C}_2 \epsilon, & \sigma \geq 4, \end{cases}$$

where $\mathbf{C}_1 = C(\sigma, c_-)$, $\mathbf{C}_2 = C(\sigma, c_-, \lambda_1) > 0$ for $\sigma > 0$.

Since $\mathscr{S}_\alpha(t)$ is not bounded linear operator on $L^2(\Omega)$ at time $t = 0$, it means that problem (5.1) is not stable on $L^\infty(0, T; L^2(\Omega))$, and it can lead to the general ill-posed problem, in the sequel, we define a family of regularizing operators $\mathscr{S}_\alpha^\epsilon(t)$ with the main idea of a general filter regularization method by

$$\mathscr{S}_\alpha^\epsilon(t) v = \sum_{k=1, k\in\Pi}^{\infty} \mathbf{R}(\epsilon, \lambda_k) \frac{\mathscr{E}_\alpha(-\lambda_k t^\alpha)}{\mathscr{E}_\alpha(-\lambda_k T^\alpha)} (v, \phi_k) \phi_k.$$

Then we can obtain the following regularized solution by the same way as in above theorems

$$\mathbf{u}^\epsilon(t) = \mathscr{S}_\alpha^\epsilon(t) g^\epsilon - (\mathscr{T}_\alpha^\epsilon f)(t, \mathbf{u}^\epsilon) + (\mathscr{Q}_\alpha f)(t, \mathbf{u}^\epsilon),$$

where g^ϵ is a noisy final data and $\epsilon > 0$ is a noise level which is assumed to satisfy

$$\|g^\epsilon - g\| \le \epsilon, \tag{5.21}$$

and hence we can rewrite it as follows:

$$\begin{aligned}
\mathbf{u}^\epsilon(t,x) = & \sum_{k=1,k\in\Pi}^{\infty} \mathbf{R}(\epsilon,\lambda_k)\frac{\mathscr{E}_\alpha(-\lambda_k t^\alpha)}{\mathscr{E}_\alpha(-\lambda_k T^\alpha)} g_k^\epsilon \phi_k(x) \\
& - \sum_{k=1,k\in\Pi}^{\infty} \mathbf{R}(\epsilon,\lambda_k)\frac{\mathscr{E}_\alpha(-\lambda_k t^\alpha)}{\mathscr{E}_\alpha(-\lambda_k T^\alpha)} \\
& \times \left[\int_0^T (T-\tau)^{\alpha-1}\mathscr{E}_{\alpha,\alpha}(-\lambda_k(T-\tau)^\alpha) f_k(\tau,\mathbf{u}^\epsilon(\tau))d\tau\right]\phi_k(x) \\
& + \sum_{k=1,k\in\Pi}^{\infty} \left[\int_0^t (t-\tau)^{\alpha-1}\mathscr{E}_{\alpha,\alpha}(-\lambda_k(T-\tau)^\alpha) f_k(\tau,\mathbf{u}^\epsilon(\tau))d\tau\right]\phi_k(x).
\end{aligned}$$

Let us introduce the function u_α by

$$\begin{aligned}
u_\alpha(t,x) = & \sum_{k=1,k\in\Pi}^{\infty} \mathbf{R}(\epsilon,\lambda_k)\frac{\mathscr{E}_\alpha(-\lambda_k t^\alpha)}{\mathscr{E}_\alpha(-\lambda_k T^\alpha)} g_k \phi_k(x) \\
& - \sum_{k=1,k\in\Pi}^{\infty} \mathbf{R}(\epsilon,\lambda_k)\frac{\mathscr{E}_\alpha(-\lambda_k t^\alpha)}{\mathscr{E}_\alpha(-\lambda_k T^\alpha)} \\
& \times \left[\int_0^T (T-\tau)^{\alpha-1}\mathscr{E}_{\alpha,\alpha}(-\lambda_k(T-\tau)^\alpha) f_k(\tau,u(\tau))d\tau\right]\phi_k(x) \\
& + \sum_{k=1,k\in\Pi}^{\infty} \left[\int_0^t (t-\tau)^{\alpha-1}\mathscr{E}_{\alpha,\alpha}(-\lambda_k(T-\tau)^\alpha) f_k(\tau,u(\tau))d\tau\right]\phi_k(x).
\end{aligned}$$

Theorem 5.6 *Assume that $\psi(x) := u(0,x) \in \mathscr{H}^\sigma(\Omega)$ for any $\sigma > 0$ satisfies an a priori bound condition*

$$\|\psi\|_{\mathscr{H}^\sigma(\Omega)} \le \mathbf{M},$$

for a positive constant **M**. *Furthermore, let f satisfy that i) f is continuous with respect to u and is measurable with respect to t, ii) there exists a positive function $\vartheta(\cdot)$ satisfying ($H\rho$) such that (5.16) and the following condition hold*

$$\|f(t,u) - f(t,v)\| \le \vartheta(t)\|u-v\|, \quad \forall u,v \in L^2(\Omega), \ \ t \in (0,T].$$

If $\left(\frac{c_+^2}{c_-} + c_+T^\alpha\right)\Lambda_\vartheta < 1$, *then*

$$\|\mathbf{u}^\epsilon - u\|_{L^\infty(0,T;L^2(\Omega))} \le \left[1 - \left(\frac{c_+^2}{c_-} + c_+T^\alpha\right)\Lambda_\vartheta\right]^{-1}\left[\frac{c_+^2\sqrt{\epsilon}}{2c_-} + c_+\mathbf{M}C_\epsilon\right].$$

Proof According to the assumptions of f, it is easy to check that there is a unique solution $\mathbf{u}^\epsilon \in L^\infty(0, T; L^2(\Omega))$. We next show that $\mathbf{u}^\epsilon - u \in L^\infty(0, T; L^2(\Omega))$ for each $\epsilon > 0$ and its exact upper bound.

By the triangle inequality, we have

$$\|\mathbf{u}^\epsilon(t) - u(t)\| \le \|\mathbf{u}^\epsilon(t) - u_\alpha(t)\| + \|u_\alpha(t) - u(t)\|.$$

Firstly, we estimate $\|\mathbf{u}^\epsilon(t) - u_\alpha(t)\|$.

Indeed, in view of the inequalities in Lemma 5.3, we have

$$\frac{1}{|\mathscr{E}_\alpha(-\lambda_k T^\alpha)|^2 + \epsilon\lambda_k^2} \le \frac{1}{\frac{c_-^2}{\lambda_k^2} + \epsilon\lambda_k^2} \le \frac{1}{2c_-\sqrt{\epsilon}}.$$

Thus, by virtue of inequality $z/(z + a) \le 1$ for any $z, a \ge 0$, it yields that

$$|\mathbf{R}(\epsilon, \lambda_k)| \le 1, \qquad \left|\mathbf{R}(\epsilon, \lambda_k)\frac{\mathscr{E}_\alpha(-\lambda_k t^\alpha)}{\mathscr{E}_\alpha(-\lambda_k T^\alpha)}\right| \le \frac{c_+^2}{2c_-\sqrt{\epsilon}},$$

which implies from (5.16) and (5.21) that

$$\begin{aligned}\|\mathbf{u}^\epsilon(t) - u_\alpha(t)\| \le& \frac{c_+^2}{2c_-\sqrt{\epsilon}}\|g^\epsilon - g\| + \frac{c_+^2}{c_-}\int_0^T (T-\tau)^{-1} \\ &\times \|f(\tau, \mathbf{u}^\epsilon(\tau)) - f(\tau, u(\tau))\|d\tau \\ &+ c_+\int_0^t (t-\tau)^{\alpha-1}\|f(\tau, \mathbf{u}^\epsilon(\tau)) - f(\tau, u(\tau))\|d\tau \\ \le& \frac{c_+^2}{2c_-\sqrt{\epsilon}}\|g^\epsilon - g\| + \frac{2c_+^2}{c_-}\int_0^T (T-\tau)^{-1}\vartheta(\tau)d\tau \\ &+ 2c_+\int_0^t (t-\tau)^{\alpha-1}\vartheta(\tau)d\tau \\ \le& \frac{c_+^2\sqrt{\epsilon}}{2c_-} + \left(\frac{2c_+^2}{c_-} + 2c_+T^\alpha\right)\Lambda_\vartheta.\end{aligned}$$

Next, we estimate $\|u_\alpha(t) - u(t)\|$.

Obviously, from the initial value $u_k(0)$, we know

$$u_k(0) = \frac{1}{\mathscr{E}_\alpha(-\lambda_k T^\alpha)}\left(g_k - \int_0^T (T-\tau)^{\alpha-1}\mathscr{E}_{\alpha,\alpha}(-\lambda_k(T-\tau)^\alpha) f_k(\tau, u)d\tau\right),$$

and it yields

$$\begin{aligned}
\|u_\alpha(t) - u(t)\| = &\sqrt{\sum_{k=1,k\in\Pi}^{\infty}\left|(\mathbf{R}(\epsilon,\lambda_k)-1)\frac{\mathscr{E}_\alpha(-\lambda_k t^\alpha)}{\mathscr{E}_\alpha(-\lambda_k T^\alpha)}\right|^2} \\
&\overline{\times\left[g_k - \int_0^T (T-\tau)^{\alpha-1}\mathscr{E}_{\alpha,\alpha}(-\lambda_k(T-\tau)^\alpha) f_k(\tau, u(\tau))d\tau\right]^2} \\
\le & c_+\sqrt{\sum_{k=1,k\in\Pi}^{\infty}\left|\frac{\epsilon\lambda_k^2}{|\mathscr{E}_\alpha(-\lambda_k T^\alpha)|^2+\epsilon\lambda_k^2}\right|^2 |u_k(0)|^2} \\
\le & c_+ \sup_{k\in\mathbb{N}^+\cap\Pi} A(k)\|\psi\|_{\mathscr{H}^\sigma(\Omega)},
\end{aligned}$$

where

$$A(k) = \frac{\epsilon\lambda_k^{2-\sigma}}{|\mathscr{E}_\alpha(-\lambda_k T^\alpha)|^2+\epsilon\lambda_k^2}.$$

It follows from [225, Lemma 2.5] and Lemma 5.3 that

$$A(k) \le \frac{\epsilon\lambda_k^{4-\sigma}}{c_-^2+\epsilon\lambda_k^4} \le \begin{cases} \mathbf{C}_1\,\epsilon^{\frac{\sigma}{4}}, & 0<\sigma<4, \\ \mathbf{C}_2\,\epsilon, & \sigma\ge 4. \end{cases}$$

Therefore, $\mathbf{u}^\epsilon - u \in L^\infty(0,T;L^2(\Omega))$ for each $\epsilon>0$ follow. In particular, by the same way as the proof in $\|\mathbf{u}^\epsilon(t) - u_\alpha(t)\|$, we have

$$\begin{aligned}
\|\mathbf{u}^\epsilon(t) - u_\alpha(t)\| \le & \frac{c_+^2}{2c_-\sqrt{\epsilon}}\|g^\epsilon - g\| + \frac{c_+^2}{c_-}\int_0^T (T-\tau)^{-1}\vartheta(\tau)\|\mathbf{u}^\epsilon(\tau) - u(\tau)\|d\tau \\
& + c_+\int_0^t (t-\tau)^{\alpha-1}\vartheta(\tau)\|\mathbf{u}^\epsilon(\tau) - u(\tau)\|d\tau \\
\le & \frac{c_+^2\sqrt{\epsilon}}{2c_-} + \left(\frac{c_+^2}{c_-} + c_+T^\alpha\right)\Lambda_\vartheta\|\mathbf{u}^\epsilon - u\|_{L^\infty(0,T;L^2(\Omega))}.
\end{aligned}$$

Therefore, we have

$$\|\mathbf{u}^\epsilon - u\|_{L^\infty(0,T;L^2(\Omega))} \le \frac{c_+^2\sqrt{\epsilon}}{2c_-} + \left(\frac{c_+^2}{c_-} + c_+T^\alpha\right) \Lambda_\vartheta \|\mathbf{u}^\epsilon - u\|_{L^\infty(0,T;L^2(\Omega))}$$
$$+ c_+\mathbf{M}C_\epsilon.$$

Consequently, since $\left(\frac{c_+^2}{c_-} + c_+T^\alpha\right)\Lambda_\vartheta < 1$, we deduce the desired conclusions. The proof is completed.

5.2 Initial Inverse Problem

5.2.1 Introduction

Consider the time fractional wave equation

$$\begin{cases} {}_0^C D_t^\beta u(t,x) - \mathscr{L}u(t,x) = G(t,x), & (t,x) \in (0,T_0) \times \Omega, \\ u(t,x) = 0, & (t,x) \in (0,T_0) \times \partial\Omega, \\ u(0,x) = u_0(x), & x \in \Omega, \\ u_t(0,x) = 0, & x \in \Omega, \end{cases} \tag{5.22}$$

where the domain Ω is a subset of d-dimensional space $\mathbb{R}^d$, which is a bounded domain with sufficient smooth boundary $\partial\Omega$; $d = 1, 2, 3$ is the dimension of the domain Ω, $T_0 > 0$ is fixed (we let $I_{T_0} := [0, T_0]$), and $\beta \in (1, 2)$ is the fractional order of the time derivative. Here, ${}_0^C D_t^\beta$ refers to the (left-sided) Caputo fractional derivative of order β with respect to $t \in (0, T_0]$:

$${}_0^C D_t^\beta u(t,x) := \frac{\partial^\beta u}{\partial t^\beta} = \begin{cases} \frac{1}{\Gamma(2-\beta)} \int_0^t (t-s)^{1-\beta} \frac{\partial^2}{\partial s^2} u(s,x) ds, & 1 < \beta < 2, \\ \frac{\partial^2 u(t,x)}{\partial t^2} = 0, & \beta = 2. \end{cases} \tag{5.23}$$

The operator $-\mathscr{L}$ in (5.22) represents the unbounded uniformly elliptic operator with domain $D(-\mathscr{L}) = H_0^1(\Omega) \cap H^2(\Omega)$ defined in [188] (see p.427).

If the initial data u_0 and the source term G are given, the problem (5.22) is called the direct problem. In [188], Yamamoto proved that the direct problem (5.22) has a unique weak solution, and for other results, see for example [4, 52, 180, 188, 226]. The inverse problem for (5.22) is less known. Inverse problems occur when we do not know all the given data (the initial data, boundary value, diffusion coefficient, or source term). By adding some given data, we can have inverse problems such as the backward problem (recovering the initial data) and the source identification problem (recovering the source function), see, for example, [96, 147, 206, 207, 209, 223] and the references therein.

In this section, we consider the backward problem (the initial inverse problem) of the inhomogeneous time fractional wave equation:

$$\begin{cases} {}_0^C D_t^\beta u(t,x) - \mathscr{L} u(t,x) = G(t,x), & (t,x) \in (0,T_0] \times \Omega, \\ u(t,x) = 0, & (t,x) \in I_{T_0} \times \partial\Omega, \\ u(T_0,x) = h(x), & x \in \Omega, \\ u_t(T_0,x) = 0, & x \in \Omega. \end{cases} \tag{5.24}$$

Our goal is to construct the initial data $f(x) = u(0,x)$ from given data (h, G). We show that this problem is ill-posed (see Sect. 5.2.2). Note that we cannot observe the data (h, G), so we only get approximate data (h^ϵ, G^ϵ) such that

$$\|h - h^\epsilon\|_{L^2(\Omega)} + \|G - G^\epsilon\|_{L^1(0,T_0;L^2(\Omega))} \leq \epsilon, \tag{5.25}$$

where $\epsilon > 0$ is the noise level (in this section we will also let $\|\cdot\|$ denote the $L^2(\Omega)$-norm). A regularization method is required for constructing approximations of stability for a sought solution. There are only a few results on inverse problems for fractional wave equation such as inverse source problems on a bounded domain [28, 146, 196, 208, 226], and recently, Tuan et al. in [208] considered existence and regularity of final value problems for time fractional wave equation.

We use the fractional Landweber method to find a regularized solution. This method was introduced by Klann and Ramlau [119] to consider linear ill-posed problem. The main idea of the fractional Landweber method is based on iterative sequences, which is similar to the classical iterative method. Using this method, some authors established a fractional method for solving some linear ill-posed models, see, for example, [88, 167, 229]. We will consider regularized solutions and regularity for the regularized solutions. Also we will present an error estimate of the fractional Landweber regularized solution to the exact solution under an a priori assumption using a priori and a posteriori regularization parameter choice rules.

The rest of the section is organized as follows. In the next subsection, some preliminaries are presented, and a mild solution of our backward problem is discussed. In Sect. 5.2.3, we present a regularized problem and consider the well-posedness of the regularized solutions. In Sect. 5.2.4, error estimates under two parameter choice rules are considered.

5.2.2 Preliminaries

Consider the operator $-\mathscr{L}$ on $L^2(\Omega)$ with domain $D(-\mathscr{L}) = H^2(\Omega) \cap H_0^1(\Omega)$. Assume that $-\mathscr{L}$ has an eigenvalue $\widetilde{a}_k$ with corresponding eigenfunction $e_k \in H^2(\Omega) \cap H_0^1(\Omega)$. Note

$$0 < \widetilde{a}_1 \leq \widetilde{a}_2 \leq \widetilde{a}_3 \leq \ldots \leq \widetilde{a}_k \leq \ldots$$

and $\widetilde{a}_k \to \infty$ as $k \to \infty$. Moreover

$$\begin{cases} \mathscr{L} e_k(x) = -\widetilde{a}_k e_k(x), & x \in \Omega, \\ e_k(x) = 0, & x \in \partial\Omega, \end{cases}$$

and we note that there exists a positive constant C such that $\widetilde{a}_k \geq C k^{\frac{2}{d}}$ for $k \in \mathbb{N}^+$, where d is the dimension of the domain Ω, see [42].

For $r \geq 0$, consider the Hilbert scale space (see [159])

$$\mathscr{H}^r(\Omega) = \left\{ v \in L^2(\Omega) : \sum_{k=1}^{\infty} \widetilde{a}_k^r |\langle v, e_k \rangle|^2 < +\infty \right\} \tag{5.26}$$

with norm

$$\|v\|_{\mathscr{H}^r(\Omega)} = \left(\sum_{k=1}^{\infty} \widetilde{a}_k^r |\langle v, e_k \rangle|^2 \right)^{\frac{1}{2}}.$$

If $r = 0$, we have $\mathscr{H}^0(\Omega) = L^2(\Omega)$.

For a given real number $1 \leq p < \infty$, let $L^p(0, T_0; L^2(\Omega))$ be the space of all functions such that

$$\|v\|_{L^p(0,T_0;L^2(\Omega))} := \left(\int_0^{T_0} \|v(t)\|_{L^2(\Omega)}^p dt \right)^{\frac{1}{p}} < +\infty.$$

Let $E_{\alpha,\beta}(z)$ be the Mittag–Leffler function as in Definition 1.7. Note that $E_{\alpha,\beta}(z)$ is an entire function in $z \in \mathbb{C}$.

Lemma 5.6 ([180]) *For $c > 0$, $\beta > 0$, and a positive integer $n \in \mathbb{N}^+$, we have*

$$\frac{d^n}{dt^n} E_{\beta,1}(-ct^\beta) = -ct^{\beta-n} E_{\beta,\beta-n+1}(-ct^\beta), \quad t > 0.$$

Lemma 5.7 ([141, 213]) *For $0 < \lambda < 1$, $r > 0$, and $n \in \mathbb{N}^+$, we have*

$$(1 - \lambda)^n \lambda^r \leq r^r (n+1)^{-r} < r^r n^{-r}.$$

From result in [188], the direct problem (5.22) exists a unique weak solution $u \in C\big(I_{T_0}, L^2(\Omega)\big) \cap C\big((0, T_0]; H^2(\Omega) \cap H_0^1(\Omega)\big)$ with ${}_0^C D_t^\beta u \in C\big((0, T_0]; L^2(\Omega)\big)$. Using the Fourier series expansion, the formal solution of the direct problem (5.22) can be constructing in the following form [188]:

$$u(t,x)=\sum_{k=1}^{\infty}\Big[E_{\beta,1}(-\widetilde{a}_k t^{\beta})f_k+\mathscr{F}_{\beta,k}(t)\Big]e_k(x), \tag{5.27}$$

where $f_k = \langle f, e_k\rangle$, $G_k(t) = \langle G(t,\cdot), e_k\rangle$, $f(x) := u(0,x)$, and $\mathscr{F}_{\beta,k}(t) = \big\langle \mathscr{F}_{\beta}(t), e_k\big\rangle$ with

$$\mathscr{F}_{\beta}(t):=\sum_{k=1}^{\infty}\int_0^t (t-\tau)^{\beta-1}E_{\beta,\beta}(-\widetilde{a}_k(t-\tau)^{\beta})G_k(\tau)d\tau e_k(x). \tag{5.28}$$

Using Proposition 1.16, we have the following lemma.

Lemma 5.8 *Let $1<\beta<2$, for all $0\le t\le T_0$ and $k\in\mathbb{N}^+$. There exist two positive constants $\mathscr{M}$, $\mathscr{M}_{\beta,+}$ such that*

$$\Big|E_{\beta,1}(-\widetilde{a}_k t^{\beta})\Big|\le\frac{\mathscr{M}_{\beta,+}}{1+\widetilde{a}_k t^{\beta}},\qquad \Big|E_{\beta,\beta}(-\widetilde{a}_k t^{\beta})\Big|\le\frac{\mathscr{M}}{1+\widetilde{a}_k t^{\beta}}.$$

Next, we will consider the mild solution of the problem (5.24). Assume the problem (5.24) has a unique solution u.

Let $t=T_0$ in (5.27), and we obtain

$$h_k=E_{\beta,1}(-\widetilde{a}_k T_0^{\beta})f_k+\mathscr{F}_{\beta,k}(T_0), \tag{5.29}$$

where $h_k=\langle h, e_k\rangle$, with $k\in\mathbb{N}^+$. From Lemma 3.2 in [226], there exists a positive constant $L_0>0$ which does not depend on k such that

$$E_{\beta,1}(-\widetilde{a}_k T_0^{\beta})\le\frac{1}{2\Gamma(1-\beta)\widetilde{a}_k T_0^{\beta}}<0,\quad \widetilde{a}_k T_0^{\beta}>L_0,$$

for $1<\beta<2$. Therefore, if T_0 is large enough such that

$$T_0^{\beta}\ge L_0(\widetilde{a}_1)^{-1}, \tag{5.30}$$

then $T_0^{\beta}\ge L_0(\widetilde{a}_k)^{-1}$, for all $k\in\mathbb{N}^+$. Therefore

$$E_{\beta,1}(-\widetilde{a}_k T_0^{\beta})\neq 0, \tag{5.31}$$

for all $k\in\mathbb{N}^+$. In this section, we assume that T_0 always satisfies (5.31).

From Lemma 2.5 in [115], Lemma 5.8, and (5.31), we have the following lemma.

Lemma 5.9 *Let $1<\beta<2$ and $T_0^{\beta}\ge L_0(\widetilde{a}_k)^{-1}$, for all $k\in\mathbb{N}^+$. Then there exist positive constants $\overline{\mathscr{M}}_{\beta,-}$, $\overline{\mathscr{M}}_{\beta,+}$ depending on β, T_0, $\widetilde{a}_k$ such that:*

$$\frac{\overline{\mathscr{M}}_{\beta,-}}{\widetilde{a}_k} \leq |E_{\beta,1}(-\widetilde{a}_k T_0^\beta)| \leq \frac{\overline{\mathscr{M}}_{\beta,+}}{\widetilde{a}_k}.$$

It follows from (5.29) that

$$f_k = \frac{1}{E_{\beta,1}(-\widetilde{a}_k T_0^\beta)}\Big[h_k - \mathscr{F}_{\beta,k}(T_0)\Big]. \tag{5.32}$$

By substituting f_k into (5.27), we obtain

$$u(t,x) = \sum_{k=1}^{\infty}\Big[\frac{E_{\beta,1}(-\widetilde{a}_k t^\beta)}{E_{\beta,1}(-\widetilde{a}_k T_0^\beta)}\Big[h_k - \mathscr{F}_{\beta,k}(T_0)\Big] + \mathscr{F}_{\beta,k}(t)\Big]e_k(x). \tag{5.33}$$

Lemma 5.10 *Let $G \in L^1(0,T_0;L^2(\Omega))$ and $\mathscr{F}_\beta(t)$ is defined as in (5.28). Then there exists a positive constant Λ such that*

$$\begin{aligned}\|\mathscr{F}_\beta(t)\|_{L^2(\Omega)} =& \Big\|\sum_{k=1}^{\infty}\int_0^t (t-\tau)^{\beta-1}E_{\beta,\beta}(-\widetilde{a}_k(t-\tau)^\beta)G_k(\tau)d\tau e_k(x)\Big\|_{L^2(\Omega)}\\ \leq& \Lambda\|G\|_{L^1(0,T_0;L^2(\Omega))}.\end{aligned}$$

Proof First note

$$\begin{aligned}\|\mathscr{F}_\beta(t)\|_{L^2(\Omega)} =& \Big\|\sum_{k=1}^{\infty}\int_0^t (t-\tau)^{\beta-1}E_{\beta,\beta}(-\widetilde{a}_k(t-\tau)^\beta)G_k(\tau)d\tau e_k(x)\Big\|_{L^2(\Omega)}\\ \leq& \int_0^t \sqrt{\sum_{k=1}^{\infty}(t-\tau)^{2(\beta-1)}\Big|E_{\beta,\beta}(-\widetilde{a}_k(t-\tau)^\beta)G_k(\tau)\Big|^2}\,d\tau.\end{aligned}$$

Using Lemma 5.8, we obtain

$$\begin{aligned}\|\mathscr{F}_\beta(t)\|_{L^2(\Omega)} \leq& \mathscr{M}\int_0^t (t-\tau)^{\beta-1}\|G(\tau,\cdot)\|_{L^2(\Omega)}d\tau\\ \leq& \mathscr{M}T_0^{\beta-1}\|G\|_{L^1(0,T_0;L^2(\Omega))}\end{aligned} \tag{5.34}$$

and so

$$\|\mathscr{F}_\beta(t)\|_{L^2(\Omega)} \leq \Lambda\|G\|_{L^1(0,T_0;L^2(\Omega))}$$

with $\Lambda = \mathscr{M}T_0^{\beta-1}$.

The solution $u(t,x)$ of the problem (5.24) for $1 < \beta < 2$ is given by (5.33). For $t = 0$, we get

$$f(x) := u(0,x) = \sum_{k=1}^{\infty} \frac{\Psi_k}{E_{\beta,1}(-\widetilde{a}_k T_0^{\beta})} e_k(x), \tag{5.35}$$

where

$$\Psi(x) := \sum_{k=1}^{\infty} \Psi_k e_k(x), \quad \text{with} \quad \Psi_k := h_k - \mathscr{F}_{\beta,k}(T_0). \tag{5.36}$$

Our main goal is to find the initial data $f(x)$ from the final data $h(x)$ and the source term $G(t,x)$. To find $f(x)$, we need to solve an operator equation:

$$\mathscr{K} f = \Psi,$$

where $\mathscr{K} : L^2(\Omega) \to L^2(\Omega)$ is the integral operator defined by

$$(\mathscr{K} f)(x) := \sum_{k=1}^{\infty} E_{\beta,1}(-\widetilde{a}_k T_0^{\beta}) \langle f, e_k \rangle e_k = \int_{\Omega} \kappa(\zeta, x) f(\zeta) d\zeta$$

with kernel $\kappa(\cdot,\cdot)$, i.e.,

$$\kappa(x,\zeta) := \sum_{k=1}^{\infty} E_{\beta,1}(-\widetilde{a}_k T_0^{\beta}) e_k(\zeta) e_k(x).$$

Since $\kappa(x,\zeta) = \kappa(\zeta,x)$, it is easy to see that the operator $\mathscr{K}$ is self-adjoint.

Lemma 5.11 *Let $h \in L^2(\Omega)$ and $G \in L^1(0,T_0;L^2(\Omega))$. Then Ψ as in* (5.36) *belongs to $L^2(\Omega)$ and*

$$\|\Psi\|_{L^2(\Omega)} \le \|h\|_{L^2(\Omega)} + \Lambda \|G\|_{L^1(0,T_0;L^2(\Omega))}.$$

Proof We have

$$\begin{aligned} \|\Psi\|_{L^2(\Omega)} &= \Big\| \sum_{k=1}^{\infty} \Big(h_k - \mathscr{F}_{\beta,k}(T_0) \Big) e_k(x) \Big\|_{L^2(\Omega)} \\ &\le \Big\| \sum_{k=1}^{\infty} h_k e_k(x) \Big\|_{L^2(\Omega)} + \Big\| \sum_{k=1}^{\infty} \mathscr{F}_{\beta,k}(T_0) e_k(x) \Big\|_{L^2(\Omega)}. \end{aligned}$$

Using Lemma 5.10 we obtain

$$\|\Psi\|_{L^2(\Omega)} \leq \|h\|_{L^2(\Omega)} + \Lambda\|G\|_{L^1(0,T_0;L^2(\Omega))}.$$

This completes the proof.

One can show that $\kappa(\cdot,\cdot) \in L^2(\Omega \times \Omega)$. Therefore, $\mathscr{K} : L^2(\Omega) \to L^2(\Omega)$ is compact operator of finite rank. Hence $\mathscr{K}$ does not have a continuous inverse [169].

To illustrate the ill-posedness of the backward problem, we give an example. Let $(h, G) = (0, 0)$ and $(\overline{h}, \overline{G}) = \left(\frac{1}{\sqrt{\widetilde{a}_l}} e_l, \frac{1}{\sqrt{\widetilde{a}_l}} e_l\right)$, where $l \in \mathbb{N}^+$. It is easy to see that

$$\|\overline{h} - h\|_{L^2(\Omega)} = \frac{1}{\sqrt{\widetilde{a}_l}} \text{ and } \|\overline{G} - G\|_{L^2(\Omega)} = \frac{1}{\sqrt{\widetilde{a}_l}}.$$

Hence

$$\lim_{l\to\infty} \|\overline{h} - h\|_{L^2(\Omega)} = 0 \text{ and } \lim_{l\to\infty} \|\overline{G} - G\|_{L^2(\Omega)} = 0, \tag{5.37}$$

so $(\overline{h}, \overline{G})$ is an approximation of (h, G) when l is large enough. Using $(\overline{h}, \overline{G})$, we get the corresponding initial data $\overline{f}$ and the equation $\overline{\Psi}$ as follows:

$$\overline{\Psi}(x) := \sum_{k=1}^{\infty} \left(\overline{h}_k - \int_0^{T_0} (T_0 - \tau)^{\beta-1} E_{\beta,\beta}(-\widetilde{a}_k (T_0 - \tau)^{\beta}) \overline{G}_k(\tau) d\tau\right) e_k(x),$$

$$\overline{f}(x) := \sum_{k=1}^{\infty} \frac{1}{E_{\beta,1}(-\widetilde{a}_k T_0^{\beta})} \times \left(\overline{h}_k - \int_0^{T_0} (T_0 - \tau)^{\beta-1} E_{\beta,\beta}(-\widetilde{a}_k (T_0 - \tau)^{\beta}) \overline{G}_k(\tau) d\tau\right) e_k(x).$$

From the Parseval's equality and Lemma 5.6, we obtain

$$\begin{aligned}
&\|\overline{\Psi} - \Psi\|^2_{L^2(\Omega)} \\
=&\sum_{k=1}^{\infty} \left[\langle \overline{h} - h, e_k\rangle - \int_0^{T_0} (T_0 - \tau)^{\beta-1} E_{\beta,\beta}(-\widetilde{a}_k (T_0 - \tau)^{\beta}) \langle \overline{G} - G, e_k\rangle d\tau\right]^2 \\
=&\left[\frac{1}{\sqrt{\widetilde{a}_l}} - \frac{1}{\sqrt{\widetilde{a}_l}} \int_0^{T_0} (T_0 - \tau)^{\beta-1} E_{\beta,\beta}(-\widetilde{a}_l (T_0 - \tau)^{\beta}) d\tau\right]^2 \\
=&\frac{1}{\widetilde{a}_l}\left[1 - \int_0^{T_0} \tau^{\beta-1} E_{\beta,\beta}(-\widetilde{a}_l \tau^{\beta}) d\tau\right]^2 \\
=&\frac{1}{\widetilde{a}_l}\left[1 - \frac{1}{\widetilde{a}_l} + \frac{1}{\widetilde{a}_l} E_{\beta,1}(-\widetilde{a}_l T_0^{\beta})\right]^2.
\end{aligned} \tag{5.38}$$

This gives

$$\lim_{l\to\infty} \|\overline{\Psi} - \Psi\|_{L^2(\Omega)} = 0. \tag{5.39}$$

On the other hand, we have

$$\begin{aligned}
\|\overline{f} - f\|^2_{L^2(\Omega)} =& \sum_{k=1}^{\infty} \frac{1}{E^2_{\beta,1}(-\widetilde{a}_k T_0^\beta)} \Big[\langle \overline{h} - h, e_k \rangle \\
& - \int_0^{T_0} (T_0 - \tau)^{\beta-1} E_{\beta,\beta}(-\widetilde{a}_k (T_0 - \tau)^\beta) \langle \overline{G} - G, e_k \rangle d\tau \Big]^2 \\
=& \frac{1}{E^2_{\beta,1}(-\widetilde{a}_l T_0^\beta)} \Big[\frac{1}{\sqrt{\widetilde{a}_l}} - \frac{1}{\sqrt{\widetilde{a}_l}} \int_0^{T_0} (T_0 - \tau)^{\beta-1} \\
& \times E_{\beta,\beta}(-\widetilde{a}_l (T_0 - \tau)^\beta) d\tau \Big]^2 \\
\geq& \frac{\widetilde{a}_l^2}{\overline{\mathscr{M}}^2_{\beta,+}} \frac{1}{\widetilde{a}_l} \Big[1 - \int_0^{T_0} \tau^{\beta-1} E_{\beta,\beta}(-\widetilde{a}_l \tau^\beta) d\tau \Big]^2 \\
\geq& \frac{\widetilde{a}_l}{\overline{\mathscr{M}}^2_{\beta,+}} \Big[1 - \frac{1}{\widetilde{a}_l} + \frac{1}{\widetilde{a}_l} E_{\beta,1}(-\widetilde{a}_l T_0^\beta) \Big]^2 .
\end{aligned}$$

Therefore

$$\lim_{l\to\infty} \|\overline{f} - f\|_{L^2(\Omega)} = +\infty. \tag{5.40}$$

We conclude that the backward problem is ill-posed in the sense of Hadamard. Hence a regularization method is necessary.

Now we give a stability estimate.

Theorem 5.7 *Suppose $f \in \mathscr{H}^r(\Omega)$ satisfies $\|f\|_{\mathscr{H}^r(\Omega)} \leq P$ for any $r > 0$. Then*

$$\|f\|_{L^2(\Omega)} \leq \overline{\mathscr{M}}_{\beta,-}^{\frac{-r}{r+2}} \|\Psi\|_{L^2(\Omega)}^{\frac{r}{r+2}} P^{\frac{2}{r+2}},$$

where P is a positive constant.

Proof Using (5.35) and the Hölder's inequality, we get

$$\|f\|^2_{L^2(\Omega)} = \sum_{k=1}^{\infty} \frac{\Psi_k^2}{E^2_{\beta,1}(-\widetilde{a}_k T_0^\beta)} \leq \Big[\sum_{k=1}^{\infty} \Psi_k^2 \Big]^{\frac{r}{r+2}} \Big[\sum_{k=1}^{\infty} \frac{\Psi_k^2}{E^{r+2}_{\beta,1}(-\widetilde{a}_k T_0^\beta)} \Big]^{\frac{2}{r+2}} .$$

Hence

$$\|f\|_{L^2(\Omega)} \leq \|\Psi\|_{L^2(\Omega)}^{\frac{r}{r+2}} \Upsilon^{\frac{1}{r+2}}, \tag{5.41}$$

where

$$\Upsilon := \sum_{k=1}^{\infty} \frac{\Psi_k^2}{E_{\beta,1}^{r+2}(-\widetilde{a}_k T_0^{\beta})}.$$

Using Lemma 5.9 and (5.35), we get

$$\Upsilon = \sum_{k=1}^{\infty} \frac{\Psi_k^2}{E_{\beta,1}^{r+2}(-\widetilde{a}_k T_0^{\beta})} \leq \sum_{k=1}^{\infty} \frac{\widetilde{a}_k^r}{\overline{\mathscr{M}}_{\beta,-}^r} f_k^2 \leq \frac{1}{\overline{\mathscr{M}}_{\beta,-}^r} \|f\|_{\mathscr{H}^r(\Omega)}^2.$$

This implies that

$$\|f\|_{L^2(\Omega)} \leq \overline{\mathscr{M}}_{\beta,-}^{\frac{-r}{r+2}} \|\Psi\|_{L^2(\Omega)}^{\frac{r}{r+2}} P^{\frac{2}{r+2}}.$$

5.2.3 Regularization Method

In this subsection, we introduce the fractional Landweber regularization method, and we also analyze convergence properties of regularization methods under two parameter choice rules.

From [118], the operator equation $\mathscr{K} f = \Psi$ is equivalent to the following equation:

$$f = \left(I - a\mathscr{K}^* \mathscr{K}\right) f + a\mathscr{K}^* \Psi, \tag{5.42}$$

for any $a > 0$. Here, $\mathscr{K}^*$ is the adjoint operator of $\mathscr{K}$, and a satisfies $0 < a < \frac{1}{\|\mathscr{K}\|_{\mathfrak{L}(L^2(\Omega))}^2}$. The iterative implementation of the fractional Landweber method was considered in [119]. Denote the fractional Landweber regularization solution by

$$f_{n,\varrho}(x) = \sum_{k=1}^{\infty} \frac{1}{E_{\beta,1}(-\widetilde{a}_k T_0^{\beta})} \left[1 - \left(1 - aE_{\beta,1}^2(-\widetilde{a}_k T_0^{\beta})\right)^n\right]^{\varrho} \langle \Psi, e_k \rangle e_k, \tag{5.43}$$

and the fractional Landweber regularization solution with the noisy data by

$$f_{n,\varrho}^{\epsilon}(x) = \sum_{k=1}^{\infty} \frac{1}{E_{\beta,1}(-\widetilde{a}_k T_0^{\beta})} \left[1 - \left(1 - aE_{\beta,1}^2(-\widetilde{a}_k T_0^{\beta})\right)^n\right]^{\varrho} \left\langle \Psi^{\epsilon}, e_k \right\rangle e_k, \tag{5.44}$$

where $\varrho \in (\frac{1}{2}, 1]$ is called the fractional parameter, and $n = 1, 2, \ldots$ is a regularization parameter. When $\varrho = 1$, this is the classical Landweber method.

Hence, we get the fractional Landweber regularization solution of the problem (5.24)

$$\begin{aligned} & u_{n,\varrho}(t,x) \\ =& \sum_{k=1}^{\infty}\left[\frac{E_{\beta,1}(-\widetilde{a}_k t^{\beta})}{E_{\beta,1}(-\widetilde{a}_k T_0^{\beta})}\left[1-\left(1-aE_{\beta,1}^2(-\widetilde{a}_k T_0^{\beta})\right)^n\right]^{\varrho}\langle\Psi, e_k\rangle + \mathscr{F}_{\beta,k}(t)\right]e_k(x) \end{aligned}$$

and the fractional Landweber regularization solution of the problem (5.24) with the noisy data

$$\begin{aligned} & u_{n,\varrho}^{\epsilon}(t,x) \\ =& \sum_{k=1}^{\infty}\left[\frac{E_{\beta,1}(-\widetilde{a}_k t^{\beta})}{E_{\beta,1}(-\widetilde{a}_k T_0^{\beta})}\left[1-\left(1-aE_{\beta,1}^2(-\widetilde{a}_k T_0^{\beta})\right)^n\right]^{\varrho}\langle\Psi^{\epsilon}, e_k\rangle + \mathscr{F}_{\beta,k}^{\epsilon}(t)\right]e_k(x) \\ =& \sum_{k=1}^{\infty}\left[\frac{E_{\beta,1}(-\widetilde{a}_k t^{\beta})}{E_{\beta,1}(-\widetilde{a}_k T_0^{\beta})}\left[1-\left(1-aE_{\beta,1}^2(-\widetilde{a}_k T_0^{\beta})\right)^n\right]^{\varrho}h_k^{\epsilon}\right. \\ & \left. -\frac{E_{\beta,1}(-\widetilde{a}_k t^{\beta})}{E_{\beta,1}(-\widetilde{a}_k T_0^{\beta})}\left[1-\left(1-aE_{\beta,1}^2(-\widetilde{a}_k T_0^{\beta})\right)^n\right]^{\varrho}\mathscr{F}_{\beta,k}^{\epsilon}(T_0) + \mathscr{F}_{\beta,k}^{\epsilon}(t)\right]e_k(x) \\ =:& \phi_1(t,x) - \phi_2(t,x) + \phi_3(t,x), \end{aligned}$$

where

$$\phi_1(t,x) = \sum_{k=1}^{\infty}\frac{E_{\beta,1}(-\widetilde{a}_k t^{\beta})}{E_{\beta,1}(-\widetilde{a}_k T_0^{\beta})}\left[1-\left(1-aE_{\beta,1}^2(-\widetilde{a}_k T_0^{\beta})\right)^n\right]^{\varrho}h_k^{\epsilon}e_k(x),$$

$$\phi_2(t,x) = \sum_{k=1}^{\infty}\frac{E_{\beta,1}(-\widetilde{a}_k t^{\beta})}{E_{\beta,1}(-\widetilde{a}_k T_0^{\beta})}\left[1-\left(1-aE_{\beta,1}^2(-\widetilde{a}_k T_0^{\beta})\right)^n\right]^{\varrho}\mathscr{F}_{\beta,k}^{\epsilon}(T_0)e_k(x),$$

$$\phi_3(t,x) = \sum_{k=1}^{\infty}\mathscr{F}_{\beta,k}^{\epsilon}(t)e_k(x)$$

with

$$\mathscr{F}_{\beta,k}^{\epsilon}(t) := \int_0^t (t-\tau)^{\beta-1}E_{\beta,\beta}(-\widetilde{a}_k(t-\tau)^{\beta})G_k^{\epsilon}(\tau)d\tau.$$

Next, we consider the regularity of the solution $u_{n,\varrho}^{\epsilon}$.

Theorem 5.8 *Let $h^\epsilon \in L^2(\Omega)$, for $1 < \beta < 2$, and assume* (5.30) *holds. Furthermore, suppose that $G^\epsilon \in L^1(0, T_0; L^2(\Omega))$. Then (the regularized solution) $u^\epsilon_{n,\varrho} \in L^\infty(0, T_0; L^2(\Omega))$ and*

$$\begin{aligned}\|u^\epsilon_{n,\varrho}\|_{L^\infty(0,T_0;L^2(\Omega))} \leq& \mathscr{M}_{\beta,+} a^{\frac{1}{2}} n^{\frac{1}{2}} \|h^\epsilon\|_{L^2(\Omega)}\\ &+ \left(\mathscr{M}_{\beta,+} a^{\frac{1}{2}} n^{\frac{1}{2}} + 1\right) \Lambda \|G^\epsilon\|_{L^1(0,T_0;L^2(\Omega))}.\end{aligned}$$

Proof From Lemma 5.8, we get

$$\begin{aligned}&\frac{E_{\beta,1}(-\widetilde{a}_k t^\beta)}{E_{\beta,1}(-\widetilde{a}_k T_0^\beta)}\left[1 - \left(1 - aE^2_{\beta,1}(-\widetilde{a}_k T_0^\beta)\right)^n\right]^\varrho\\ \leq& a^{\frac{1}{2}} \mathscr{M}_{\beta,+} \left(a^{\frac{1}{2}} E_{\beta,1}(-\widetilde{a}_k T_0^\beta)\right)^{-1} \left[1 - \left(1 - aE^2_{\beta,1}(-\widetilde{a}_k T_0^\beta)\right)^n\right]^\varrho.\end{aligned} \tag{5.45}$$

Let $\vartheta := a^{\frac{1}{2}} E_{\beta,1}(-\widetilde{a}_k T_0^\beta)$ and

$$\phi(\vartheta) = \vartheta^{-2}\left[1 - \left(1 - \vartheta^2\right)^n\right]^{2\varrho}.$$

Since $0 < a < \dfrac{1}{\|\mathscr{K}\|^2_{\mathfrak{L}(L^2(\Omega))}}$, we have $0 < aE^2_{\beta,1}(-\widetilde{a}_k T_0^\beta) < 1$. Hence, the function is continuous when $\vartheta \in (0, 1)$.

For $\varrho \in (\frac{1}{2}, 1)$ and $\vartheta \in (0, 1)$, using Lemma 3.3 in [119] :

$$\phi(\vartheta) \leq n. \tag{5.46}$$

Combining (5.45) and (5.46), we deduce that

$$\sup_{\widetilde{a}_k > 0} \frac{E_{\beta,1}(-\widetilde{a}_k t^\beta)}{E_{\beta,1}(-\widetilde{a}_k T_0^\beta)}\left[1 - \left(1 - aE^2_{\beta,1}(-\widetilde{a}_k T_0^\beta)\right)^n\right]^\varrho \leq \mathscr{M}_{\beta,+} a^{\frac{1}{2}} n^{\frac{1}{2}}. \tag{5.47}$$

First note

$$\begin{aligned}&\|\phi_1(t,\cdot)\|_{L^2(\Omega)}\\ =& \sqrt{\sum_{k=1}^{\infty}\left(\frac{E_{\beta,1}(-\widetilde{a}_k t^\beta)}{E_{\beta,1}(-\widetilde{a}_k T_0^\beta)}\left[1 - \left(1 - aE^2_{\beta,1}(-\widetilde{a}_k T_0^\beta)\right)^n\right]^\varrho h^\epsilon_k\right)^2}\\ \leq& \sup_{\widetilde{a}_k > 0}\left[\frac{E_{\beta,1}(-\widetilde{a}_k t^\beta)}{E_{\beta,1}(-\widetilde{a}_k T_0^\beta)}\left[1 - \left(1 - aE^2_{\beta,1}(-\widetilde{a}_k T_0^\beta)\right)^n\right]^\varrho\right]\sqrt{\sum_{k=1}^{\infty}\left(h^\epsilon_k\right)^2}\\ \leq& \mathscr{M}_{\beta,+} a^{\frac{1}{2}} n^{\frac{1}{2}} \|h^\epsilon\|_{L^2(\Omega)}.\end{aligned} \tag{5.48}$$

By a similar method, we obtain

$$\begin{aligned}
&\|\phi_2(t,\cdot)\|_{L^2(\Omega)}\\
=&\left\|\sum_{k=1}^{\infty}\frac{E_{\beta,1}(-\widetilde{a}_k t^{\beta})}{E_{\beta,1}(-\widetilde{a}_k T_0^{\beta})}\left[1-\left(1-aE_{\beta,1}^2(-\widetilde{a}_k T_0^{\beta})\right)^n\right]^{\varrho}\mathscr{F}_{\beta,k}^{\epsilon}(T_0)e_k(x)\right\|_{L^2(\Omega)}\\
\le&\mathscr{M}_{\beta,+}a^{\frac{1}{2}}n^{\frac{1}{2}}\|\mathscr{F}_{\beta}^{\epsilon}(T_0)\|_{L^2(\Omega)}.
\end{aligned} \tag{5.49}$$

Using Lemma 5.10, we get

$$\|\phi_2(t,\cdot)\|_{L^2(\Omega)}\le\mathscr{M}_{\beta,+}a^{\frac{1}{2}}n^{\frac{1}{2}}\Lambda\|G^{\epsilon}\|_{L^1(0,T_0;L^2(\Omega))}. \tag{5.50}$$

On the other hand, from Lemma 5.10, we obtain

$$\|\phi_3(t,\cdot)\|_{L^2(\Omega)}=\|\mathscr{F}_{\beta}^{\epsilon}(t)\|_{L^2(\Omega)}\le\Lambda\|G^{\epsilon}\|_{L^1(0,T_0;L^2(\Omega))}. \tag{5.51}$$

Combining (5.48), (5.50), and (5.51), we have

$$\begin{aligned}
\|u_{n,\varrho}^{\epsilon}\|_{L^\infty(0,T_0;L^2(\Omega))}\le&\mathscr{M}_{\beta,+}a^{\frac{1}{2}}n^{\frac{1}{2}}\|h^{\epsilon}\|_{L^2(\Omega)}\\
&+\left(\mathscr{M}_{\beta,+}a^{\frac{1}{2}}n^{\frac{1}{2}}+1\right)\Lambda\|G^{\epsilon}\|_{L^1(0,T_0;L^2(\Omega))}.
\end{aligned}$$

Now, we give the regularity of the solution at $t=0$.

Theorem 5.9 *Let $f_{n,\varrho}^{\epsilon}(x):=u_{n,\varrho}^{\epsilon}(0,x)$ and assume that $h^{\epsilon}\in\mathscr{H}^{r+2}(\Omega)$ and $G^{\epsilon}\in L^{\infty}(0,T_0;\mathscr{H}^{r+1}(\Omega))$. Then*

$$\|f_{n,\varrho}^{\epsilon}\|_{\mathscr{H}^r(\Omega)}\le\frac{1}{\mathscr{M}_{\beta,-}}\|h^{\epsilon}\|_{\mathscr{H}^{r+2}(\Omega)}+\frac{\mathscr{M}T_0^{\frac{\beta-1}{2}}}{2\mathscr{M}_{\beta,-}\sqrt{\beta-1}}\|G^{\epsilon}\|_{L^\infty(0,T_0;\mathscr{H}^{r+1}(\Omega))}.$$

Proof Since $\varrho\in(\frac{1}{2},1]$ and $0<aE_{\beta,1}^2(-\widetilde{a}_k T_0^{\beta})<1$, it is easy to see that

$$\begin{aligned}
&\|f_{n,\varrho}^{\epsilon}\|_{\mathscr{H}^r(\Omega)}\\
=&\left\|\sum_{k=1}^{\infty}\frac{1}{E_{\beta,1}(-\widetilde{a}_k T_0^{\beta})}\left[1-\left(1-aE_{\beta,1}^2(-\widetilde{a}_k T_0^{\beta})\right)^n\right]^{\varrho}\left\langle\Psi^{\epsilon},e_k\right\rangle e_k\right\|_{\mathscr{H}^r(\Omega)}\\
\le&\left\|\sum_{k=1}^{\infty}\frac{\widetilde{a}_k}{\mathscr{M}_{\beta,-}}\left\langle h^{\epsilon}-\mathscr{F}_{\beta}^{\epsilon}(T_0),e_k\right\rangle e_k\right\|_{\mathscr{H}^r(\Omega)}
\end{aligned}$$

$$\leq \left\| \sum_{k=1}^{\infty} \frac{\widetilde{a}_k}{\overline{\mathscr{M}}_{\beta,-}} h_k^\epsilon e_k \right\|_{\mathscr{H}^r(\Omega)} + \left\| \sum_{k=1}^{\infty} \frac{\widetilde{a}_k}{\overline{\mathscr{M}}_{\beta,-}} \mathscr{F}_{\beta,k}^\epsilon(T_0) e_k \right\|_{\mathscr{H}^r(\Omega)}.$$

A direct calculation gives

$$\left\| \sum_{k=1}^{\infty} \frac{\widetilde{a}_k}{\overline{\mathscr{M}}_{\beta,-}} h_k^\epsilon e_k \right\|_{\mathscr{H}^r(\Omega)} = \frac{1}{\overline{\mathscr{M}}_{\beta,-}} \sqrt{\sum_{k=1}^{\infty} \widetilde{a}_k^2 \widetilde{a}_k^r \left(h_k^\epsilon\right)^2} = \frac{1}{\overline{\mathscr{M}}_{\beta,-}} \|h^\epsilon\|_{\mathscr{H}^{r+2}(\Omega)}.$$

On the other hand

$$\begin{aligned}
&\left\| \sum_{k=1}^{\infty} \frac{\widetilde{a}_k}{\overline{\mathscr{M}}_{\beta,-}} \mathscr{F}_{\beta,k}^\epsilon(T_0) e_k \right\|_{\mathscr{H}^r(\Omega)} \\
=&\left\| \sum_{k=1}^{\infty} \frac{\widetilde{a}_k}{\overline{\mathscr{M}}_{\beta,-}} \int_0^{T_0} (T_0-\tau)^{\beta-1} E_{\beta,\beta}(-\widetilde{a}_k (T_0-\tau)^\beta) G_k^\epsilon(\tau) d\tau e_k \right\|_{\mathscr{H}^r(\Omega)} \\
\leq& \frac{1}{\overline{\mathscr{M}}_{\beta,-}} \sqrt{\sum_{k=1}^{\infty} \widetilde{a}_k^{r+2} \int_0^{T_0} (T_0-\tau)^{2(\beta-1)} |E_{\beta,\beta}(-\widetilde{a}_k (T_0-\tau)^\beta)|^2 |G_k^\epsilon(\tau)|^2 d\tau}\ .
\end{aligned}$$

From Lemma 5.8 and the inequality $(a+b)^2 \geq 4ab$ for any $a, b \in \mathbb{R}$, we obtain

$$\begin{aligned}
&\left\| \sum_{k=1}^{\infty} \frac{\widetilde{a}_k}{\overline{\mathscr{M}}_{\beta,-}} \mathscr{F}_{\beta,k}^\epsilon(T_0) e_k \right\|_{\mathscr{H}^r(\Omega)} \\
\leq& \frac{1}{\overline{\mathscr{M}}_{\beta,-}} \sqrt{\sum_{k=1}^{\infty} \widetilde{a}_k^{r+2} \int_0^{T_0} (T_0-\tau)^{2(\beta-1)} \left(\frac{\mathscr{M}}{1+\widetilde{a}_k (T_0-\tau)^\beta} \right)^2 |G_k^\epsilon(\tau)|^2 d\tau} \\
\leq& \frac{\mathscr{M}}{2\overline{\mathscr{M}}_{\beta,-}} \sqrt{\sum_{k=1}^{\infty} \widetilde{a}_k^{r+2} \int_0^{T_0} (T_0-\tau)^{2(\beta-1)} \frac{1}{\widetilde{a}_k (T_0-\tau)^\beta} |G_k^\epsilon(\tau)|^2 d\tau} \\
\leq& \frac{\mathscr{M}}{2\overline{\mathscr{M}}_{\beta,-}} \|G^\epsilon\|_{L^\infty(0,T_0;\mathscr{H}^{r+1}(\Omega))} \sqrt{\int_0^{T_0} (T_0-\tau)^{\beta-2} d\tau} \\
=& \frac{\mathscr{M} T_0^{\frac{\beta-1}{2}}}{2\overline{\mathscr{M}}_{\beta,-} \sqrt{\beta-1}} \|G^\epsilon\|_{L^\infty(0,T_0;\mathscr{H}^{r+1}(\Omega))},
\end{aligned}$$

where we note that for $0 \leq t \leq T_0$ and

$$\widetilde{a}_k^{r+1} |G_k^\epsilon(t)|^2 \leq \sum_{k=1}^{\infty} \widetilde{a}_k^{r+1} \left| \left\langle G^\epsilon(t,\cdot), e_k \right\rangle \right|^2 \leq \|G^\epsilon\|_{L^\infty(0,T_0;\mathscr{H}^{r+1}(\Omega))}.$$

Thus

$$\|f^{\epsilon}_{n,\varrho}\|_{\mathscr{H}^r(\Omega)} \le \frac{1}{\mathscr{M}_{\beta,-}}\|h^{\epsilon}\|_{\mathscr{H}^{r+2}(\Omega)} + \frac{\mathscr{M} T_0^{\frac{\beta-1}{2}}}{2\mathscr{M}_{\beta,-}\sqrt{\beta-1}}\|G^{\epsilon}\|_{L^{\infty}(0,T_0;\mathscr{H}^{r+1}(\Omega))}.$$

5.2.4 Convergence Analysis and Error Estimate

In this subsection, we choose a regularization parameter $n := n(\epsilon)$ such that $\|u - u^{\epsilon}_{n,\varrho}\|_{L^2(\Omega)} \to 0$ as $\epsilon \to 0$, and we also consider the convergence analysis between the regularized solution $u^{\epsilon}_{n,\varrho}$ and the exact solution u.

Theorem 5.10 *Let $h \in L^2(\Omega)$ and $G \in L^1(0, T_0; L^2(\Omega))$. Assume the a priori bound condition $\|f\|_{\mathscr{H}^r(\Omega)} \le P$ holds. If we choose the regularization parameter*

$$n = \left\lfloor \left(\frac{P}{\epsilon}\right)^{\frac{4}{r+2}} \right\rfloor,$$

then we get the following error estimate between the exact solution and its regularization solution with noisy data

$$\|u^{\epsilon}_{n,\varrho} - u\|_{L^2(\Omega)} \le \mathscr{M}_{\beta,+} a^{\frac{1}{2}}\Big(1+\Lambda\Big) P^{\frac{2}{r+2}}\epsilon^{\frac{r}{r+2}} + \Lambda\epsilon$$
$$+ a^{-\frac{r}{4}} \frac{\mathscr{M}_{\beta,+}}{\mathscr{M}^{\frac{r}{2}}_{\beta,-}}\left(\frac{r}{4}\right)^{\frac{r}{4}} P^{\frac{2}{r+2}}\epsilon^{\frac{r}{r+2}},$$

where $\lfloor x \rfloor$ denotes the largest integer less than or equal to x.

Proof From the triangle inequality, we get

$$\|u - u^{\epsilon}_{n,\varrho}\|_{L^2(\Omega)} \le \|u_{n,\varrho} - u^{\epsilon}_{n,\varrho}\|_{L^2(\Omega)} + \|u - u_{n,\varrho}\|_{L^2(\Omega)}.$$

Using result (i) in Theorem 5.8:

$$\begin{aligned}\|u_{n,\varrho} - u^{\epsilon}_{n,\varrho}\|_{L^2(\Omega)} \le& \|u_{n,\varrho} - u^{\epsilon}_{n,\varrho}\|_{L^{\infty}(0,T_0;L^2(\Omega))} \\ \le& \mathscr{M}_{\beta,+} a^{\frac{1}{2}} n^{\frac{1}{2}}\|h^{\epsilon} - h\|_{L^2(\Omega)} \\ &+ \Big(\mathscr{M}_{\beta,+} a^{\frac{1}{2}} n^{\frac{1}{2}} + 1\Big)\Lambda\|G^{\epsilon} - G\|_{L^1(0,T_0;L^2(\Omega))} \\ \le& \mathscr{M}_{\beta,+} a^{\frac{1}{2}} n^{\frac{1}{2}}\epsilon\Big(1+\Lambda\Big) + \Lambda\epsilon.\end{aligned} \tag{5.52}$$

On the other hand, we have

$$u(t,x)=\sum_{k=1}^{\infty}\left[\frac{E_{\beta,1}(-\widetilde{a}_k t^{\beta})}{E_{\beta,1}(-\widetilde{a}_k T_0^{\beta})}\langle\Psi,e_k\rangle+\mathscr{F}_{\beta,k}(t)\right]e_k(x).$$

Note $\varrho\in(\frac{1}{2},1]$ and $0<aE^2_{\beta,1}(-\widetilde{a}_k T_0^{\beta})<1$, so it follows that

$$\begin{aligned}
&\|u-u_{n,\varrho}\|_{L^2(\Omega)}\\
=&\left\|\sum_{k=1}^{\infty}\frac{E_{\beta,1}(-\widetilde{a}_k t^{\beta})}{E_{\beta,1}(-\widetilde{a}_k T_0^{\beta})}\left(1-\left[1-\left(1-aE^2_{\beta,1}(-\widetilde{a}_k T_0^{\beta})\right)^n\right]^{\varrho}\right)\langle\Psi,e_k\rangle\, e_k(x)\right\|_{L^2(\Omega)}\\
\le&\left\|\sum_{k=1}^{\infty}\frac{E_{\beta,1}(-\widetilde{a}_k t^{\beta})}{E_{\beta,1}(-\widetilde{a}_k T_0^{\beta})}\left(1-aE^2_{\beta,1}(-\widetilde{a}_k T_0^{\beta})\right)^n\langle\Psi,e_k\rangle\, e_k(x)\right\|_{L^2(\Omega)}.
\end{aligned}$$

From the definition of f in (5.35) and using Lemma 5.8, we deduce that

$$\begin{aligned}
&\|u-u_{n,\varrho}\|_{L^2(\Omega)}\\
\le&\left\|\sum_{k=1}^{\infty}E_{\beta,1}(-\widetilde{a}_k t^{\beta})\left(1-aE^2_{\beta,1}(-\widetilde{a}_k T_0^{\beta})\right)^n\langle f,e_k\rangle\, e_k(x)\right\|_{L^2(\Omega)}\\
\le&\sqrt{\sum_{k=1}^{\infty}\left(\mathscr{M}_{\beta,+}\right)^2\left(1-aE^2_{\beta,1}(-\widetilde{a}_k T_0^{\beta})\right)^{2n}(\widetilde{a}_k)^{-r}(\widetilde{a}_k)^{r}\,|\langle f,e_k\rangle|^2}\\
\le&\sqrt{\sum_{k=1}^{\infty}\frac{\mathscr{M}^2_{\beta,+}}{\mathscr{M}^r_{\beta,-}}\left(1-aE^2_{\beta,1}(-\widetilde{a}_k T_0^{\beta})\right)^{2n}E^r_{\beta,1}(-\widetilde{a}_k T_0^{\beta})(\widetilde{a}_k)^{r}\,|\langle f,e_k\rangle|^2}\\
\le&\frac{\mathscr{M}_{\beta,+}}{\mathscr{M}^{\frac{r}{2}}_{\beta,-}}\sup_{\widetilde{a}_k>0}\left(1-aE^2_{\beta,1}(-\widetilde{a}_k T_0^{\beta})\right)^n E^{\frac{r}{2}}_{\beta,1}(-\widetilde{a}_k T_0^{\beta})\|f\|_{\mathscr{H}^r(\Omega)}.
\end{aligned}$$

Using Lemma 5.7, we have

$$\begin{aligned}
&\|u-u_{n,\varrho}\|_{L^2(\Omega)}\\
\le&a^{-\frac{r}{4}}\frac{\mathscr{M}_{\beta,+}}{\mathscr{M}^{\frac{r}{2}}_{\beta,-}}\sup_{\widetilde{a}_k>0}\left(1-aE^2_{\beta,1}(-\widetilde{a}_k T_0^{\beta})\right)^n\left(aE^2_{\beta,1}(-\widetilde{a}_k T_0^{\beta})\right)^{\frac{r}{4}}\|f\|_{\mathscr{H}^r(\Omega)}\\
\le&a^{-\frac{r}{4}}\frac{\mathscr{M}_{\beta,+}}{\mathscr{M}^{\frac{r}{2}}_{\beta,-}}\left(\frac{r}{4}\right)^{\frac{r}{4}}(n+1)^{-\frac{r}{4}}P.
\end{aligned}\tag{5.53}$$

Combining the above two inequalities (5.52) and (5.53), we obtain

$$\|u - u^{\epsilon}_{n,\varrho}\|_{L^2(\Omega)} \leq \mathscr{M}_{\beta,+} a^{\frac{1}{2}} n^{\frac{1}{2}} \epsilon \Big(1 + \Lambda\Big) + \Lambda\epsilon + a^{-\frac{r}{4}} \frac{\mathscr{M}_{\beta,+}}{\mathscr{M}^{\frac{r}{2}}_{\beta,-}} \left(\frac{r}{4}\right)^{\frac{r}{4}} (n+1)^{-\frac{r}{4}} P.$$

Choosing the regularization parameter $n := \lfloor (\frac{P}{\epsilon})^{\frac{4}{r+2}} \rfloor$, we obtain the error estimate

$$\begin{aligned}\|u - u^{\epsilon}_{n,\varrho}\|_{L^2(\Omega)} &\leq \mathscr{M}_{\beta,+} a^{\frac{1}{2}} \Big(1 + \Lambda\Big) P^{\frac{2}{r+2}} \epsilon^{\frac{r}{r+2}} + \Lambda\epsilon \\ &\quad + a^{-\frac{r}{4}} \frac{\mathscr{M}_{\beta,+}}{\mathscr{M}^{\frac{r}{2}}_{\beta,-}} \left(\frac{r}{4}\right)^{\frac{r}{4}} P^{\frac{2}{r+2}} \epsilon^{\frac{r}{r+2}}.\end{aligned}$$

In the above result, we obtained an error estimate between the exact solution and its regularization solution with noisy data by choosing the a priori parameter n, and this n depends on the noise level ϵ and the a priori bound condition P. Now, from results in Morozov's discrepancy principal [58], we choose the regularization parameter n by using an a posteriori choice rule.

The general a posteriori rule can be formulated as follows:

$$\|\mathscr{K} f^{\epsilon}_{n,\varrho} - \Psi^{\epsilon}\|_{L^2(\Omega)} \leq \eta\epsilon \leq \|\mathscr{K} f^{\epsilon}_{n-1,\varrho} - \Psi^{\epsilon}\|_{L^2(\Omega)}, \tag{5.54}$$

where $\|\Psi^{\epsilon}\|_{L^2(\Omega)} \geq \eta\epsilon$, η is a constant independent of ϵ, and $n > 0$ is the regularization parameter which makes (5.54) hold at the first iteration time.

Choosing $\eta > 1$, the following lemma gives a bound for n in terms of ϵ and P.

Lemma 5.12 *Let $\eta > 1 + \Lambda$ and n satisfies* (5.54). *Assume the a priori bound condition $\|f\|_{\mathscr{H}^r(\Omega)} \leq P$ holds. Then*

$$n \leq \frac{r+2}{4a} \left(\frac{1}{\mathscr{M}^{\frac{r}{2}}_{\beta,-}(\eta - \Lambda - 1)} \right)^{\frac{4}{r+2}} \left(\frac{P}{\epsilon}\right)^{\frac{4}{r+2}}.$$

Proof From the definition of n, we get

$$\begin{aligned}&\|\mathscr{K} f^{\epsilon}_{n-1,\varrho} - \Psi^{\epsilon}\|_{L^2(\Omega)} \\ &= \left\| \sum_{k=1}^{\infty} \left(\Big[1 - \Big(1 - aE^2_{\beta,1}(-\widetilde{a}_k T_0^{\beta})\Big)^{n-1}\Big]^{\varrho} - 1 \right) \langle \Psi^{\epsilon}, e_k \rangle e_k \right\|_{L^2(\Omega)} \\ &\leq \left\| \sum_{k=1}^{\infty} \left(\Big[1 - \Big(1 - aE^2_{\beta,1}(-\widetilde{a}_k T_0^{\beta})\Big)^{n-1}\Big]^{\varrho} - 1 \right) \langle \Psi^{\epsilon} - \Psi, e_k \rangle e_k \right\|_{L^2(\Omega)}\end{aligned}$$

$$+\left\|\sum_{k=1}^{\infty}\left(\left[1-\left(1-aE_{\beta,1}^2(-\widetilde{a}_kT_0^{\beta})\right)^{n-1}\right]^{\varrho}-1\right)\langle\Psi,e_k\rangle\,e_k\right\|_{L^2(\Omega)}.$$

Since $\varrho\in(\frac{1}{2},1]$ and $0<aE_{\beta,1}^2(-\widetilde{a}_kT_0^{\beta})<1$, we get

$$\begin{aligned}&\|\mathscr{K}f_{n-1,\varrho}^{\epsilon}-\Psi^{\epsilon}\|_{L^2(\Omega)}\\ \le&\left\|\Psi^{\epsilon}-\Psi\right\|_{L^2(\Omega)}+\left\|\sum_{k=1}^{\infty}\left(1-aE_{\beta,1}^2(-\widetilde{a}_kT_0^{\beta})\right)^{n-1}\langle\Psi,e_k\rangle\,e_k\right\|_{L^2(\Omega)}.\end{aligned}$$

Using Lemma 5.7, we have

$$\begin{aligned}&\left\|\sum_{k=1}^{\infty}\left(1-aE_{\beta,1}^2(-\widetilde{a}_kT_0^{\beta})\right)^{n-1}\langle\Psi,e_k\rangle\,e_k\right\|_{L^2(\Omega)}\\ \le&\sqrt{\sum_{k=1}^{\infty}\left(1-aE_{\beta,1}^2(-\widetilde{a}_kT_0^{\beta})\right)^{2(n-1)}(\widetilde{a}_k)^{-r}(\widetilde{a}_k)^{r}\,|\,\langle\Psi,e_k\rangle\,|^2}\\ \le&\sqrt{\sum_{k=1}^{\infty}\frac{1}{\mathscr{M}_{\beta,-}^{r}}\left(1-aE_{\beta,1}^2(-\widetilde{a}_kT_0^{\beta})\right)^{2(n-1)}E_{\beta,1}^{r+2}(-\widetilde{a}_kT_0^{\beta})(\widetilde{a}_k)^{r}\,|\,\langle f,e_k\rangle\,|^2}\\ \le&\frac{a^{-\frac{r+2}{4}}}{\mathscr{M}_{\beta,-}^{\frac{r}{2}}}\sup_{\widetilde{a}_k>0}\left(1-aE_{\beta,1}^2(-\widetilde{a}_kT_0^{\beta})\right)^{n-1}\left(aE_{\beta,1}^2(-\widetilde{a}_kT_0^{\beta})\right)^{\frac{r+2}{4}}\|f\|_{\mathscr{H}^r(\Omega)}\\ \le&\frac{a^{-\frac{r+2}{4}}}{\mathscr{M}_{\beta,-}^{\frac{r}{2}}}\left(\frac{r+2}{4}\right)^{\frac{r+2}{4}}n^{-\frac{r+2}{4}}\|f\|_{\mathscr{H}^r(\Omega)}.\end{aligned}$$

Using Lemma 5.11, we get

$$\begin{aligned}\|\mathscr{K}f_{n-1,\varrho}^{\epsilon}-\Psi^{\epsilon}\|_{L^2(\Omega)}\le&\|h^{\epsilon}-h\|_{L^2(\Omega)}+\Lambda\|G^{\epsilon}-G\|_{L^1(0,T_0;L^2(\Omega))}\\ &+\frac{a^{-\frac{r+2}{4}}}{\mathscr{M}_{\beta,-}^{\frac{r}{2}}}\left(\frac{r+2}{4}\right)^{\frac{r+2}{4}}n^{-\frac{r+2}{4}}\|f\|_{\mathscr{H}^r(\Omega)}.\end{aligned}$$

This implies that

$$\eta\epsilon\le(1+\Lambda)\epsilon+\frac{1}{\mathscr{M}_{\beta,-}^{\frac{r}{2}}}\left(\frac{r+2}{4a}\right)^{\frac{r+2}{4}}n^{-\frac{r+2}{4}}P.$$

Thus

$$n \leq \frac{r+2}{4a}\left(\frac{1}{\overline{\mathscr{M}}^{\frac{r}{2}}_{\beta,-}(\eta-\Lambda-1)}\right)^{\frac{4}{r+2}}\left(\frac{P}{\epsilon}\right)^{\frac{4}{r+2}}.$$

Theorem 5.11 *Let n be as in Lemma 5.12. Assume the a priori bound condition $\|f\|_{\mathscr{H}^r(\Omega)} \leq P$ holds. Then*

$$\|u-u^\epsilon_{n,\varrho}\|_{L^2(\Omega)} \leq \Big(I_1+I_2\Big)\epsilon^{\frac{r}{r+2}}P^{\frac{2}{r+2}}+\Lambda\epsilon,$$

where

$$I_1 := \mathscr{M}_{\beta,+}a^{\frac{1}{2}}\Big(1+\Lambda\Big)\left(\frac{r+2}{4a}\right)^{\frac{1}{2}}\left(\frac{1}{\overline{\mathscr{M}}^{\frac{r}{2}}_{\beta,-}(\eta-\Lambda-1)}\right)^{\frac{2}{r+2}},$$

$$I_2 := \mathscr{M}_{\beta,+}\left(\frac{\eta+1+\Lambda}{\overline{\mathscr{M}}_{\beta,-}}\right)^{\frac{r}{r+2}}.$$

Proof From the triangle inequality, we get

$$\|u-u^\epsilon_{n,\varrho}\|_{L^2(\Omega)} \leq \|u_{n,\varrho}-u^\epsilon_{n,\varrho}\|_{L^2(\Omega)}+\|u-u_{n,\varrho}\|_{L^2(\Omega)}.$$

Using the result of Theorem 5.10, we obtain

$$\|u_{n,\varrho}-u^\epsilon_{n,\varrho}\|_{L^2(\Omega)} \leq \mathscr{M}_{\beta,+}a^{\frac{1}{2}}n^{\frac{1}{2}}\epsilon(1+\Lambda)+\Lambda\epsilon.$$

From Lemma 5.12, we deduce that

$$\begin{aligned}
&\|u_{n,\varrho}-u^\epsilon_{n,\varrho}\|_{L^2(\Omega)}\\
\leq&\mathscr{M}_{\beta,+}a^{\frac{1}{2}}n^{\frac{1}{2}}\epsilon\Big(1+\Lambda\Big)+\Lambda\epsilon\\
\leq&\mathscr{M}_{\beta,+}a^{\frac{1}{2}}\Big(1+\Lambda\Big)\left(\frac{r+2}{4a}\right)^{\frac{1}{2}}\left(\frac{1}{\overline{\mathscr{M}}^{\frac{r}{2}}_{\beta,-}(\eta-\Lambda-1)}\right)^{\frac{2}{r+2}}P^{\frac{2}{r+2}}\epsilon^{\frac{r}{r+2}}+\Lambda\epsilon.
\end{aligned}$$

Using the Hölder's inequality, we get

$$\begin{aligned}
&\|u-u_{n,\varrho}\|_{L^2(\Omega)}\\
\leq&\left\|\sum_{k=1}^{\infty}\frac{E_{\beta,1}(-\widetilde{a}_k t^\beta)}{E_{\beta,1}(-\widetilde{a}_k T_0^\beta)}\left(1-\left[1-\left(1-aE^2_{\beta,1}(-\widetilde{a}_k T_0^\beta)\right)^n\right]^\varrho\right)\langle\Psi,e_k\rangle\, e_k(x)\right\|_{L^2(\Omega)}
\end{aligned}$$

$$
\begin{aligned}
&\leq \mathscr{M}_{\beta,+}\Bigg\|\sum_{k=1}^{\infty}\frac{1}{E_{\beta,1}(-\widetilde{a}_k T_0^{\beta})} \\
&\times\left(1-\left[1-\left(1-aE_{\beta,1}^2(-\widetilde{a}_k T_0^{\beta})\right)^n\right]^{\varrho}\right)\langle\Psi, e_k\rangle\, e_k(x)\Bigg\|_{L^2(\Omega)}^{\frac{r}{r+2}} \\
&\times\left\|\sum_{k=1}^{\infty}\left(1-\left[1-\left(1-aE_{\beta,1}^2(-\widetilde{a}_k T_0^{\beta})\right)^n\right]^{\varrho}\right)\langle f, e_k\rangle\, e_k(x)\right\|_{L^2(\Omega)}^{\frac{2}{r+2}}.
\end{aligned}
$$

On the other hand, since $\varrho \in (\frac{1}{2}, 1]$ and $0 < aE_{\beta,1}^2(-\widetilde{a}_k T_0^{\beta}) < 1$, we deduce that

$$
\begin{aligned}
&\left\|\sum_{k=1}^{\infty}\left(1-\left[1-\left(1-aE_{\beta,1}^2(-\widetilde{a}_k T_0^{\beta})\right)^n\right]^{\varrho}\right)\langle f, e_k\rangle\, e_k(x)\right\|_{L^2(\Omega)}^{\frac{2}{r+2}} \\
&\leq\left\|\sum_{k=1}^{\infty}\widetilde{a}_k^{-\frac{r}{2}}\widetilde{a}_k^{\frac{r}{2}}\langle f, e_k\rangle\, e_k(x)\right\|_{L^2(\Omega)}^{\frac{2}{r+2}} \\
&\leq \sup_{\widetilde{a}_k>0}\left(\frac{1}{\widetilde{a}_k}\right)^{\frac{r}{r+2}}\|f\|_{\mathscr{H}^r(\Omega)}^{\frac{2}{r+2}}.
\end{aligned}
$$

This implies that

$$
\begin{aligned}
&\|u-u_{n,\varrho}\|_{L^2(\Omega)} \\
&\leq \mathscr{M}_{\beta,+}\sup_{\widetilde{a}_k>0}\left(\frac{1}{\widetilde{a}_k E_{\beta,1}(-\widetilde{a}_k T_0^{\beta})}\right)^{\frac{r}{r+2}} P^{\frac{2}{r+2}}\Bigg(\left\|\sum_{k=1}^{\infty}\left\langle\Psi-\Psi^{\epsilon}, e_k\right\rangle e_k(x)\right\|_{L^2(\Omega)} \\
&\quad+\left\|\sum_{k=1}^{\infty}\left(1-\left[1-\left(1-aE_{\beta,1}^2(-\widetilde{a}_k T_0^{\beta})\right)^n\right]^{\varrho}\right)\left\langle\Psi^{\epsilon}, e_k\right\rangle e_k(x)\right\|_{L^2(\Omega)}\Bigg)^{\frac{r}{r+2}} \\
&\leq \mathscr{M}_{\beta,+}\sup_{\widetilde{a}_k>0}\left(\frac{1}{\widetilde{a}_k E_{\beta,1}(-\widetilde{a}_k T_0^{\beta})}\right)^{\frac{r}{r+2}}\left(\eta+1+\Lambda\right)^{\frac{r}{r+2}}\epsilon^{\frac{r}{r+2}}P^{\frac{2}{r+2}} \\
&\leq \mathscr{M}_{\beta,+}\left(\frac{\eta+1+\Lambda}{\overline{\mathscr{M}}_{\beta,-}}\right)^{\frac{r}{r+2}}\epsilon^{\frac{r}{r+2}}P^{\frac{2}{r+2}}.
\end{aligned}
$$

From the above arguments, we deduce the desired inequality.

5.3 Terminal Value Problem

5.3.1 Introduction

5.3.1.1 Statement of the Problem

In this section, we consider the following fractional wave equation:

$$\begin{cases} \partial_t^\alpha u(x,t) = -\mathscr{A}u(x,t) + G(t,u(x,t)), & x \in \Omega,\ 0 < t \le T, \\ u(x,t) = 0, & x \in \partial\Omega,\ 0 < t < T, \\ u_t(x,0) = 0, & x \in \Omega, \end{cases} \tag{5.55}$$

where G is called a source function which will be defined later. The time fractional derivative ∂_t^α, $1 < \alpha < 2$, is understood as the left-sided Caputo fractional derivative of order α with respect to t, which is defined by

$$\partial_t^\alpha v(x,t) = \frac{1}{\Gamma(2-\alpha)} \int_0^t (t-s)^{1-\alpha} \frac{\partial^2}{\partial s^2} v(x,s)ds,$$

where $\Gamma(\cdot)$ is the Gamma function. For $\alpha = 2$, we recover the usual time derivative of second order $\frac{\partial^2}{\partial t^2}$. Let us assume that Ω is a non-empty open set and possesses a Lipschitz continuous boundary in $\mathbb{R}^N$, $N \ge 1$, $T > 0$, and let $\mathscr{A}$ be a symmetric and uniformly elliptic operator on Ω defined by

$$\mathscr{A}v(x) = -\sum_{m=1}^N \frac{\partial}{\partial x_m} \left(\sum_{n=1}^N a_{mn}(x) \frac{\partial}{\partial x_n} v(x) \right) + q(x)v(x), \quad x \in \overline{\Omega},$$

where $a_{ij} \in C^1(\overline{\Omega})$, $q \in C(\overline{\Omega}; [0,+\infty))$, and $a_{mn} = a_{nm}$, $1 \le m,n \le N$. We also assume that there exists a constant $b_0 > 0$ such that, for $x \in \overline{\Omega}$, $y = (y_1, y_2, ..., y_N) \in \mathbb{R}^N$,

$$\sum_{1 \le m,n \le N} a_{mn}(x) y_m y_n \ge b_0 |y|^2.$$

This section considers the inverse problem of determining the initial value $u(x,0) = u_0(x)$ from its final value $u(x,T)$. We focus to study existence, uniqueness, and regularity of mild solutions of the problem (5.55) associated with the final value condition

$$u(x,T) = f(x), \quad x \in \Omega, \tag{5.56}$$

where f belongs to an appropriate space.

The study of (5.55) is mainly motivated by problems arising in anomalous diffusion phenomena. Anomalous diffusion and wave equations are of great interest in physics. They are frequently used for the superdiffusive models of anomalous diffusion such as diffusion in heterogeneous media. These fractional differential equations have another important issue in the probability theory related to non-Markovian diffusion processes with memory. Fractional wave equation also describes evolution processes intermediate between diffusion and wave propagation [150–152]. In [151], it has been shown that the fractional wave equation governs the propagation of mechanical diffusion waves in viscoelastic media. Such waves are relevant in acoustics, seismology, medical imaging, etc. The physical background for a time-space fractional diffusion-wave equation can be seen in [35].

5.3.1.2 Motivations

If the condition (5.56) is replaced by

$$u(x,0) = \bar{f}(x), \quad x \in \Omega, \tag{5.57}$$

then we have the direct problem or initial value problem (IVP) of (5.55). Some quasi-linear equations of the form (5.55) and (5.57) with standard time derivative ($\alpha = 2$) have been extensively studied in the published literature. The global well-posedness has been proved both in the subcritical case by Ginibre and Velo [70] and in the critical case by Grillarkis [74], Shatah and Struwe [193, 194], and references therein.

In fractional derivative cases, such as the Caputo or Riemann–Liouville derivative, the problem (5.55) and (5.57) has been considered with $G = 0$ or $G = G(x,t)$ by some authors, see, e.g., [26, 52, 55, 126, 180, 187] and also [61, 76, 125].

To our knowledge, the study of the initial value problem for the fractional wave equation in the nonlinear case is still limited. Recently, Kian and Yamamoto [112] studied the problem (5.55) and (5.57) with an inhomogeneous source, i.e., $G = G(x,t)$, and then further investigated local solutions with a nonlinear source. Warma et al. [10] considered the existence and regularity of local and global weak solutions with a suitable growth assumption on the nonlinearity G. Very recently, the authors have studied the uniqueness of inverse problems for a fractional equation with a single measurement in [113].

In practice, there are some physical models which are not subjected to initial value problem. Some phenomena cannot be observed at the time $t = 0$ and only can be measured at a terminal time $t = T > 0$. Hence, a final value condition appears instead of the respectively initial value one. It has great importance in engineering areas and aimed at detecting the previous state of a physical field from its present information. In a few sentences, we explain the presence of the equation $u_t(x,0) = 0$. By Yang and Liu [232], the system (5.55) and (5.56) in the two-dimensional case can be considered as description for an imaging process, namely, to recover an exact picture from its blurry form. The condition $u_t(x,0) = 0$ means that the distribution does not change on the interval $(0, t_0)$ when t_0 is near zero. Therefore, the necessity of studying terminal value problems or final value problems (FVPs) or backward problems is out of any doubt.

The FVP (5.55) and (5.56) with derivatives of integer orders has been treated for a long time, e.g., see [16, 29, 195]. In [29], Carasso considered the following final value problem for classical wave equation (i.e., $\alpha = 2$)

$$\begin{cases} u_{tt} = (\Delta + k)^2 u, & x \in \Omega,\ 0 < t < T, \\ u = \Delta u = 0, & x \in \partial\Omega,\ 0 < t < T, \\ u_t(x, 0) = g(x), & x \in \Omega, \\ u(x, T) = f(x), & x \in \Omega, \end{cases}$$

where k is a given positive number which may equal several eigenvalues of $-\Delta$, and f, g are given functions. Up to date, little research has been done on the inverse problems of time-space fractional diffusion equations. FVPs for fractional PDEs can be roughly divided into two topics. The first one contains problems related to the ill-posedness and proposes some regularization methods for approximating a sought solution. We can list some well-known results, for example, Jia et al. [101], Wang and Liu [220], some papers of Yamamoto and his group, see [102, 127, 130, 145, 166], Kaltenbacher and Rundell [107, 108], Rundell and Zhang [183, 184], Janno and Kinash, see [97, 98], etc. The second topic contains problems concerning the existence and regularity of solutions such as [147]. Investigating the existence and regularity of solutions of ODE/PDE models plays an important role in both the development of the ODE/PDE theory and their applications in real-life problems. Furthermore, studying regularity helps to improve the smoothness and stability of solutions in different spaces and hence makes the numerical simulations valuable. This second topic has not been treated well in the literature.

As far as we know, there are only a few works analyzing the problem (5.55) and (5.56), which provides existence, uniqueness, and some regularity estimates. The main difficulty in the analysis of the problem (5.55) and (5.56) and the essential difference from the traditional problems come from the nonlocality of the time fractional derivative ∂_t^α. The major question for this work in our mind is: What is the regularity of the corresponding solution u (output data) if the given data (input data) f, G are regular?

Our goal in this section is to find suitable Banach spaces for the given data (f, G) in order to obtain regularity results for the corresponding solution. The regularity estimates are important in the analysis of time discretization schemes for the problems (5.55) and (5.56) in the future.

The difficulties of a final value problem can be briefly described as follows. Firstly, since the fractional derivative ∂_t^α is nonlocally defined on the time interval $(0, t)$, we cannot convert a final value problem for fractional wave equation into an initial value problem by using some substitution methods. Secondly, the formulation of mild solutions of a final value problem is more complex than the corresponding initial problem. This positively promotes us to construct new solution techniques to deal with problem (5.55) and (5.56). Some more details can be found in Sect. 5.3.3, where the explicit representation of solutions relies on the eigenfunctions expansion and the Mittag–Leffler functions.

Let us describe the main results of this section in two cases as follows. The first case is related to the properties of solutions under a globally Lipschitz (GL) assumption on the nonlinearity corresponding to two theorems, while the second one concerns a critical nonlinearity corresponding to the third theorem. In the first theorem, we obtain the regularity results of solutions and their derivatives of first and fractional orders under the (GL) assumption $(\mathscr{H}_1)$. The key idea is based on a Picard iteration argument and techniques to find appropriate spaces for f. Choosing spaces of f and G is a difficult and nontrivial task when we study the regularity of the solution. Although applications of our problem under $(\mathscr{H}_1)$ are not wide, the analysis and techniques here are helpful tools to study the next result. Moreover, the existence of a mild solution in the space L^∞ may not be obtained by considering $(\mathscr{H}_1)$. This can be overcome by considering the (GL) assumption $(\mathscr{H}_2)$ of the nonlinearity which is presented in the second theorem. The third theorem uses the contraction mapping principle to prove the existence of a mild solution in the critical case. As we know, nonlinear PDEs with critical nonlinearities are an interesting topic. We can mention [47] and references therein. Studying the IVP for (5.55) in the critical case is also a challenging problem. Therefore, investigating the regularity of the mild solution and its derivatives is very difficult.

The outline of this section is as follows. In Sect. 5.3.2, we introduce some terminology used throughout this work. Moreover, we obtain a precise representation of solutions by using the Mittag–Leffler functions. In Sect. 5.3.3, we investigate the well-posedness, and regularity of the mild solutions of the problem (5.55) and (5.56). Three main results on the existence, uniqueness (in some suitable class of functions), regularity of the mild solution, and its derivative are proved under some suitable assumptions on the terminal data and nonlinearity. In Sect. 5.3.4, we apply the theoretical results to some typical time fractional diffusion equation: time fractional Ginzburg–Landau equation and Burger equation. Finally, in Sect. 5.3.5, we provide full proofs of the main theorems established in Sect. 5.3.3.

5.3.2 Preliminaries

In this subsection we will recall some properties that will be useful for the study of the well-posedness of the problem (5.55) and (5.56). We start by introducing some functional spaces. Then we will recall some properties of Mittag–Leffler functions. Let the operator $\mathscr{L}$ be considered on $L^2(\Omega)$ with respect to domain $D(\mathscr{L}) = W_0^{1,2}(\Omega) \cap W^{2,2}(\Omega)$, where $L^2(\Omega)$, $W_0^{1,2}(\Omega)$, $W^{2,2}(\Omega)$ are usual Sobolev spaces. Then the spectrum of $\mathscr{L}$ is a nondecreasing sequence of positive real numbers $\{\lambda_j\}_{j=1,2,\ldots}$ satisfying $\lim_{j\to\infty} \lambda_j = \infty$. Moreover, there exists a positive constant $c_{\mathscr{L}}$ such that $\lambda_j \geq c_{\mathscr{L}} j^{2/d}$, for all $j \geq 1$, see [42]. Let us denote by $\{\varphi_j\}_{j=1,2,...} \subset D(\mathscr{L})$ the set of eigenfunctions of $\mathscr{L}$, i.e., $\mathscr{L}\varphi_j = \lambda_j \varphi_j$, and $\varphi_j = 0$ on $\partial\Omega$, for all $j \geq 1$. The sequence $\{\varphi_k\}_{k=1,2,...}$ forms an orthonormal basis of $L^2(\Omega)$, see, e.g., [109]. For a given real number $\gamma \geq 0$, the Hilbert scale space

$$\mathbb{H}^\gamma(\Omega) := \left\{ v \in L^2(\Omega) : \sum_{j=1}^{\infty} \lambda_j^{2\gamma} |\langle v, \varphi_j \rangle|^2 < \infty \right\}$$

endows with the norm

$$\|v\|^2_{\mathbb{H}^\gamma(\Omega)} := \sum_{j=1}^{\infty} \lambda_j^{2\gamma} \left|\langle v, \varphi_j \rangle\right|^2,$$

where $\langle \cdot, \cdot \rangle$ is the usual product of $L^2(\Omega)$. We have $\mathbb{H}^0(\Omega) = L^2(\Omega)$ if $\gamma = 0$, and $\mathbb{H}^{\frac{1}{2}}(\Omega) = W_0^{1,2}(\Omega)$. We denote by $\mathbb{H}^{-\gamma}(\Omega)$ the dual space of $\mathbb{H}^\gamma(\Omega)$ provided that the dual space of $L^2(\Omega)$ is identified with itself, e.g., see [159]. The space $\mathbb{H}^{-\gamma}(\Omega)$ is a Hilbert space with respect to the norm $\|v\|^2_{\mathbb{H}^{-\gamma}(\Omega)} = \sum_{j=1}^{\infty} \lambda_j^{-2\gamma} \langle v, \varphi_j \rangle^2_{-\gamma,\gamma}$, for $v \in \mathbb{H}^{-\gamma}(\Omega)$, where $\langle \cdot, \cdot \rangle_{-\gamma,\gamma}$ is the dual product between $\mathbb{H}^{-\gamma}(\Omega)$ and $\mathbb{H}^\gamma(\Omega)$. We note that

$$\langle \tilde{v}, v \rangle_{-\gamma,\gamma} = \langle \tilde{v}, v \rangle, \quad \text{for } \tilde{v} \in L^2(\Omega), v \in \mathbb{H}^\gamma(\Omega). \tag{5.58}$$

By identifying $L^2(\Omega)$ with its dual space, and making use of the inclusion $\mathbb{H}^\gamma(\Omega) \hookrightarrow L^2(\Omega)$, the embedding $\mathbb{H}^\gamma(\Omega) \hookrightarrow L^2(\Omega) \hookrightarrow \mathbb{H}^{-\gamma}(\Omega)$ holds for $\gamma \geq 0$. Hence, it is suitable to call the space $\mathbb{H}^s(\Omega)$, $s \in \mathbb{R}$, by a Hilbert scale space. For given numbers $p \geq 1$ and $\nu \in \mathbb{R}$, let $L^p(0, T; \mathbb{H}^\nu(\Omega))$ be the space of all functions $w : (0, T) \to \mathbb{H}^\nu(\Omega)$ such that

$$\|w\|_{L^p(0,T;\mathbb{H}^\nu(\Omega))} := \left(\int_0^T \|w(t)\|^p_{\mathbb{H}^\nu(\Omega)} dt \right)^{1/p} < \infty.$$

We denote by $C([0, T]; \mathbb{H}^\nu(\Omega))$ the space of all continuous functions from $[0, T]$ to $\mathbb{H}^\nu(\Omega)$ corresponding to the usual supremum norm $\|w\|_{C([0,T];\mathbb{H}^\nu(\Omega))} := \sup_{0 \leq t \leq T} \|w(t)\|_{\mathbb{H}^\nu(\Omega)}$; and denote by $C^\delta([0, T]; \mathbb{H}^\nu(\Omega))$, $\delta \in (0, 1)$, the space of all Hölder continuous functions from $[0, T]$ to $\mathbb{H}^\nu(\Omega)$ with exponent δ, namely, $w \in C([0, T]; \mathbb{H}^\nu(\Omega))$ satisfies that

$$\|w\|_{C^\delta([0,T];\mathbb{H}^\nu(\Omega))} := \sup_{0 \leq t, s \leq T, t \neq s} \frac{\|w(t) - w(s)\|_{\mathbb{H}^\nu(\Omega)}}{|t - s|^\delta} < \infty.$$

Let us denote by $C((0, T]; \mathbb{H}^\nu(\Omega))$ the set of all continuous functions which map $(0, T]$ into $\mathbb{H}^\nu(\Omega)$. For a given number $\eta > 0$, we denote by $C^\eta((0, T]; \mathbb{H}^\nu(\Omega))$

the space of all functions w in $C((0,T];\mathbb{H}^\nu(\Omega))$ such that $\|w\|_{C^\eta((0,T];\mathbb{H}^\nu(\Omega))} := \sup_{0<t\le T} t^\eta \|w(t)\|_{\mathbb{H}^\nu(\Omega)} < \infty$, see [43].

We recall some Sobolev embeddings as follows. Let Ω be a non-empty open set with a Lipschitz continuous boundary in $\mathbb{R}^N$, $N \ge 1$. Let us recall that the notation $W^{s,p}(\Omega)$, $s \in \{0, 1, 2, ...\}$, $p \ge 1$, denotes by the standard Sobolev space, e.g., see [3]. In the case $0 \le s \le 1$, the intermediary space $W^{s,p}(\Omega) = \big[L^p(\Omega); W^{1,p}(\Omega)\big]_s$ can be defined by

$$W^{s,p}(\Omega) = \left\{ u \in L^p(\Omega) : \frac{|u(x) - u(x')|}{|x - x'|^{\frac{N}{p}+s}} \in L^p(\Omega \times \Omega) \right\}.$$

Since Ω is a non-empty open set and possesses a Lipschitz continuous boundary in $\mathbb{R}^N$, then the following Sobolev embedding holds

$$W^{\sigma,p}(\Omega) \hookrightarrow W^{\gamma,q}(\Omega) \quad \text{if} \quad \begin{cases} 1 \le p, q < \infty, \\ 0 \le \gamma \le \sigma < \infty, \\ \sigma - \gamma \ge \dfrac{N}{p} - \dfrac{N}{q}. \end{cases} \tag{5.59}$$

By letting $p = 2$, $\gamma = 0$ in (5.59), one obtains that $W^{\sigma,2}(\Omega) \hookrightarrow L^q(\Omega)$ with $1 \le q < \infty$, $0 \le \sigma < \infty$, and $\sigma \ge \frac{N}{2} - \frac{N}{q}$. Henceforth, setting $0 \le \sigma < \frac{N}{2}$ infers that $1 \le q \le \frac{2N}{N-2\sigma}$. Summarily, we obtain the following embedding:

$$W^{\sigma_1,2}(\Omega) \hookrightarrow L^{q_1}(\Omega) \quad \text{if} \quad 0 \le \sigma_1 < \frac{N}{2},\ 1 \le q_1 \le \frac{2N}{N - 2\sigma_1}. \tag{5.60}$$

This implies $W_0^{\sigma_1,2}(\Omega) \hookrightarrow L^{q_1}(\Omega)$, and so $L^{q_1^*}(\Omega) = [L^{q_1}(\Omega)]^* \hookrightarrow [W_0^{\sigma_1,2}(\Omega)]^* = W^{-\sigma_1,2}(\Omega)$ with respect to $-\frac{N}{2} < -\sigma_1 \le 0$ and $q_1^* \ge (\frac{2N}{N-2\sigma_1})\{(\frac{2N}{N-2\sigma_1}) - 1\}^{-1} = \frac{2N}{N+2\sigma_1}$. Thus,

$$L^{q_2}(\Omega) \hookrightarrow W^{\sigma_2,2}(\Omega) \quad \text{if} \quad -\frac{N}{2} < \sigma_2 \le 0,\ q_2 \ge \frac{2N}{N - 2\sigma_2}. \tag{5.61}$$

On the other hand, the Hilbert scale spaces and the fractional Sobolev space are related to each other by the following embeddings:

$$\mathbb{H}^s(\Omega) \hookrightarrow W^{2s,2}(\Omega) \hookrightarrow L^2(\Omega), \quad \text{if} \quad s \ge 0. \tag{5.62}$$

Let $E_{\alpha,\beta}(z)$ be the Mittag–Leffler function as in Definition 1.7. In this section, we always consider T satisfies the following assumption (5.63).

Lemma 5.13 ([226]) *Given $1 < \alpha < 2$, if the number T is large enough, then*

$$E_{\alpha,1}(-\lambda_j T^\alpha) \neq 0, \quad \textit{for all } j \in \mathbb{N}^+, \tag{5.63}$$

and there exist two positive constants m_α and M_α such that

$$\frac{m_\alpha}{1+\lambda_j T^\alpha} \leq \left| E_{\alpha,1}(-\lambda_j T^\alpha) \right| \leq \frac{M_\alpha}{1+\lambda_j T^\alpha}.$$

Lemma 5.14 *Given $1 < \alpha < 2$ and $0 < \theta < 1$, for $j \in \mathbb{N}^+$ and $t \in (0, T]$, there hold that:*

(a) $z^{\alpha-1} E_{\alpha,\alpha}(-\lambda_j z^\alpha) \leq M_\alpha \lambda_j^{-\theta} z^{\alpha(1-\theta)-1}$.

(b) $\mathscr{E}_{\alpha,T}(-\lambda_j t^\alpha) := \dfrac{E_{\alpha,1}(-\lambda_j t^\alpha)}{E_{\alpha,1}(-\lambda_j T^\alpha)} \leq M_\alpha m_\alpha^{-1} T^{\alpha(1-\theta)} \Big(T^{\alpha\theta} + \lambda_1^{-\theta}\Big) \lambda_j^\theta t^{-\alpha(1-\theta)}$.

Proof The first inequality is estimated as follows:

$$\begin{aligned}
z^{\alpha-1} E_{\alpha,\alpha}(-\lambda_j z^\alpha) &\leq M_\alpha z^{\alpha-1} \frac{1}{1+\lambda_j z^\alpha} \\
&= M_\alpha z^{\alpha-1} \Big(\frac{1}{1+\lambda_j z^\alpha}\Big)^\theta \Big(\frac{1}{1+\lambda_j z^\alpha}\Big)^{1-\theta} \\
&\leq M_\alpha \lambda_j^{-\theta} z^{\alpha(1-\theta)-1},
\end{aligned} \tag{5.64}$$

and the second inequality is shown as follows:

$$\begin{aligned}
\mathscr{E}_{\alpha,T}(-\lambda_j t^\alpha) = \left| \frac{E_{\alpha,1}(-\lambda_j t^\alpha)}{E_{\alpha,1}(-\lambda_j T^\alpha)} \right| &\leq \frac{M_\alpha}{m_\alpha} \left(\frac{1+\lambda_j T^\alpha}{1+\lambda_j t^\alpha}\right)^{1-\theta} \left(\frac{1+\lambda_j T^\alpha}{1+\lambda_j t^\alpha}\right)^\theta \\
&\leq \frac{M_\alpha}{m_\alpha} T^{\alpha(1-\theta)} (1+\lambda_j T^\alpha)^\theta t^{-\alpha(1-\theta)} \\
&\leq M_\alpha m_\alpha^{-1} T^{\alpha(1-\theta)} \Big(T^{\alpha\theta} + \lambda_1^{-\theta}\Big) \lambda_j^\theta t^{-\alpha(1-\theta)}.
\end{aligned}$$

The proof is completed.

From now on, we will use $a \lesssim b$ to denote the existence of a constant $C > 0$, which may depend only on α, T such that $a \leq Cb$.

5.3.3 Existence and Regularity

5.3.3.1 Mild Solutions

Solutions of partial differential equations can be considered in the classical, weak, or mild sense. In this chapter, we will study mild solutions of FVP (5.55)–(5.56).

There are many works considering the precise formulation of mild solutions to IPVs for time fractional wave equation, such as [10, 35, 46, 112, 114, 128, 150, 172, 188], by using complex integral representations on Banach spaces or spectral representations on Hilbert scale spaces of the Mittag–Leffler operators. In studying FVPs for time fractional wave equation, the precise formulation of mild solutions can be derived by using spectral representations of the inverse Mittag–Leffler operators, such as [48, 97, 101, 107, 145, 184, 208, 220]. In what follows, we give a definition of mild solutions to FVP (5.55)–(5.56) where the precise formulation can be obtained by some simple computations.

In extra, for a given two-variables function $w = w(x, t)$, we will write $w(t)$ instead of $w(\cdot, t)$ and understand $w(t)$ as a function of the spatial variable x.

Definition 5.2 A function u in $L^p(0, T; \mathbb{H}^\nu(\Omega))$ or $C^\eta((0, T]; \mathbb{H}^\nu(\Omega))$ (with some suitable numbers $p \geq 1$, $\nu \geq 0$ and $\eta > 0$) is called a mild solution of the problem (5.55)–(5.56) if it satisfies the following equation:

$$
\begin{aligned}
u(t) =& \mathbf{B}_\alpha(t, T)f + \int_0^t \mathbf{P}_\alpha(t-r)G(r, u(r))dr \\
&- \int_0^T \mathbf{B}_\alpha(t, T)\mathbf{P}_\alpha(T-r)G(r, u(r))dr
\end{aligned}
\tag{5.65}
$$

in the sense of $\mathbb{H}^\nu(\Omega)$, where, for $0 \leq t \leq T$, the solution operators $\mathbf{B}_\alpha$, $\mathbf{P}_\alpha$ are given by

$$
\begin{aligned}
\mathbf{B}_\alpha(t, T)v &:= \sum_{j=1}^{\infty} \frac{E_{\alpha,1}(-\lambda_j t^\alpha)}{E_{\alpha,1}(-\lambda_j T^\alpha)} \langle v, \varphi_j \rangle \varphi_j, \\
\mathbf{P}_\alpha(t)v &:= \sum_{j=1}^{\infty} t^{\alpha-1} E_{\alpha,\alpha}(-\lambda_j t^\alpha) \langle v, \varphi_j \rangle \varphi_j,
\end{aligned}
\tag{5.66}
$$

where $v = \sum_{j=1}^{\infty} \langle v, \varphi_j \rangle \varphi_j$.

Remark 5.3 A mild formulation of this IVP (5.55), (5.57) is given by

$$
u(t) = \mathscr{B}_\alpha(t)u_0 + \int_0^t \mathbf{P}_\alpha(t-r)G(r, u(r))dr, \tag{5.67}
$$

where $\mathscr{B}_\alpha(t)w := \sum_{j=1}^{\infty} E_{\alpha,1}(-\lambda_j t^\alpha)\langle w, \varphi_j \rangle \varphi_j$, see [35, 46, 112, 114, 134, 150, 172, 188], etc. It should be pointed out some core differences between the IVP (5.55), (5.57), and the FVP (5.55)–(5.56) for fractional wave equation as what follows.

- The solution operator $\mathbf{B}_\alpha(t, T)$ is weaker than $\mathscr{B}_\alpha(t)$. Indeed, one can see that if $v \in L^2(\Omega)$, then $\mathscr{B}_\alpha(t)v \in L^\infty(0, T; L^2(\Omega))$ and

$$\mathbf{B}_\alpha(t,T)v \notin L^\infty(0,T;L^2(\Omega)) \cup C([0,T];L^2(\Omega)).$$

Therefore, it is actually difficult to establish the existence of mild solutions, especially in the critical nonlinear case.

- Mild formulation of the FVP (5.55)–(5.56) contains more terms than the IVP (5.55), (5.57). In particular, estimating the last term of (5.65) requires very clever techniques in acting $\mathbf{B}_\alpha(t,T)$, $\mathbf{P}_\alpha(t-r)$ on $G(r,u(r))$. In the critical nonlinear case, it is very difficult to determine where the quantity

$$\mathbf{B}_\alpha(t,T)\mathbf{P}_\alpha(t-r)G(r,u(r))$$

belongs to and also how to bound this quantity such that its integration on the whole interval $(0,T)$ is convergent.

- The Gronwall's inequality is available to apply when we estimate solutions of the IVP (5.55), (5.57). However, it is not available when we estimate solutions of the FVP (5.55)–(5.56) since (5.65) contains the integral on $(0,T)$.

Hence, studying FVP (5.55)–(5.56) is a difficult task.

5.3.3.2 Well-Posedness of the Problem (5.55) and (5.56) Under Globally Lipschitz Case

In the following, we study the well-posedness of the problem (5.55) and (5.56) and regularity of the solution when we consider the following globally Lipschitz assumptions on G:

$(\mathscr{H}_1)$ The function $G:[0,T]\times \mathbb{H}^\nu(\Omega)\to \mathbb{H}^\nu(\Omega)$ satisfies $G(t,0)=0$, and there exists a nonnegative function $L_1\in L^\infty(0,T)$ such that

$$\|G(t,w_1)-G(t,w_2)\|_{\mathbb{H}^\nu(\Omega)} \le L_1(t)\|w_1-w_2\|_{\mathbb{H}^\nu(\Omega)}, \tag{5.68}$$

for all $0\le t\le T$, and $w_1,w_2\in \mathbb{H}^\nu(\Omega)$.

$(\mathscr{H}_2)$ The function $G:[0,T]\times C([0,T];\mathbb{H}^\nu(\Omega))\cap L^q(0,T;\mathbb{H}^\sigma(\Omega))\to \mathbb{H}^{\nu+1}(\Omega)$ satisfies $G(t,0)=0$, and there exists a nonnegative function $L_2\in L^\infty(0,T)$ such that

$$\|G(w_1)-G(w_2)\|_{\mathbb{H}^{\nu+1}(\Omega)} \le L_2\Big\|w_1-w_2\Big\|_{C([0,T];\mathbb{H}^\nu(\Omega))\cap L^q(0,T;\mathbb{H}^\sigma(\Omega))}, \tag{5.69}$$

for all $0\le t\le T$, and $w_1,w_2\in C([0,T];\mathbb{H}^\nu(\Omega))\cap L^q(0,T;\mathbb{H}^\sigma(\Omega))$, where

$$\Big\|w_1-w_2\Big\|_{C([0,T];\mathbb{H}^\nu(\Omega))\cap L^q(0,T;\mathbb{H}^\sigma(\Omega))} = \Big\|w_1-w_2\Big\|_{C([0,T];\mathbb{H}^\nu(\Omega))} + \Big\|w_1-w_2\Big\|_{L^q(0,T;\mathbb{H}^\sigma(\Omega))},$$

and $\nu \geq 0$, $q \geq 1$, $\sigma \geq 0$.

In order to establish our main results, it is useful to note that

$$0 < \frac{\alpha - 1}{\alpha} < \frac{1}{2} < \frac{1}{\alpha} < 1, \quad \text{as} \quad 1 < \alpha < 2.$$

Besides, we recall that the Sobolev embedding $\mathbb{H}^{\nu+\theta}(\Omega) \hookrightarrow \mathbb{H}^{\nu}(\Omega)$ holds as $\nu \geq 0$ and $\theta \geq 0$, so there exists a positive constant $C_1(\nu, \theta)$ such that

$$\left\|\bar{v}\right\|_{\mathbb{H}^{\nu}(\Omega)} \leq C_1(\nu, \theta)\left\|\bar{v}\right\|_{\mathbb{H}^{\nu+\theta}(\Omega)}, \tag{5.70}$$

for all $\bar{v} \in \mathbb{H}^{\nu+\theta}(\Omega)$. In addition, for the convenience purpose, the important constants (which may appear in some proofs) are summarily given by **(AP.)** in the Appendix.

The first result in Theorem 5.12 ensures the existence of a mild solution in $L^p(0, T; \mathbb{H}^{\nu}(\Omega))$ under appropriate assumptions on p, the final value data f, and the assumption $(\mathscr{H}_1)$ on the nonlinearity G. The idea is to construct a Cauchy sequence in $L^p(0, T; \mathbb{H}^{\nu}(\Omega))$ which will be bounded by a power function and converge to a mild solution of the problem (5.55)–(5.56). The solution is then bounded by the power function. After that, time continuity and spatial regularities can be consequently derived. Furthermore, we also discuss on the existence of the derivatives ∂_t, ∂_t^{α} of the mild solution in some appropriate spaces.

Theorem 5.12 *Assume that $f \in \mathbb{H}^{\nu+\theta}(\Omega)$ and G satisfies $(\mathscr{H}_1)$ such that $\|L_1\|_{L^{\infty}(0,T)} \in \left(0, \mathscr{U}_1^{-1}\right)$ with $\nu \geq 0$ and θ satisfies that $\frac{\alpha-1}{\alpha} < \theta < 1$, where the constant $\mathscr{U}_1$ is given by* **(AP.)** *in the Appendix. Then the problem (5.55)–(5.56) has a unique mild solution*

$$u \in L^p(0, T; \mathbb{H}^{\nu}(\Omega)) \cap C^{\alpha(1-\theta)}\big((0, T]; \mathbb{H}^{\nu}(\Omega)\big),$$

for all $p \in \left[1, \frac{1}{\alpha(1-\theta)}\right)$, which corresponds to the estimate

$$\|u(t)\|_{\mathbb{H}^{\nu}(\Omega)} \lesssim t^{-\alpha(1-\theta)}\|f\|_{\mathbb{H}^{\nu+\theta}(\Omega)}. \tag{5.71}$$

We also have the following spatial and time regularities:

(a) *Let θ' satisfy that $\dfrac{\alpha-1}{\alpha} < \theta' \leq \theta$. Then $u \in L^p(0, T; \mathbb{H}^{\nu+\theta-\theta'}(\Omega))$, for all $p \in \left[1, \frac{1}{\alpha(1-\theta')}\right)$, which corresponds to the estimate*

$$\|u(t)\|_{\mathbb{H}^{\nu+\theta-\theta'}(\Omega)} \lesssim t^{-\alpha(1-\theta')}\|f\|_{\mathbb{H}^{\nu+\theta}(\Omega)}.$$

(b) *Let $1 - \theta < \nu' \leq 2 - \theta$. Then $u \in C^{\eta_{glo}}\left([0, T]; \mathbb{H}^{\nu-\nu'}(\Omega)\right)$ and*

$$\left\| u(\widetilde{t}) - u(t) \right\|_{\mathbb{H}^{\nu-\nu'}(\Omega)} \lesssim (\widetilde{t} - t)^{\eta_{glo}} \|f\|_{\mathbb{H}^{\nu+\theta}(\Omega)},$$

and here η_{glo} is defined in the Appendix.

(c) *Let $0 \le \nu_1 < \min\left\{1-\theta, \frac{\alpha-1}{\alpha}\right\}$. Then $\partial_t u \in L^p(0,T;\mathbb{H}^{\nu-\nu_1-\frac{1}{\alpha}}(\Omega))$, for all $p \in \left[1, \frac{1}{\alpha(1-\theta-\nu_1)}\right)$, which corresponds to the estimate*

$$\|\partial_t u(t)\|_{\mathbb{H}^{\nu-\nu_1-\frac{1}{\alpha}}(\Omega)} \lesssim t^{-\alpha(1-\theta-\nu_1)} \|f\|_{\mathbb{H}^{\nu+\theta}(\Omega)}.$$

(d) *Let $\frac{2(\alpha-1)}{\alpha} - \theta < \nu_\alpha < 2 - \frac{1}{\alpha} - \theta$. Then $\partial_t^\alpha u \in L^p(0,T;\mathbb{H}^{\nu-\nu_\alpha-\frac{1}{\alpha}}(\Omega))$, for any $p \in \left[1, \frac{1}{\alpha \max\left\{2-\theta-\nu_\alpha-\frac{1}{\alpha}, 1-\theta\right\}}\right)$, which corresponds to the estimate*

$$\|\partial_t^\alpha u(t)\|_{\mathbb{H}^{\nu-\nu_\alpha-\frac{1}{\alpha}}(\Omega)} \lesssim t^{-\alpha \max\left\{2-\theta-\nu_\alpha-\frac{1}{\alpha}, 1-\theta\right\}} \|f\|_{\mathbb{H}^{\nu+\theta}(\Omega)}.$$

The hidden constants (as using the notation $\lesssim$) depend only on α, ν, θ, T in the inequality (5.71), on $\alpha, \nu, \theta, \theta', T$ in Part (a), on $\alpha, \nu, \theta, \nu', T$ in Part (b), on $\alpha, \nu, \theta, \nu_1, T$ in Part (c)*, on $\alpha, \nu, \theta, \nu_\alpha, T, \lambda_1$ in Part* (d)*.*

Proposition 5.1 *By Theorem 5.12, the smoothness of the mild solution can be summarized together as*

$$\begin{cases} u \in \left\{ \bigcup_{1 \le p < \frac{1}{\alpha(1-\theta')}} L^p(0,T;\mathbb{H}^{\nu+\theta-\theta'}(\Omega)) \right\} \bigcap C^{\alpha(1-\theta)}\big((0,T];\mathbb{H}^{\nu}(\Omega)\big), \\ \partial_t u \in \bigcup_{1 \le p < \frac{1}{\alpha(1-\theta-\nu_1)}} L^p(0,T;\mathbb{H}^{\nu-\nu_1-\frac{1}{\alpha}}(\Omega)), \\ \partial_t^\alpha u \in \bigcup_{1 \le p < \frac{1}{\alpha \max\left\{2-\theta-\nu_\alpha-\frac{1}{\alpha}, 1-\theta\right\}}} L^p(0,T;\mathbb{H}^{\nu-\nu_\alpha-\frac{1}{\alpha}}(\Omega)), \end{cases}$$

where values of the parameters are given by Theorem 5.12. Moreover, the spatial regularity in Part (b) shows how the best spatial regularity of the mild solution u can achieve. Then, by using some suitable Sobolev embeddings, one can derive the Gradient and Laplacian estimates for the solution on L^q spaces.

Remark 5.4 In fact, one can investigate the continuity of the first order derivative $\partial_t u$ which is established in Part (d) of the above theorem. Moreover, if the nonlinearity G is continuous in the time variable t, for instance, G verifies that

$$\|G(t_1, v_1) - G(t_2, v_2)\|_{\mathbb{H}^\nu(\Omega)} \lesssim |t_1 - t_2|^{\text{positive exponent}} + \|v_1 - v_2\|_{\mathbb{H}^\nu(\Omega)},$$

then one can establish the continuity of the fractional derivative ∂_t^α of the solution.

In Theorem 5.12, under assumption $(\mathscr{H}_1)$, we do not get the regularity results of u in $C([0,T];\mathbb{H}^\nu(\Omega))$ or $L^\infty(0,T;\mathbb{H}^\nu(\Omega))$. The main reason is that the information at the initial time $u(0)$ does not actually exist on $\mathbb{H}^\nu(\Omega)$. To overcome this restriction, we are going to consider the existence of a mild solution in the spaces $C([0,T];\mathbb{H}^\nu(\Omega))$ or $L^\infty(0,T;\mathbb{H}^\nu(\Omega))$ by imposing the assumption $(\mathscr{H}_2)$ on the nonlinearity G. In addition, it is necessary to suppose a smoother assumption on the final value data f. In the following theorem, we will build up this existence and also a regularity result for the mild solution by using the Banach fixed point theorem. Let us recall the fact that the embedding $\mathbb{H}^{\nu+1}(\Omega) \hookrightarrow \mathbb{H}^\sigma(\Omega)$ holds as $0 \le \sigma \le \nu+1$. So, there exists a positive constant $C_2(\nu,\sigma)$ such that

$$\left\|\bar{v}\right\|_{\mathbb{H}^\sigma(\Omega)} \le C_2(\nu,\sigma)\left\|\bar{v}\right\|_{\mathbb{H}^{\nu+1}(\Omega)}, \tag{5.72}$$

for all $\bar{v} \in \mathbb{H}^{\nu+1}(\Omega)$.

Theorem 5.13 *Let* $\frac{\alpha-1}{\alpha} < \theta < 1$, $0 \le \nu \le \sigma \le \nu+1$, *and* $1 \le q < \frac{1}{\alpha(1-\theta)}$. *Assume that* $f \in \mathbb{H}^{\nu+\theta+1}(\Omega)$, *and* G *satisfies* $(\mathscr{H}_2)$ *with* $L_2 \in \left(0, \mathscr{U}_2^{-1}\right)$ *where* $\mathscr{U}_2$ *is given by* **(AP.)** *in the Appendix. Then, the problem (5.55)–(5.56) has a unique mild solution*

$$u \in C\left([0,T];\mathbb{H}^\nu(\Omega)\right) \cap L^q(0,T;\mathbb{H}^\sigma(\Omega)).$$

Moreover, we have

$$\sup_{0\le t\le T}\left\|u(t)\right\|_{\mathbb{H}^\nu(\Omega)} + \left(\int_0^T \left\|u(t)\right\|^q_{\mathbb{H}^\sigma(\Omega)} dt\right)^{1/q} \lesssim \|f\|_{\mathbb{H}^{\nu+\theta+1}(\Omega)}. \tag{5.73}$$

5.3.3.3 Well-Posedness of the Problem (5.55)–(5.56) Under Critical Nonlinearities Case

Theorems 5.12 and 5.13 state the results in the globally Lipschitz case, and they cannot virtually be applied in many models such as time fractional Ginzburg–Landau, Allen–Cahn, Burger, Navier–Stokes, Schrödinger, etc., equations. In the following, we state the well-posedness of the problem (5.55)–(5.56) under the critical nonlinearities case.

Theorem 5.14 *Assume that* $\alpha \in (1,2)$, $\sigma \in (-1,0)$, $0 < \nu < 1+\sigma$, *and* $s > 0$. *Let* ϑ *such that* $\vartheta \in (\nu-\sigma, 1)$ *and set* $\mu = \nu - \sigma$. *Let* ζ *satisfy that*

$$\zeta < \min\left\{\alpha^{-1} - (1+s)\vartheta, \vartheta(1-s) - \nu + \sigma\right\}. \tag{5.74}$$

The function G *satisfies the next assumption* $(\mathscr{H}_3)$, *that is,* $G : [0,T] \times \mathbb{H}^\nu(\Omega) \to \mathbb{H}^\sigma(\Omega)$, $G(0) = 0$ *and*

$$\|G(v_1) - G(v_2)\|_{\mathbb{H}^\sigma(\Omega)} \le L_3(t)\Big(1 + \|v_1\|^s_{\mathbb{H}^\nu(\Omega)} + \|v_2\|^s_{\mathbb{H}^\nu(\Omega)}\Big)\|v_1 - v_2\|_{\mathbb{H}^\nu(\Omega)}, \tag{5.75}$$

where L_3 satisfies that $L_3(t)t^{\alpha\zeta} \in L^\infty(0, T)$. Set

$$\mathfrak{X}_{\alpha,\vartheta,\nu,T}(\mathscr{R}) := \Big\{w \in C^{\alpha\vartheta}((0, T]; \mathbb{H}^\nu(\Omega)),\ \ \|w\|_{C^{\alpha\vartheta}((0,T];\mathbb{H}^\nu(\Omega))} \le \mathscr{R}\Big\}.$$

*If $f \in \mathbb{H}^{\nu+(1-\vartheta)}(\Omega)$ and $K_0T^{s\alpha\vartheta} \in \big(0, \min\big\{\frac{1}{2}\overline{\mathfrak{N}_2}^{-1}, \mathscr{N}_f\big\}\big)$ with $K_0 = \|L_3(t)t^{\alpha\zeta}\|_{L^\infty(0,T)}$, where the constants are formulated by **(AP.)** in the Appendix, then the problem (5.55)–(5.56) has a unique mild solution $u \in \mathfrak{X}_{\alpha,\vartheta,\nu,T}(\widehat{\mathscr{R}})$ with respect to the estimate*

$$\|u(t)\|_{\mathbb{H}^\nu(\Omega)} \lesssim t^{-\alpha\vartheta}\ \|f\|_{\mathbb{H}^{\nu+(1-\vartheta)}(\Omega)}\,. \tag{5.76}$$

Moreover, we get the following spatial and time regularities:

(a) Let $\vartheta \le \vartheta' \le 1$ and $\alpha\vartheta - 1 \le \beta \le \alpha\vartheta$; then $t^\beta u \in L^p(0, T; \mathbb{H}^{\nu+(\vartheta'-\vartheta)}(\Omega))$ for all $p \in \Big[1, \frac{1}{\alpha\vartheta-\beta}\Big)$, with respect to the estimate

$$\Big\|u(t)\Big\|_{\mathbb{H}^{\nu+(\vartheta'-\vartheta)}(\Omega)} \lesssim t^{-\alpha\vartheta'}\ \|f\|_{\mathbb{H}^{\nu+(1-\vartheta)}(\Omega)}\,.$$

(b) Let $\vartheta < \eta \le \vartheta + 1$ and then $u \in C^{\eta_{cri}}\big([0, T]; \mathbb{H}^{\nu-\eta}(\Omega)\big)$ and

$$\Big\|u(\widetilde{t}) - u(t)\Big\|_{\mathbb{H}^{\nu-\eta}(\Omega)} \lesssim (\widetilde{t} - t)^{\eta_{cri}}\ \|f\|_{\mathbb{H}^{\nu+(1-\vartheta)}(\Omega)}\,,$$

where η_{cri} is defined in the Appendix.

The hidden constants (as using the notation $\lesssim$) depend only on $\alpha, \mu, \vartheta, \zeta, s, T$ in the inequality (5.76), on $\alpha, \mu, \vartheta, \vartheta', \zeta, s, T$ in Part (a), on $\alpha, \mu, \vartheta, \eta, \zeta, s, T$ in Part (b).

Remark 5.5 One can actually establish the existence of the derivatives $\partial_t u$ and ∂_t^α of the solution as follows.

(i) Assume that $\vartheta < \frac{\mu+1}{2}$ and let $\vartheta_1 \in \Big[\vartheta, \frac{\nu-\sigma+1}{2}\Big)$, and then $\partial_t u(t) \in \mathbb{H}^{\sigma+\vartheta_1-1}(\Omega)$ for each $t > 0$, which corresponds to the estimate

$$\|\partial_t u(t)\|_{\mathbb{H}^{\sigma+\vartheta_1-1}(\Omega)} \lesssim t^{-\alpha\left(2\vartheta_1-\mu-\frac{\alpha-1}{\alpha}\right)}\ \|f\|_{\mathbb{H}^{\nu+(1-\vartheta)}(\Omega)}\,. \tag{5.77}$$

(ii) Assume that $\vartheta \le \frac{2\nu-2\sigma+1}{3}$ and let $\vartheta_\alpha \in \Big[\frac{\nu-\sigma+2}{3}, \frac{\nu-\sigma+5}{3}\Big)$, and then $\partial_t^\alpha u(t) \in \mathbb{H}^{\sigma+\vartheta_\alpha-2}(\Omega)$ for each $t > 0$, which corresponds to the estimate

$$\Big\|\partial_t^\alpha u(t)\Big\|_{\mathbb{H}^{\sigma+\vartheta_\alpha-2}(\Omega)} \lesssim t^{-\max\left\{\alpha\left(\vartheta_\alpha-\frac{\mu+2}{3}\right),\alpha(2\vartheta-\mu)\right\}}\|f\|_{\mathbb{H}^{\nu+(1-\vartheta)}(\Omega)}. \tag{5.78}$$

The above results are weaker than Parts (c), (d) of Theorem 5.12 since the powers in (5.77), (5.78) are really less than -1. So, we cannot obtain the existence of $\partial_t u$, $\partial_t^\alpha u$ in the L^p space with respect to the time variable. This obviously comes from the critical property of the nonlinearity.

5.3.4 Applications

In this subsection we apply the theory developed in this work to some well-known equations. The classes of time fractional Ginzburg–Landau equation and time fractional Burger equation are studied in L^q ($q \geq 1$) settings via interpolation–extrapolation scales and dual interpolation–extrapolation scales of Sobolev spaces. We will discuss both time and spatial regularity of solutions by considering:

- The time continuities of solutions on L^q spaces with respect to the intervals $(0, T]$, $[0, T]$
- The Gradient and Laplacian estimates for the solutions on L^q spaces

5.3.4.1 Time Fractional Ginzburg–Landau Equation

We discuss now an application of our methods to a final value problem for a time fractional Ginzburg–Landau equation which is stated as follows:

$$\begin{cases} \partial_t^\alpha u(x,t) + \mathscr{L}u(x,t) = \rho(t)|u(x,t)|^s u(x,t), & x \in \Omega, t \in (0,T), \\ u(x,t) = 0, & x \in \partial\Omega, t \in (0,T), \\ \partial_t u(x,0) = 0, & x \in \Omega \end{cases} \tag{5.79}$$

associated with the final value data (5.56) and where $s > 0$ is a given number.

Theorem 5.15 *Assume that* $2 \leq N \leq 4$*:*

(a) *The case* $0 < s < \dfrac{4}{N}$*.*

Let the numbers κ, ν, σ, μ, ϑ, ϑ', *respectively, satisfy that*

$$\kappa \in \left(0, \frac{Ns}{8}\right], \quad \nu \in \left[\frac{N}{4} - \frac{\kappa}{s}, \frac{N}{4}\right), \quad \mu \in \left(\mu_0, \frac{N+4}{8}\right),$$

$$\vartheta \in \left(\mu, \frac{N+4}{8}\right), \quad \vartheta' \in \left[\vartheta + \frac{4-N}{8}, 1\right),$$

where $\mu = \nu - \sigma$ *and* $\mu_0 := \max\left\{\nu, s\left(\frac{N}{4} - \nu\right)\right\}$. *If* $f \in \mathbb{H}^{\nu+(1-\vartheta)}(\Omega)$, *and* $\rho(t) \leq C_\rho t^{\alpha b}$ *with* $b > -\min\left\{\frac{1}{\alpha} - (1+s)\vartheta, (\vartheta - \mu) - s\vartheta\right\}$ *and* C_ρ *is small enough, then the problem (5.79) has a unique mild solution* u *such that:*

(Time regularity) $u \in C^{\alpha\vartheta}((0,T]; L^4(\Omega)) \cap C^{\eta_{cri}}\left([0,T]; \mathbb{H}^{\nu-\eta}(\Omega)\right)$, *where* $\vartheta < \eta \leq \vartheta + 1$. *This solution satisfies the estimate*

$$t^{\alpha\vartheta} \|u(t)\|_{L^4(\Omega)} \mathbf{1}_{t>0} + \frac{\|u(t+\gamma) - u(t)\|_{\mathbb{H}^{\nu-\eta}(\Omega)}}{\gamma^{\eta_{cri}}} \mathbf{1}_{t\geq 0} \lesssim \|f\|_{\mathbb{H}^{\nu+(1-\vartheta)}(\Omega)}. \tag{5.80}$$

(Spatial regularity) *For each* $t > 0$, $u(t)$ *belongs to* $W^{1,\frac{4N}{3N-8\nu}}(\Omega)$ *and verifies the estimate*

$$t^{\alpha\vartheta'} \left\|\nabla u(t)\right\|_{L^{\frac{4N}{3N-8\nu}}(\Omega)} + t^{\alpha\vartheta'} \left\|(-\Delta)^{\vartheta'-\vartheta} u(t)\right\|_{L^4(\Omega)} \lesssim \|f\|_{\mathbb{H}^{\nu+(1-\vartheta)}(\Omega)}. \tag{5.81}$$

(b) *The case* $s \geq \dfrac{4}{N}$.

Let the numbers $\nu, \sigma, \mu, \vartheta, \vartheta'$, *respectively, satisfy that*

$$\nu \in \left[\frac{N}{4} - \frac{1}{2s}, \frac{N}{4}\right), \quad \mu \in \left(\mu_0, \frac{N+4}{8}\right),$$
$$\vartheta \in \left(\mu, \frac{N+4}{8}\right), \quad \vartheta' \in \left[\vartheta + \frac{4-N}{8}, 1\right),$$

whereupon $\mu = \nu - \sigma$ *and* $\mu_0 := \max\left\{\nu, s\left(\frac{N}{4} - \nu\right)\right\}$. *If* $f \in \mathbb{H}^{\nu+(1-\vartheta)}(\Omega)$, *and* $\rho(t) \leq C_\rho t^{\alpha b}$ *such that* $b > -\min\left\{\frac{1}{\alpha} - (1+s)\vartheta, (\vartheta - \mu) - s\vartheta\right\}$ *and* C_ρ *is small enough, then the problem (5.79) has a unique mild solution* u *such that:*

(Time regularity) $u \in C^{\alpha\vartheta}((0,T]; L^{Ns}(\Omega)) \cap C^{\eta_{cri}}\left([0,T]; \mathbb{H}^{\nu-\eta}(\Omega)\right)$, *where* $\vartheta < \eta \leq \vartheta + 1$. *This solution satisfies the estimate*

$$t^{\alpha\vartheta} \|u(t)\|_{L^{Ns}(\Omega)} \mathbf{1}_{t>0} + \frac{\|u(t+\gamma) - u(t)\|_{\mathbb{H}^{\nu-\eta}(\Omega)}}{\gamma^{\eta_{cri}}} \mathbf{1}_{t\geq 0} \lesssim \|f\|_{\mathbb{H}^{\nu+(1-\vartheta)}(\Omega)}. \tag{5.82}$$

(Spatial regularity) *For each* $t > 0$, $u(t)$ *belongs to* $W^{1,\frac{4N}{3N-8\nu}}(\Omega)$ *and verifies the estimate*

$$t^{\alpha\vartheta'} \left\|\nabla u(t)\right\|_{L^{\frac{4N}{3N-8\nu}}(\Omega)} + t^{\alpha\vartheta'} \left\|(-\Delta)^{\vartheta'-\vartheta} u(t)\right\|_{L^{Ns}(\Omega)} \lesssim \|f\|_{\mathbb{H}^{\nu+(1-\vartheta)}(\Omega)}. \tag{5.83}$$

Proof

(a) This proof will be based on applying and improving Theorem 5.14 actually. We firstly give some explanations that the assumptions in this part are suitable:

- $0 < \kappa < \frac{1}{2}$ since $\frac{Ns}{8} < \frac{N\left(\frac{4}{N}\right)}{8} = \frac{1}{2}$ as the assumption $s < \frac{4}{N}$.
- $0 < \kappa \leq \frac{3N-4}{8}s$ since $\frac{Ns}{8} \leq \frac{Ns}{\left(\frac{8N}{3N-4}\right)} = \frac{3N-4}{8}$ by the fact that $8 \geq \frac{8N}{3N-4}$.
- $\nu \geq \frac{N}{8}$ since $\nu \geq \frac{N}{4} - \frac{\kappa}{s} \geq \frac{N}{4} - \frac{N}{8} = \frac{N}{8}$ as the assumption $\nu \geq \frac{N}{4} - \frac{\kappa}{s}$ and $\kappa \leq \frac{Ns}{8}$.
- The interval $\left(\mu_0, \frac{N+4}{8}\right)$ is not really empty. Indeed, it is easy to see from $\nu < \frac{N}{4}$ and $\frac{N}{4} < \frac{N+4}{8}$ that $\nu < \frac{N+4}{8}$. Moreover, we have
$$s\left(\frac{N}{4} - \nu\right) \leq s\left(\frac{N}{4} - \left(\frac{N}{4} - \frac{\kappa}{s}\right)\right) = \kappa < \frac{1}{2} < \frac{N+4}{8},$$
by using the assumption $\nu > \frac{N}{4} - \frac{\kappa}{s}$ and noting that $\kappa < \frac{1}{2}$.
- The interval $\left[\vartheta + \frac{4-N}{8}, 1\right)$ is also not empty as $\vartheta < \frac{N+4}{8}$.
- The number σ belongs to $\left(-\frac{N}{4}, 0\right)$ since $\sigma = \nu - \mu \leq \mu_0 - \mu < 0$, and furthermore
$$\nu - \mu > \frac{N}{4} - \frac{\kappa}{s} - \frac{N+4}{8} \geq \frac{N}{4} - \frac{3N-4}{8} - \frac{N+4}{8} = -\frac{N}{4},$$
by using the assumption $\nu \geq \frac{N}{4} - \frac{\kappa}{s}$ and the fact that $\kappa < \frac{3N-4}{8}s$.

Secondly, we obtain some important Sobolev embeddings which help to establish the existence of a mild solution. By applying the embeddings (5.61)–(5.62), and the dualities $\left[\mathbb{H}^{-\sigma}(\Omega)\right]^* = \mathbb{H}^{\sigma}(\Omega)$, $\left[W^{-2\sigma,2}(\Omega)\right]^* = W^{2\sigma,2}(\Omega)$, one can see that:

- The Sobolev embedding $L^{\frac{2N}{N-4\sigma}}(\Omega) \hookrightarrow W^{2\sigma,2}(\Omega)$ holds as $-\frac{N}{2} < 2\sigma < 0$, $\frac{2N}{N-4\sigma} = \frac{2N}{N-2(2\sigma)}$.
- The Sobolev embedding $\mathbb{H}^{-\sigma}(\Omega) \hookrightarrow W^{-2\sigma,2}(\Omega)$ holds as $-\sigma > 0$, which implies that $W^{2\sigma,2}(\Omega) \hookrightarrow \mathbb{H}^{\sigma}(\Omega)$ holds.

As a consequence of the above embeddings, we obtain the following Sobolev embedding:

$$L^{\frac{2N}{N-4\sigma}}(\Omega) \hookrightarrow \mathbb{H}^{\sigma}(\Omega). \tag{5.84}$$

By the assumption $s\left(\frac{N}{4} - \nu\right) \leq \mu_0 < \mu$, we have

$$\frac{2N(1+s)}{N-4\sigma} = \frac{2N(1+s)}{N-4(\nu-\mu)} \leq \frac{2N(1+s)}{N-4\nu+4s\left(\frac{N}{4}-\nu\right)} = \frac{2N}{N-4\nu}.$$

Therefore, using (5.61) yields that

$$W^{2\nu,2}(\Omega) \hookrightarrow L^{\frac{2N(1+s)}{N-4\sigma}}(\Omega) \quad \text{since} \quad 0 \leq 2\nu < \frac{N}{2}, \ \frac{2N(1+s)}{N-4\sigma} \leq \frac{2N}{N-4\nu}.$$

Besides, using (5.62) invokes that $\mathbb{H}^{\nu}(\Omega) \hookrightarrow W^{2\nu,2}(\Omega)$, as $\nu \geq 0$, which consequently infers the embedding

$$\mathbb{H}^{\nu}(\Omega) \hookrightarrow L^{\frac{2N(1+s)}{N-4\sigma}}(\Omega). \tag{5.85}$$

Thirdly, let us set the nonlinearity $G(v) := \rho(t)|v|^s v$, and show that G satisfies $\mathscr{H}_3$. Indeed, it is obvious that $|G(v_1) - G(v_2)|$ is pointwise bounded by $(1+s)\left(|v_1|^s + |v_2|^s\right)|v_1 - v_2|$, and so one can derive the following chain of the estimates:

$$\begin{aligned}
&\|G(v_1) - G(v_2)\|_{\mathbb{H}^{\sigma}(\Omega)} \\
&\lesssim \|G(v_1) - G(v_2)\|_{L^{\frac{2N}{N-4\sigma}}(\Omega)} \\
&\lesssim \rho(t)\left[\left\| |v_1|^s |v_1 - v_2| \right\|_{L^{\frac{2N}{N-4\sigma}}(\Omega)} + \left\| |v_2|^s |v_1 - v_2| \right\|_{L^{\frac{2N}{N-4\sigma}}(\Omega)}\right] \\
&\lesssim \rho(t)\left(\|v_1\|^s_{L^{\frac{2N(1+s)}{N-4\sigma}}(\Omega)} + \|v_2\|^s_{L^{\frac{2N(1+s)}{N-4\sigma}}(\Omega)} \right) \|v_1 - v_2\|_{L^{\frac{2N(1+s)}{N-4\sigma}}(\Omega)} \\
&\lesssim \rho(t)\left(\|v_1\|^s_{\mathbb{H}^{\nu}(\Omega)} + \|v_2\|^s_{\mathbb{H}^{\nu}(\Omega)} \right) \|v_1 - v_2\|_{\mathbb{H}^{\nu}(\Omega)},
\end{aligned}$$

where the embedding (5.84) has been used in the first estimate, the pointwise boundedness in the second estimate, the Hölder's inequality in the third estimate, and the embedding (5.85) in the last estimate. Therefore, we can take the Lipschitz coefficient in the form $K(t) = K_{\text{Gin}}\rho(t)$ with some positive constant K_{Gin}. Furthermore, the assumption $b > -\min\left\{\frac{1}{\alpha} - (1+s)\vartheta, (\vartheta-\mu) - s\vartheta\right\}$ ensures that there always exists a real constant ζ such that

$$-b < \zeta < \min\left\{\frac{1}{\alpha} - (1+s)\vartheta, (\vartheta-\mu) - s\vartheta\right\},$$

and then one derives $K(t) \leq C_\rho K_{\text{Gin}} t^{-\alpha\zeta} t^{\alpha(b+\zeta)} \leq K_0 t^{-\alpha\zeta}$, where $K_0 = C_\rho K_{\text{Gin}} T^{\alpha(b+\zeta)}$. We conclude that ζ and G satisfy the assumptions of Theorem 5.14. It is obvious that all assumptions of this theorem also fulfill the assumptions of Theorem 5.14. Thus, applying Theorem 5.14 invokes that the problem (5.79) has a unique mild solution $u \in C^{\alpha\vartheta}((0,T];\mathbb{H}^\nu(\Omega)) \cap C^{\eta_{cri}}\left([0,T];\mathbb{H}^{\nu-\eta}(\Omega)\right)$ with C_ρ is small enough. Now, the assumption $\nu \geq \frac{N}{4} - \frac{\kappa}{s}$ invokes that

$$\frac{2N}{N-4\nu} \geq \frac{2N}{N-4\left(\frac{N}{4}-\frac{\kappa}{s}\right)} = \frac{2N}{\left(4\frac{\kappa}{s}\right)} \geq \frac{2N}{\left(4\frac{N}{8}\right)} = 4, \tag{5.86}$$

where $\frac{\kappa}{s} \leq \frac{N}{8}$. Hence, we infer from Ω is a bounded domain that $L^{\frac{2N}{N-4\nu}}(\Omega) \hookrightarrow L^4(\Omega)$. Besides, applying the embedding (5.60) again combined with the above embedding to allow that

$$W^{2\nu,2}(\Omega) \hookrightarrow L^{\frac{2N}{N-4\nu}}(\Omega) \hookrightarrow L^4(\Omega),$$

where we note that $0 < 2\nu < \frac{N}{2}$, $\frac{2N}{N-4\nu} = \frac{2N}{N-2(2\nu)}$. Therefore, we deduce that

$$u \in C^{\alpha\vartheta}((0,T];L^4(\Omega)) \cap C^{\eta_{cri}}\left([0,T];\mathbb{H}^{\nu-\eta}(\Omega)\right),$$

where $\vartheta < \eta \leq \vartheta + 1$ as in Part (b) of Theorem 5.14 and

$$t^{\alpha\vartheta}\,\|u(t)\|_{L^4(\Omega)}\,\mathbf{1}_{t>0} + \frac{\|u(t+\gamma)-u(t)\|_{\mathbb{H}^{\nu-\eta}(\Omega)}}{\gamma^{\eta_{cri}}}\mathbf{1}_{t\geq 0} \lesssim \|f\|_{\mathbb{H}^{\nu+(1-\vartheta)}(\Omega)}.$$

This shows the inequality (5.80). Finally, we need to prove the inequality (5.81). Indeed, we have

$$\vartheta' - \vartheta \geq \frac{4-N}{8} \quad \text{as} \quad \vartheta' \in \left[\vartheta + \frac{4-N}{8}, 1\right), \tag{5.87}$$

which associates with $\nu \geq \frac{N}{8}$ that $\nu + (\vartheta' - \vartheta) \geq \frac{N}{8} + \frac{4-N}{8} = \frac{1}{2}$. Therefore, we obtain:

- The Sobolev embedding $\mathbb{H}^{\nu+(\vartheta'-\vartheta)}(\Omega) \hookrightarrow W^{2\nu+2(\vartheta'-\vartheta),2}(\Omega)$ holds as $\nu + (\vartheta' - \vartheta) > 0$.
- The Sobolev embedding $W^{2\nu+2(\vartheta'-\vartheta),2}(\Omega) \hookrightarrow W^{1,\frac{4N}{3N-8\nu}}(\Omega)$ holds by using the embedding (5.59) as $\frac{4N}{3N-8\nu} \geq 1$ and $2\nu + 2(\vartheta' - \vartheta) \geq 1$, where we note from (5.87) that

$$2\nu + 2(\vartheta' - \vartheta) - 1 \geq 2\nu + 2\left(\frac{4-N}{8}\right) - 1 = 2\nu - \frac{N}{4} = \frac{N}{2} - \frac{N}{\left(\frac{4N}{3N-8\nu}\right)}.$$

Two above embeddings consequently infer that the Sobolev embedding $\mathbb{H}^{\nu+(\vartheta'-\vartheta)}(\Omega) \hookrightarrow W^{1,\frac{4N}{3N-8\nu}}(\Omega)$ holds. Hence, we deduce from Part (a) of Theorem 5.14 that $u(t) \in W^{1,\frac{4N}{3N-8\nu}}(\Omega)$ with respect to the estimate

$$t^{\alpha\vartheta'} \left\|\nabla u(t)\right\|_{L^{\frac{4N}{3N-8\nu}}(\Omega)} + t^{\alpha\vartheta'} \left\|(-\Delta)^{\vartheta'-\vartheta} u(t)\right\|_{L^4(\Omega)} \lesssim \|f\|_{\mathbb{H}^{\nu+(1-\vartheta)}(\Omega)},$$

which finalizes the proof of Part (a) of this theorem.

(b) We note from Part (a) that the mediate number κ belongs to the interval $\left(0, \frac{1}{2}\right)$. In this part, we try to extend the method in Part (a) with $\kappa = \frac{1}{2}$. It is important to explain the similarities and differences between the numbers in this part from Part (a) as follows:

- $\nu \geq \frac{N}{8}$ since $\nu \geq \frac{N}{4} - \frac{1}{2s} \geq \frac{N}{4} - \frac{1}{2\left(\frac{4}{N}\right)} = \frac{N}{8}$ by employing $\nu \geq \frac{N}{4} - \frac{1}{2s}$ and $s \geq \frac{4}{N}$.
- The interval $\left(\mu_0, \frac{N+4}{8}\right)$ is not really empty since

$$s\left(\frac{N}{4} - \nu\right) \leq s\left(\frac{N}{4} - \left(\frac{N}{4} - \frac{1}{2s}\right)\right) = \frac{1}{2} < \frac{N+4}{8}.$$

- The number σ belongs to $\left(-\frac{N}{4}, 0\right)$ since

$$\nu - \mu > \frac{N}{4} - \frac{1}{2s} - \frac{N+4}{8} \geq \frac{N}{4} - \frac{1}{2\left(\frac{4}{N}\right)} - \frac{N+4}{8} = -\frac{1}{2} \geq -\frac{N}{4},$$

by also employing $\nu \geq \dfrac{N}{4} - \dfrac{1}{2s}$ and $s \geq \dfrac{4}{N}$.

By using the same methods as Part (a), one can establish the existence and uniqueness of a mild solution u to the problem (5.79) in $C^{\alpha\vartheta}((0,T];\mathbb{H}^{\nu}(\Omega)) \cap C^{\eta_{cri}}([0,T];\mathbb{H}^{\nu-\eta}(\Omega))$ with C_ρ is small enough. Next, the inequality (5.86) can be modified as

$$\frac{2N}{N-4\nu} \geq \frac{2N}{N-4\left(\frac{N}{4}-\frac{1}{2s}\right)} = Ns.$$

Hence, we obtain the Sobolev embedding

$$W^{2\nu,2}(\Omega) \hookrightarrow L^{\frac{2N}{N-4\nu}}(\Omega) \hookrightarrow L^{Ns}(\Omega),$$

which deduces the inequality (5.82). Moreover, we also have $\nu + (\vartheta' - \vartheta) \geq \frac{1}{2}$ by noting the assumption $\vartheta' \in \left[\vartheta + \frac{4-N}{8}, 1\right)$ and the fact that $\nu \geq \frac{N}{8}$. Then, we obtain the Sobolev embedding

$$\mathbb{H}^{\nu+(\vartheta'-\vartheta)}(\Omega) \hookrightarrow W^{1,\frac{4N}{3N-8\nu}}(\Omega),$$

and the inequality (5.83) also holds. We finally complete the proof.

5.3.4.2 Time Fractional Burgers Equation

In the following, we deal with a terminal value problem for a time fractional Burgers equation which is given by

$$\begin{cases} \partial_t^\alpha u(x,t) + \rho(t)(u\cdot\nabla)u(x,t) = \Delta u(x,t), & x \in \Omega,\ 0 < t < T, \\ u(x,t) = 0, & x \in \partial\Omega,\ 0 < t < T, \\ \partial_t u(x,0) = 0, & x \in \Omega \end{cases} \tag{5.88}$$

associated with the final value data (5.56). Here f and ρ are given functions, and the operator $\mathscr{A}$ is $-\Delta$ which acts on $L^2(\Omega)$ with its domain $W_0^{1,2}(\Omega) \cap W^{2,2}(\Omega)$. In the following, we will apply Theorem 5.14 to obtain a mild solution of the problem (5.88) and then obtain the spatial regularity with an $L^q(\Omega)$-estimate for ∇u and an $\mathbb{H}^\nu(\Omega)$-estimate for $(-\Delta)^{\vartheta'-\vartheta}u$.

Theorem 5.16 *Assume that* $3 \leq N \leq 4$ *(N is the dimension of Ω). Let the numbers* $\nu, \sigma, \mu, \vartheta, \vartheta'$*, respectively, satisfy that*

$$\nu \in \left[\frac{1}{2}, \frac{N}{4}\right),$$

$$\mu \in \left[\mu_2, \frac{N+4}{8}\right), \quad \vartheta \in \left(\mu, \frac{N+4}{8}\right), \quad \vartheta' \in \left[\vartheta + \frac{4-N}{8}, 1\right),$$

where $\mu = \nu - \sigma$ *and* $\mu_2 := \max\left\{\nu, \frac{N+2}{4} - \nu\right\}$.

If $\rho(t) \le C_\rho t^b$ *with* $b > -\min\left\{-\mu, \frac{1}{\alpha} - 2\vartheta\right\}$, $f \in \mathbb{H}^{\nu+(1-\vartheta)}(\Omega)$, *and* C_ρ *is small enough, then the problem (5.88) has a unique mild solution* u *such that the following conclusions hold:*

(a) (Time regularity) *Let* $\vartheta < \eta \le \vartheta + 1$. *Then we have*

$$u \in C^{\alpha\vartheta}((0,T]; L^{\frac{2N}{4\mu-2}}(\Omega)) \cap C^{\eta_{cri}}\left([0,T]; \mathbb{H}^{\nu-\eta}(\Omega)\right),$$

and time regularity result for u *holds*

$$t^{\alpha\vartheta} \|u(t)\|_{L^{\frac{2N}{4\mu-2}}(\Omega)} + \frac{\|u(t+\gamma) - u(t)\|_{\mathbb{H}^{\nu-\eta}(\Omega)}}{\gamma^{\eta_{cri}}} \lesssim \|f\|_{\mathbb{H}^{\nu+(1-\vartheta)}(\Omega)}. \tag{5.89}$$

(b) (Spatial regularity) *For each* $t > 0$, $u(t)$ *belongs to* $W^{1,\frac{4N}{3N-4}}(\Omega)$ *and satisfies the following estimate:*

$$t^{\alpha\vartheta'} \left\|\nabla u(t)\right\|_{L^{\frac{4N}{3N-4}}(\Omega)} + t^{\alpha\vartheta'} \left\|(-\Delta)^{\vartheta'-\vartheta} u(t)\right\|_{\mathbb{H}^{\nu}(\Omega)} \lesssim \|f\|_{\mathbb{H}^{\nu+(1-\vartheta)}(\Omega)}. \tag{5.90}$$

Proof In order to prove this theorem, we will apply Theorem 5.14 and then improve the time and spatial regularities of the mild solution. Let us set $G(t, v) = \rho(t)(v \cdot \nabla)v$ and show that G satisfies the assumption $(\mathscr{H}_3)$ corresponding to $s = 1$. Firstly, we analyze the values of the numbers $\nu, \sigma, \mu, \vartheta, \vartheta'$ as follows:

- The interval $\left[\mu_2, \frac{N+4}{8}\right)$ is not empty since $\nu < \frac{N+4}{8}$ as $\nu < \frac{N}{4} \le \frac{N+4}{8}$ (here $N \le 4$), and $\frac{N+2}{4} - \nu < \frac{N+2}{4} - \frac{1}{2} < \frac{N+2}{4} - \frac{N}{8} = \frac{N+4}{8}$.
- The numbers $\frac{N-4\sigma}{4\mu-2}, \frac{N-4\sigma}{N+2-4\nu}$ are greater than 1. Indeed, $\mu > \mu_2 \ge \nu \ge \frac{1}{2}$ and

$$\frac{N-4\sigma}{4\mu-2} = \frac{N+2-4\nu}{4\mu-2} + 1 > \frac{N+2-4\left(\frac{N}{4}\right)}{4\mu-2} + 1 > 1,$$

$$\frac{N-4\sigma}{N+2-4\nu} = \frac{4\mu-2}{N+2-4\nu} + 1 > 1.$$

Moreover, these are the dual numbers of each other.

Therewith, one can obtain the following chains of the Sobolev embeddings by applying (5.59), (5.60), and (5.61):

- The Sobolev embedding $L^{\frac{2N}{N-4\sigma}}(\Omega) \hookrightarrow W^{2\sigma,2}(\Omega)$ holds as $-\frac{N}{2} < 2\sigma \leq 0$, $\frac{2N}{N-4\sigma} = \frac{2N}{N-2(2\sigma)}$, and $W^{2\sigma,2}(\Omega) \hookrightarrow \mathbb{H}^{\sigma}(\Omega)$ as $\sigma \leq 0$, and so that

$$L^{\frac{2N}{N-4\sigma}}(\Omega) \hookrightarrow W^{2\sigma,2}(\Omega) \hookrightarrow \mathbb{H}^{\sigma}(\Omega). \tag{5.91}$$

- The Sobolev embedding $\mathbb{H}^{\nu}(\Omega) \hookrightarrow W^{2\nu,2}(\Omega)$ holds as $\nu \geq 0$, and $W^{2\nu,2}(\Omega) \hookrightarrow W^{1,\frac{2N}{N+2-4\nu}}(\Omega)$ as $\nu \geq \frac{1}{2}$, $2\nu - 1 = \frac{N}{2} - N/\left(\frac{2N}{N+2-4\nu}\right)$, which implies that the following Sobolev embedding

$$\mathbb{H}^{\nu}(\Omega) \hookrightarrow W^{2\nu,2}(\Omega) \hookrightarrow W^{1,\frac{2N}{N+2-4\nu}}(\Omega). \tag{5.92}$$

- The Sobolev embedding $W^{2\nu,2}(\Omega) \hookrightarrow L^{\frac{2N}{4\mu-2}}(\Omega)$ holds as $0 < 2\nu < \frac{N}{2}$, $1 \leq \frac{2N}{4\mu-2} \leq \frac{2N}{N-4\nu}$ $(\nu < \mu)$, and henceforth

$$\mathbb{H}^{\nu}(\Omega) \hookrightarrow W^{2\nu,2}(\Omega) \hookrightarrow L^{\frac{2N}{4\mu-2}}(\Omega). \tag{5.93}$$

- The Sobolev embedding $\mathbb{H}^{\nu+(\vartheta'-\vartheta)}(\Omega) \hookrightarrow W^{2\nu+2(\vartheta'-\vartheta),2}(\Omega) \hookrightarrow W^{2-\frac{N}{4},2}(\Omega)$ holds since $\nu + (\vartheta' - \vartheta) \geq \frac{1}{2} + \frac{4-N}{8} = 1 - \frac{N}{8}$ by using the assumption $\vartheta' \in \left[\vartheta + \frac{4-N}{8}, 1\right)$. In addition, $W^{2-\frac{N}{4},2}(\Omega) \hookrightarrow W^{1,\frac{4N}{3N-4}}(\Omega)$ as $2-\frac{N}{4} \geq 1$, $2 - \frac{N}{4} - 1 = \frac{N}{2} - N/\left(\frac{4N}{3N-4}\right)$. Therefore, we obtain the following Sobolev embedding:

$$\mathbb{H}^{\nu+(\vartheta'-\vartheta)}(\Omega) \hookrightarrow W^{2\nu+2(\vartheta'-\vartheta),2}(\Omega) \hookrightarrow W^{2-\frac{N}{4},2}(\Omega) \hookrightarrow W^{1,\frac{4N}{3N-4}}(\Omega). \tag{5.94}$$

On account of the above embeddings, and the Hölder's inequality, we deduce the following chain of the estimates:

$$\begin{aligned}
&\|G(v_1) - G(v_2)\|_{\mathbb{H}^{\sigma}(\Omega)} \\
&\lesssim \rho(t)\left(\|(v_1 \cdot \nabla)(v_1 - v_2)\|_{\mathbb{H}^{\sigma}(\Omega)} + \|((v_1 - v_2) \cdot \nabla)v_2\|_{\mathbb{H}^{\sigma}(\Omega)}\right)
\end{aligned}$$

$$
\begin{aligned}
&\lesssim \rho(t)\left(\|(v_1\cdot\nabla)(v_1-v_2)\|_{L^{\frac{2N}{N-4\sigma}}(\Omega)} + \|((v_1-v_2)\cdot\nabla)v_2\|_{L^{\frac{2N}{N-4\sigma}}(\Omega)} \right) \\
&\lesssim \rho(t)\left(\|v_1\|_{L^{\frac{2N}{N-4\sigma}\frac{N-4\sigma}{4\mu-2}}(\Omega)} \|\nabla(v_1-v_2)\|_{L^{\frac{2N}{N-4\sigma}\frac{N-4\sigma}{N+2-4\nu}}(\Omega)} \right. \\
&\quad \left. + \|v_1-v_2\|_{L^{\frac{2N}{N-4\sigma}\frac{N-4\sigma}{4\mu-2}}(\Omega)} \|\nabla v_2\|_{L^{\frac{2N}{N-4\sigma}\frac{N-4\sigma}{N+2-4\nu}}(\Omega)} \right) \\
&= \rho(t)\left(\|v_1\|_{L^{\frac{2N}{4\mu-2}}(\Omega)} \|\nabla(v_1-v_2)\|_{L^{\frac{2N}{N+2-4\nu}}(\Omega)} \right. \\
&\quad \left. + \|v_1-v_2\|_{L^{\frac{2N}{4\mu-2}}(\Omega)} \|\nabla v_2\|_{L^{\frac{2N}{N+2-4\nu}}(\Omega)} \right) \\
&\lesssim \rho(t)\left(\|v_1\|_{\mathbb{H}^\nu(\Omega)} + \|v_2\|_{\mathbb{H}^\nu(\Omega)} \right) \|v_1-v_2\|_{\mathbb{H}^\nu(\Omega)},
\end{aligned}
$$

where the chain (5.91) has been used in the first estimate, the triangle inequality in the second estimate, the Hölder's inequality with the dual numbers $\frac{N-4\sigma}{4\mu-2}$, $\frac{N-4\sigma}{N+2-4\nu}$ in the third one, and the chains (5.92) and (5.93) in the last one. This means that G is really a critical nonlinearity from $\mathbb{H}^\nu(\Omega)$ to $\mathbb{H}^\sigma(\Omega)$ with respect to $s = 1$ and $\mathfrak{N}(v_1, v_2) = \|v_1\|_{\mathbb{H}^\nu(\Omega)} + \|v_2\|_{\mathbb{H}^\nu(\Omega)}$. Furthermore, we can write $K(t) = \rho(t)K_{\text{Bur}}$ with a positive constant K_{Bur}. Let us take ζ satisfies that

$$
-b < \zeta < \min\left\{-\mu, \frac{1}{\alpha} - 2\vartheta\right\},
$$

and then one has $K(t) \le K_0 t^{-\alpha\zeta}$, where $K_0 = C_\rho K_{\text{Bur}} T^{\alpha(b+\zeta)}$. Due to the above arguments, we consequently conclude that G fulfills the assumption ($\mathscr{H}_3$). One can check that all numbers in this theorem obviously satisfy the assumptions of Theorem 5.14. Hence, we can apply Theorem 5.14, and the chain (5.93) ensures that the problem (5.88) has a unique mild solution

$$
u \in C^{\alpha\vartheta}((0,T]; L^{\frac{2N}{4\mu-2}}(\Omega)) \cap C^{\eta_{cri}}\left([0,T]; \mathbb{H}^{\nu-\eta}(\Omega)\right),
$$

with C_ρ is small enough. Besides, the boundedness (5.76) and Part (b) of Theorem 5.14 can be combined to allow the following estimate:

$$
t^{\alpha\vartheta}\|u(t)\|_{L^{\frac{2N}{4\mu-2}}(\Omega)} + \frac{\|u(t+\gamma)-u(t)\|_{\mathbb{H}^{\nu-\eta}(\Omega)}}{\gamma^{\eta_{cri}}} \lesssim \|f\|_{\mathbb{H}^{\nu+(1-\vartheta)}(\Omega)},
$$

i.e., the inequality (5.89) is easily obtained. We now prove the spatial regularity. Indeed, Part (a) of Theorem 5.14 can be rewritten as $(-\Delta)^{\vartheta'-\vartheta}u(t) \in \mathbb{H}^\nu(\Omega)$ with respect to the estimate

$$
\left\|(-\Delta)^{\vartheta'-\vartheta}u(t)\right\|_{L^{\frac{2N}{4\mu-2}}(\Omega)} \lesssim \left\|(-\Delta)^{\vartheta'-\vartheta}u(t)\right\|_{\mathbb{H}^\nu(\Omega)} \lesssim t^{-\alpha\vartheta'}\|f\|_{\mathbb{H}^{\nu+(1-\vartheta)}(\Omega)}.
$$

On the other hand, by using the chain (5.94), we deduce that

$$\left\|\nabla u(t)\right\|_{L^{\frac{4N}{3N-4}}(\Omega)} \lesssim \left\|(-\Delta)^{\vartheta'-\vartheta} u(t)\right\|_{\mathbb{H}^\nu(\Omega)} \lesssim t^{-\alpha\vartheta'} \|f\|_{\mathbb{H}^{\nu+(1-\vartheta)}(\Omega)},$$

which implies the inequality (5.90).

5.3.5 *Proof of Theorems*

In this subsection, we give full proofs for Theorem 5.12, Theorem 5.13, and Theorem 5.14. We prove Theorem 5.12 by using some new techniques of the Picard approximation method. We show Theorem 5.13 by applying Banach fixed point theorem. And we end the subsection by proving Theorem 5.14. For the sake of convenience, some important constants, which will be used in the proofs, will be listed in part **(AP.)** of the Appendix.

5.3.5.1 Proof of Theorem 5.12

Let us begin with the proof of Theorem 5.12 by using Picard approximation method. We construct a Picard sequence defined by Lemma 5.15. With some appropriate assumptions, we will bound the sequence by a power function. Then, we can prove it is a Cauchy sequence in a Banach space as Lemma 5.16. Now, we consider two following lemmas.

Lemma 5.15 *Let the Picard sequence $\{w^{(k)}\}_{k=1,2,\dots}$ be defined by $w^{(1)}(t) = f$, and*

$$\begin{aligned} w^{(k+1)}(t) =& \mathbf{B}_\alpha(t,T)f + \int_0^t \mathbf{P}_\alpha(t-r)G(r, w^{(k)}(r))dr \\ &- \int_0^T \mathbf{B}_\alpha(t,T)\mathbf{P}_\alpha(T-r)G(r, w^{(k)}(r))dr, \quad 0 \le t \le T. \end{aligned} \tag{5.95}$$

Then, for all $t > 0$, $k \in \mathbb{N}^+$ it holds

$$\left\|w^{(k)}(t)\right\|_{\mathbb{H}^\nu(\Omega)} \le \mathscr{N}_1 t^{-\alpha(1-\theta)} \|f\|_{\mathbb{H}^{\theta+\nu}(\Omega)}, \tag{5.96}$$

*where $\mathscr{N}_1$ is given by **(AP.)** in the Appendix.*

Lemma 5.16 *Let $\{u^{(k)}\}_{k=1,2,\dots}$ be the sequence defined by Lemma 5.15, and then it is a bounded and Cauchy sequence in the Banach space $L^p(0,T;\mathbb{H}^\nu(\Omega))$ with $p \in \left[1, \frac{1}{\alpha(1-\theta)}\right)$.*

Proof of Lemma 5.15 Let us consider the case $k = 1$. First, $\|f\|_{\mathbb{H}^\nu(\Omega)}$ is bounded by $C_1(\nu,\theta)\|f\|_{\mathbb{H}^{\nu+\theta}(\Omega)}$ upon the embedding (5.70). Furthermore, it is obvious to see from **(AP.)** in the Appendix that $\mathscr{N}_1 \geq C_1(\nu,\theta)t^{\alpha(1-\theta)}$ by noting the number $\alpha(1-\theta)$ be contained in the interval $(0, 1)$. These easily imply the desired inequality (5.96) for $k = 1$. Assume that (5.96) holds for $k = n$. This means that

$$\left\|w^{(n)}(t)\right\|_{\mathbb{H}^\nu(\Omega)} \leq \mathscr{N}_1 t^{-\alpha(1-\theta)}\|f\|_{\mathbb{H}^{\nu+\theta}(\Omega)}. \tag{5.97}$$

We show that (5.96) holds for $k = n + 1$. Thanks to the definition (5.66), using the fact that $\{\varphi_j\}$ is an orthonormal basis of $L^2(\Omega)$, and then using Lemma 5.14, one arrives at

$$\begin{aligned}
&\left\|\mathbf{B}_\alpha(t,T)f\right\|_{\mathbb{H}^\nu(\Omega)} \\
\leq& M_\alpha m_\alpha^{-1}T^{\alpha(1-\theta)}(T^{\alpha\theta}+\lambda_1^{-\theta})t^{-\alpha(1-\theta)}\left(\sum_{j=1}^{\infty}\lambda_j^{2(\nu+\theta)}\langle f,\varphi_j\rangle^2\right)^{1/2} \\
=& M_\alpha m_\alpha^{-1}T^{\alpha(1-\theta)}\left(T^{\alpha\theta}+\lambda_1^{-\theta}\right)t^{-\alpha(1-\theta)}\|f\|_{\mathbb{H}^{\nu+\theta}(\Omega)}.
\end{aligned}$$

Next, let us estimate the integrals by using the assumption $(\mathscr{H}_1)$. The idea is to try to bound them by the convergent improper integrals. Indeed, one can show that

$$\begin{aligned}
&\left\|\int_0^t \mathbf{P}_\alpha(t-r)G(r,w^{(n)}(r))dr\right\|_{\mathbb{H}^\nu(\Omega)} \\
\leq& \int_0^t \left\|\sum_{j=1}^{\infty}(t-r)^{\alpha-1}E_{\alpha,\alpha}(-\lambda_j(t-r)^\alpha)G_j(r,w^{(n)}(r))\varphi_j\right\|_{\mathbb{H}^\nu(\Omega)} dr \\
\leq& M_\alpha\lambda_1^{-\theta}\int_0^t (t-r)^{\alpha(1-\theta)-1}\left\|G(r,w^{(n)}(r))\right\|_{\mathbb{H}^\nu(\Omega)} dr \\
\leq& \|L_1\|_{L^\infty(0,T)}M_\alpha\lambda_1^{-\theta}\int_0^t (t-r)^{\alpha(1-\theta)-1}\left\|w^{(n)}(r)\right\|_{\mathbb{H}^\nu(\Omega)} dr.
\end{aligned} \tag{5.98}$$

On the other hand, the quantity $\mathscr{E}_{\alpha,T}(-\lambda_j t^\alpha)(T-r)^{\alpha-1}E_{\alpha,\alpha}(-\lambda_j(T-r)^\alpha)$ is obviously bounded by $\mathscr{M}_1(T-r)^{\alpha(1-\theta)-1}t^{-\alpha(1-\theta)}$ due to applying Lemma 5.14. Here, the constant $\mathscr{M}_1$ is given by **(AP.)** in the Appendix. We then get the following estimate

$$\left\| \int_0^T \mathbf{B}_\alpha(t,T)\mathbf{P}_\alpha(T-r)G(r,w^{(n)}(r))dr \right\|_{\mathbb{H}^\nu(\Omega)}$$
$$\leq \int_0^T \left\| \sum_{j=1}^{\infty} (T-r)^{\alpha-1}\mathscr{E}_{\alpha,T}(-\lambda_j t^\alpha)E_{\alpha,\alpha}(-\lambda_j(T-r)^\alpha)G_j(r,w^{(n)}(r))\varphi_j \right\|_{\mathbb{H}^\nu(\Omega)} dr$$
$$\leq \|L_1\|_{L^\infty(0,T)}\mathscr{M}_1 t^{-\alpha(1-\theta)} \int_0^T (T-r)^{\alpha(1-\theta)-1} \left\| w^{(n)}(r) \right\|_{\mathbb{H}^\nu(\Omega)} dr. \tag{5.99}$$

According to the above inequalities, we need to estimate the integral

$$\int_0^t (t-r)^{\alpha(1-\theta)-1}\|w^{(n)}(r)\|_{\mathbb{H}^\nu(\Omega)}dr.$$

To do this, we will apply the inductive hypothesis (5.97). Moreover, by also using the facts that $1 \leq T^{\alpha(1-\theta)}t^{-\alpha(1-\theta)}$ for all $0 < t \leq T$, and $\int_0^t (t-r)^{\alpha(1-\theta)-1}r^{-\alpha(1-\theta)}dr$ is equal to $\pi/\sin(\pi\alpha(1-\theta))$, we consequently obtain the following estimates:

$$\begin{aligned}
&\int_0^t (t-r)^{\alpha(1-\theta)-1}\|w^{(n)}(r)\|_{\mathbb{H}^\nu(\Omega)}dr \\
\leq &\mathscr{N}_1\|f\|_{\mathbb{H}^{\nu+\theta}(\Omega)} \int_0^t (t-r)^{\alpha(1-\theta)-1}r^{-\alpha(1-\theta)}dr \\
\leq &\mathscr{N}_1\|f\|_{\mathbb{H}^{\nu+\theta}(\Omega)} \frac{\pi T^{\alpha(1-\theta)}}{\sin(\pi\alpha(1-\theta))} t^{-\alpha(1-\theta)}.
\end{aligned} \tag{5.100}$$

Here, in the last inequality, we use **(AP.1)** in the Appendix. By similar arguments as above, we get

$$\begin{aligned}
&\left\| \int_0^t \mathbf{P}_\alpha(t-r)G(r,w^{(n)}(r))dr \right\|_{\mathbb{H}^\nu(\Omega)} \\
&+ \left\| \int_0^T \mathbf{B}_\alpha(t,T)\mathbf{P}_\alpha(T-r)G(r,w^{(n)}(r))dr \right\|_{\mathbb{H}^\nu(\Omega)} \\
\leq &\|L_1\|_{L^\infty(0,T)}\mathscr{U}_1\mathscr{N}_1 t^{-\alpha(1-\theta)}\|f\|_{\mathbb{H}^{\nu+\theta}(\Omega)}.
\end{aligned} \tag{5.101}$$

From some preceding estimates and by some simple computations, we can find that

$$\|w^{(n+1)}(t)\|_{\mathbb{H}^\nu(\Omega)} \leq \left\| \mathbf{B}_\alpha(t,T)f \right\|_{\mathbb{H}^\nu(\Omega)} + \left\| \int_0^t \mathbf{P}_\alpha(t-r)G(r,w^{(n)}(r))dr \right\|_{\mathbb{H}^\nu(\Omega)}$$

$$+\left\| \int_0^T \mathbf{B}_\alpha(t,T)\mathbf{P}_\alpha(T-r)G(r,w^{(n)}(r))dr \right\|_{\mathbb{H}^\nu(\Omega)}$$

$$\leq \Big(\mathscr{M}_1 M_\alpha^{-1} + \|L_1\|_{L^\infty(0,T)} \mathscr{U}_1 \mathscr{N}_1 \Big) t^{-\alpha(1-\theta)} \|f\|_{\mathbb{H}^{\nu+\theta}(\Omega)}$$

$$\leq \mathscr{N}_1 t^{-\alpha(1-\theta)} \|f\|_{\mathbb{H}^{\nu+\theta}(\Omega)}.$$

By inductive method, we deduce that (5.97) holds for any $k = n \in \mathbb{N}^+$.

Proof of Lemma 5.16 Since $p \in [1, \frac{1}{\alpha(1-\theta)})$, we know that the function $t \mapsto t^{-\alpha(1-\theta)}$ is $L^p(0,T)$-integrable which implies that $\{u^{(n)}\}$ is a bounded sequence in $L^p(0,T;\mathbb{H}^\nu(\Omega))$. Hence, it is necessary to prove $\{u^{(n)}\}$ is a Cauchy sequence. By using the notation $\mathbf{u}^{(n,k)} := u^{(n+k)} - u^{(n)}$ and using triangle inequality, one has the following estimate:

$$\left\| \mathbf{u}^{(n+1,k)}(t) \right\|_{\mathbb{H}^\nu(\Omega)}$$

$$\leq \|L_1\|_{L^\infty(0,T)} M_\alpha \lambda_1^{-\theta} \int_0^t (t-r)^{\alpha(1-\theta)-1} \left\| \mathbf{u}^{(n,k)}(r) \right\|_{\mathbb{H}^\nu(\Omega)} dr$$

$$+ \|L_1\|_{L^\infty(0,T)} \mathscr{M}_1 t^{-\alpha(1-\theta)} \int_0^T (T-r)^{\alpha(1-\theta)-1} \left\| \mathbf{u}^{(n,k)}(r) \right\|_{\mathbb{H}^\nu(\Omega)} dr.$$

Similarly as the proof of Lemma 5.15, one can use the inductive hypothesis to estimate the above right-hand side. Then by iterating the same computations in Lemma 5.15, we can bound $\|\mathbf{u}^{(n+1,k)}(\cdot,t)\|_{\mathbb{H}^\nu(\Omega)}$ by the quantity $2\mathscr{N}_1 (\|L_1\|_{L^\infty(0,T)} \mathscr{U}_1)^n \, t^{-\alpha(1-\theta)} \|f\|_{\mathbb{H}^{\nu+\theta}(\Omega)}$. Summarily, one can obtain the following conclusion by the inductive method:

$$\left\| \mathbf{u}^{(n,k)}(t) \right\|_{\mathbb{H}^\nu(\Omega)} \leq 2\mathscr{N}_1 \Big(\|L_1\|_{L^\infty(0,T)} \mathscr{U}_1 \Big)^{n-1} t^{-\alpha(1-\theta)} \|f\|_{\mathbb{H}^{\nu+\theta}(\Omega)}, \tag{5.102}$$

which completed the proof by letting $n \to \infty$.

Proof of Theorem 5.12 We firstly prove the existence of a mild solution u in the space $L^p(0,T;\mathbb{H}^\nu(\Omega))$ and then obtain its continuity in the following steps (1 and 2). After that, we will present the proofs the Parts (a)–(d) in the sequel.

Step 1. Prove the existence of a mild solution u in the space $L^p(0,T;\mathbb{H}^\nu(\Omega))$.

Since $L^p(0,T;\mathbb{H}^\nu(\Omega))$ is a Banach space and $\{u^{(n)}\}$ is a Cauchy sequence in $L^p(0,T;\mathbb{H}^\nu(\Omega))$, thanks to Lemmas 5.15, 5.16, we deduce that there exists a function $u \in L^p(0,T;\mathbb{H}^\nu(\Omega))$ such that $\lim_{n\to+\infty} u^{(n)} = u$. Now, we show that u is a mild solution of the problem (5.55)–(5.56) by showing that $u = \overline{\mathbf{J}}u$, where

$$\begin{aligned}\bar{\mathbf{J}}u(t) :=&\mathbf{B}_\alpha(t,T)f + \int_0^t \mathbf{P}_\alpha(t-r)G(r,u(r))dr\\ &- \int_0^T \mathbf{B}_\alpha(t,T)\mathbf{P}_\alpha(T-r)G(r,u(r))dr.\end{aligned}\tag{5.103}$$

Since the sequence $\{u^{(n)}\}$ converges to u in $L^p(0,T;\mathbb{H}^\nu(\Omega))$-norm, there exists a sub-sequence $\{u^{(n_m)}\}$ that pointwise converges to u, i.e., $u^{(n_m)}(t) \to u(t)$ in $\mathbb{H}^\nu(\Omega)$-norm for almost everywhere t in $(0,T)$. Let us denote by $\mathbf{v}^{(n_m)} := u^{(n_m)} - u$. This fact and taking $k \to \infty$ in the estimate (5.102) allow us to obtain

$$\left\|\mathbf{v}^{(n_m)}(t)\right\|_{\mathbb{H}^\nu(\Omega)} \le 2\mathscr{N}_1\Big(\|L_1\|_{L^\infty(0,T)}\mathscr{U}_1\Big)^{n_m-1} t^{-\alpha(1-\theta)}\|f\|_{\mathbb{H}^{\nu+\theta}(\Omega)},\tag{5.104}$$

for almost everywhere t in $(0,T)$. Moreover, we note from Lemma 5.15 that this sub-sequence is also bounded by the power function $t \mapsto t^{-\alpha(1-\theta)}$. These help us to apply dominated convergence theorem as follows. Indeed, it is obvious that the quantities

$$\int_0^t (t-r)^{\alpha(1-\theta)-1}\left\|\mathbf{v}^{(n_m)}(r)\right\|_{\mathbb{H}^\nu(\Omega)}dr,$$

$$t^{-\alpha(1-\theta)}\int_0^T (T-r)^{\alpha(1-\theta)-1}\left\|\mathbf{v}^{(n_m)}(r)\right\|_{\mathbb{H}^\nu(\Omega)}dr$$

pointwise converge to zero by (5.104) as $n_m \to \infty$ and are bounded by $L^p(0,T)$-integrable functions. Thus, the same computations as (5.98) show

$$\begin{aligned}&\left\|\int_0^t \mathbf{P}_\alpha(t-r)\Big(G(r,u^{(n_m)}(r)) - G(r,u(r))\Big)dr\right\|^p_{L^p(0,T;\mathbb{H}^\nu(\Omega))}\\ &= \int_0^T \left\|\int_0^t \mathbf{P}_\alpha(t-r)\Big(G(r,u^{(n_m)}(r)) - G(r,u(r))\Big)dr\right\|^p_{\mathbb{H}^\nu(\Omega)} dt\\ &\le \int_0^T \left\{\int_0^t (t-r)^{\alpha(1-\theta)-1}\left\|\mathbf{v}^{(n_m)}(r)\right\|_{\mathbb{H}^\nu(\Omega)}dr\right\}^p dt,\end{aligned}\tag{5.105}$$

and by using a similar way as in (5.99), we have the following bound:

$$\begin{aligned}&\left\|\int_0^T \mathbf{B}_\alpha(t,T)\mathbf{P}_\alpha(T-r)\Big(G(r,u^{(n_m)}(r)) - G(r,u(r))\Big)dr\right\|^p_{L^p(0,T;\mathbb{H}^\nu(\Omega))}\\ &= \int_0^T \left\|\int_0^T \mathbf{B}_\alpha(t,T)\mathbf{P}_\alpha(T-r)\Big(G(r,u^{(n_m)}(r)) - G(r,u(r))\Big)dr\right\|^p_{\mathbb{H}^\nu(\Omega)} dt\\ &\le \int_0^T \left\{t^{-\alpha(1-\theta)}\int_0^T (T-r)^{\alpha(1-\theta)-1}\left\|\mathbf{v}^{(n_m)}(r)\right\|_{\mathbb{H}^\nu(\Omega)}dr\right\}^p dt.\end{aligned}\tag{5.106}$$

The right-hand sides of (5.105) and (5.106) tend to zero when n_m goes to positive infinity. The above arguments conclude that $u = \overline{\mathbf{J}}u$, and so that is a mild solution of the problem (5.55)–(5.56) in $L^p(0, T; \mathbb{H}^\nu(\Omega))$. By taking the limit of the left-hand side of (5.96), we obtain

$$\|u(t)\|_{\mathbb{H}^\nu(\Omega)} \lesssim t^{-\alpha(1-\theta)} \|f\|_{\mathbb{H}^{\nu+\theta}(\Omega)}. \tag{5.107}$$

Step 2. Prove $u \in C^{\alpha(1-\theta)}\big((0, T]; \mathbb{H}^\nu(\Omega)\big)$.

We need to estimate $u(\widetilde{t}) - u(t)$ in $\mathbb{H}^\nu$-norm, for all $0 < t \le \widetilde{t} \le T$. By the formulation (5.65), the triangle inequality yields that

$$\begin{aligned}
&\left\|u(\widetilde{t}) - u(t)\right\|_{\mathbb{H}^\nu(\Omega)} \\
\le&\left\|\Big(\mathbf{B}_\alpha(\widetilde{t}, T) - \mathbf{B}_\alpha(t, T)\Big) f\right\|_{\mathbb{H}^\nu(\Omega)} \\
&+\left\|\int_0^t \Big(\mathbf{P}_\alpha(\widetilde{t} - r) - \mathbf{P}_\alpha(t - r)\Big) G(r, u(r)) dr\right\|_{\mathbb{H}^\nu(\Omega)} \\
&+\left\|\int_t^{\widetilde{t}} \mathbf{P}_\alpha(\widetilde{t} - r) G(r, u(r)) dr\right\|_{\mathbb{H}^\nu(\Omega)} \\
&+\left\|\int_0^T \Big(\mathbf{B}_\alpha(\widetilde{t}, T) - \mathbf{B}_\alpha(t, T)\Big) \mathbf{P}_\alpha(T - r) G(r, u(r)) dr\right\|_{\mathbb{H}^\nu(\Omega)} \\
=:&\|\mathfrak{M}_1\|_{\mathbb{H}^\nu(\Omega)} + \|\mathfrak{M}_2\|_{\mathbb{H}^\nu(\Omega)} + \|\mathfrak{M}_3\|_{\mathbb{H}^\nu(\Omega)} + \|\mathfrak{M}_4\|_{\mathbb{H}^\nu(\Omega)}.
\end{aligned} \tag{5.108}$$

In what follows, we will estimate the terms $\mathfrak{M}_j$ for $1 \le j \le 4$.

Estimate of $\mathfrak{M}_1$. Using the fact that $\dfrac{d}{dt} E_{\alpha,1}(-\lambda_j t^\alpha) = -\lambda_j t^{\alpha-1} E_{\alpha,\alpha}(-\lambda_j t^\alpha)$, we find that

$$\frac{E_{\alpha,1}(-\lambda_j \widetilde{t}^\alpha) - E_{\alpha,1}(-\lambda_j t^\alpha)}{E_{\alpha,1}(-\lambda_j T^\alpha)} = \int_t^{\widetilde{t}} -\lambda_j r^{\alpha-1} \frac{E_{\alpha,\alpha}(-\lambda_j r^\alpha)}{E_{\alpha,1}(-\lambda_j T^\alpha)} dr.$$

This leads to

$$\begin{aligned}
&\left\|\Big(\mathbf{B}_\alpha(\widetilde{t}, T) - \mathbf{B}_\alpha(t, T)\Big) f\right\|_{\mathbb{H}^\nu(\Omega)} \\
\lesssim&\int_t^{\widetilde{t}} \left\|\sum_{j \in \mathbb{N}} \lambda_j r^{\alpha-1} \frac{E_{\alpha,\alpha}(-\lambda_j r^\alpha)}{E_{\alpha,1}(-\lambda_j T^\alpha)} f_j \varphi_j\right\|_{\mathbb{H}^\nu(\Omega)} dr \\
\lesssim&\left(\int_t^{\widetilde{t}} r^{\alpha(\theta-1)-1} dr\right) \|f\|_{\mathbb{H}^{\nu+\theta}(\Omega)} \\
=&\frac{\widetilde{t}^{\alpha(1-\theta)} - t^{\alpha(1-\theta)}}{\alpha(1-\theta) t^{\alpha(1-\theta)} \widetilde{t}^{\alpha(1-\theta)}} \|f\|_{\mathbb{H}^{\nu+\theta}(\Omega)},
\end{aligned} \tag{5.109}$$

where we have used the fact that

$$\lambda_j r^{\alpha-1}\frac{E_{\alpha,\alpha}(-\lambda_j r^\alpha)}{E_{\alpha,1}(-\lambda_j T^\alpha)} \lesssim \lambda_j r^{\alpha-1}\frac{1+\lambda_j T^\alpha}{1+(\lambda_j r^\alpha)^2} \lesssim \lambda_j^\theta r^{\alpha(\theta-1)-1}. \tag{5.110}$$

By noting that $0 < \alpha(1-\theta) < 1$, we now have $\widetilde{t}^{\alpha(1-\theta)} - t^{\alpha(1-\theta)} \leq (\widetilde{t}-t)^{\alpha(1-\theta)}$ and furthermore $\alpha(1-\theta)t^{\alpha(1-\theta)}\widetilde{t}^{\alpha(1-\theta)} \geq \alpha(1-\theta)t^{2\alpha(1-\theta)}$. Consequently, employing the estimate (5.109), we arrive at

$$\|\mathfrak{M}_1\|_{\mathbb{H}^\nu(\Omega)} \lesssim t^{-2\alpha(1-\theta)}(\widetilde{t}-t)^{\alpha(1-\theta)}\|f\|_{\mathbb{H}^{\nu+\theta}(\Omega)}. \tag{5.111}$$

Estimate of $\mathfrak{M}_2$. In virtue of $\frac{d}{d\rho}\left(\rho^{\alpha-1}E_{\alpha,\alpha}(-\lambda_j\rho^\alpha)\right) = \rho^{\alpha-2}E_{\alpha,\alpha-1}(-\lambda_j\rho^\alpha)$, for all $\rho > 0$ (see Proposition 1.14), we obtain that

$$\begin{aligned}
&\int_0^t \left((\widetilde{t}-r)^{\alpha-1}E_{\alpha,\alpha}(-\lambda_j(\widetilde{t}-r)^\alpha) - (t-r)^{\alpha-1}E_{\alpha,\alpha}(-\lambda_j(t-r)^\alpha)\right) \\
&\quad\times G_j(r,u(r))dr \\
&= \int_0^t \int_{t-r}^{\widetilde{t}-r} \rho^{\alpha-2}E_{\alpha,\alpha-1}(-\lambda_j\rho^\alpha)G_j(r,u(r))d\rho dr.
\end{aligned} \tag{5.112}$$

This implies that

$$\begin{aligned}
\|\mathfrak{M}_2\|_{\mathbb{H}^\nu(\Omega)} &\lesssim \int_0^t \int_{t-r}^{\widetilde{t}-r} \rho^{\alpha-2}\,\|G(r,u(r))\|_{\mathbb{H}^\nu(\Omega)}\,d\rho dr \\
&\lesssim (\widetilde{t}-t)^{\alpha-1}\int_0^t \|u(r)\|_{\mathbb{H}^\nu(\Omega)}\,dr \\
&\lesssim (\widetilde{t}-t)^{\alpha-1}\|u\|_{L^p(0,T;\mathbb{H}^\nu(\Omega))} \\
&\lesssim (\widetilde{t}-t)^{\alpha-1}\|f\|_{\mathbb{H}^{\nu+\theta}(\Omega)},
\end{aligned} \tag{5.113}$$

where we note that

$$\int_{t-r}^{\widetilde{t}-r}\rho^{\alpha-2}d\rho = \frac{(\widetilde{t}-r)^{\alpha-1}-(t-r)^{\alpha-1}}{\alpha-1} \leq \frac{(\widetilde{t}-t)^{\alpha-1}}{\alpha-1},$$

$$\rho^{\alpha-2}E_{\alpha,\alpha-1}(-\lambda_j\rho^\alpha) \lesssim \rho^{\alpha-2}\frac{1}{1+\lambda_j\rho^\alpha} \lesssim \rho^{\alpha-2}.$$

Estimate of $\mathfrak{M}_3$. For all $t < r < \widetilde{t}$, using the inequality $|E_{\alpha,\alpha}(-\lambda_j(\widetilde{t}-r)^\alpha)| \leq M_\alpha$ and the fact that $(\widetilde{t}-r)^{\alpha-1} \leq (\widetilde{t}-t)^{\alpha-1}$, we have

$$\begin{aligned}\left\|\mathfrak{M}_3\right\|_{\mathbb{H}^\nu(\Omega)} \leq & M_\alpha(\widetilde{t}-t)^{\alpha-1}\int_t^{\widetilde{t}}\|G(r,u(r))\|_{\mathbb{H}^\nu(\Omega)}\,dr \\ \lesssim & (\widetilde{t}-t)^{\alpha-1}\|f\|_{\mathbb{H}^{\nu+\theta}(\Omega)}.\end{aligned} \tag{5.114}$$

Estimate of $\mathfrak{M}_4$. Using (5.110), we get

$$\left\|\sum_{j=1}^{\infty}(T-r)^{\alpha-1}\lambda_j\rho^{\alpha-1}\frac{E_{\alpha,\alpha}(-\lambda_j\rho^\alpha)}{E_{\alpha,1}(-\lambda_jT^\alpha)}E_{\alpha,\alpha}(-\lambda_j(T-r)^\alpha)G_j(r,u(r))\varphi_j\right\|_{\mathbb{H}^\nu(\Omega)}$$
$$\lesssim(T-r)^{\alpha(1-\theta)-1}\rho^{\alpha(\theta-1)-1}\,\|G(r,u(r))\|_{\mathbb{H}^\nu(\Omega)}\,.$$

By using the fact that $\widetilde{t}^{\alpha(1-\theta)}-t^{\alpha(1-\theta)}\leq(\widetilde{t}-t)^{\alpha(1-\theta)}$ and $t^{\alpha(1-\theta)}\widetilde{t}^{\alpha(1-\theta)}\geq t^{2\alpha(1-\theta)}$, we get that

$$\begin{aligned}&\|\mathfrak{M}_4\|_{\mathbb{H}^\nu(\Omega)} \\ \lesssim&\int_0^T\int_t^{\widetilde{t}}(T-r)^{\alpha(1-\theta)-1}\xi^{\alpha(\theta-1)-1}\,\|G(r,u(r))\|_{\mathbb{H}^\nu(\Omega)}\,d\xi dr \\ =&\frac{\widetilde{t}^{\alpha(1-\theta)}-t^{\alpha(1-\theta)}}{\alpha(1-\theta)t^{\alpha(1-\theta)}\widetilde{t}^{\alpha(1-\theta)}}\int_0^T(T-r)^{\alpha(1-\theta)-1}\,\|G(r,u(r))\|_{\mathbb{H}^\nu(\Omega)}\,dr \\ \lesssim& t^{-2\alpha(1-\theta)}(\widetilde{t}-t)^{\alpha(1-\theta)}\int_0^T(T-r)^{\alpha(1-\theta)-1}\,\|G(r,u(r))\|_{\mathbb{H}^\nu(\Omega)}\,dr.\end{aligned} \tag{5.115}$$

It is easy to see that

$$\begin{aligned}&\int_0^T(T-r)^{\alpha(1-\theta)-1}\,\|G(u(r))\|_{\mathbb{H}^\nu(\Omega)} \\ &\quad\lesssim\int_0^T(T-r)^{\alpha(1-\theta)-1}\,\|u(r)\|_{\mathbb{H}^\nu(\Omega)}\,dr \\ &\quad\lesssim\|f\|_{\mathbb{H}^{\nu+\theta}(\Omega)}\int_0^T(T-r)^{\alpha(1-\theta)-1}r^{-\alpha(1-\theta)}dr \\ &\quad\lesssim\|f\|_{\mathbb{H}^{\nu+\theta}(\Omega)}.\end{aligned}$$

The latter estimate together with (5.115) leads to

$$\|\mathfrak{M}_4\|_{\mathbb{H}^\nu(\Omega)}\lesssim t^{-2\alpha(1-\theta)}(\widetilde{t}-t)^{\alpha(1-\theta)}\|f\|_{\mathbb{H}^{\nu+\theta}(\Omega)}. \tag{5.116}$$

Estimate $u(\widetilde{t})-u(t)$. Combining with (5.108), (5.111), (5.113), (5.114), and (5.116), we deduce that

$$\|u(\widetilde{t})-u(t)\|_{\mathbb{H}^\nu(\Omega)}\lesssim\left[t^{-2\alpha(1-\theta)}(\widetilde{t}-t)^{\alpha(1-\theta)}+(\widetilde{t}-t)^{\alpha-1}\right]\|f\|_{\mathbb{H}^{\nu+\theta}(\Omega)},$$

which implies that $u \in C^{\alpha(1-\theta)}\big((0,T];\mathbb{H}^{\nu}(\Omega)\big)$, and so that the inequality (5.71) can be obtained by using (5.107) .

In what follows, we give the proofs of Part (a), (b), (c), (d).

Part (a). Prove $u \in L^p(0,T;\mathbb{H}^{\nu+\theta-\theta'}(\Omega))$, for any $p \in \Big[1, \frac{1}{\alpha(1-\theta')}\Big)$.

Lemma 5.14 yields the estimate

$$\begin{aligned}\Big\|\mathbf{B}_\alpha(t,T)f\Big\|^2_{\mathbb{H}^{\nu+\theta-\theta'}(\Omega)} &= \sum_{j=1}^{\infty}\lambda_j^{2\nu+2\theta-2\theta'}\Big|\mathscr{E}_{\alpha,T}(-\lambda_j t^\alpha)\Big|^2\langle f,\varphi_j\rangle^2 \\ &\lesssim t^{-2\alpha(1-\theta')}\sum_{j=1}^{\infty}\lambda_j^{2\nu+2\theta-2\theta'}\lambda_j^{2\theta'}\langle f,\varphi_j\rangle^2.\end{aligned} \tag{5.117}$$

Therefore, $\mathbf{B}_\alpha(t,T)f \in L^p(0,T;\mathbb{H}^{\nu+\theta-\theta'}(\Omega))$. Further, from assumption $(\mathscr{H}_1)$ and the estimate $(t-r)^{\alpha-1}E_{\alpha,\alpha}(-\lambda_j(t-r)^\alpha) \lesssim \lambda_j^{-(\theta-\theta')}(t-r)^{\alpha(1-(\theta-\theta'))-1}$ as in Lemma 5.14, we infer that

$$\begin{aligned}&\left\|\int_0^t \mathbf{P}_\alpha(t-r)G(r,u(r))dr\right\|_{\mathbb{H}^{\nu+\theta-\theta'}(\Omega)} \\ &\lesssim \int_0^t (t-r)^{\alpha(1-\theta+\theta')-1}\,\|G(r,u(r))\|_{\mathbb{H}^{\nu}(\Omega)}\,dr \\ &\lesssim \int_0^t (t-r)^{\alpha(1-\theta+\theta')-1}r^{-\alpha(1-\theta)}dr\,\|f\|_{\mathbb{H}^{\nu+\theta}(\Omega)} \\ &\lesssim t^{-\alpha(1-\theta')}\|f\|_{\mathbb{H}^{\nu+\theta}(\Omega)},\end{aligned} \tag{5.118}$$

where we used the definition of the Beta function as **(AP.1)** in the Appendix, and the fact that $t^{\alpha\theta'} = t^{-\alpha(1-\theta')}t^\alpha \lesssim t^{-\alpha(1-\theta')}$. Now, we proceed to estimate the last term of u. Lemma 5.14 yields that

$$\mathscr{E}_{\alpha,T}(-\lambda_j t^\alpha)(T-r)^{\alpha-1}E_{\alpha,\alpha}(-\lambda_j(T-r)^\alpha) \lesssim \lambda_j^{\theta'-\theta}t^{-\alpha(1-\theta')}(T-r)^{\alpha(1-\theta)-1}.$$

This invokes from assumption $(\mathscr{H}_1)$ that

$$\begin{aligned}&\left\|\int_0^T \mathbf{B}_\alpha(t,T)\mathbf{P}_\alpha(T-r)G(r,u(r))dr\right\|_{\mathbb{H}^{\nu+\theta-\theta'}(\Omega)} \\ &\lesssim t^{-\alpha(1-\theta')}\int_0^T (T-r)^{\alpha(1-\theta)-1}\,\|G(r,u(r))\|_{\mathbb{H}^{\nu}(\Omega)}\,dr \\ &\lesssim t^{-\alpha(1-\theta')}\int_0^T (T-r)^{\alpha(1-\theta)-1}r^{-\alpha(1-\theta)}dr\,\|f\|_{\mathbb{H}^{\nu+\theta}(\Omega)} \\ &\lesssim t^{-\alpha(1-\theta')}\|f\|_{\mathbb{H}^{\nu+\theta}(\Omega)}.\end{aligned} \tag{5.119}$$

Taking the above estimates (5.117)–(5.119) together, we obtain that $u \in L^p(0, T; \mathbb{H}^{\nu+\theta-\theta'}(\Omega))$, for all $p \in \left[1, \frac{1}{\alpha(1-\theta')}\right)$, and complete this part.

Part (b). Show that $u \in C^{\eta_{glo}}\left([0, T]; \mathbb{H}^{\nu-\nu'}(\Omega)\right)$.

Let t and $\widetilde{t}$ satisfy that $0 \le t < \widetilde{t} \le T$. By using the equation (5.108) and the embedding $\mathbb{H}^{\nu}(\Omega) \hookrightarrow \mathbb{H}^{\nu-\nu'}(\Omega)$, we obtain

$$\begin{aligned}
&\|u(\widetilde{t}) - u(t)\|_{\mathbb{H}^{\nu-\nu'}(\Omega)} \\
\le&\|\mathfrak{M}_1\|_{\mathbb{H}^{\nu-\nu'}(\Omega)} + \|\mathfrak{M}_2\|_{\mathbb{H}^{\nu-\nu'}(\Omega)} + \|\mathfrak{M}_3\|_{\mathbb{H}^{\nu-\nu'}(\Omega)} + \|\mathfrak{M}_4\|_{\mathbb{H}^{\nu-\nu'}(\Omega)} \\
\lesssim&\|\mathfrak{M}_1\|_{\mathbb{H}^{\nu-\nu'}(\Omega)} + |\mathfrak{M}_4\|_{\mathbb{H}^{\nu-\nu'}(\Omega)} + \|\mathfrak{M}_2\|_{\mathbb{H}^{\nu}(\Omega)} + \|\mathfrak{M}_3\|_{\mathbb{H}^{\nu}(\Omega)}.
\end{aligned} \tag{5.120}$$

For the right-hand side of (5.108), we thus need to estimate the terms $\|\mathfrak{M}_1\|_{\mathbb{H}^{\nu-\nu'}(\Omega)}$, $|\mathfrak{M}_4\|_{\mathbb{H}^{\nu-\nu'}(\Omega)}$. We now continue to consider the following estimates.

Estimate $\|\mathfrak{M}_1\|_{\mathbb{H}^{\nu-\nu'}(\Omega)}$. It is easy to show that

$$\|\mathfrak{M}_1\|_{\mathbb{H}^{\nu-\nu'}(\Omega)} \lesssim (\widetilde{t} - t)^{\alpha(\theta+\nu'-1)} \|f\|_{\mathbb{H}^{\nu+\theta}(\Omega)}.$$

Estimate $\|\mathfrak{M}_4\|_{\mathbb{H}^{\nu-\nu'}(\Omega)}$. By a similar argument as in (5.115), we obtain

$$\|\mathfrak{M}_4\|_{\mathbb{H}^{\nu-\nu'}(\Omega)} \lesssim \int_0^T \int_t^{\widetilde{t}} (T-r)^{\alpha(1-\theta)-1} \rho^{\alpha(\theta+\nu'-1)-1} \|G(r, u(r))\|_{\mathbb{H}^{\nu}(\Omega)}\, d\rho dr. \tag{5.121}$$

Since $0 < \alpha(\theta + \nu' - 1) < 2$, we divide into two following cases.

Case 1. If $0 < \alpha(\theta + \nu' - 1) \le 1$, then applying $(a+b)^{\sigma} \le a^{\sigma} + b^{\sigma},\ a, b \ge 0, 0 < \sigma < 1$, we get

$$(\widetilde{t})^{\alpha(\theta+\nu'-1)} - t^{\alpha(\theta+\nu'-1)} \le (\widetilde{t} - t)^{\alpha(\theta+\nu'-1)}. \tag{5.122}$$

From two latter observations, we find that

$$\begin{aligned}
&\|\mathfrak{M}_4\|_{\mathbb{H}^{\nu-\nu'}(\Omega)} \\
\lesssim&\int_0^T \int_t^{\widetilde{t}} (T-r)^{\alpha(1-\theta)-1} \rho^{\alpha(\theta+\nu'-1)-1} \|G(r, u(r))\|_{\mathbb{H}^{\nu}(\Omega)}\, d\rho dr \\
=&\frac{(\widetilde{t})^{\alpha(\theta+\nu'-1)} - t^{\alpha(\theta+\nu'-1)}}{\alpha(\theta+\nu'-1)} \int_0^T (T-r)^{\alpha(1-\theta)-1} \|G(r, u(r))\|_{\mathbb{H}^{\nu}(\Omega)}\, dr \\
\lesssim&(\widetilde{t} - t)^{\alpha(\theta+\nu'-1)} \int_0^T (T-r)^{\alpha(1-\theta)-1} \|G(r, u(r))\|_{\mathbb{H}^{\nu}(\Omega)}\, dr \\
\lesssim&(\widetilde{t} - t)^{\alpha(\theta+\nu'-1)} \|f\|_{\mathbb{H}^{\nu+\theta}(\Omega)}.
\end{aligned} \tag{5.123}$$

Case 2. If $1 < \alpha(\theta + \nu' - 1) \le 2$, for $0 \le t \le \tilde{t} \le T$, we get

$$
\begin{aligned}
&(\tilde{t})^{\alpha(\theta+\nu'-1)} - t^{\alpha(\theta+\nu'-1)} \\
=&\Big[(\tilde{t})^{\alpha(\theta+\nu'-1)} - t^{\alpha(\theta+\nu'-1)-1}\tilde{t}\Big] + \Big[t^{\alpha(\theta+\nu'-1)-1}\tilde{t} - t^{\alpha(\theta+\nu'-1)}\Big] \\
=&\tilde{t}\Big[(\tilde{t})^{\alpha(\theta+\nu'-1)-1} - t^{\alpha(\theta+\nu'-1)-1}\Big] + (\tilde{t}-t)t^{\alpha(\theta+\nu'-1)-1} \\
\le&\max\Big\{T, T^{\alpha(\theta+\nu'-1)-1}\Big\}\Big[(\tilde{t}-t)^{\alpha(\theta+\nu'-1)-1} + (\tilde{t}-t)\Big].
\end{aligned}
\tag{5.124}
$$

Therefore

$$
\|\mathfrak{M}_4\|_{\mathbb{H}^{\nu-\nu'}(\Omega)} \lesssim \left\{ \begin{array}{l} (\tilde{t}-t)^{\alpha(\theta+\nu'-1)}\mathbf{1}_{0<\alpha(\theta+\nu'-1)\le 1} \\ \Big((\tilde{t}-t)^{\alpha(\theta+\nu'-1)-1} + (\tilde{t}-t)\Big)\mathbf{1}_{1<\alpha(\theta+\nu'-1)<2} \end{array} \right\} \|f\|_{\mathbb{H}^{\nu+\theta}(\Omega)}.
\tag{5.125}
$$

Collecting the results (5.113), (5.114), (5.120), (5.121),(5.123), and (5.125), we deduce that $u \in C^{\eta_{glo}}([0,T]; \mathbb{H}^{\nu-\nu'}(\Omega))$ and finish the desired inequality.

Part (c). Show that

$$
\|\partial_t u(t)\|_{\mathbb{H}^{\nu-\nu_1-\frac{1}{\alpha}}(\Omega)} \lesssim t^{-\alpha(1-\theta-\nu_1)}\|f\|_{\mathbb{H}^{\nu+\theta}(\Omega)}.
$$

In order to establish the result, let us define the following projection operator, for any $v = \sum_{j=1}^{\infty}\langle v, \varphi_j\rangle\varphi_j$ and any $M > 0$,

$$
\mathscr{P}_M v := \sum_{j=1}^{M}\langle v, \varphi_j\rangle\varphi_j,
$$

and two following operators:

$$
\mathscr{D}_{1,\alpha}(t,T)v := \sum_{j=1}^{\infty}\frac{-\lambda_j t^{\alpha-1}E_{\alpha,\alpha}(-\lambda_j t^{\alpha})}{E_{\alpha,1}(-\lambda_j T^{\alpha})}\langle v, \varphi_j\rangle\varphi_j,
\tag{5.126}
$$

$$
\mathscr{D}_{2,\alpha}(t)v := \sum_{j=1}^{\infty} t^{\alpha-2}E_{\alpha,\alpha-1}(-\lambda_j t^{\alpha})\langle v, \varphi_j\rangle\varphi_j.
\tag{5.127}
$$

Noting that $\mathscr{P}_M$ has the finite rank, we have the following equality after some simple computations

$$\begin{aligned}\partial_t \mathscr{P}_M u(t) =& \mathscr{D}_{1,\alpha}(t,T)\mathscr{P}_M f + \int_0^t \mathscr{D}_{2,\alpha}(t-r)\mathscr{P}_M G(r,u(r))dr \\ &- \int_0^T \mathscr{D}_{1,\alpha}(t,T)\mathbf{P}_\alpha(T-r)\mathscr{P}_M G(r,u(r))dr.\end{aligned} \tag{5.128}$$

One can infer from $0 \le \nu_1 \le 1-\theta$ that $0 \le \nu_1 < \frac{2\alpha-1}{\alpha} - \theta$. Hence, this can be associated with $\frac{\alpha-1}{\alpha} < \theta < 1$ that $1 < \theta + \nu_1 + \frac{1}{\alpha} < 2$, and this implies that

$$\begin{aligned} t^{\alpha-1}\left|\frac{E_{\alpha,\alpha}(-\lambda_j t^\alpha)}{E_{\alpha,1}(-\lambda_j T^\alpha)}\right| &\lesssim t^{\alpha-1}\left(\frac{1+\lambda_j T^\alpha}{1+\lambda_j t^\alpha}\right)^{2-\theta-\nu_1-\frac{1}{\alpha}}\left(\frac{1+\lambda_j T^\alpha}{1+\lambda_j t^\alpha}\right)^{\frac{1}{\alpha}+\nu_1+\theta-1} \\ &\lesssim t^{-\alpha(1-\theta-\nu_1)}\lambda_j^{\frac{1}{\alpha}+\nu_1+\theta-1}. \end{aligned} \tag{5.129}$$

It is easy to see that

$$\begin{aligned} &\left\|\mathscr{D}_{1,\alpha}(t,T)\Big(\mathscr{P}_{M'} - \mathscr{P}_M\Big)f\right\|^2_{\mathbb{H}^{\nu-\nu_1-\frac{1}{\alpha}}(\Omega)} \\ &= \sum_{j=M+1}^{M'} \lambda_j^{2\nu-2\nu_1-\frac{2}{\alpha}}\left|\frac{-\lambda_j t^{\alpha-1}E_{\alpha,\alpha}(-\lambda_j t^\alpha)}{E_{\alpha,1}(-\lambda_j T^\alpha)}\langle f,\varphi_j\rangle\right|^2 \\ &\lesssim t^{-2\alpha(1-\theta-\nu_1)}\sum_{j=M+1}^{M'}\lambda_j^{2\nu+2\theta}\left|\langle f,\varphi_j\rangle\right|^2. \end{aligned} \tag{5.130}$$

On the other hand, it is certain that

$$\begin{aligned} &\left\|\int_0^t \mathscr{D}_{2,\alpha}(t-r)\Big(\mathscr{P}_{M'} - \mathscr{P}_M\Big)G(r,u(r))dr\right\|_{\mathbb{H}^{\nu-\nu_1-\frac{1}{\alpha}}(\Omega)} \\ &\lesssim \int_0^t (t-r)^{\alpha-2}\left(\sum_{j=M+1}^{M'}\lambda_j^{2\nu-2\nu_1-\frac{2}{\alpha}}G_j^2(r,u(r))\right)^{1/2}dr \\ &\lesssim \int_0^t (t-r)^{\alpha-2}\left(\sum_{j=M+1}^{M'}\lambda_j^{2\nu}G_j^2(r,u(r))\right)^{1/2}dr. \end{aligned} \tag{5.131}$$

Now, let us estimate the third term on the right-hand side of (5.128). We see that

$$\left\| \int_0^T \mathscr{D}_{1,\alpha}(t,T)\mathbf{P}_\alpha(T-r)\Big(\mathscr{P}_{M'}-\mathscr{P}_M\Big)G(r,u(r))dr \right\|_{\mathbb{H}^{\nu-\nu_1-\frac{1}{\alpha}}(\Omega)}$$

$$\lesssim \int_0^T (T-r)^{\alpha-1}\Bigg(\sum_{j=M+1}^{M'} \lambda_j^{2\nu-2\nu_1-\frac{2}{\alpha}}$$

$$\times \left| t^{-\alpha(1-\theta-\nu_1)}\lambda_j^{\frac{1}{\alpha}+\nu_1+\theta}\lambda_j^{-\theta}(T-r)^{-\alpha\theta}G_j(r,u(r))\right|^2\Bigg)^{\frac{1}{2}} dr \tag{5.132}$$

$$\lesssim t^{-\alpha(1-\theta-\nu_1)}\int_0^T (T-r)^{\alpha(1-\theta)-1}\Bigg(\sum_{j=M+1}^{M'} \lambda_j^{2\nu}G_j^2(r,u(r))\Bigg)^{\frac{1}{2}} dr,$$

where we have used the estimates $\Big|E_{\alpha,\alpha}(-\lambda_j(T-r)^\alpha)\Big| \lesssim \lambda_j^{-\theta}(T-r)^{-\alpha\theta}$ and

$$\left|\lambda_j^{\nu-\nu_1-\frac{1}{\alpha}}\frac{-\lambda_j t^{\alpha-1}E_{\alpha,\alpha}(-\lambda_j t^\alpha)}{E_{\alpha,1}(-\lambda_j T^\alpha)}E_{\alpha,\alpha}(-\lambda_j(T-\tau)^\alpha)G_j(r,u(r))\right|$$

$$\lesssim (T-r)^{-\alpha\theta}t^{-\alpha(1-\theta-\nu_1)}\lambda_j^{\nu}\Big|G_j(r,u(r))\Big|. \tag{5.133}$$

Applying the Lebesgue's dominated convergence theorem, we deduce that three terms

$$\mathscr{D}_{1,\alpha}(t,T)\mathscr{P}_M f, \quad \int_0^t \mathscr{D}_{2,\alpha}(t-r)\mathscr{P}_M G(r,u(r))dr$$

and

$$\int_0^T \mathscr{D}_{1,\alpha}(t,T)\mathbf{P}_\alpha(T-r)\mathscr{P}_M G(r,u(r))dr$$

are the Cauchy sequences in the space $\mathbb{H}^{\nu-\nu_1-\frac{1}{\alpha}}(\Omega)$. Then, we obtain three convergences in the space $\mathbb{H}^{\nu-\nu_1-\frac{1}{\alpha}}(\Omega)$ as follows:

$$\lim_{M\to\infty} \mathscr{D}_{1,\alpha}(t,T)\mathscr{P}_M f = \mathscr{D}_{1,\alpha}(t,T)f,$$

$$\lim_{M\to\infty}\int_0^t \mathscr{D}_{2,\alpha}(t-r)\mathscr{P}_M G(r,u(r))dr = \int_0^t \mathscr{D}_{2,\alpha}(t-r)G(r,u(r))dr,$$

$$\lim_{M\to\infty}\int_0^T \mathscr{D}_{1,\alpha}(t,T)\mathbf{P}_\alpha(T-r)\mathscr{P}_M G(r,u(r))dr$$

$$= \int_0^T \mathscr{D}_{1,\alpha}(t,T)\mathbf{P}_\alpha(T-r)G(r,u(r))dr. \tag{5.134}$$

The above equalities imply that $\partial_t \mathscr{P}_M u(t)$ consequently converges to $\partial_t u(t)$ in $\mathbb{H}^{\nu-\nu_1-\frac{1}{\alpha}}(\Omega)$. Further, the following estimates also hold

$$\|\mathscr{D}_{1,\alpha}(t,T)f\|_{\mathbb{H}^{\nu-\nu_1-\frac{1}{\alpha}}(\Omega)} \lesssim t^{-\alpha(1-\theta-\nu_1)}\|f\|_{\mathbb{H}^{\nu+\theta}(\Omega)},$$

$$\begin{aligned}\left\|\int_0^t \mathscr{D}_{2,\alpha}(t-r)G(r,u(r))dr\right\|_{\mathbb{H}^{\nu-\nu_1-\frac{1}{\alpha}}(\Omega)} &\lesssim \int_0^t (t-r)^{\alpha-2}\|G(r,u(r))\|_{\mathbb{H}^{\nu}(\Omega)}\,dr\\ &\lesssim \|f\|_{\mathbb{H}^{\nu+\theta}} \int_0^t (t-r)^{\alpha-2} r^{-\alpha(1-\theta)}dr\\ &\lesssim t^{-\alpha(1-\theta-\nu_1)}\|f\|_{\mathbb{H}^{\nu+\theta}}\end{aligned}$$

and

$$\begin{aligned}&\left\|\int_0^T \mathscr{D}_{1,\alpha}(t,T)\mathbf{P}_\alpha(T-r)G(r,u(r))dr\right\|_{\mathbb{H}^{\nu-\nu_1-\frac{1}{\alpha}}(\Omega)}\\ &\lesssim t^{-\alpha(1-\theta-\nu_1)}\int_0^T (T-r)^{\alpha(1-\theta)-1}\|G(r,u(r))\|_{\mathbb{H}^{\nu}(\Omega)}\,dr\\ &\lesssim t^{-\alpha(1-\theta-\nu_1)}\|f\|_{\mathbb{H}^{\nu+\theta}(\Omega)}\int_0^T (T-r)^{\alpha(1-\theta)-1} r^{-\alpha(1-\theta)}dr\\ &\lesssim t^{-\alpha(1-\theta-\nu_1)}\|f\|_{\mathbb{H}^{\nu+\theta}(\Omega)}.\end{aligned}$$

Here, we note that

$$\int_0^t (t-r)^{\alpha-2} r^{-\alpha(1-\theta)}dr \lesssim t^{\alpha\theta-1}.$$

By using **(AP.1)** and noting that $1-\nu_1-\frac{1}{\alpha} \geq 0$ as $\nu_1 \leq \frac{\alpha-1}{\alpha}$, we have

$$t^{\alpha\theta-1} = t^{-\alpha(1-\theta-\nu_1)} t^{\alpha\left(1-\nu_1-\frac{1}{\alpha}\right)} \lesssim t^{-\alpha(1-\theta-\nu_1)}.$$

Consolidating all the above arguments, we obtain that

$$\begin{aligned}\left\|\partial_t u(t)\right\|_{\mathbb{H}^{\nu-\nu_1-\frac{1}{\alpha}}(\Omega)} \leq &\left\|\mathscr{D}_{1,\alpha}(t,T)f\right\|_{\mathbb{H}^{\nu-\nu_1-\frac{1}{\alpha}}(\Omega)}\\ &+\left\|\int_0^T \mathscr{D}_{1,\alpha}(t,T)\mathbf{P}_\alpha(T-r)G(r,u(r))dr\right\|_{\mathbb{H}^{\nu-\nu_1-\frac{1}{\alpha}}(\Omega)}\end{aligned}$$

$$+\left\| \int_0^t \mathscr{D}_{2,\alpha}(t-r)G(r,u(r))dr \right\|_{\mathbb{H}^{\nu-\nu_1-\frac{1}{\alpha}}(\Omega)}$$
$$\lesssim t^{-\alpha(1-\theta-\nu_1)}\|f\|_{\mathbb{H}^{\nu+\theta}(\Omega)}.$$

Since $\nu_1 \le 1-\theta$ and $1 < \theta+\nu_1+\frac{1}{\alpha}$, these straightforwardly imply that $0 \le \alpha(1-\theta-\nu_1) < 1$ and $\partial_t u \in L^p(0,T;\mathbb{H}^{\nu-\nu_1-\frac{1}{\alpha}}(\Omega))$, for all $p \in \left[1, \frac{1}{\alpha(1-\theta-\nu_1)}\right)$. This ends up Part (c).

Part (d). Show that

$$\|\partial_t^\alpha u(t)\|_{\mathbb{H}^{\nu-\nu_\alpha-\frac{1}{\alpha}}(\Omega)} \lesssim t^{-\alpha\max\left\{2-\theta-\nu_\alpha-\frac{1}{\alpha},1-\theta\right\}}\|f\|_{\mathbb{H}^{\nu+\theta}(\Omega)}.$$

To study the fractional derivative of order α of the mild solution u, let us denote some operators by

$$\mathscr{D}_{3\alpha}(t,T)w := \sum_{j=1}^{\infty} \frac{-\lambda_j E_{\alpha,1}(-\lambda_j t^\alpha)}{E_{\alpha,1}(-\lambda_j T^\alpha)}\langle w,\varphi_j\rangle\varphi_j,$$

$$\mathscr{D}_{4\alpha}(t)w := -\sum_{j=1}^{\infty} \lambda_j t^{\alpha-1}E_{\alpha,\alpha}(-\lambda_j t^\alpha)\langle w,\varphi_j\rangle\varphi_j.$$

By applying the projection $\mathscr{P}_M$ to the solution u, and then calculating the fractional differentiation ∂_t^α, one can get that

$$\begin{aligned}\partial_t^\alpha \mathscr{P}_M u(t) =&\mathscr{D}_{3,\alpha}(t,T)\mathscr{P}_M f + \int_0^t \mathscr{D}_{4,\alpha}(t-r)\mathscr{P}_M G(r,u(r))dr \\ &- \int_0^T \mathscr{D}_{3,\alpha}(t,T)\mathbf{P}_\alpha(T-r)\mathscr{P}_M G(r,u(r))dr + \mathscr{P}_M G(t,u(t)).\end{aligned} \tag{5.135}$$

From the assumption $\frac{2(\alpha-1)}{\alpha} - \theta < \nu_\alpha < 2 - \frac{1}{\alpha} - \theta$, we find that $1 < \nu_\alpha + \theta + \frac{1}{\alpha} < 2$. Therewith, the same techniques as (5.130) invoke that $\mathscr{D}_{3\alpha}(t,T)f$ exists in the space $\mathbb{H}^{\nu-\nu_\alpha-\frac{1}{\alpha}}(\Omega)$ if $f \in \mathbb{H}^{\nu+\theta}(\Omega)$, and

$$\|\mathscr{D}_{3\alpha}(t,T)f\|_{\mathbb{H}^{\nu-\nu_\alpha-\frac{1}{\alpha}}(\Omega)} \lesssim t^{-\alpha\left(2-\theta-\nu_\alpha-\frac{1}{\alpha}\right)}\|f\|_{\mathbb{H}^{\nu+\theta}(\Omega)}. \tag{5.136}$$

The proof for integrals

$$\int_0^t \mathscr{D}_{4,\alpha}(t-r)G(r,u(r))dr$$

and

$$\int_0^T \mathscr{D}_{3,\alpha}(t,T)\mathbf{P}_\alpha(T-r)G(r,u(r))dr$$

in the space $\mathbb{H}^{\nu-\nu_\alpha-\frac{1}{\alpha}}(\Omega)$ can be done by using the same arguments of (5.131) and (5.132) by using assumption $(\mathscr{H}_1)$ and the argument of Cauchy sequences. Aside from the above existence results, we can also verify the following estimates:

$$\begin{aligned}\left\| \int_0^t \mathscr{D}_{4,\alpha}(t-r)G(r,u(r))dr \right\|_{\mathbb{H}^{\nu-\nu_\alpha-\frac{1}{\alpha}}(\Omega)} &\lesssim \int_0^t (t-r)^{\alpha-2} \|G(r,u(r))\|_{\mathbb{H}^\nu(\Omega)} dr \\ &\lesssim t^{\alpha\theta-1}\|f\|_{\mathbb{H}^{\nu+\theta}(\Omega)},\end{aligned} \tag{5.137}$$

where $\nu_\alpha - 1 + \frac{2}{\alpha} > 0$ as $\nu_\alpha > \frac{2(\alpha-1)}{\alpha} - \theta$ and $\theta < 1$, and by a similar argument, we get

$$\begin{aligned}&\left\| \int_0^T \mathscr{D}_{3,\alpha}(t,T)\mathbf{P}_\alpha(T-r)G(r,u(r))dr \right\|_{\mathbb{H}^{\nu-\nu_\alpha-\frac{1}{\alpha}}(\Omega)} \\ &\lesssim t^{-\alpha\left(2-\theta-\nu_\alpha-\frac{1}{\alpha}\right)}\|f\|_{\mathbb{H}^{\nu+\theta}(\Omega)}.\end{aligned} \tag{5.138}$$

Since $\nu_\alpha < 2 - \frac{1}{\alpha} - \theta$, we have $\alpha\left(2-\theta-\nu_\alpha-\frac{1}{\alpha}\right) > 0$. In addition, it can be deduced from $\frac{2(\alpha-1)}{\alpha} - \theta < \nu_\alpha$ that $\alpha\left(2-\theta-\nu_\alpha-\frac{1}{\alpha}\right) < 1$. Hence, we straightforwardly infer that $0 < \alpha\left(2-\theta-\nu_\alpha-\frac{1}{\alpha}\right) < 1$. Moreover, it results from $\nu_\alpha > \frac{2(\alpha-1)}{\alpha} - \theta$ that $\nu - \nu_\alpha - \frac{1}{\alpha} < \nu - (1-\frac{1}{\alpha}) - (1-\theta) < \nu$, and then Sobolev embedding $\mathbb{H}^\nu(\Omega) \hookrightarrow \mathbb{H}^{\nu-\nu_\alpha-\frac{1}{\alpha}}(\Omega)$ holds. This implies that

$$\|G(t,u(t))\|_{\mathbb{H}^{\nu-\nu_\alpha-\frac{1}{\alpha}}(\Omega)} \lesssim \|G(t,u(t))\|_{\mathbb{H}^\nu(\Omega)} \lesssim t^{-\alpha(1-\theta)}\|f\|_{\mathbb{H}^{\nu+\theta}(\Omega)}.$$

Combining the above inequalities and $\alpha\theta - 1 > -\alpha(1-\theta)$, we finally show that

$$\|\partial_t^\alpha u(t)\|_{\mathbb{H}^{\nu-\nu_\alpha-\frac{1}{\alpha}}(\Omega)} \lesssim t^{-\alpha\max\left\{2-\theta-\nu_\alpha-\frac{1}{\alpha},1-\theta\right\}}\|f\|_{\mathbb{H}^{\nu+\theta}(\Omega)}$$

and $\partial_t^\alpha u \in L^p(0,T;\mathbb{H}^{\nu-\nu_\alpha-\frac{1}{\alpha}}(\Omega))$, for all $p \in \left[1, \frac{1}{\alpha\max\left\{2-\theta-\nu_\alpha-\frac{1}{\alpha},1-\theta\right\}}\right)$. The proof is accomplished.

5.3.5.2 Proofs of Theorem 5.13

The proof of Theorem 5.13 relies on a contraction mapping principle. In order to prove this, we have to prove the following lemma.

Lemma 5.17 *Let us take* $\frac{\alpha-1}{\alpha} < \theta < 1$, $0 \le \nu \le \sigma \le \nu + 1$, *and* $1 \le q < \frac{1}{\alpha(1-\theta)}$. *Assume that* $f \in \mathbb{H}^{\nu+\theta+1}(\Omega)$ *and* G *sastisfies* $(\mathscr{H}_2)$ *with* $\|L_2\|_{L^\infty(0,T)} \in \left(0, \mathscr{U}_2^{-1}\right)$. *Set*

$$\begin{aligned} \mathscr{T}v(t) :=& \mathbf{B}_\alpha(t,T)f + \int_0^t \mathbf{P}_\alpha(t-r)G(r,v(r))dr \\ &- \int_0^T \mathbf{B}_\alpha(t,T)\mathbf{P}_\alpha(T-r)G(r,v(r))dr, \end{aligned} \tag{5.139}$$

and then for any $v \in C([0,T];\mathbb{H}^\nu(\Omega)) \cap L^q(0,T;\mathbb{H}^\sigma(\Omega))$, *it holds that*

$$\mathscr{T}v \in C\left([0,T];\mathbb{H}^\nu(\Omega)\right) \cap L^q(0,T;\mathbb{H}^\sigma(\Omega)).$$

Proof of Lemma 5.17 We divide this proof into the following steps as follows. *Prove* $\mathscr{T}v \in C([0,T];\mathbb{H}^\nu(\Omega))$. Namely, we need to estimate the norm $\|\mathscr{T}v(\tilde{t}) - \mathscr{T}v(t)\|_{\mathbb{H}^\nu(\Omega)}$ for all $0 \le t < \tilde{t} \le T$. For more convenience, we will use notation $\mathfrak{M}_j$, $1 \le j \le 4$ as (5.108) again. However, the estimates for $\mathfrak{M}_j$ in Step 2 of the proof of Theorem (5.12) will be modified suitably to fit the assumptions of f and G in this theorem. Indeed, a slight modification of the techniques in the estimates (5.109) and (5.110) invokes that

$$\begin{aligned} \|\mathfrak{M}_1\|_{\mathbb{H}^\nu(\Omega)} &\le \int_t^{\tilde{t}} \left\| \sum_{j=1}^\infty \lambda_j r^{\alpha-1} \frac{E_{\alpha,\alpha}(-\lambda_j r^\alpha)}{E_{\alpha,1}(-\lambda_j T^\alpha)} f_j \varphi_j \right\|_{\mathbb{H}^\nu(\Omega)} d\tau \\ &\lesssim \left(\int_t^{\tilde{t}} r^{\alpha\theta-1} dr \right) \|f\|_{\mathbb{H}^{\nu+\theta+1}(\Omega)} \\ &\lesssim \begin{Bmatrix} (\tilde{t}-t)^{\alpha\theta} \mathbf{1}_{0<\alpha\theta\le 1} \\ \left((\tilde{t}-t)^{\alpha\theta-1} + (\tilde{t}-t)\right) \mathbf{1}_{1<\alpha\theta<2} \end{Bmatrix} \|f\|_{\mathbb{H}^{\nu+\theta+1}(\Omega)}, \end{aligned}$$

where we note that $\frac{1-\theta}{2}$ belongs to $(0,1)$, and it notes that $0 < \alpha - 1 < \alpha\theta < \alpha < 2$ as $\frac{\alpha-1}{\alpha} < \theta < 1$. Next, estimates for the terms $\mathfrak{M}_j$, $2 \le j \le 4$ will be based on the assumption $(\mathscr{H}_2)$ of the nonlinearity G. We see that

$$
\begin{aligned}
\|\mathfrak{M}_2\|_{\mathbb{H}^\nu(\Omega)} &\lesssim \int_0^t \int_{t-r}^{\widetilde{t}-r} \rho^{\alpha-2} \|G(r,u(r))\|_{\mathbb{H}^\nu(\Omega)}\, d\rho dr \\
&\lesssim \int_0^t \int_{t-r}^{\widetilde{t}-r} \rho^{\alpha-2} \|G(r,v(r))\|_{\mathbb{H}^{\nu+1}(\Omega)}\, d\rho dr \\
&\lesssim \|v\|_{C([0,T];\mathbb{H}^\nu(\Omega))\cap L^q(0,T;\mathbb{H}^\sigma(\Omega))} \int_0^t \int_{t-r}^{\widetilde{t}-r} \rho^{\alpha-2} d\rho dr,
\end{aligned}
$$

where the norm $\|G(r,u(r))\|_{\mathbb{H}^\nu(\Omega)}$ is certainly $\lesssim$-bounded by $\|G(r,v(r))\|_{\mathbb{H}^{\nu+1}(\Omega)}$ due to the embedding $\mathbb{H}^{\nu+1}(\Omega) \hookrightarrow \mathbb{H}^\nu(\Omega)$. Observation from the above estimate that the last right-hand side clearly tends to zero as $\widetilde{t}$ tends to t. Hence, the preceding estimate implies the continuity of the term $\mathfrak{M}_2$ on $\mathbb{H}^\nu(\Omega)$. In addition, the continuity of the term $\mathfrak{M}_3$ is obvious by using similar arguments as in (5.114) and the assumption $(\mathscr{H}_2)$. Precisely,

$$
\|\mathfrak{M}_3\|_{\mathbb{H}^\nu(\Omega)} \lesssim \int_t^{\widetilde{t}} \|G(r,v(r))\|_{\mathbb{H}^{\nu+1}(\Omega)}\, dr \lesssim (\widetilde{t}-t)\|v\|_{C([0,T];\mathbb{H}^\nu(\Omega))\cap L^q(0,T;\mathbb{H}^\sigma(\Omega))}.
$$

Finally, we consider the term $\|\mathfrak{M}_4\|_{\mathbb{H}^\nu(\Omega)}$. The idea is combining similar arguments as in Step 2 and the modification in the above estimates for $\mathfrak{M}_1$. Here, the maximum of the spatial smoothness of G should be estimated in the space $\mathbb{H}^{\nu+1}$. Indeed, the following chain of the estimates can be checked

$$
\begin{aligned}
&\|\mathfrak{M}_4\|_{\mathbb{H}^\nu(\Omega)} \\
&\leq \int_0^T \int_t^{\widetilde{t}} \Bigg(\sum_{j=1}^{\infty} \Big| \lambda_j^\nu (T-\tau)^{\alpha-1} \lambda_j \rho^{\alpha-1} \mathscr{E}_{\alpha,T}(-\lambda_j \rho^\alpha) \\
&\qquad \times E_{\alpha,\alpha}(-\lambda_j (T-r)^\alpha) G_j(r,u(r)) \Big|^2 \Bigg)^{\frac{1}{2}} d\rho dr \\
&\lesssim \int_0^T \int_t^{\widetilde{t}} (T-r)^{\alpha(1-\theta)-1} \rho^{\alpha\theta-1} \|G(r,v(r))\|_{\mathbb{H}^{\nu+1}(\Omega)} d\rho dr \\
&\lesssim \|v\|_{C([0,T];\mathbb{H}^\nu(\Omega))\cap L^q(0,T;\mathbb{H}^\sigma(\Omega))} \left[(\widetilde{t})^{\alpha\theta} - t^{\alpha\theta} \right] \\
&\lesssim \|v\|_{C([0,T];\mathbb{H}^\nu(\Omega))\cap L^q(0,T;\mathbb{H}^\sigma(\Omega))} \left\{ \begin{array}{l} (\widetilde{t}-t)^{\alpha\theta} \mathbf{1}_{0<\alpha\theta\leq 1} \\ \left((\widetilde{t}-t)^{\alpha\theta-1} + (\widetilde{t}-t) \right) \mathbf{1}_{1<\alpha\theta<2} \end{array} \right\}.
\end{aligned}
$$

The preceding estimates lead to $\mathscr{T}v \in C([0,T];\mathbb{H}^\nu(\Omega))$.
Prove $\mathscr{T}v \in L^q(0,T;\mathbb{H}^\sigma(\Omega))$. We observe that

$$\begin{aligned}
\left\|\mathbf{B}_\alpha(t,T)f\right\|^2_{\mathbb{H}^\sigma(\Omega)} &= \sum_{j=1}^{\infty} \lambda_j^{2\sigma} \left| \frac{E_{\alpha,1}(-\lambda_j t^\alpha)}{E_{\alpha,1}(-\lambda_j T^\alpha)} \right|^2 \langle f, \varphi_j \rangle^2 \\
&\lesssim t^{-2\alpha(1-\theta)} \sum_{j=1}^{\infty} \lambda_j^{2\theta+2\sigma} \langle f, \varphi_j \rangle^2 \\
&\lesssim t^{-2\alpha(1-\theta)} \sum_{j=1}^{\infty} \lambda_j^{2\theta+2\nu+2} \langle f, \varphi_j \rangle^2 \\
&= t^{-2\alpha(1-\theta)} \| f \|^2_{\mathbb{H}^{\nu+\theta+1}}.
\end{aligned} \tag{5.140}$$

Thus, $\mathbf{B}_\alpha(t,T)f \in L^q(0,T;\mathbb{H}^\sigma(\Omega))$ since $t^{-\alpha(1-\theta)} \in L^q(0,T;\mathbb{R})$.

Next, will estimate the second term of $\mathscr{T}v$ where we will bound the operator norm of $\mathbf{P}_\alpha(t-r)$ on $\mathbb{H}^\sigma(\Omega)$ by $M_\alpha(t-r)^{\alpha-1}$, and then we estimate $\|G(r,v(r))\|_{\mathbb{H}^\sigma(\Omega)}$ by $\|G(r,v(r))\|_{\mathbb{H}^{\nu+1}(\Omega)}$ upon the assumption $(\mathscr{H}_2)$ and the embedding $\mathbb{H}^{\nu+1}(\Omega) \hookrightarrow \mathbb{H}^\sigma(\Omega)$ as $0 \le \sigma \le \nu+1$, see (5.72). Precisely, these arguments can be performed as follows:

$$\begin{aligned}
&\left\| \int_0^t \mathbf{P}_\alpha(t-r)G(r,v(r))dr \right\|^q_{L^q(0,T;\mathbb{H}^\sigma(\Omega))} \\
=& \int_0^T \left\| \int_0^t \mathbf{P}_\alpha(t-r)G(r,v(r))dr \right\|^q_{\mathbb{H}^\sigma(\Omega)} dt \\
\le& (C_2(\nu,\sigma)M_\alpha)^q \int_0^T \left(\int_0^t (t-r)^{\alpha-1} \|G(r,v(r))\|_{\mathbb{H}^{\nu+1}(\Omega)} dr \right)^q dt \\
\le& \left(\|L_2\|_{L^\infty(0,T)} \overline{\mathscr{M}}_1 \right)^q \|v\|^q_{C([0,T];\mathbb{H}^\nu(\Omega)) \cap L^q(0,T;\mathbb{H}^\sigma(\Omega))},
\end{aligned} \tag{5.141}$$

where the constant $\overline{\mathscr{M}}_1$ is given by **(AP.)** in the Appendix.

Next, we will estimate the $L^q(0,T;\mathbb{H}^\sigma(\Omega))$-norm of the last term of $\mathscr{T}v$. The idea is trying to combine estimates for the operators $\mathbf{B}_\alpha(t,T)$ and $\mathbf{P}_\alpha(T-r)$. We can estimate that the operator $\mathbf{B}_\alpha(t,T)$ maps $\mathbb{H}^{\sigma+\theta}$ into $\mathbb{H}^\sigma(\Omega)$, and the operator $\mathbf{P}_\alpha(T-r)$ maps $\mathbb{H}^\sigma(\Omega)$ into $\mathbb{H}^{\sigma+\theta}(\Omega)$. Therefore, by using the assumption $(\mathscr{H}_2)$, this term can be estimated on the space $\mathbb{H}^\sigma$. In the technical aspect, we also note that the assumption $1-1/(\alpha q) < \theta < 1$ guarantees that $(\alpha-1)/\alpha < \theta < 1$, and so $\alpha(1-\theta) \in (0,1)$. Moreover, the power function $t^{-\alpha(1-\theta)q}$ is clearly integrable on $(0,T)$. Indeed, one can show the following chain of the estimates:

$$
\begin{aligned}
&\left\| \int_0^T \mathbf{B}_\alpha(t,T)\mathbf{P}_\alpha(T-r)G(r,v(r))dr \right\|^q_{L^q(0,T;\mathbb{H}^\sigma(\Omega))} \\
&= \int_0^T \left\| \int_0^T \mathbf{B}_\alpha(t,T)\mathbf{P}_\alpha(T-r)G(r,v(r))dr \right\|^q_{\mathbb{H}^\sigma(\Omega)} dt \\
&\le \mathscr{M}^q_{\alpha,T,\theta} \int_0^T t^{-\alpha(1-\theta)q} \left(\int_0^T \|\mathbf{P}_\alpha(T-r)G(r,v(r))\|_{\mathbb{H}^{\sigma+\theta}(\Omega)} dr \right)^q dt \\
&\le \overline{\mathscr{M}}^q_{\alpha,T,\theta} \int_0^T t^{-\alpha(1-\theta)q} \left(\int_0^T (T-\tau)^{\alpha(1-\theta)-1} \|G(r,v(r))\|_{\mathbb{H}^\sigma(\Omega)} dr \right)^q dt \\
&\le \left(\|L_2\|_{L^\infty(0,T)} \overline{\mathscr{M}}_2 \right)^q \|v\|^q_{C([0,T];\mathbb{H}^\nu(\Omega)) \cap L^q(0,T;\mathbb{H}^\sigma(\Omega))},
\end{aligned}
\tag{5.142}
$$

where $\mathscr{M}_{\alpha,T,\theta} := T^{\alpha(1-\theta)}(T^{\alpha\theta} + \lambda_1^{-\theta})$, $\overline{\mathscr{M}}_{\alpha,T,\theta} = \mathscr{M}_{\alpha,T,\theta} M_\alpha$, and the constant $\overline{\mathscr{M}}_2$ is given by **(AP.)** in the Appendix. A collection of the derived estimates (5.140), (5.141), (5.142) reveals $\mathscr{T}v \in L^q(0,T;\mathbb{H}^\sigma(\Omega))$. Finally, we wrap up the proof.

Proof of Theorem 5.13 In order to show that the problem (5.55)–(5.56) has a unique mild solution, we will prove the operator $\mathscr{T}$ has a unique fixed point in $C([0,T];\mathbb{H}^\nu(\Omega)) \cap L^q(0,T;\mathbb{H}^\sigma(\Omega))$. The proof is based on the Banach contraction principle. We have

$$
\begin{aligned}
&\left\| \mathscr{T}v_1 - \mathscr{T}v_2 \right\|_{C([0,T];\mathbb{H}^\nu(\Omega)) \cap L^q(0,T;\mathbb{H}^\sigma(\Omega))} \\
&= \|\mathscr{T}v_1 - \mathscr{T}v_2\|_{L^q(0,T;\mathbb{H}^\sigma(\Omega))} + \|\mathscr{T}v_1 - \mathscr{T}v_2\|_{C([0,T];\mathbb{H}^\nu(\Omega))} \\
&=: \mathfrak{M}_5 + \mathfrak{M}_6.
\end{aligned}
\tag{5.143}
$$

For estimating $\mathfrak{M}_5$, we applied the previous results in estimating (5.141) and (5.142) to get

$$
\begin{aligned}
\mathfrak{M}_5 \le& \left\| \int_0^t \mathbf{P}_\alpha(t-r)(G(r,v_1(r)) - G(r,v_2(r)))dr \right\|_{L^q(0,T;\mathbb{H}^\sigma(\Omega))} \\
&+ \left\| \int_0^T \mathbf{B}_\alpha(t,T)\mathbf{P}_\alpha(T-r)(G(r,v_1(r)) - G(r,v_2(r)))dr \right\|_{L^q(0,T;\mathbb{H}^\sigma(\Omega))} \\
\le& \|L_2\|_{L^\infty(0,T)} \left(\overline{\mathscr{M}}_1 + \overline{\mathscr{M}}_2 \right) \|v_1 - v_2\|_{C([0,T];\mathbb{H}^\nu(\Omega)) \cap L^q(0,T;\mathbb{H}^\sigma(\Omega))}.
\end{aligned}
\tag{5.144}
$$

On the other hand, to estimate the term $\mathfrak{M}_6$, we estimate the operator norm of $\mathbf{P}_\alpha(t-r)$ acting on $\mathbb{H}^\nu(\Omega)$ by $M_\alpha(t-r)^{\alpha-1}$, and of $\mathbf{B}_\alpha(t,T)\mathbf{P}_\alpha(T-r)$ acting from $\mathbb{H}^\nu(\Omega)$ to $\mathbb{H}^{\nu+1}(\Omega)$ by $M_\alpha^2 m_\alpha^{-1}(T^\alpha + \lambda_1^{-1})$. Applying the embedding $\mathbb{H}^\nu(\Omega) \hookrightarrow \mathbb{H}^{\nu+1}(\Omega)$ and the assumption $(\mathscr{H}_2)$ consequently deduces the following estimates:

$$
\begin{aligned}
\mathfrak{M}_6 \le & \left\| \int_0^t \mathbf{P}_\alpha(t-r)\,(G(r,v_1(r)) - G(r,v_2(r)))\,dr \right\|_{C([0,T];\mathbb{H}^\nu(\Omega))} \\
& + \left\| \int_0^T \mathbf{B}_\alpha(t,T)\mathbf{P}_\alpha(T-r)\,(G(v_1(r)) - G(v_2(r)))\,dr \right\|_{C([0,T];\mathbb{H}^\nu(\Omega))} \\
\le & M_\alpha \sup_{0\le t\le T} \left(\int_0^t (t-r)^{\alpha-1} \left\| G(r,v_1(r)) - G(r,v_2(r)) \right\|_{H^\nu(\Omega)}\,dr \right) \\
& + \frac{M_\alpha^2}{m_\alpha}(T^\alpha + \lambda_1^{-1}) \int_0^T (T-r)^{\alpha-1} \left\| G(r,v_1(r)) - G(r,v_2(r)) \right\|_{\mathbb{H}^{\nu+1}(\Omega)}\,dr \\
\le & \|L_2\|_{L^\infty(0,T)} \overline{\mathscr{M}}_3 \left\| v_1 - v_2 \right\|_{C([0,T];\mathbb{H}^\nu(\Omega)) \cap L^q(0,T;\mathbb{H}^\sigma(\Omega))}.
\end{aligned}
\tag{5.145}
$$

A collection of the estimates (5.143), (5.144), (5.145) reveals that

$$
\begin{aligned}
& \left\| \mathscr{T} v_1 - \mathscr{T} v_2 \right\|_{C([0,T];\mathbb{H}^\nu(\Omega)) \cap L^q(0,T;\mathbb{H}^\sigma(\Omega))} \\
& \le \|L_2\|_{L^\infty(0,T)} \mathscr{U}_2 \|v_1 - v_2\|_{C([0,T];\mathbb{H}^\nu(\Omega)) \cap L^q(0,T;\mathbb{H}^\sigma(\Omega))}.
\end{aligned}
$$

Since $\|L_2\|_{L^\infty(0,T)} \mathscr{U}_2 < 1$, we conclude that $\mathscr{T}$ is a contraction in $C([0,T]; \mathbb{H}^\nu(\Omega)) \cap L^q(0,T; \mathbb{H}^\sigma(\Omega))$, which ensures the existence and uniqueness of a fixed point. The desired inequality is easy to obtain. Hence, we finalize the proof.

5.3.5.3 Proof of Theorem 5.14

To start with, let us proceed the following lemma.

Lemma 5.18 *Assume that all assumptions of Theorem 5.14 are fulfilled:*

(a) *For $t > 0$, and $\mathscr{N}_2$ is given by (AP.) in the Appendix, then*

$$
\|\mathbf{B}_\alpha(t,T) f\|_{\mathbb{H}^\nu(\Omega)} \le \mathscr{N}_2 t^{-\alpha\vartheta} \|f\|_{\mathbb{H}^{\nu+(1-\vartheta)}(\Omega)}. \tag{5.146}
$$

Moreover, the following convergence holds

$$
\mathbf{B}_\alpha(\tilde{t},T) f \xrightarrow{\tilde{t}\to t} \mathbf{B}_\alpha(t,T) f, \quad \text{in } \mathbb{H}^\nu(\Omega). \tag{5.147}
$$

(b) *For $t > 0$, $w \in \mathfrak{X}_{\alpha,\vartheta,\nu,T}$, and $\mathfrak{N}_2$ is given by **(AP.)** in the Appendix, then*

$$\left\| \int_0^t \mathbf{P}_\alpha(t-r)G(r,w(r))dr \right\|_{\mathbb{H}^\nu(\Omega)} \tag{5.148}$$
$$\leq \mathfrak{N}_2 K_0 \big(T^{s\alpha\vartheta} + \mathscr{R}^s\big) t^{-\alpha\vartheta} \|w\|_{C^{\alpha\vartheta}((0,T];\mathbb{H}^\nu(\Omega))}.$$

Moreover, the following convergence holds

$$\int_0^{\widetilde{t}} \mathbf{P}_\alpha(\widetilde{t}-r)G(r,w(r))dr \tag{5.149}$$
$$\xrightarrow{\widetilde{t}\to t} \int_0^t \mathbf{P}_\alpha(t-r)G(r,w(r))dr, \quad \text{in } \mathbb{H}^\nu(\Omega).$$

(c) *For $t>0$, $w \in \mathfrak{X}_{\alpha,\vartheta,\nu,T}$, then*

$$\left\| \int_0^t \mathbf{P}_\alpha(t-r)G(r,w(r))dr \right\|_{\mathbb{H}^{\nu+1-\vartheta}(\Omega)} \tag{5.150}$$
$$\leq \mathfrak{N}_2 K_0 \big(T^{s\alpha\vartheta} + \mathscr{R}^s\big) t^{-\alpha\vartheta} \|w\|_{C^{\alpha\vartheta}((0,T];\mathbb{H}^\nu(\Omega))}.$$

Proof Proof of Part (a). By applying the first part of Lemma 5.14, we obtain

$$\left\| \mathbf{B}_\alpha(t,T)f \right\|^2_{\mathbb{H}^\nu(\Omega)} = \sum_{j=1}^{\infty} \left| \frac{E_{\alpha,1}(-\lambda_j t^\alpha)}{E_{\alpha,1}(-\lambda_j T^\alpha)} \right|^2 f_j^2 \leq \mathscr{N}_2^2 t^{-2\alpha\vartheta} \|f\|^2_{\mathbb{H}^{\nu+(1-\vartheta)}(\Omega)},$$

where $\mathscr{N}_2$ is given by **(AP.)** in the Appendix. This directly implies the inequality (5.146). Let us proceed to prove the convergence (5.147). By the fact that $E_{\alpha,\alpha}(-z) \lesssim (1+z^2)^{-1}$ for all $z \geq 0$, see, e.g., [52, 180, 187], one can apply the same techniques as (5.110) to show the following inequalities:

$$\left| \frac{E_{\alpha,\alpha}(-\lambda_j r^\alpha)}{E_{\alpha,1}(-\lambda_j T^\alpha)} \right| \lesssim \frac{1+\lambda_j T^\alpha}{\left[1+(\lambda_j r^\alpha)^2\right]^{1-\frac{1-\vartheta}{2}}} \lesssim \lambda_j \left[(\lambda_j r^\alpha)^2\right]^{\frac{1-\vartheta}{2}-1}, \tag{5.151}$$

where $0 < (1-\vartheta)/2 < (1-\mu)/2 < 1$. Hence, we derive that

$$\left\| \mathbf{B}_\alpha(\widetilde{t},T)f - \mathbf{B}_\alpha(t,T)f \right\|_{\mathbb{H}^\nu(\Omega)} \leq \int_t^{\widetilde{t}} r^{\alpha-1} \left\| \sum_{j=1}^{\infty} \lambda_j \frac{E_{\alpha,\alpha}(-\lambda_j r^\alpha)}{E_{\alpha,1}(-\lambda_j T^\alpha)} f_j \varphi_j \right\|_{\mathbb{H}^\nu(\Omega)} dr,$$
$$\lesssim \|f\|_{\mathbb{H}^{\nu+(1-\vartheta)}(\Omega)} \int_t^{\widetilde{t}} r^{-\alpha\vartheta-1} dr.$$

Since the integral in the above inequality tends to zero as t approaches $\widetilde{t}$ from the right, we obtain (5.147) and finish the proof of Part (a).

Proof of Part (b). We divide this proof into two parts as follows.
Step 1. Prove the inequality (5.148).
It follows from $E_{\alpha,\alpha}(-\lambda_j(t-r)^\alpha) \le M_\alpha \lambda_j^{-\mu}(t-r)^{-\alpha\mu}$ that

$$\begin{aligned}
&\left\| \int_0^t \mathbf{P}_\alpha(t-r)G(r,w(r))dr \right\|_{\mathbb{H}^\nu(\Omega)} \\
&\le \int_0^t (t-r)^{\alpha-1} \left\| \sum_{j=1}^\infty E_{\alpha,\alpha}(-\lambda_j(t-r)^\alpha)G_j(w(r))\varphi_j \right\|_{\mathbb{H}^\nu(\Omega)} dr \\
&\le M_\alpha \int_0^t (t-r)^{\alpha(1-\mu)-1} \|G(r,w(r))\|_{\mathbb{H}^\sigma(\Omega)} dr, \qquad (5.152)
\end{aligned}$$

where $\sigma = \nu - \mu$. Since $w \in \mathfrak{X}_{\alpha,\vartheta,\nu,T}(\mathscr{R})$, we see that $\|w(r)\|_{\mathbb{H}^\nu(\Omega)} \le \mathscr{R} r^{-\alpha\vartheta}$. Thus, we have in view of (5.75) that

$$\|G(r,w(r))\|_{\mathbb{H}^\sigma(\Omega)} \le L_3(r)\Big(1 + \|w(r)\|^s_{\mathbb{H}^\nu(\Omega)}\Big)\|w(r)\|_{\mathbb{H}^\nu(\Omega)}.$$

It follows from (5.152) that

$$\left\| \int_0^t \mathbf{P}_\alpha(t-r)G(r,w(r))dr \right\|_{\mathbb{H}^\nu(\Omega)} \le M_\alpha \|w\|_{C^{\alpha\vartheta}((0,T];\mathbb{H}^\nu(\Omega))} \widehat{L}_3(t), \qquad (5.153)$$

where

$$\widehat{L}_3(t) := \int_0^t (t-r)^{\alpha(1-\mu)-1} \left(r^{-\alpha\vartheta} + \mathscr{R}^s r^{-(1+s)\alpha\vartheta}\right) L_3(r)dr. \qquad (5.154)$$

Our next purpose is to find an upper bound of $\widehat{L}_3(t)$. In order to control this term, we observe from $0 < r < T$ that $r^{-\alpha\vartheta} \le T^{s\alpha\vartheta} r^{-(1+s)\alpha\vartheta}$, and from $K_0 = \|L_3(t)t^{\alpha\zeta}\|_{L^\infty(0,T)}$ that $L_3(r) \le K_0 r^{-\alpha\zeta}$, which yields the following estimates:

$$\begin{aligned}
\widehat{L}_3(t) &\le \left(T^{s\alpha\vartheta} + \mathscr{R}^s\right) \int_0^t (t-r)^{\alpha(1-\mu)-1} r^{-(1+s)\alpha\vartheta} L_3(r)dr \\
&\le K_0\big(T^{s\alpha\vartheta} + \mathscr{R}^s\big) \int_0^t (t-r)^{\alpha(1-\mu)-1} r^{-\alpha((1+s)\vartheta+\zeta)} dr.
\end{aligned}$$

By noting $\min\{\alpha(1-\mu)-1, \ -\alpha((1+s)\vartheta+\zeta)\} > -1$ as $0 < \mu < 1$, $\zeta < \alpha^{-1} - (1+s)\vartheta$, and using **(AP.1)**, we find that

$$\int_0^t (t-r)^{\alpha(1-\mu)-1} r^{-\alpha((1+s)\vartheta+\zeta)} dr \le \mathfrak{N}_1 t^{\alpha((1-\mu)-(1+s)\vartheta-\zeta)}.$$

This implies that

$$\widehat{L}_3(t) \leq K_0\big(T^{s\alpha\vartheta} + \mathscr{R}^s\big)\mathfrak{N}_1 T^{\alpha((1-\mu)-s\vartheta-\zeta)}\, t^{-\alpha\vartheta},$$

where we have noted that $\zeta \leq (1-\mu)-s\vartheta$ since $\zeta < \alpha^{-1} - \vartheta - s\vartheta \leq (1-\mu)-s\vartheta$ as $\alpha^{-1} < 1$ and $\vartheta > \mu$. The latter estimate together with (5.153) and (5.154) that

$$\left\| \int_0^t \mathbf{P}_\alpha(t-r)G(r,w(r))dr \right\|_{\mathbb{H}^\nu(\Omega)} \leq \mathfrak{N}_2 K_0\big(T^{s\alpha\vartheta} + \mathscr{R}^s\big)t^{-\alpha\vartheta} \|w\|_{C^{\alpha\vartheta}((0,T];\mathbb{H}^\nu(\Omega))},$$

where we recall that $\mathfrak{N}_2$ is given by **(AP.)** in the Appendix.

Step 2. Show that (5.149) holds.

By dealing with $\|G(r,w(r))\|_{\mathbb{H}^\sigma(\Omega)}$ as the same arguments in Step 1, we derive that

$$\begin{aligned}
\|\mathfrak{M}_2\|_{\mathbb{H}^\nu(\Omega)} &\leq \int_0^t \int_{t-r}^{\widetilde{t}-r} \rho^{\alpha-2} \left\| \sum_{j=1}^\infty E_{\alpha,\alpha-1}(-\lambda_j \rho^\alpha) G_j(r,u(r))\varphi_j \right\|_{\mathbb{H}^\nu(\Omega)} d\rho dr \\
&\lesssim \int_0^t \int_{t-r}^{\widetilde{t}-r} \rho^{\alpha(1-\mu)-2} \|G(r,u(r))\|_{\mathbb{H}^\sigma(\Omega)} d\rho dr \\
&\lesssim \int_0^t \int_{t-r}^{\widetilde{t}-r} \rho^{\alpha(1-\mu)-2} \left(\mathscr{R} r^{-\alpha\vartheta} + \mathscr{R}^{1+s} r^{-(1+s)\alpha\vartheta} \right) L_3(r) d\rho dr \\
&\lesssim \left| \int_0^t \left((\widetilde{t}-r)^{\alpha(1-\mu)-1} - (t-r)^{\alpha(1-\mu)-1} \right) r^{-\alpha((1+s)\vartheta+\zeta)} dr \right|,
\end{aligned}$$

where $\mathfrak{M}_2$ is formulated by (5.108). By the fact that $\alpha(1-\mu) > 0$ and $1 - \alpha((1+s)\vartheta + \zeta) > 0$ and using **(AP.2)** in Appendix, we know that the right-hand side of the latter inequality tends to zero, as $\widetilde{t}$ approaches t. Hence, $\|\mathfrak{M}_2\|_{\mathbb{H}^\nu(\Omega)} \xrightarrow{\widetilde{t}\to t} 0$. Now, by the same way as the above arguments, we get

$$\begin{aligned}
\|\mathfrak{M}_3\|_{\mathbb{H}^\nu(\Omega)} &\leq \int_t^{\widetilde{t}} (\widetilde{t}-r)^{\alpha-1} \left\| \sum_{j=1}^\infty E_{\alpha,\alpha}(-\lambda_j(\widetilde{t}-\tau)^\alpha) G_j(r,u(r))\varphi_j \right\|_{\mathbb{H}^\nu(\Omega)} dr \\
&\lesssim \int_t^{\widetilde{t}} (\widetilde{t}-r)^{\alpha(1-\mu)-1} \|G(r,u(r))\|_{\mathbb{H}^\sigma(\Omega)} dr \\
&\lesssim \int_t^{\widetilde{t}} (\widetilde{t}-r)^{\alpha(1-\mu)-1} r^{-\alpha((1+s)\vartheta+\zeta)} dr, \qquad (5.155)
\end{aligned}$$

where $\mathfrak{M}_3$ is formulated by (5.108). From that $(\widetilde{t}-r)^{\alpha(1-\vartheta)} \leq (\widetilde{t}-t)^{\alpha(1-\vartheta)}$ as $t \leq r \leq \widetilde{t}$, we bound the right-hand side of the (5.155) as follows:

$$\begin{aligned}\text{(RHS) of (5.155)} &\leq (\widetilde{t}-t)^{\alpha(1-\vartheta)} \int_0^{\widetilde{t}} (\widetilde{t}-r)^{\alpha(\vartheta-\mu)-1} r^{-\alpha((1+s)\vartheta+\zeta)} dr \\ &\lesssim (\widetilde{t}-t)^{\alpha(1-\vartheta)} \int_0^{\widetilde{t}} (\widetilde{t}-r)^{\alpha(\vartheta-\mu)-1} r^{-\alpha((1+s)\vartheta+\zeta)} dr.\end{aligned}$$

Noting that $\alpha(\vartheta-\mu) > 0$ and $1-\alpha\left((1+s)\vartheta+\zeta\right) > 0$, we ensure that

$$\int_0^{\widetilde{t}} (\widetilde{t}-r)^{\alpha(\vartheta-\mu)-1} r^{-\alpha((1+s)\vartheta+\zeta)} dr$$

is convergent. The above observations imply that $\|\mathfrak{M}_3\|_{\mathbb{H}^\nu(\Omega)} \xrightarrow{\widetilde{t}\to t} 0$. Since the fact that

$$\int_0^{\widetilde{t}} \mathbf{P}_\alpha(\widetilde{t}-r)G(r,w(r))dr - \int_0^t \mathbf{P}_\alpha(t-r)G(r,w(r))dr = \mathfrak{M}_2 + \mathfrak{M}_3,$$

we finish this step.
Proof of Part (c).

In view of $0 \leq 1 + [(\nu-\sigma)-\vartheta] \leq 1$, one can see that

$$\begin{aligned}&\left\| \int_0^t \mathbf{P}_\alpha(t-r)G(r,w(r))dr \right\|_{\mathbb{H}^{\nu+(1-\vartheta)}(\Omega)} \\ \leq & M_\alpha \int_0^t (t-r)^{\alpha(\vartheta-\mu)-1} \|G(r,w(r))\|_{\mathbb{H}^\sigma(\Omega)}\, dr \\ \leq & M_\alpha K_0(T^{s\alpha\vartheta} + \mathscr{R}^s)\|w\|_{C^{\alpha\vartheta}((0,T];\mathbb{H}^\nu(\Omega))} \int_0^t (t-r)^{\alpha(\vartheta-\mu)-1} r^{-\alpha((1+s)\vartheta+\zeta)} dr \\ \leq & M_\alpha K_0(T^{s\alpha\vartheta} + \mathscr{R}^s)\|w\|_{C^{\alpha\vartheta}((0,T];\mathbb{H}^\nu(\Omega))} \mathfrak{N}_1 t^{\alpha((\vartheta-\mu)-(1+s)\vartheta-\zeta)} \\ \leq & \mathfrak{N}_2 K_0\big(T^{s\alpha\vartheta} + \mathscr{R}^s\big) t^{-\alpha\vartheta} \|w\|_{C^{\alpha\vartheta}((0,T];\mathbb{H}^\nu(\Omega))},\end{aligned}$$

where we also recall that $\mathfrak{N}_2$ is given by **(AP.)** in the Appendix. This completes the proof.

Proof of Theorem 5.14 The proof will be based on contraction mapping theorem on the Banach space. For this purpose, let us define the mapping

$$\mathscr{Q} : \mathfrak{X}_{\alpha,\vartheta,\nu,T}(\mathscr{R}) \longrightarrow \mathfrak{X}_{\alpha,\vartheta,\nu,T}(\mathscr{R})$$

given by

$$\mathscr{Q}w = \mathbf{B}_\alpha(t,T)f + \int_0^t \mathbf{P}_\alpha(t-r)G(r,w(r))dr$$
$$- \int_0^T \mathbf{B}_\alpha(t,T)\mathbf{P}_\alpha(T-r)G(r,w(r))dr. \tag{5.156}$$

Since $f \in \mathbb{H}^{\nu+(1-\vartheta)}(\Omega)$, the convergence (5.147) in Part (a) of Lemma 5.18 yields that the first term of $\mathscr{Q}$ is time continuous for all $0 < t \leq T$. The estimate (5.146) deduces that this term belongs to $C^{\alpha\vartheta}((0,T];\mathbb{H}^\nu(\Omega))$. Similarly, we observe from G satisfies assumption $(\mathscr{H}_3)$ and the estimate (5.148), the convergence (5.149) in Part (b) of Lemma 5.18 that the second term of $\mathscr{Q}$ belongs to $C^{\alpha\vartheta}((0,T];\mathbb{H}^\nu(\Omega))$. On the other hand, using in Part (c) of Lemma 5.18 shows that the integral $\int_0^T \mathbf{P}_\alpha(T-r)G(r,w(r))dr$ belongs to $\mathbb{H}^{\nu+(1-\vartheta)}(\Omega)$, so we deduce from Part (a) of Lemma 5.18 that

$$\mathbf{B}_\alpha(t,T)\int_0^T \mathbf{P}_\alpha(T-r)G(r,w(r))dr \quad \text{belongs to} \quad C^{\alpha\vartheta}((0,T];\mathbb{H}^\nu(\Omega)). \tag{5.157}$$

Therefore, the last term of $\mathscr{Q}$ also belongs to $C^{\alpha\vartheta}((0,T];\mathbb{H}^\nu(\Omega))$.

In the following, we prove $\mathscr{Q}$ maps $\mathfrak{X}_{\alpha,\vartheta,\nu,T}(\mathscr{R})$ into itself. Indeed, let $w^\dagger, w^\ddagger \in \mathfrak{X}_{\alpha,\vartheta,\nu,T}(\mathscr{R})$. By using the formula (5.156), we have

$$\begin{aligned}
&t^{\alpha\vartheta}\left\|\mathscr{Q}w^\dagger(t) - \mathscr{Q}w^\ddagger(t)\right\|_{\mathbb{H}^\nu(\Omega)} \\
\leq & t^{\alpha\vartheta}\left\|\int_0^t \mathbf{P}_\alpha(t-r)\Big(G(r,w^\dagger(r)) - G(r,w^\ddagger(r))\Big)dr\right\|_{\mathbb{H}^\nu(\Omega)} \\
&+ t^{\alpha\vartheta}\left\|\mathbf{B}_\alpha(t,T)\int_0^T \mathbf{P}_\alpha(T-r)\Big(G(r,w^\dagger(r)) - G(r,w^\ddagger(r))\Big)dr\right\|_{\mathbb{H}^\nu(\Omega)} \\
\leq & \mathfrak{N}_2 K_0(T^{s\alpha\vartheta} + \mathscr{R}^s)\left\|w^\dagger - w^\ddagger\right\|_{C^{\alpha\vartheta}((0,T];\mathbb{H}^\nu(\Omega))} \\
&+ \mathscr{N}_2\left\|\int_0^T \mathbf{P}_\alpha(T-r)\Big(G(r,w^\dagger(r)) - G(r,w^\ddagger(r))\Big)dr\right\|_{\mathbb{H}^{\nu+(1-\vartheta)}(\Omega)} \\
\leq & \overline{\mathfrak{N}_2} K_0(T^{s\alpha\vartheta} + \mathscr{R}^s)\|w^\dagger - w^\ddagger\|_{C^{\alpha\vartheta}((0,T];\mathbb{H}^\nu(\Omega))},
\end{aligned}$$

where on the right-hand side of (5.156), we have used the inequalities (5.146) of Lemma 5.18, (5.148) of Lemma 5.18 in the first estimate, and the inequality (5.150) of Lemma 5.18 in the second estimate. This implies that

$$\left\|\mathscr{Q}w^\dagger - \mathscr{Q}w^\ddagger\right\|_{C^{\alpha\vartheta}((0,T];\mathbb{H}^\nu(\Omega))} \leq \overline{\mathfrak{N}_2} K_0(T^{s\alpha\vartheta} + \mathscr{R}^s)\|w^\dagger - w^\ddagger\|_{C^{\alpha\vartheta}((0,T];\mathbb{H}^\nu(\Omega))}. \tag{5.158}$$

By letting $w^{\ddagger} = 0$ into the latter inequality and noting that $\mathscr{Q}\overline{w}^{\ddagger}(t) = \mathbf{B}_\alpha(t, T)f$ if $w^{\ddagger} = 0$, we derive

$$\left\| \mathscr{Q}w^{\dagger} - \mathbf{B}_\alpha(t, T)f \right\|_{C^{\alpha\vartheta}((0,T];\mathbb{H}^\nu(\Omega))} \le \overline{\mathfrak{N}_2} K_0 (T^{s\alpha\vartheta} + \mathscr{R}^s) \| w^{\dagger} \|_{C^{\alpha\vartheta}((0,T];\mathbb{H}^\nu(\Omega))}.$$

From (5.146) and using the triangle inequality, we know that

$$\begin{aligned}\left\| \mathscr{Q}w^{\dagger} \right\|_{C^{\alpha\vartheta}((0,T];\mathbb{H}^\nu(\Omega))} \le & \left\| \mathscr{Q}w^{\dagger} - \mathbf{B}_\alpha(t, T)f \right\|_{C^{\alpha\vartheta}((0,T];\mathbb{H}^\nu(\Omega))} \\ & + \sup_{0 \le t \le T} t^{\alpha\vartheta} \| \mathbf{B}_\alpha(t, T)f \|_{\mathbb{H}^\nu(\Omega)} \\ \le & \overline{\mathfrak{N}_2} K_0 (T^{s\alpha\vartheta} + \mathscr{R}^s) \| w^{\dagger} \|_{C^{\alpha\vartheta}((0,T];\mathbb{H}^\nu(\Omega))} \\ & + \mathscr{N}_2 \| f \|_{\mathbb{H}^{\nu+(1-\vartheta)}(\Omega)}.\end{aligned}$$

Since $w^{\dagger} \in \mathfrak{X}_{\alpha,\vartheta,\nu,T}(\mathscr{R})$, we have $\| w^{\dagger} \|_{C^{\alpha\vartheta}((0,T];\mathbb{H}^\nu(\Omega))} \le \mathscr{R}$. It implies that

$$\begin{aligned}\left\| \mathscr{Q}w^{\dagger} \right\|_{C^{\alpha\vartheta}((0,T];\mathbb{H}^\nu(\Omega))} & \le \overline{\mathfrak{N}_2} K_0 (T^{s\alpha\vartheta} + \mathscr{R}^s)\mathscr{R} + \mathscr{N}_2 \| f \|_{\mathbb{H}^{\nu+(1-\vartheta)}(\Omega)} \\ & =: \pi(\mathscr{R}). \end{aligned} \tag{5.159}$$

Due to the assumption $K_0 T^{s\alpha\vartheta} \in \left(0, \min\left\{\frac{1}{2}\overline{\mathfrak{N}_2}^{-1}, \mathscr{N}_f\right\}\right)$, we now show that there exists $0 < \mathscr{R} < \widehat{\mathscr{R}}$ which is a solution to the equation $\pi(\mathscr{R}) = \mathscr{R}$, where we denote by the constant

$$\widehat{\mathscr{R}} := \left(\frac{1 - \overline{\mathfrak{N}_2} K_0 T^{s\alpha\vartheta}}{(1+s)\overline{\mathfrak{N}_2} K_0} \right)^{1/s}.$$

We note that the function $\mathscr{R} \mapsto \widehat{\pi}(\mathscr{R}) := \pi(\mathscr{R}) - \mathscr{R}$ is continuous on $(0, \widehat{\mathscr{R}})$ with the terminal values $\widehat{\pi}(0) = \mathscr{N}_2 \| f \|_{\mathbb{H}^{\nu+(1-\vartheta)}(\Omega)}$ and

$$\begin{aligned}\widehat{\pi}(\widehat{\mathscr{R}}) & = \overline{\mathfrak{N}_2} K_0 (T^{s\alpha\vartheta} + \widehat{\mathscr{R}}^s)\widehat{\mathscr{R}} + \mathscr{N}_2 \| f \|_{\mathbb{H}^{\nu+(1-\vartheta)}(\Omega)} - \widehat{\mathscr{R}} \\ & = \left(\overline{\mathfrak{N}_2} K_0 \widehat{\mathscr{R}}^s - (1 - \overline{\mathfrak{N}_2} K_0 T^{s\alpha\vartheta}) \right)\widehat{\mathscr{R}} + \mathscr{N}_2 \| f \|_{\mathbb{H}^{\nu+(1-\vartheta)}(\Omega)} \\ & = \left(1 - \overline{\mathfrak{N}_2} K_0 T^{s\alpha\vartheta}\right) \left(\frac{1}{1+s} - 1 \right) \widehat{\mathscr{R}} + \mathscr{N}_2 \| f \|_{\mathbb{H}^{\nu+(1-\vartheta)}(\Omega)} \\ & = \mathscr{N}_2 \| f \|_{\mathbb{H}^{\nu+(1-\vartheta)}(\Omega)} - \frac{s}{1+s} \left(1 - \overline{\mathfrak{N}_2} K_0 T^{s\alpha\vartheta}\right) \widehat{\mathscr{R}} \\ & = \mathscr{N}_2 \| f \|_{\mathbb{H}^{\nu+(1-\vartheta)}(\Omega)} - \frac{s}{1+s} \frac{\left(1 - \overline{\mathfrak{N}_2} K_0 T^{s\alpha\vartheta}\right)^{1+1/s}}{(1+s)^{1/s} (\overline{\mathfrak{N}_2} K_0)^{1/s}}\end{aligned}$$

$$< \mathcal{N}_2 \|f\|_{\mathbb{H}^{\nu+(1-\vartheta)}(\Omega)} \left(1 - \frac{s}{1+s} \frac{(1/2)^{1+1/s}}{(1+s)^{1/s} s} (2(1+s))^{1+1/s}\right)$$
$$= 0,$$

where we note that $1 - \overline{\mathfrak{N}_2} K_0 T^{s\alpha\vartheta} > \frac{1}{2}$. Therefore, there exists $0 < \mathscr{R} < \widehat{\mathscr{R}}$ such that $\pi(\mathscr{R}) = \mathscr{R}$. So it follows from (5.159) that $\mathscr{Q}$ maps $\mathfrak{X}_{\alpha,\vartheta,\nu,T}(\mathscr{R})$ into itself.

In the following, we prove $\mathscr{Q}$ is a contraction mapping and then establish the existence of the mild solution. We note that

$$\begin{aligned}
\overline{\mathfrak{N}_2} K_0 (T^{s\alpha\vartheta} + \widehat{\mathscr{R}}^s) &= \overline{\mathfrak{N}_2} K_0 \left(T^{s\alpha\vartheta} + \frac{1 - \overline{\mathfrak{N}_2} K_0 T^{s\alpha\vartheta}}{(1+s)\overline{\mathfrak{N}_2} K_0}\right) \\
&= \frac{1 - \overline{\mathfrak{N}_2} K_0 T^{s\alpha\vartheta}}{1+s} - \left(1 - \overline{\mathfrak{N}_2} K_0 T^{s\alpha\vartheta}\right) + 1 \\
&= 1 - \frac{s}{1+s}\left(1 - \overline{\mathfrak{N}_2} K_0 T^{s\alpha\vartheta}\right) < \frac{2+s}{2+2s}.
\end{aligned}$$

Hence, we can imply from (5.158) that

$$\begin{aligned}
\left\|\mathscr{Q}w^\dagger - \mathscr{Q}w^\ddagger\right\|_{C^{\alpha\vartheta}((0,T];\mathbb{H}^\nu(\Omega))} &\le \overline{\mathfrak{N}_2} K_0 (T^{s\alpha\vartheta} + \widehat{\mathscr{R}}^s) \|w^\dagger - w^\ddagger\|_{C^{\alpha\vartheta}((0,T];\mathbb{H}^\nu(\Omega))} \\
&\le \frac{2+s}{2+2s} \|w^\dagger - w^\ddagger\|_{C^{\alpha\vartheta}((0,T];\mathbb{H}^\nu(\Omega))}.
\end{aligned}$$

We imply that $\mathscr{Q}$ is a contraction mapping on $\mathfrak{X}_{\alpha,\vartheta,\nu,T}(\mathscr{R})$ which has a unique fixed point u in this space. This fixed point is the unique mild solution of the problem (5.55)–(5.56). In addition, the inequality (5.76) can be easily obtained. The remain of the proof is split as the following steps.

Part (a). Show that $u \in L^p(0,T;\mathbb{H}^{\nu+(\vartheta'-\vartheta)}(\Omega))$ for all $1 \le p < \frac{1}{\alpha\vartheta'}$. It is easy to see that the estimate $\|\mathbf{B}_\alpha(t,T)f\|_{\mathbb{H}^{\nu+(\vartheta'-\vartheta)}(\Omega)} \lesssim t^{-\alpha\vartheta'} \|f\|_{\mathbb{H}^{\nu+(1-\vartheta)}(\Omega)}$ for all $t > 0$. Moreover, using this estimate combines with Lemma 5.18, the inequality (5.76) to invoke that

$$\begin{aligned}
&\left\|\mathbf{B}_\alpha(t,T)\int_0^T \mathbf{P}_\alpha(T-r)G(r,u(r))dr\right\|_{\mathbb{H}^{\nu+(\vartheta'-\vartheta)}(\Omega)} \\
\lesssim & t^{-\alpha\vartheta'} \left\|\int_0^T \mathbf{P}_\alpha(T-r)G(r,u(r))dr\right\|_{\mathbb{H}^{\nu+(1-\vartheta)}(\Omega)} \\
\lesssim & t^{-\alpha\vartheta'} \|f\|_{\mathbb{H}^{\nu+(1-\vartheta)}(\Omega)}.
\end{aligned}$$

On the other hand, it follows from $\nu + (\vartheta' - \vartheta) \le \nu + (1 - \vartheta)$ that the Sobolev embedding $\mathbb{H}^{\nu+(1-\vartheta)}(\Omega) \hookrightarrow \mathbb{H}^{\nu+(\vartheta'-\vartheta)}(\Omega)$ holds. Hence, we can infer from Lemma 5.18 that

$$\begin{aligned}
&\left\| \int_0^t \mathbf{P}_\alpha(t-r)G(r,u(r))dr \right\|_{\mathbb{H}^{\nu+(\vartheta'-\vartheta)}(\Omega)} \\
&\lesssim \left\| \int_0^t \mathbf{P}_\alpha(t-r)G(r,u(r))dr \right\|_{\mathbb{H}^{\nu+(1-\vartheta)}(\Omega)} \\
&\lesssim t^{-\alpha\vartheta} \, \|f\|_{\mathbb{H}^{\nu+(1-\vartheta)}(\Omega)} \\
&\lesssim t^{-\alpha\vartheta'} \, \|f\|_{\mathbb{H}^{\nu+(1-\vartheta)}(\Omega)} .
\end{aligned}$$

Summarily, the solution $u \in L^p(0,T;\mathbb{H}^{\nu+(\vartheta'-\vartheta)}(\Omega))$ for all $1 \le p < \frac{1}{\alpha\vartheta'}$ since $t^{-\alpha\vartheta'}$ clearly belongs to $L^p(0,T;\mathbb{R})$ for all $1 \le p < \frac{1}{\alpha\vartheta'}$. The proof is finalized.

Part (b). Show that $u \in C\left([0,T];\mathbb{H}^{\nu-\eta}(\Omega)\right)$. Let $t, \widetilde{t}$ such that $0 \le t \le \widetilde{t} \le T$. Our purpose here is to find a upper bound of the norm $\left\| u(\widetilde{t}) - u(t) \right\|_{\mathbb{H}^{\nu-\eta}(\Omega)}$. Since $\vartheta < \eta \le \vartheta + 1$ and $0 < \vartheta < 1$, the number $\frac{1+\vartheta-\eta}{2}$ consequently belongs to $[0,1]$. Hence, replacing $1-\frac{1-\vartheta}{2}$ by $\frac{1+\vartheta-\eta}{2}$ helps to improve the inequalities (5.151). Indeed, we have

$$\left| \frac{E_{\alpha,\alpha}(-\lambda_j r^\alpha)}{E_{\alpha,1}(-\lambda_j T^\alpha)} \right| \lesssim r^{\alpha(\eta-\vartheta-1)} \lambda_j^{\eta-\vartheta}. \tag{5.160}$$

As a consequence of the above inequality, we have

$$\begin{aligned}
&\left\| \mathbf{B}_\alpha(\widetilde{t},T)f - \mathbf{B}_\alpha(t,T)f \right\|_{\mathbb{H}^{\nu-\eta}(\Omega)} \\
&\lesssim \|f\|_{\mathbb{H}^{\nu+(1-\vartheta)}(\Omega)} \int_t^{\widetilde{t}} r^{\alpha(\eta-\vartheta)-1} dr \\
&\lesssim \|f\|_{\mathbb{H}^{\nu+(1-\vartheta)}(\Omega)} \left\{ \begin{array}{l} (\widetilde{t}-t)^{\alpha(\eta-\vartheta)} \mathbf{1}_{0<\alpha(\eta-\vartheta)\le 1} \\ \left((\widetilde{t}-t)^{\alpha(\eta-\vartheta)-1} + (\widetilde{t}-t)\right) \mathbf{1}_{1<\alpha(\eta-\vartheta)<2} \end{array} \right\},
\end{aligned}$$

where the number $\alpha(\eta-\vartheta)$ includes in $(0,2)$. Employing Lemma 5.18 allows that

$$\begin{aligned}
\left\| \mathfrak{M}_4 \right\|_{\mathbb{H}^{\nu-\eta}(\Omega)} &\lesssim \left\| \int_0^T \mathbf{P}_\alpha(T-r)G(r,u(r))dr \right\|_{\mathbb{H}^{\nu+(1-\vartheta)}(\Omega)} \\
&\quad \times \left\{ \begin{array}{l} (\widetilde{t}-t)^{\alpha(\eta-\vartheta)} \mathbf{1}_{0<\alpha(\eta-\vartheta)\le 1} \\ \left((\widetilde{t}-t)^{\alpha(\eta-\vartheta)-1} + (\widetilde{t}-t)\right) \mathbf{1}_{1<\alpha(\eta-\vartheta)<2} \end{array} \right\} \\
&\lesssim \|f\|_{\mathbb{H}^{\nu+(1-\vartheta)}(\Omega)} \left\{ \begin{array}{l} (\widetilde{t}-t)^{\alpha(\eta-\vartheta)} \mathbf{1}_{0<\alpha(\eta-\vartheta)\le 1} \\ \left((\widetilde{t}-t)^{\alpha(\eta-\vartheta)-1} + (\widetilde{t}-t)\right) \mathbf{1}_{1<\alpha(\eta-\vartheta)<2} \end{array} \right\},
\end{aligned}$$

provided that notation $\mathfrak{M}_4$ is given by (5.108). Now, let us consider the terms $\mathfrak{M}_2$ and $\mathfrak{M}_3$. It indicates from $\mu < \vartheta$ and $\vartheta < \eta$ that $0 \le -\sigma \le \eta - \nu$, and it results $\mathbb{H}^\sigma(\Omega) \hookrightarrow \mathbb{H}^{\nu-\eta}(\Omega)$. This suggests to estimate the term $\mathfrak{M}_2$. In fact, we have

$$\begin{aligned}
\|\mathfrak{M}_2\|_{\mathbb{H}^{\nu-\eta}(\Omega)} &\lesssim \int_0^t \int_{t-r}^{t'-r} \rho^{\alpha-2} \|G(r, u(r))\|_{\mathbb{H}^\sigma(\Omega)} d\rho dr \\
&\lesssim \|f\|_{\mathbb{H}^{\nu+(1-\vartheta)}(\Omega)} \int_0^t \int_{t-r}^{\tilde{t}-r} \rho^{\alpha-2} \left(r^{-\alpha\vartheta} + r^{-(1+s)\alpha\vartheta}\right) L_3(r) d\rho dr \\
&\lesssim \|f\|_{\mathbb{H}^{\nu+(1-\vartheta)}(\Omega)} (\tilde{t} - t)^{\alpha-1} \int_0^t \left(r^{-\alpha(\vartheta+\zeta)} + r^{-\alpha((1+s)\vartheta+\zeta)}\right) dr,
\end{aligned}$$

provided that $L_3(t) \le K_0 t^{-\alpha\zeta}$ as G satisfies (5.75). Since $\zeta < \dfrac{1}{\alpha} - (1+s)\vartheta$, we derive that the integral on the right-hand side of the previous express is convergent. Hence, we get immediately the estimate

$$\|\mathfrak{M}_2\|_{\mathbb{H}^{\nu-\eta}(\Omega)} \lesssim \|f\|_{\mathbb{H}^{\nu+(1-\vartheta)}(\Omega)} (\tilde{t} - t)^{\alpha-1},$$

and from the locally property of G as in (5.75), we find that

$$\begin{aligned}
\|\mathfrak{M}_3\|_{\mathbb{H}^{\nu-\eta}(\Omega)} &\lesssim \int_t^{\tilde{t}} (\tilde{t} - r)^{\alpha-1} \|G(r, u(r))\|_{\mathbb{H}^\sigma(\Omega)} dr \\
&\lesssim \|f\|_{\mathbb{H}^{\nu+(1-\vartheta)}(\Omega)} \int_t^{\tilde{t}} (\tilde{t} - r)^{\alpha-1} \left(r^{-\alpha\vartheta} + r^{-(1+s)\alpha\vartheta}\right) L_3(r) dr \\
&\lesssim \|f\|_{\mathbb{H}^{\nu+(1-\vartheta)}(\Omega)} \int_t^{\tilde{t}} (\tilde{t} - r)^{\alpha-1} \left(r^{-\alpha(\vartheta+\zeta)} + r^{-\alpha((1+s)\vartheta+\zeta)}\right) dr \\
&\lesssim \|f\|_{\mathbb{H}^{\nu+(1-\vartheta)}(\Omega)} (\tilde{t} - t)^{\alpha-1}.
\end{aligned}$$

The above explanations deduce $u \in C\left([0, T]; \mathbb{H}^{\nu-\eta}(\Omega)\right)$. This part is accomplished.

Appendix

(AP.) List of Important Constants

Here, we list some important constants appeared in this section, where some of them contain the constant $C_1(\nu, \theta)$, $C_2(\nu, \sigma)$ in the embeddings (5.70) and (5.72).

$$
\begin{cases}
\mathscr{M}_1 = \mathscr{M}_1(\alpha, \theta, T) := M_\alpha^2 m_\alpha^{-1} T^{\alpha(1-\theta)}(T^{\alpha\theta} + \lambda_1^{-\theta}), \\
\mathscr{M}_2 = \mathscr{M}_2(\alpha, \theta, T) := \frac{M_\alpha^2}{m_\alpha} \frac{T^{\alpha(1-\theta)}}{\alpha^2\theta(1-\theta)}(T^\alpha + \lambda_1^{-1}), \\
\mathscr{U}_1 = \mathscr{U}_1(\alpha, \theta, T) := \frac{\pi M_\alpha \lambda_1^{-\theta} T^{\alpha(1-\theta)} + \pi \mathscr{M}_1}{\sin(\pi\alpha(1-\theta))}, \\
\overline{\mathscr{M}}_1 = \overline{\mathscr{M}}_1(q, \alpha, \nu, \sigma, T) := C_2(\nu, \sigma) M_\alpha \frac{T^{\alpha+1/q}}{\alpha(\alpha q+1)^{1/q}}, \\
\overline{\mathscr{M}}_2 = \overline{\mathscr{M}}_2(q, \alpha, \theta, T) := \frac{T^{\alpha(1-\theta)}(T^{\alpha\theta+1/q} + \lambda_1^{-\theta})}{(1-\alpha(1-\theta)q)^{1/q}\alpha(1-\theta)}, \\
\overline{\mathscr{M}}_3 = \overline{\mathscr{M}}_3(\alpha, T) := \frac{M_\alpha T^\alpha}{\alpha}\left(1 + \frac{M_\alpha}{m_\alpha}(T^\alpha + \lambda_1^{-1})\right), \\
\mathscr{U}_2 = \mathscr{U}_2(q, \alpha, \nu, \sigma, T) := \|L_2\|_{L^\infty(0,T)} \sum_{1 \le j \le 3} \overline{\mathscr{M}}_j, \\
\mathscr{N}_1 = \mathscr{N}_1(\alpha, \theta, \nu, T) := \frac{\mathscr{M}_1 M_\alpha^{-1} \max\{1, (T^{\alpha\theta} + \lambda_1^{-\theta})^{-1}\} C_1(\nu, \theta)}{1 - \|L_1\|_{L^\infty(0,T)} \mathscr{U}_1}, \\
\mathscr{N}_2 = \mathscr{N}_2(\alpha, \vartheta, T) := M_\alpha m_\alpha^{-1} T^{\alpha\vartheta}(T^{\alpha(1-\vartheta)} + \lambda_1^{\vartheta-1}), \\
\mathfrak{N}_1 = \mathfrak{N}_1(\alpha, \mu, \vartheta, \zeta) := \max\{\mathbf{B}(\alpha z_j, 1 - \alpha(1+s)\vartheta - \alpha\zeta), j = 1, 2\}, \\
\qquad \begin{cases} z_1 = \vartheta - \mu, \\ z_2 = 1 - \mu, \end{cases} \\
\mathfrak{N}_2 = \mathfrak{N}_2(\alpha, \mu, \vartheta, \zeta, s, T) := M_\alpha \mathfrak{N}_1 \max\{T^{\alpha(z_j - s\vartheta - \zeta)}, j = 1, 2\}, \\
\overline{\mathfrak{N}_2} = \overline{\mathfrak{N}_2}(\alpha, \mu, \vartheta, \zeta, s, T) := \mathfrak{N}_2(1 + \mathscr{N}_2 T^{-\alpha\vartheta}), \\
\mathscr{N}_f = \mathscr{N}_f(\alpha, \nu, \vartheta, T, s) := \left(\frac{s}{\mathscr{N}_2 \|f\|_{\mathbb{H}^{\nu+(1-\vartheta)}(\Omega)}}\right)^s \frac{1}{(2+2s)^{1+s}}, \\
\widehat{\mathscr{R}} = \widehat{\mathscr{R}}(\alpha, \mu, \vartheta, \zeta, s, T) := \left(\frac{1 - \overline{\mathfrak{N}_2} K_0 T^{s\alpha\vartheta}}{(1+s)\overline{\mathfrak{N}_2} K_0}\right)^{1/s}, \\
\eta_{glo} = \eta_{glo}(\alpha, \theta, \nu') := \begin{Bmatrix} \min\left\{\alpha(\theta + \nu' - 1), \alpha - 1\right\} \mathbf{1}_{0 < \alpha(\theta+\nu'-1) \le 1} \\ \min\left\{\alpha(\theta + \nu' - 1) - 1, \alpha - 1\right\} \mathbf{1}_{1 < \alpha(\theta+\nu'-1) < 2} \end{Bmatrix}, \\
\eta_{cri} = \eta_{glo}(\alpha, \eta, \vartheta) := \begin{Bmatrix} \min\left\{\alpha(\eta - \vartheta), \alpha - 1\right\} \mathbf{1}_{0 < \alpha(\eta-\vartheta) \le 1} \\ \min\left\{\alpha(\eta - \vartheta) - 1, \alpha - 1\right\} \mathbf{1}_{1 < \alpha(\eta-\vartheta) < 2} \end{Bmatrix}.
\end{cases}
$$

(AP.1) A Singular Integral

It is useful to recall some basic properties of a singular integral. For given $z_1 > 0$, $z_2 > 0$, and $0 \le a < b \le T$, we denote by

$$\mathscr{K}(z_1, z_2, a, b) := \int_a^b (b-\tau)^{z_1-1}(\tau-a)^{z_2-1}d\tau = (b-a)^{z_1+z_2-1}\mathbf{B}(z_1, z_2), \tag{5.161}$$

where $\mathbf{B}$ is the Beta function, $\mathbf{B}(z_1, z_2) := \int_0^1 t^{z_1-1}(1-t)^{z_2-1}dt$. Moreover, a special case of the Beta function is $\mathbf{B}(z, 1-z) = \pi/\sin(\pi z)$, see, e.g., [26, 52, 126, 180, 187].

(AP.2) A Useful Limit

For $a > 0, b > 0, t > 0, h > 0$, the following convergence holds

$$\int_0^t (t+h-r)^{a-1}r^{b-1}dr \xrightarrow{h\to 0^+} \int_0^t (t-r)^{a-1}r^{b-1}dr.$$

Indeed, it can be proved by noting that $\int_0^t (t-r)^{a-1}r^{b-1}dr = t^{a+b-1}\mathbf{B}(a, b)$ and

$$\int_0^t (t+h-r)^{a-1}r^{b-1}dr = (t+h)^{a+b-1}\int_0^{t/(t+h)} (1-s)^{a-1}s^{b-1}ds$$

$$\xrightarrow{h\to 0^+} t^{a+b-1}\mathbf{B}(a, b).$$

References

1. E. Affili, E. Valdinoci, Decay estimates for evolution equations with classical and fractional time-derivatives. J. Diff. Equ. **266**(7), 4027–4060 (2019)
2. M. Abramowitz, I.A. Stegun, in *Handbook of Mathematical Functions with Formulas, Graphs, and Mathematical Tables*. Applied Mathematics Series (Courier Corporation, Washington, D.C., 1967)
3. R.A. Adams, *Sobolev Spaces* (Academic, London, 1975)
4. O.P. Agrawal, Fractional variational calculus in terms of Riesz fractional derivatives. J. Phy. A: Math. Theor. **40**, 6287–6303 (2007)
5. B. Ahmad, M.S. Alhothuali, H.H. Alsulami, M. Kirane, S. Timoshin, On a time fractional reaction diffusion equation. Appl. Math. Comput. **257**, 199–204 (2015)
6. M.R. Alaimia, N.E. Tatar, Blow up for the wave equation with a fractional damping. J. Appl. Anal. **11**(1), 133–144 (2005)
7. M. Ali, S.A. Malik, An inverse problem for a family of time fractional diffusion equations. Inverse Probl. Sci. Eng. **25**(9), 1299–1322 (2017)
8. L. Aloui, S. Ibrahim, M. Khenissi, Energy decay for linear dissipative wave equations in exterior domains. J. Diff. Equ. **259**(5), 2061–2079 (2015)
9. A. Alsaedi, B. Ahmad, M. Kirane, B. Rebiai, Local and blowing-up solutions for a space-time fractional evolution system with nonlinearities of exponential growth. Math. Methods Appl. Sci. **42**(12), 4378–4393 (2019)
10. E. Alvarez, C.G. Gal, V. Keyantuo, M. Warma, Well-posedness results for a class of semi-linear super-diffusive equations. Nonlinear Anal. **181**, 24–61 (2019)
11. H. Amann, *Linear and Quasilinear Parabolic Problems* (Birkhäuser, Berlin, 1995)
12. B.D. Andrade, A. Viana, On a fractional reaction-diffusion equation. Z. Angew. Math. Phys. **68**(3), 59 (2017)
13. D. Araya, C. Lizama, Almost automorphic mild solutions to fractional differential equations. Nonlinear Anal. **69**, 3692–3705 (2008)
14. T.M. Atanacković, S. Pilipović, B. Stanković, D. Zorica, *Fractional Calculus with Applications in Mechanics* (Wiley-ISTE, New York, 2014)
15. V.V. Au, M. Kirane, N.H. Tuan, Determination of initial data for a reaction-diffusion system with variable coefficients. Discrete Contin. Dyn. Syst. **39**(2), 771–801 (2019)
16. J. Baumeister, *Stable Solution of Inverse Problems* (Springer, Berlin, 1986)
17. A. Bekkai, B. Rebiai, M. Kirane, On local existence and blowup of solutions for a time-space fractional diffusion equation with exponential nonlinearity. Math. Methods Appl. Sci. **42**, 1819–1830 (2019)
18. J. Bergh, J. Löfström, *Interpolation Spaces: An Introduction* (Springer, Berlin, 1976)

Y. Zhou, *Fractional Diffusion and Wave Equations*,
https://doi.org/10.1007/978-3-031-74031-2

19. B. Berkowitz, J. Klafter, R. Metzler, H. Scher, Physical pictures of transport in heterogeneous media: Advection-dispersion, random-walk, and fractional derivative formulations. Water Res. Res. **38**, 9–1-9-12 (2002)
20. A.V. Bobylev, C. Cercignani, The inverse Laplace transform of some analytic functions with an application to the eternal solutions of the Boltzmann equation. Appl. Math. Lett. **15**, 807–813 (2002)
21. K. Bogdan, K. Burdzy, Z-Q. Chen, Censored stable processes. Probab. Theory Related Fields **127**(1), 89–152 (2003)
22. M. Bonforte, Y. Sire, J.L. Vazquez, Optimal existence and uniqueness theory for the fractional heat equation. Nonlinear Anal. **153**, 142–168 (2017)
23. M. Bonforte, J.L. Vazquez, Quantitative local and global a priori estimates for fractional nonlinear diffusion equations. Adv. Math. **250**, 242–284 (2014)
24. V. Bögelein, F. Duzaar, P. Marcellini, S. Signoriello, Nonlocal diffusion equations. J. Math. Anal. Appl. **432**(1), 398–428 (2015)
25. V. Bögelein, F. Duzaar, P. Marcellini, C. Scheven, Doubly nonlinear equations of porous medium type. Arch. Ration. Mech. Anal. **229**(2), 503–545 (2018)
26. H. Brezis, *Functional Analysis, Sobolev Spaces and Partial Differential Equations* (Springer Science & Business Media, Berlin, 2010)
27. Z. Brzezniak, N. Rana, Local solution to an energy critical 2-D stochastic wave equation with exponential nonlinearity in a bounded domain. J. Differ. Equ. **340**, 386–462 (2022)
28. M.V. Burtsev, A.N. Zarubin, Inverse initial-boundary value problem for a fractional diffusion wave equation with a non-carleman shift. Differ. Equ. **44**(3), 390–400 (2008)
29. A. Carasso, Error bounds in the final value problem for the heat equation. SIAM J. Math. Anal. **7**, 195–199 (1976)
30. Y. Cao, J.X. Yin, C.P. Wang, Cauchy problems of semilinear pseudo-parabolic equations. J. Differ. Equ. **46**, 4568–4590 (2009)
31. C.M. Carracedo, M.S. Alix, *The Theory of Fractional Powers of Operators, North-Holland Mathematics Studies* (Elsevier, San Diego, 2001)
32. T. Cazenave, *Semilinear Schrödinger Equations* (American Mathematical Society, New York, 2003)
33. T. Cazenave, B. Weissler, The cauchy problem for the critical nonlinear Schrödinger equation in H^s. Nonlinear Anal. **14**(10), 807–836 (1990)
34. L. Chen, Nonlinear stochastic time-fractional diffusion equations on $\mathbb{R}$: moments, Hölder regularity and intermittency. Trans. Amer. Math. Soc. **369**(12), 8497–8535 (2017)
35. W. Chen, S. Holm, Physical interpretation of fractional diffusion-wave equation via lossy media obeying frequency power law. arXiv preprint math-ph/0303040 (2003)
36. J. Chen, F. Liu, V. Anh, Analytical solution for the time-fractional telegraph equation by the method of separating variables. J. Math. Anal. Appl. **338**, 1364–1377 (2008)
37. Y. Chen, H. Gao, M. Garrido-Atienza, B. Schmalfuss, Pathwise solutions of SPDEs driven by Hölder-continuous integrators with exponent larger than 1/2 and random dynamical systems. Discrete Contin. Dyn. Syst. **34**, 79–98 (2014)
38. D. Chen, B. Hofmann, J. Zou, Regularization and convergence for ill-posed backward evolution equations in Banach spaces. J. Differ. Equ. **265**, 3533–3566 (2018)
39. B. Claus, M. Warma, Realization of the fractional laplacian with nonlocal exterior conditions via forms method. J. Evol. Equ. **20**, 1597–1631 (2020)
40. Ph. Clément, G. Gripenberg, S.O. Londen, Regularity properties of solutions of fractional evolution equations, in *Evolution Equations and Their Applications in Physical and Life Sciences* (Dekker, New York, 2001)
41. Ph. Clément, S.O. Londen, G. Simonett, Quasilinear evolutionary equations and continuous interpolation spaces. J. Differ. Equ. **196**, 418–447 (2004)
42. R. Courant, D. Hilbert, *Methods of Mathematical Physics: Partial Differential Equations* (Wiley, New Jersey, 2008)
43. D.T. Dang, E. Nane, D.M. Nguyen, N.H. Tuan, Continuity of solutions of a class of fractional equations. Potential Anal. **49**, 423–478 (2018)

44. M.F. de Alemida, J.C.P. Precioso, Existence and symmetries of solutions in Besov-Morrey spaces for a semilinear heat-wave type equation. J. Math. Anal. Appl. **432**, 338–355 (2015)
45. M.F. de Alemida, A. Viana, Self-similar solutions for a superdiffusive heat equation with gradient nonlinearity. Electr. J. Diff. Equ. **2016**(250), 1–20 (2016)
46. B. de Andrade, A.N. Carvalho, P.M. Carvalho-Neto, P. Marin-Rubio, Semilinear fractional differential equations: global solutions, critical nonlinearities and comparison results. Topol. Methods Nonlinear Anal. **45**, 439–467 (2015)
47. P.M. de Carvalho-Neto, P. Gabriela, Mild solutions to the time fractional Navier-Stokes equations in $\mathbb{R}^N$. J. Differ. Equ. **259**, 2948–2980 (2015)
48. A. Deiveegan, J.J. Nieto, P. Prakash, Periasamy, The revised generalized Tikhonov method for the backward time-fractional diffusion equation. J. Appl. Math. Comput. **9**(1), 45–56 (2019)
49. D. del Castillo-Negrete, B.A. Carreras, V.E. Lynch, Fractional diffusion in plasma turbulence Phys. Plasmas **11**(8), 3854–3864 (2004)
50. D. del-Castillo-Negrete, B.A. Carreras, V.E. Lynch, Nondiffusive transport in plasma turbulene: a fractional diffusion approach. Phys. Rev. Lett. **94**, 065003 (2005)
51. H.F. Di, Y.D. Shang, X.M. Peng, Blow-up phenomena for a pseudo-parabolic equation with variable exponents. Appl. Math. Lett. **64**, 67–73 (2017)
52. K. Diethelm, *The Analysis of Fractional Differential Equations: An Application-Oriented Exposition Using Differential Operators of Caputo Type* (Springer Science and Business Media, Berlin, 2010)
53. S. Dipierro, X. Ros-Oton, E. Valdinoci, Nonlocal problems with Neumann boundary conditions, Rev. Mat. Iberoam. **33**(2), 377–416 (2017)
54. J.D. Djida, A. Fernandez, I. Area, Well-posedness results for fractional semi-linear wave equations. Discrete Contin. Dyn. Syst. Ser. B **25**(2), 569–597 (2020)
55. H. Dong, D. Kim, L_p-estimates for time fractional parabolic equations with coefficients measurable in time. Adv. Math. **345**, 289–345 (2019)
56. H. Dong, Y. Liu, Weighted mixed norm estimates for fractional wave equations with VMO coefficients. J. Differ. Equ. **337**, 168–254 (2022)
57. F. Duzaar, J. Habermann, Partial regularity for parabolic systems with non-standard growth. J. Evol. Equ. **12**(1), 203–244 (2012)
58. H.W. Engl, M. Hanke, A. Neubauer, *Regularization of Inverse Problems* (Springer Science and Business Media, Berlin, 1996)
59. L.C. Evans, *Partial Differential Equations*, 2nd edn. (American Mathematical Society, Providence, 2010)
60. L.R. Evangelista, E.K. Lenzi, *Fractional Diffusion Equations and Anomalous Diffusion* (Cambridge University Press, Cambridge, 2018)
61. W. Fan, F. Liu, X. Jiang, I. Turner, A novel unstructublack mesh finite element method for solving the time-space fractional wave equation on a two-dimensional irregular convex domain. Fract. Calc. Appl. Anal. **20**, 352–383 (2017)
62. A.Z. Fino, M. Kirane, The Cauchy problem for heat equation with fractional Laplacian and exponential nonlinearity. Commun. Pure Appl. Anal. **19**(7), 3625–3650 (2020)
63. G. Fragnelli, D. Mugnai, Stability of solutions for some classes of nonlinear damped wave equations. SIAM J. Control Optim. **47**(5), 2520–2539 (2008)
64. Y. Fujita, Integrodifferential equation which interpolates the heat equation and the wave equation. Osaka J. Math. **27**, 309–321 (1990)
65. G. Furioli, T. Kawakami, B. Ruf, E. Terraneo, Asymptotic behavior and decay estimates of the solutions for a nonlinear parabolic equation with exponential nonlinearity. J. Differ. Equ. **262**, 145–180 (2017)
66. C.G. Gal, M. Warma, Fractional in time semilinear parabolic equations and applications. HAL Id: hal-01578788 (2017)
67. T. Ghosh, M. Salo, G. Uhlmann, The Calderón problem for the fractional Schrödinger equation. Anal. PDE. **13**(2), 455–475 (2020)
68. D. Gilbarg, N.S. Trudinger, *Elliptic Partial Differential Equations of Second Order* (Springer, Berlin, 2001)

69. Y. Giga, T. Namba, Well-posedness of Hamilton-Jacobi equations with Caputo's time fractional derivative. Comm. Partial Differ. Equ. **42**(7), 1088–1120 (2017)
70. J. Ginibre, G. Velo, The global Cauchy problem for nonlinear Klein-Gordon equation. Math. Z **189**, 487–505 (1985)
71. T. Ghoula, V.T. Nguyen, H. Zaag, Blowup solutions for a reaction-diffusion system with exponential nonlinearities. J. Differ. Equ. **264**, 7523–7579 (2018)
72. I. Golding, E.C. Cox, Physical nature of bacterial cytoplasm. Phys. Rev. Lett. **96**, 098102 (2006)
73. I. Graham, U. Langer, J. Melenk, M. Sini (Eds.), *Direct and Inverse Problems in Wave Propagation and Applications* (Walter de Gruyter, Berlin, 2013)
74. M.G. Grillakis, Regularity and asymptotic behavior of the wave equation with a critical nonlinearity. Ann. Math. **132**, 485–509 (1990)
75. G. Grubb, Fractional Laplacians on domains, a development of Hörmander's theory of μ-transmission pseudo-differential operators. Adv. Math. **268**, 478–528 (2015)
76. S. Guo, L. Mei, Y. Li, An efficient Galerkin spectral method for two-dimensional fractional nonlinear reaction-diffusion-wave equation. Comput. Math. Appl. **74**, 2449–2465 (2017)
77. B.H. Guswanto, T. Suzuki, Existence and uniqueness of mild solutions for fractional semilinear differential equations. Electron. J. Differ. Equ. **2015**, 1–16 (2015)
78. H. Hajaiej, X. Yu, Z. Zhai, Fractional Gagliardo-Nirenberg and Hardy inequalities under Lorentz norms. J. Math. Anal. Appl. **396**, 569–577 (2012)
79. B. Han, K. Kim, D. Park, Weighted $L_q(L_p)$-estimate with Muckenhoupt weights for the diffusion-wave equations with time-fractional derivatives. J. Differ. Equ. **269**, 3515–3550 (2020)
80. D.N. Hao, N.V. Duc, N.V. Thang, Backward semi-linear parabolic equations with time-dependent coefficients and local Lipschitz source. Inverse Probl. **34**, 33 (2018)
81. J.W. He, L. Peng, Approximate controllability for a class of fractional stochastic wave equations. Comput. Math. Appl. **78**, 1463–1476 (2019)
82. J.W. He, Y. Zhou, On a backward problem for nonlinear time fractional wave equations. Proc. Roy. Soc. Edinburgh Sect. A. **152**(6), 1589–1612 (2022)
83. J.W. He, Y. Zhou, Local/global existence analysis of fractional wave equations with exponential nonlinearity. Bull. Sci. Math. **189**, 103357 (2023)
84. J.W. He, Y. Zhou, L. Peng, On well-posedness of semilinear Rayleigh-Stokes problem with fractional derivative on $\mathbb{R}^N$. Adv. Nonlinear Anal. **11**, 580–597 (2022)
85. J.W. He, Y. Zhou, A. Alsaedi, B. Ahmad, The existence of solutions to fractional diffusion equations with exponential growth. Nonlinear Anal. Model. Control **29**(2), 286–304 (2024)
86. R. Hilfer, *Applications of Fractional Calculus in Physics* (World Scientific, Singapore, 2000)
87. H. Hirata, C. Miao, Space-time estimates of linear flow and application to some nonlinear integro-differential equations corresponding to fractional-order time derivative. Adv. Differ. Equ. **7**(2), 217–236 (2002)
88. M. E. Hochstenbach, L. Reichel, Fractional Tikhonov regularization for linear discrete ill-posed problems. BIT **51**(1), 197–215 (2011)
89. N.Q. Hung, J.L. Vazquez, Porous medium equation with nonlocal pressure in a bounded domain. Comm. Partial Differ. Equ. **43**(10), 1502–1539 (2018)
90. J. Huang, G. Wang, J. Xiong, A maximum principle for partial information backward stochastic control problems with applications. SIAM J. Control Optim. **48**(4), 2106–2117 (2009)
91. L.N. Huynh, Y. Zhou, D. O'Regan, N.H. Tuan, Fractional Landweber method for an initial inverse problem for time-fractional wave equations. Appl. Anal. **100**(4), 860–878 (2021)
92. S. Ibrahim, M. Majdoub, N. Masmoudi, K. Nakanishi, Scattering for the two-dimensional energy-critical wave equation. Duke Math. J. **150**(2), 287–329 (2009)
93. R. Ikehata, G. Todorova, B. Yordanov, Wave equations with strong damping in Hilbert spaces. J.Differ. Equ. **254**(8), 3352–3368 (2013)
94. N. Ioku, The Cauchy problem for heat equations with exponential nonlinearity. J. Differ. Equ. **251**, 1172–1194 (2011)

95. N. Ioku, B. Ruf, E. Terraneo, Existence, non-existence, and uniqueness for a heat equation with exponential nonlinearity in $\mathbb{R}^2$. Math. Phys. Anal. Geom. **18**(29), 19 (2015)
96. M.I. Ismailov, M. Çiçek, Inverse source problem for a time-fractional diffusion equation with nonlocal boundary conditions. Appl. Math. Modell. **40**(7–8), 4891–4899 (2016)
97. J. Janno, N. Kinash, Reconstruction of an order of derivative and a source term in a fractional diffusion equation from final measurements. Inverse Probl. **34**(2), 025007(2018)
98. J. Janno, K. Kasemets, Uniqueness for an inverse problem for a semilinear time-fractional diffusion equation. Inverse Probl. Imag. **11**, 125–149 (2017)
99. J.-H. Jeon, H.M.-S. Monne, M. Javanainen, R. Metzler, Anomalous diffusion of phospholipids and cholesterols in a lipid bilayer and its origins. Phys. Rev. Lett. **109**, 188103 (2012)
100. J.-H. Jeon, N. Leijnse, L.B. Oddershede, R. Metzler, Anomalous diffusion and power-law relaxation of the time averaged mean squared displacement in worm-like micellar solutions. New J. Phys. **15**, 045011 (2013)
101. J. Jia, J. Peng, J. Gao, Y. Li, Backward problem for a time-space fractional diffusion equation. Inverse Probl. Imag. **12**(3), 773–800 (2018)
102. D. Jiang, Z. Li, Y. Liu, M. Yamamoto, Weak unique continuation property and a related inverse source problem for time-fractional diffusion-advection equations. Inverse Probl. **33**, 21 (2017)
103. Z. Jiao, Y.Q, Chen, I. Podlubny, *Distributed-Order Dynamic Systems: Stability, Simulation, Applications and Perspectives* (Springer, London, 2012)
104. B. Jin, R. Lazarov, Y. Liu, Z. Zhou, The Galerkin finite element method for a multi-term time-fractional diffusion equation. J. Comput. Phys. **281**, 825–843 (2015)
105. B. Jin, R. Lazarov, D. Sheen, Z. Zhou, Error estimates for approximations of distributed order time fractional diffusion with nonsmooth data. Fract. Calc. Appl. Anal. **19**(1), 69–93 (2016)
106. B. Jin, B. Li, Z. Zhou, Numerical analysis of nonlinear subdiffusion equations. SIAM J. Numer. Anal. **56**, 1–23 (2018)
107. B. Kaltenbacher, W. Rundell, On an inverse potential problem for a fractional reaction-diffusion equation. Inverse Probl. **35**(6), 065004 (2019)
108. B. Kaltenbacher, W. Rundell, Regularization of a backward parabolic equation by fractional operators. Inverse Probl. Imag. **13** (2), 401–430 (2019)
109. T. Kato, *Perturbation Theory for Linear Operators* (Springer, Berlin, 1995)
110. J. Kemppainen, J. Siljander, V. Vergara, R. Zacher, Decay estimates for time-fractional and other non-local in time subdiffusion equations in $\mathbb{R}^d$. Math. Ann. **366**(3), 941–979 (2016)
111. J. Kemppainen, J. Siljander, R. Zacher, Representation of solutions and large-time behavior for fully nonlocal diffusion equations. J. Differ. Equ. **263**(1), 149–201 (2017)
112. Y. Kian, M. Yamamoto, On existence and uniqueness of solutions for semilinear fractional wave equations. Fract. Calc. Appl. Anal. **20**(1), 117–138 (2017)
113. Y. Kian, Z. Li, Y. Liu, M. Yamamoto, The uniqueness of inverse problems for a fractional equation with a single measurement. Math. Ann. **308**(3), 1465–1495 (2021)
114. Y. Kian, L. Oksanen, E. Soccorsi, M. Yamamoto, Global uniqueness in an inverse problem for time fractional diffusion equations. J. Differ. Equ. **264**, 1146–1170 (2018)
115. A.A. Kilbas, H.M. Srivastava, J.J. Trujillo, *Theory and Applications of Fractional Differential Equations* (Elsevier Science B.V., Amsterdam, 2006)
116. I. Kim, K.H. Kim, S. Lim, An $L_q(L_p)$-theory for the time fractional evolution equations with variable coefficients. Adv. Math. **306**, 123–176 (2017)
117. M. Kirane, B. Ahmad, A. Alsaedi, M. Al-Yami, Non-existence of global solutions to a system of fractional diffusion equations. Acta Appl. Math. **133**(1), 235–248 (2014)
118. A. Kirsch, *An Introduction to the Mathematical Theory of Inverse Problem* (Springer Science and Business Media, Berlin, 2011)
119. E. Klann, R. Ramlau, Regularization by fractional filter methods and data smoothing. Inverse Probl. **24**(2), 025018 (2008)
120. H. Kita, On interpolation of the Fourier maximal operator in Orlicz spaces. Acta Math. Hungar. **81**(3), 175–193 (1998)

121. A. N. Kochubei, Distributed order calculus and equation of ultraslow diffusion. J. Math. Anal. Appl. **340**(1), 252–281 (2008)
122. S. Kou, Stochastic modeling in nanoscale biophysics: subdiffusion within proteins. Ann. Appl. Stat. **2**, 501–535 (2008)
123. A. Kubica, K. Ryszewska, Decay of solutions to parabolic-type problem with distributed order Caputo derivative. J. Math. Anal. Appl. **465**(1), 75–99 (2018)
124. A. Kubica, K. Ryszewska, Fractional diffusion equation with distributed-order Caputo derivative. J. Integr. Equ. Appl. **31**(2), 195–243 (2019)
125. D. Kumar, J. Singh, D. Baleanu, A new analysis for fractional model of regularized long-wave equation arising in ion acoustic plasma waves. Math. Methods Appl. Sci. **40**, 5642–5653 (2017)
126. P.D. Lax, *Functional Analysis* (Wiley Interscience, New York, 2002)
127. G. Li, D. Zhang, X. Jia, M. Yamamoto, Simultaneous inversion for the space-dependent diffusion coefficient and the fractional order in the time-fractional diffusion equation. Inverse Probl. **29**(6), 065014 (2013)
128. L. Li, J.G. Liu, L. Wang, Cauchy problems for Keller-Segel type time-space fractional diffusion equation. J. Differ. Equ. **265**(3), 1044–1096 (2018)
129. Z. Li, Y. Luchko, M. Yamamoto, Asymptotic estimates of solutions to initial-boundary-value problems for distributed order time-fractional diffusion equations. Fract. Calc. Appl. Anal. **17**(4), 1114–1136 (2014)
130. Z. Li, O.Y. Imanuvilov, M. Yamamoto, Uniqueness in inverse boundary value problems for fractional diffusion equations. Inverse Probl. **32**(1), 015004 (2015)
131. Z. Li, Y. Liu, M. Yamamoto, Initial-boundary value problems for multi-term time-fractional diffusion equations with positive constant coefficients. Appl. Math. Comput. **257**, 381–397 (2015)
132. Z. Li, Y. Luchko, M. Yamamoto, Analyticity of solutions to a distributed order time-fractional diffusion equation and its application to an inverse problem. Comput. Math. Appl. **73**, 1041–1052 (2016)
133. Y. Li, Y. Wang, W. Deng, Galerkin finite element approximations for stochastic space-time fractional wave equations. SIAM J. Numer. Anal. **55**(6), 3173–3202 (2017)
134. Z. Li, Y. Kian, E. Soccorsi, Initial-boundary value problem for distributed order time-fractional diffusion equations. Asymptot. Anal. **115**(1–2), 95–126 (2019)
135. W. Lian, J. Wang, R. Xu, Global existence and blow up of solutions for pseudo-parabolic equation with singular potential. J. Differ. Equ. **269**(6), 4914–4959 (2020)
136. C. Lin, G. Nakamura, Unique continuation property for anomalous slow diffusion equation. Comm. Partial Differ. Equ. **41**(5), 749–758 (2016)
137. C. Lin, G. Nakamura, Unique continuation property for multi-terms time fractional diffusion equations. Math. Ann. **373**(3–4), 929–952 (2019)
138. J.J. Liu, M. Yamamoto, A backward problem for the time-fractional diffusion equation. Appl. Anal. **89**(11), 1769–1788 (2010)
139. Y. Liu, W.S. Jiang, F.L. Huang, Asymptotic behaviour of solutions to some pseudo-parabolic equations. Appl. Math. Lett. **25**, 111–114 (2012)
140. W.J. Liu, J.Y. Yu, A note on blow-up of solution for a class of semilinear pseudo-parabolic equations. J. Funct. Anal. **274**, 1276–1283 (2018)
141. A.K. Louis, *Inverse Und Schlecht Gestellte Probleme* (Springer, Berlin, 2013)
142. Y. Luchko, Boundary value problems for the generalized time-fractional diffusion equation of distributed order. Fract. Calc. Appl. Anal. **12**(4), 409–422 (2009)
143. Y. Luchko, Fractional wave equation and damped waves. J. Math. Phys. **54**(3), 031505 (2013)
144. Y. Luchko, F. Mainardi, Fractional diffusion-wave phenomena. Handbook Fract. Calculus Appl. **5**, 71–98 (2019)
145. Y. Luchko, W. Rundell, M. Yamamoto, L. Zuo, Uniqueness and reconstruction of an unknown semilinear term in a time-fractional reaction-diffusion equation. Inverse Probl. **29**(6), 065019 (2013)

146. A. Lopushansky, H. Lopushansk, Inverse source Cauchy problem for a time fractional diffusion-wave equation with distributions. Electron. J. Differ. Equ. **182**, 1–14 (2017)
147. N.H. Luc, L.N. Huynh, N.H. Tuan, On a backward problem for inhomogeneous time-fractional diffusion equations. Comput. Math. Appl. **78**(5), 1317–1333 (2019)
148. O. Mahouachi, T. Saanouni, Well and ill-posedness issues for a class of 2D wave equation with exponential growth. J. Partial Differ. Equ. **24**(4), 361–384 (2011)
149. F. Mainardi, Fractional diffusive waves in viscoelastic solids, in *Nonlinear Waves in Solids, Fairfield* (1995), pp. 93–97
150. F. Mainardi, The fundamental solutions for the fractional diffusion-wave equation. Appl. Math. Lett. **9**, 23–28 (1996)
151. F. Mainardi, *Fractional Calculus and Waves in Linear Viscoelasticity, An Introduction to Mathematical Models* (Imperial College Press, London, 2010)
152. F. Mainardi, P. Paradisi, Fractional diffusive waves. J. Comput. Acoust. **9**, 1417–1436 (2001)
153. F. Mainardi, P. Paradisi, R. Gorenflo, Probability distributions generated by fractional diffusion equations. arXiv preprint arXiv 0704, 0320 (2007)
154. F. Mainardi, P. Paradisi, A model of diffusive waves in viscoelasticity based on fractional calculus, in *Proceedings of the 36th IEEE Conference on Decision and Control*, vol. 5 (1997), pp. 4961–4966
155. F. Mainardi, Y. Luchko, G. Pagnini, The fundamental solution of the space-time fractional diffusion equation. Fract. Calc. Appl. Anal. **4**, 153–192 (2001)
156. M. Majdoub, S. Tayachi, Well-posedness, Global existence and decay estimates for the heat equation with general power-exponential nonlinearities. Proc. Int. Cong. of Math. Rio de Janeiro **2**, 2379–2404 (2018)
157. M. Majdoub, S. Otsmane, S. Tayachi, Local well-posedness and global existence for the biharmonic heat equation with exponential nonlinearity. Adv. Differ. Equ. **23**, 489–522 (2018)
158. W. McLean, Regularity of solutions to a time-fractional diffusion equation. ANZIAM J. **52**, 123–138 (2010)
159. W. McLean, W.C. McLean, *Strongly Elliptic Systems and Boundary Integral Equations* (Cambridge University Press, Cambridge, 2000)
160. M.M. Meerschaert, H.P. Scheffler, Stochastic model for ultraslow diffusion. Stoch. Proc. Appl. **116**, 1215–1235 (2006)
161. R. Metzler, J. Klafter, The random walks guide to anomalous diffusion: A fractional dynamics approach. Phys. Rep. **339**(1), 1–77 (2000)
162. R. Metzler, J. Klafter, The restaurant at the end of the random walk: recent developments in the description of anomalous transport by fractional dynamics. J. Phys. A Math. Gen. **37**(31), R161 (2004)
163. R. Metzler, E. Barkai, J. Klafter, Anomalous diffusion and relaxation close to thermal equilibrium: a fractional Fokker-Planck equation approach. Phys. Rev. Lett. **82**, 3563–3567 (1999)
164. C. Miao, B. Yuan, B. Zhang, Well-posedness of the Cauchy problem for the fractional power dissipative equations. Nonlinear Anal. **68**, 461–484 (2008)
165. K.S. Miller, B. Ross, *An Introduction to the Fractional Calculus and Fractional Differential Equations* (Wiley, New York, 1993)
166. L. Miller, M. Yamamoto, Coefficient inverse problem for a fractional diffusion equation. Inverse Probl. **29**(7), 075013 (2013)
167. S. Morigi, L. Reichel, F. Sgallari, Fractional Tikhonov regularization with a nonlinear penalty term. E J. Comput. Appl. Math. **324**, 142–154 (2017)
168. J. Mu, B. Ahmad, S. Huang, Existence and regularity of solutions to time-fractional diffusion equations. Comput. Math. Appl. **73**(6), 985–996 (2017)
169. M.T. Nair, *Linear Operator Equation: Approximation and Regularization* (World Scientific, Singapore, 2009)
170. M. Nakamura, T. Ozawa, Global solutions in the critical Sobolev space for the wave equations with nonlinearity of exponential growth. Math. Z. **231**, 479–487 (1999)

171. R.R. Nigmatullin, The realization of the generalized transfer equation in a medium with fractal geometry. Phys. Stat. Sol. B **133**, 425–430 (1986)
172. R.H. Nochetto, E. Otárola, A.J. Salgado, A PDE approach to space-time fractional wave problems. SIAM J. Numer. Anal. **54**, 848–873 (2016)
173. E. Orsingher, L.Beghin, Time-fractional telegraph equations and telegraph process with Brownian time. Probab. Theory Relat. Fields **128**(1), 141–160 (2004)
174. E. Otárola, A.J. Salgado, Regularity of solutions to space-time fractional wave equations: a PDE approach. Fract. Calc. Appl. Anal. **21**(5), 1262–1293 (2018)
175. A. Pazy, *Semigroup of Linear Operators and Applications to Partial Differential Equations* (Springer, New York, 1983)
176. H. Pecher, Local solutions of semilinear wave equations in H^{s+1}. Math. Methods Appl. Sci. **19**(2), 145–170 (1996)
177. L. Peng, Y. Huang, On nonlocal backward problems for fractional stochastic diffusion equations. Comput. Math. Appl. **78**(5), 1450–1462 (2019)
178. L. Peng, Y. Zhou, The analysis of approximate controllability for distributed order fractional diffusion problems. Appl. Math. Optim. **86**(2), 22 (2022)
179. L. Peng, Y. Zhou, J.W. He, The well-posedness analysis of distributed order fractional diffusion problems on $\mathbb{R}^N$. Monatsh. Math. **198**, 445–463 (2022)
180. I. Podlubny, *Fractional Differential Equations* (Academic, San Diego, 1999)
181. Y. Povstenko, *Linear Fractional Diffusion-Wave Equation for Scientists and Engineers* (Springer, Cham, 2015)
182. J. Prüss, *Evolutionary Integral Equations and Applications* (Birkhäuser, Basel, 1993)
183. W. Rundell, Z. Zhang, Fractional diffusion: recovering the distributed fractional derivative from overposed data. Inverse Probl. **33**(3), 035008 (2017)
184. W. Rundell, Z. Zhang, Recovering an unknown source in a fractional diffusion problem. J. Comput. Phys. **368**, 299–314 (2018)
185. T. Saanouni, A blowing up wave equation with exponential type nonlinearity and arbitrary positive energy. J. Math. Anal. Appl. **421**, 444–452 (2015)
186. K. Sakamoto, M. Yamamoto, Initial value/boundary value problems for fractional diffusion-wave equations and applications to some inverse problems. J. Math. Anal. Appl. **382**(1), 426–447 (2011)
187. S.G. Samko, A.A. Kilbas, O.I. Marichev, *Fractional Integrals and Derivatives. Theory and Applications* (Gordon and Breach Science, London, 1987)
188. K. Sakamoto, M. Yamamoto, Initial value/boundary value problems for fractional diffusion-wave equations and applications to some inverse problems. J. Math. Anal. Appl. **382**(1), 426–447 (2011)
189. H. Scher, E.W. Montroll, Anomalous transit-time dispersion in amorphous solids. Phys. Rev. B, **12**, 2455 (1975)
190. W.R. Schneider, W. Wyss, Fractional diffusion and wave equations. J. Math. Phys. **30**(1), 134–144 (1989)
191. K. Seki, M. Wojcik, M. Tachiya, Fractional reaction-diffusion equation. J. Chem. Phy. **119**(4), 2165–2170 (2003)
192. R. Servadei, E. Valdinoci, On the spectrum of two different fractional operators. Proc. R. Soc. Edinburgh Sect. A **144**(4), 831–855 (2014)
193. J. Shatah, M. Struwe, Regularity results for nonlinear wave equations. Ann. of Math. **138**, 503–518 (1993)
194. J. Shatah, M. Struwe, Well-posedness in the energy space for semilinear wave equation with critical growth. IMRN **7**, 303–309 (1994)
195. R.E. Showalter, The final value problem for evolution equations. J. Math. Anal. Appl. **47**, 563–572 (1974)
196. K. Siskova, M. Slodicka, Recognition of a time-dependent source in a time-fractional wave equation. Appl. Numer. Math. **121**, 1–17 (2017)
197. T.H. Solomon, E.R. Weeks, H.L. Swinney, Observation of anomalous diffusion and Lévy flights in a two-dimensional rotating flow. Phys. Rev. Lett. **71**, 3975 (1993)

198. D. Stan, F. del Teso, J.L. Vázquez, Existence of weak solutions for a general porous medium equation with nonlocal pressure. Arch. Ration. Mech. Anal. **233**(1), 451–496 (2019)
199. M. Struwe, Global well-posedness of the Cauchy problem for a super-critical nonlinear wave equation in two space dimensions. Math. Ann. **350**, 707–719 (2011)
200. M. Stojanovic, R. Gorenflo, Nonlinear two term time fractional diffusion wave problem. Nonlinear Anal. Real World Appl. **11**(5), 3512–3523 (2010)
201. C.L. Sun, J.J. Liu, An inverse source problem for distributed order time-fractional diffusion equation. Inverse Probl. **36**(5), 055008 (2020)
202. M. Suzuki, Local existence and nonexistence for reaction-diffusion systems with coupled exponential nonlinearities. J. Math. Anal. Appl. **477**(1), 776–804 (2019)
203. J. Szymanski, M. Weiss, Elucidating the origin of anomalous diffusion in crowded fluids. Phys. Rev. Lett. **103**, 038102 (2009)
204. N.E. Tatar, A blow up result for a fractionally damped wave equation. NoDEA Nonlinear Differ. Equ. Appl. **12**(2), 215–226 (2005)
205. M. Taylor, Analysis on Morrey spaces and applications to Navier-Stokes equation. Comm. Partial Differ. Equ. **17**, 1407–1456 (1992)
206. N.H. Tuan, L.D. Long, N.V. Thinh, T. Tran, On a final value problem for the time-fractional diffusion equation with inhomogeneous source. Inverse Probl. Sci. Eng. **25**, 1367–1395 (2017)
207. N.H. Tuan, L.N. Huynh, T.B. Ngoc, Y. Zhou, On a backward problem for nonlinear fractional diffusion equations. Appl. Math. Lett. **92**, 76–84 (2019)
208. N.H. Tuan, A. Debbouche, T.B. Ngoc, Existence and regularity of final value problems for time fractional wave equations. Comput. Math. Appl. **78**(5), 1396–1414 (2019)
209. N.H. Tuan, V.A. Khoa, V.V. Au, Analysis of a quasi-reversibility method for a terminal value quasi-linear parabolic problem with measurements. SIAM J. Math. Anal. **51**, 60–85 (2019)
210. N.H. Tuan, T.B. Ngoc, Y. Zhou, D. O'Regan, On existence and regularity of a terminal value problem for the time fractional diffusion equation. Inverse Probl. **36**(5), 055011 (2020)
211. N.H. Tuan, T. Caraballo, T.B. Ngoc, Y. Zhou, Existence and regularity results for terminal value problem for nonlinear fractional wave equations. Nonlinearity **34**(3), 1448 (2021)
212. S. Umarov, *Introduction to Fractional and Pseudo-Differential Equations with Singular Symbols* (Springer, Berlin, 2015)
213. G.M. Vainikko, A.Y. Veretennikov, *Iteration Procedures in Ill-Posed Problems* (Nauka, Moscow, 1986)
214. V. Vergara, R. Zacher, Optimal decay estimates for time-fractional and other nonlocal subdiffusion equations via energy methods. SIAM J. Math. Anal. **47**(1), 210–239 (2015)
215. V. Vergara, R. Zacher, Stability, instability, and blowup for time fractional and other nonlocal in time semilinear subdiffusion equations. J. Evol. Equ. **17**(1), 599–626 (2017)
216. A. Viana, A local theory for a fractional reaction-diffusion equation. Commun. Contemp. Math. **21**(06), 1–26 (2019)
217. G.E. Viswanathan, M.G.E. da Luz, E.P. Raposo, H.E. Stanley, *The Physics of Foraging. An Introduction to Random Searches and Biological Encounters* (Cambridge University Press, Cambridge, 2011)
218. H. Vivian, J. Pym, M. Cloud, *Applications of Functional Analysis and Operator Theory* (Elsevier, Amsterdam, 2005)
219. V. Volpert, *Elliptic Partial Differential Equations. Volume 2: Reaction-Diffusion Equations. vol. 104* (Springer, Berlin, 2014)
220. L. Wang, J. Liu, Total variation regularization for a backward time-fractional diffusion problem. Inverse Probl. **29**(11), 115013 (2013)
221. W. Wang, M. Yamamoto, B. Han, Numerical method in reproducing kernel space for an inverse source problem for the fractional diffusion equation. Inverse Probl. **29**(9), 095009 (2013)
222. R.N. Wang, D.H. Chen, T.J. Xiao, Abstract fractional Cauchy problems with almost sectorial operators. J. Differ. Equ. **252**, 202–235 (2012)

223. J.G. Wang, Y.B. Zhou, T. Wei, A posteriori regularization parameter choice rule for the quasiboundary value method for the backward time-fractional diffusion problem. Appl. Math. Lett. **26**(7), 741–747 (2013)
224. T. Wei, J. Xian, Variational method for a backward problem for a time-fractional diffusion equation. ESAIM Math. Model. Numer. Anal. **53**(4), 1223–1244 (2019)
225. T. Wei, J.G. Wang, A modified quasi-boundary value method for the backward time-fractional diffusion problem. ESAIM Math. Model. Numer. Anal. **48**(2), 603–621 (2014)
226. T. Wei, Y. Zhang, The backward problem for a time-fractional diffusion-wave equation in a bounded domain. Comput. Math. Appl. **75**(10), 3632–3648 (2018)
227. M. Warma, Approximate controllability from the exterior of space-time fractional diffusive equations. SIAM J. Control Optim. **57**(3), 2037–2063 (2019)
228. J. Xian, T. Wei, Determination of the initial data in a time-fractional diffusion-wave problem by a final time data. Comput. Math. Appl. **78**(8), 2525–2540 (2019)
229. X. Xiong, X. Xue, Z. Qian, A modified iterative regularization method for ill-posed problems. Appl. Numer. Math. **122**, 108–128 (2017)
230. R.Z. Xu, J. Su, Global existence and finite time blow-up for a class of semilinear pseudo-parabolic equations. J. Funct. Anal. **264**, 2732–2763 (2013)
231. A. Yagi, *Abstract Parabolic Evolution Equations and Their Applications* (Springer Science & Business Media, Berlin, 2009)
232. M. Yang, J. Liu, Solving a final value fractional diffusion problem by boundary condition regularization. Appl. Numer. Math. **66**, 45–58 (2013)
233. H. Ye, J. Gao, Y. Ding, A generalized Gronwall inequality and its application to a fractional differential equation. J. Math. Anal. Appl. **328**(2), 1075–1081 (2007)
234. R. Zacher, A De Giorgi-Nash type theorem for time fractional diffusion equations. Math. Ann. **356**, 99–146 (2013)
235. G.M. Zaslavsky, Chaos, fractional kinetics, and anomalous transport. Phys. Rep. **371**, 461–580 (2002)
236. E. Zeidler, *Nonlinear Functional Analysis and Its Application II/A* (Springer, Berlin, 1990)
237. Q.G. Zhang, H.R. Sun, The blow-up and global existence of solutions of Cauchy problems for a time fractional diffusion equation. Topol. Methods Nonlinear Anal. **46**(1), 69–92 (2015)
238. Q. Zhang, Y. Li, Global well-posedness and blow-up solutions of the Cauchy problem for a time-fractional superdiffusion equation. J. Evol. Equ. **19**, 271–303 (2019)
239. A. Zhokh, A. Trypolskyi, P. Strizhak, Relationship between the anomalous diffusion and the fractal dimension of the environment. Chem. Phys. **503**, 71–76 (2018)
240. Y. Zhou, *Basic Theory of Fractional Differential Equations* (World Scientific, Singapore, 2014)
241. Y. Zhou, *Fractional Evolution Equations and Inclusions: Analysisand Control* (Academic, London, 2016)
242. Y. Zhou, J.W. He, Well-posedness and regularity for fractional damped wave equations. Monatsh. Math. **194**(2), 425–458 (2021)
243. G.H. Zhou, Z.B. Guo, Boundary feedback stabilization for an unstable time fractional reaction diffusion equation. SIAM J. Control Optim. **56**, 75–10 (2018)
244. L. Zhou, H.M. Selim, Application of the fractional advection-dispersion equation in porous media. Soil Sci. Soc. Amer. J. **67**(4), 1079–1084 (2003)
245. Y. Zhou, J.W. He, B. Ahmad, N.H. Tuan, Existence and regularity results of a backward problem for fractional diffusion equations. Math. Methods Appl. Sci. **42**, 6775–6790 (2019)
246. Y. Zhou, J.W. He, A. Alsaedi, B. Ahmad, The well-posedness for semilinear time fractional wave equations on $\mathbb{R}^n$. Elec. Res. Arch. **30**(8), 2981–3003 (2022)

Index

Y. Zhou, *Fractional Diffusion and Wave Equations*,
https://doi.org/10.1007/978-3-031-74031-2

The manufacturer's authorised representative in the EU is Springer Nature Customer Service Centre GmbH, Europaplatz 3, 69115 Heidelberg, Germany. If you have any concerns regarding our products, please contact ProductSafety@springernature.com

Printed and bound by CPI Group (UK) Ltd, Croydon, CR0 4YY
30/09/2026
02225599-0014